T0245178

CAMBRIDGE TRACTS IN MATHEMATICS

General Editors

138 Random Walks on Infinite Graphs and Groups

Wolfgang Woess
University of Milan

Random Walks on Infinite Graphs and Groups

CAMBRIDGE UNIVERSITY PRESS
Cambridge, New York, Melbourne, Madrid, Cape Town, Singapore, São Paulo

Cambridge University Press
The Edinburgh Building, Cambridge CB2 8RU, UK

Published in the United States of America by Cambridge University Press, New York

www.cambridge.org
Information on this title: www.cambridge.org/9780521552929

First published 2000
This digitally printed version (with corrections) 2008

A catalogue record for this publication is available from the British Library

ISBN 978-0-521-55292-9 hardback
ISBN 978-0-521-06172-8 paperback

CONTENTS

PREFACE

"Random walks" is a topic situated somewhere in between probability, potential theory, harmonic analysis, geometry, graph theory, and algebra. The beauty of the subject stems from this linkage, both in the way of thinking and in the methods employed, of different fields.

Let me briefly declare what - in my viewpoint - random walks are. These are time-homogeneous Markov chains whose transition probabilities are in some way (to be specified more precisely in each case) adapted to a given structure of the underlying state space. This structure may be geometric or algebraic; here it will be discrete and infinite. Typically, we shall use locally finite graphs to view the structure. This also includes groups via their Cayley graphs. From the probabilistic viewpoint, the question is what impact the particular type of structure has on various aspects of the behaviour of the random walk, such as transience/recurrence, decay and asymptotic behaviour of transition probabilities, rate of escape, convergence to a boundary at infinity and harmonic functions. Vice versa, random walks may also be seen as a nice tool for classifying, or at least describing the structure of graphs, groups and related objects.

Of course, random walks on *finite* graphs and groups are a fascinating topic as well, and have had an enormous renaissance in the last decade: a book written by two major experts, D. Aldous and J. Fill, is about to appear.

Some might object that any countable Markov chain may be viewed on a directed graph, so that our notion of random walks coincides with arbitrary Markov chains. However, our point of view is reversed: what we have in mind is to start with a graph, group, etc., and investigate the interplay between the behaviour of random walks on these objects on one hand and properties of the underlying structure itself on the other.

Historically, I believe that this spirit of approaching the theory of random walks on infinite graphs has its roots in the 1921 paper by Pólya [269], whose nice title - translated into English - is "On an exercise in probability concerning the random walk in the road network". There, Pólya shows that simple random walk in the two-dimensional Euclidean grid is recurrent, while it is transient in higher dimensions. This change of behaviour between plane and space provided inspiration for much further work. However, it took 38 years until what I (personal opinion !) consider the next "milestones". In 1959, Nash-Williams published his paper "Random walks and electric currents in networks" [245], the first to link recurrence and structural properties of networks (i.e., reversible Markov chains). This paper -

not written in the style of the mainstream of mathematics at that time - remained more or less forgotten until the 80s, when it was rediscovered by T. Lyons, Doyle and Snell, Gerl, and others. The second 1959 milestone was Kesten's "Symmetric random walks on groups" [198], founding the theory of random walks on (infinite) groups and also opening the door from random walks to amenability and other topics of harmonic and spectral analysis.

Another direct line of extension of Pólya's result is to consider sums of i.i.d. random variables taking their values in $\mathbb{Z}^d$ - this was done to perfection in Spitzer's beautiful "Principles of Random walk" [307] (first edition in 1964), which is still the most authoritative and elegant source available. Spitzer's book also contains a considerable amount of potential theory. Note that Markov chains and discrete potential theory were born more or less simultaneously (while classical potential theory had already been very well developed before its connection with Brownian motion was revealed, and one still encounters analysts who deeply mistrust the so-called probabilistic proofs of results in potential theory - probably they believe that the proofs themselves hold only almost surely). Although not being directly concerned with the type of structural considerations that are inherent to random walks, I consider the third 1959 milestone to be Doob's "Discrete potential theory and boundaries" [101]. In the sixties, potential and boundary theory of denumerable Markov chains had a strong impetus promoted by Doob, Hunt, Kemeny, Snell, Knapp and others, before being somewhat "buried" under the burden of abstract potential theory. Doob's article immediately led to considerations in the same spirit that we have in mind here, the next milestone being the note of 1961 by Dynkin and Malyutov [111]. This contains the first structural description of the Martin boundary of a class of random walks and is also - together with Kesten [198] - the first paper where one finds the principal ingredients for computations regarding nearest neighbour random walks on free groups and homogeneous trees. Indeed, it is amusing to see how many people have been redoing these computations for trees in the belief of being the first to do so.

It was in a paper on boundaries that Kesten [201] indicated a problem which then became known as "Kesten's conjecture": classify those (finitely generated) groups which carry a recurrent random walk, the conjecture (not stated explicitly by Kesten) being that such a group must grow polynomially with degree at most two. It is noteworthy that the analogous problem was first settled in the 70s for connected Lie groups, see Baldi [17]. The Lie case is not easier, but there were more analytical and structural tools available at the time. The solution in the discrete case became possible by Gromov's celebrated classification of groups with polynomial growth [149] and was carried out in a remarkable series of papers by Varopoulos, who gave the final answer in [325]. In the 80s, random walks on graphs have been repopularized, owing much to the beautiful little book by Doyle and Snell [103]. However, this discussion of selected "milestones" is bringing me too close to the present, with many of the actors still on stage and the future to judge. Other important work from the late 50s and the 60s should also be mentioned here, such as that of Choquet and Deny [74] and - in particular - Furstenberg [124].

Let me return from this "historical" excursion. This book grew out of a long survey paper that I published in 1994 [348]. It is organized in a similar way, although here, less material is covered in more detail.

Each of the four chapters is built around one specific type of question concerning the behaviour of random walks, and answers to this question are then presented for various different structures, such as integer lattices, trees, free groups, plane tilings, Gromov-hyperbolic graphs, and so on. At the beginning, I briefly considered using the "orthogonal" approach, namely to order by types of structures, for example, saying first "everything" about random walks on integer lattices, then nilpotent groups and graphs with polynomial growth, trees, hyperbolic graphs, and so on. Some thought convinced me that this was not feasible. Thus, the same classes of structures will be encountered several times in this book. For example, the reader who is interested in results concerning random walks and trees will find these in paragraphs/sections 1.D, 5, 6.B, 10.C, 12.C, 19, 21.A and 26.A, tilings and circle packings are considered in 6.C-D, 10.C and 23, and the integer grids and their generalizations appear in 1.A, 6.A, 8.B, 13 and 25. Regarding the latter, I obviously did not aim at an exposition as complete as that of Spitzer had been in its time. Most likely, every reader will find a favorite among the topics in random walk theory that are not covered here (such as random walks on direct limits of finite groups, ratio limit theorems, or random walks in random environment).

A short word on notation. Instead of using further exotic alphabets, I decided not to reserve a different symbol for each different object. For example, the symbol Φ has different meanings in Sections 6, 9 and 12, and this should be clear from the context.

I started writing this book at the beginning of 1995 (one chapter per year). Thus, Chapter I is the oldest one among the material presented here, and so on. I decided not to make a complete updating of this material to the state of the art of today (1999) - otherwise I could never stop writing. In particular, the 90s saw the emergence of a new, very strong group of random walkers (and beyond) in Israel and the US (I. Benjamini, R. Lyons, Y. Peres, O. Schramm, ...) whose work is somewhat underrepresented here by this reason. On the other hand (serving as an excuse for me), two of them (Lyons and Peres) are currently writing their own book on "Probability on Trees and Networks" that can be expected to be quite exciting.

Many mathematical monographs of today start with two claims. One is to be self-contained. This book is *not* self-contained by the nature of its topic. The other claim is to be usable for graduate students. It has been my experience that usually, this must be taken with caution and is mostly true only in the presence of a guiding hand that is acquainted with the topic. I think that this is true here as well. Proofs are sometimes a bit condensed, and it may be that even readers above the student level will need pen and paper when they want to work through them seriously - in particular because of the variety of different methods and techniques that I have tried to unite in this text. This does not mean that parts of this book could not be used for graduate or even undergraduate courses. Indeed, I have taught parts of this material on several occasions, and at various levels.

Anyone who has written a book will have experienced the mysterious fact that a text of finite length may contain an infinity of misprints and mistakes, which apparently were not there during your careful proof-reading. In this sense, I beg excuse for all those flaws whose mysterious future appearance is certain.

In conclusion, let me say that I have learned a lot in working on this book, and also had fun, and I hope that this fun will "infect" some of the readers too.

Milano, July 1999 W.W.

For the 2008 paperback edition, I only had the opportunity to add a few pages at the end of the book pointing out some corrections, without modifying the main text. Now I have finally incorporated them at the right places. Comments on those corrections can be found after the Acknowledgements at the end of Chapter IV. It is unforgivable that in the above Preface from 1999 there was no reference to the following book.

Guivarc'h, Y., Keane, M., and Roynette, B.: *Marches Aléatoires sur les Groupes de Lie*, Lect. Notes in Math. **624**, Springer, Berlin, 1977.

Indeed, while I did not use any specific material from that volume in the present monograph, it documents an important phase in the development of the theory of random walks on groups – not primarily discrete ones, but Lie groups.

Apart from that, I only have updated the "preprint" citations in the bibliography. That is, I made no attempt to expand the bibliography. As a matter of fact, since the first printing in the year 2000, the field has exploded beyond the possibility of inserting just some additional notes on recent development. But maybe this monograph can serve as a useful signpost towards what had been the future of the field 20 years ago.

Graz, November 2020 W.W.

CHAPTER I

THE TYPE PROBLEM

1. Basic facts

Before embarking on a review of the basic material concerning Markov chains, graphs, groups, etc., let us warm up by considering the classical example.

A. Pólya's walk

The d-dimensional grid, denoted briefly by $\mathbb{Z}^d$, is the graph whose vertices are integer points in d dimensions, and where two points are linked by an edge if they are at distance 1. A walker wanders randomly from point to point; at each "crossroad" (point) he chooses with equal probability the one among the $2d$ neighbouring points where his next step will take him, see Figure 1. Starting from the origin, what is the probability $p^{(2n)}(0,0)$ that the walker will be back at the $2n$th step? This is the number of closed paths of length $2n$ starting at the origin, divided by $(2d)^{2n}$. (The walker cannot be back after an odd number of steps.) For small dimensions, the solutions of this combinatorial exercise are as follows.

Figure 1: the grids $\mathbb{Z}$ and $\mathbb{Z}^2$

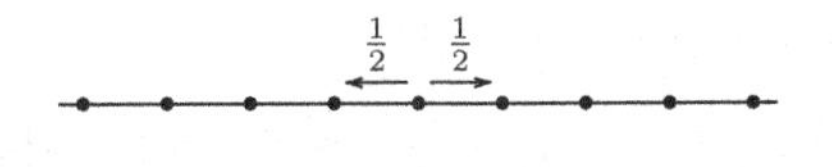

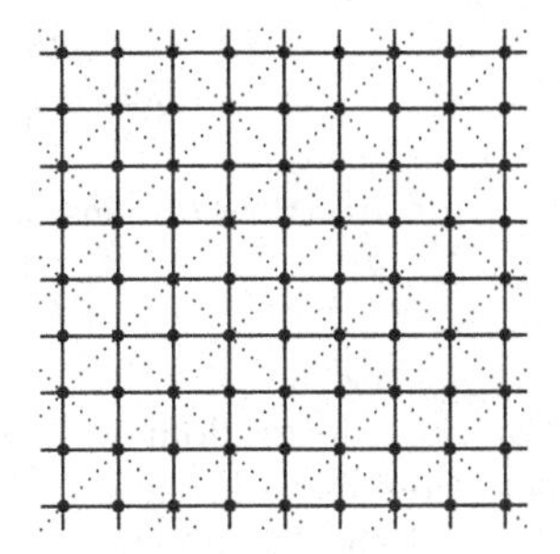

$d = 1$. Among the $2n$ steps, the walker has to make n to the left and n to the right. Hence

$$p^{(2n)}(0,0) = \frac{1}{2^{2n}}\binom{2n}{n} \sim C_1\, n^{-1/2}\,. \tag{1.1}$$

$d = 2$. Let two walkers perform the one-dimensional random walk simultaneously and independently. Their joint trajectory, viewed in $\mathbb{Z}^2$, visits only the set of points (i,j) with $i+j$ even. However, the graph with this set of vertices, and with two points neighbours if they differ by ± 1 in each component, is isomorphic with the grid $\mathbb{Z}^2$ and probabilities are preserved under this isomorphism. Hence

$$p^{(2n)}(0,0) = \left(\frac{1}{2^{2n}}\binom{2n}{n}\right)^2 \sim C_2\, n^{-1}. \tag{1.2}$$

$d = 3$. It is no longer possible to represent the random walk in terms of three independent random walks on $\mathbb{Z}$. In a path of length $2n$ starting and ending at the origin, n steps have to go north, east, or up. There are $\binom{2n}{n}$ possibilities to assign the n steps of these three types; the other n go south, west, or down. For each of these choices, i steps go north and i go south, j steps go east and j go west, $n-i-j$ steps go up and $n-i-j$ go down. Hence

$$p^{(2n)}(0,0) = \frac{1}{6^{2n}}\binom{2n}{n}\sum_{i+j\le n}\left(\frac{n!}{i!j!(n-i-j)!}\right)^2.$$

Consider the function $(x,y,z) \mapsto x!y!z!$ for $x,y,z \ge 0$. Under the condition $x+y+z=n$, it assumes its minimum for $x=y=z=n/3$, when n is sufficiently large. Hence

$$\begin{aligned} p^{(2n)}(0,0) &\le \frac{1}{6^{2n}}\binom{2n}{n}\frac{n!}{(n/3)!^3}\sum_{i+j\le n}\left(\frac{n!}{i!j!(n-i-j)!}\right) \\ &= \frac{1}{6^{2n}}\binom{2n}{n}\frac{n!}{(n/3)!^3}\,3^n \sim C_3\, n^{-3/2}. \end{aligned} \tag{1.3}$$

Indeed, for arbitrary dimension d, there are various ways to show that

$$p^{(2n)}(0,0) \sim C_d\, n^{-d/2}. \tag{1.4}$$

Now for the random walk starting at the origin, $\sum_n p^{(2n)}(0,0)$ is the expected number of visits of the walker back to the origin: this is infinite for $d = 1,2$ and finite for $d \ge 3$. This drastic change of behaviour from two to three dimensions stands at the origin of our investigations.

B. Irreducible Markov chains

A *Markov chain* is (in principle) given by a finite or countable *state space* X and a stochastic *transition matrix* (or *transition operator*) $P = \big(p(x,y)\big)_{x,y\in X}$. In addition, one has to specify the starting point (or a starting distribution on X). The matrix element $p(x,y)$ is the probability of moving from x to y in one step. Thus, we have a sequence of X-valued random variables Z_n, $n \ge 0$, with Z_n representing the random position in X at time n. To model Z_n, the usual choice of probability space is the *trajectory space* $\Omega = X^{\mathbb{N}_0}$, equipped with the product σ-algebra arising from the discrete one on X. Then Z_n is the nth projection $\Omega \to X$. This describes the Markov chain starting at x, when Ω is equipped with the probability measure given via the Kolmogorov extension theorem by

$$\mathbb{P}_x[Z_0 = x_0, Z_1 = x_1, \ldots, Z_n = x_n] = \delta_x(x_0)p(x_0, x_1)\cdots p(x_{n-1}, x_n).$$

The associated expectation is denoted by $\mathbb{E}_x$. Alternatively, we shall call a Markov chain (random walk) the pair (X, P) or the sequence of random variables $(Z_n)_{n\geq 0}$. We write

$$p^{(n)}(x, y) = \mathbb{P}_x[Z_n = y].$$

This is the (x, y)-entry of the matrix power P^n, with $P^0 = I$, the identity matrix over X. Throughout this book, we shall always require that all states communicate:

(1.5) Basic assumption. *(X, P) is irreducible, that is, for every $x, y \in X$ there is some $n \in \mathbb{N}$ such that $p^{(n)}(x, y) > 0$.*

Next, we define the *Green function* as the power series

$$G(x, y|z) = \sum_{n=0}^{\infty} p^{(n)}(x, y)\, z^n, \quad x, y \in X, z \in \mathbb{C}. \tag{1.6}$$

(1.7) Lemma. *For real $z > 0$, the series $G(x, y|z)$ either diverge or converge simultaneously for all $x, y \in X$.*

Proof. Given $x_1, y_1, x_2, y_2 \in X$, by irreducibility there are $k, \ell \in \mathbb{N}$ such that $p^{(k)}(x_1, x_2) > 0$ and $p^{(\ell)}(y_2, y_1) > 0$. We have

$$p^{(k+n+\ell)}(x_1, y_1) \geq p^{(k)}(x_1, x_2)p^{(n)}(x_2, y_2)p^{(\ell)}(y_2, y_1)$$

and hence, for $z > 0$,

$$G(x_1, y_1|z) \geq p^{(k)}(x_1, x_2)p^{(\ell)}(y_2, y_1)z^{k+\ell}G(x_2, y_2|z). \qquad \square$$

As a consequence, all the $G(x, y|z)$ (where $x, y \in X$) have the same radius of convergence $\mathtt{r}(P) = 1/\rho(P)$, given by

$$\rho(P) = \limsup_{n\to\infty} p^{(n)}(x, y)^{1/n} \in (0, 1]. \tag{1.8}$$

This number is often called the *spectral radius* of P.

The *period* of P is the number $\mathtt{d} = \mathtt{d}(P) = \gcd\{n \geq 1 : p^{(n)}(x, x) > 0\}$. It is well known and easy to check that it is independent of x by irreducibility. If $\mathtt{d}(P) = 1$ then the chain is called *aperiodic*. Choose $o \in X$ and define $Y_j = \{x \in X : p^{(n\mathtt{d}+j)}(o, x) > 0 \text{ for some } n \geq 0\}$, $j = 0, \ldots, \mathtt{d} - 1$. This defines a partition of X, and x, y are in the same class if and only if $p^{(n\mathtt{d})}(x, y) > 0$ for some n. These are the periodicity classes of (X, P), visited by the chain $(Z_n)_{n\geq 0}$ in cyclical order. The restriction of $P^{\mathtt{d}}$ to each class is irreducible and aperiodic. See e.g. Chung [75] for these facts.

(1.9) Lemma. $p^{(n)}(x,x) \le \rho(P)^n$, *and* $\lim_{n\to\infty} p^{(n\mathrm{d})}(x,x)^{1/n\mathrm{d}} = \rho(P)$.

Proof. Write $a_n = p^{(n\mathrm{d})}(x,x)$. Then $0 \le a_n \le 1$ and $\gcd N(x) = 1$, where $N(x) = \{n : a_n > 0\}$. The crucial property is $a_m a_n \le a_{m+n}$.

We first show that there is n_0 such that $a_n > 0$ for all $n \ge n_0$. If $m, n \in N(x)$ then $m+n \in N(x)$. Recall that the greatest common divisor of a set of integers can always be written as a finite linear combination with integer coefficients of elements of that set. Therefore we can write $1 = \gcd N(x) = n_1 - n_2$, where $n_1, n_2 \in N(x) \cup \{0\}$. If $n_2 = 0$ we are done ($n_0 = 1$). Otherwise, set $n_0 = n_2^2$ and decompose $n \ge n_0$ as $n = q\,n_2 + r = (q-r)n_2 + r\,n_1$, where $0 \le r < n_2$. It must be that $q \ge n_2 > r$, so that $n \in N(x)$. Next, fix $m \in N(x)$, let $n \ge n_0 + m$, and decompose $n = q_n m + r_n$, where $n_0 \le r_n < n_0 + m$. Write $b = b(m) = \min\{a_r : n_0 \le r < n_0 + m\}$. Then $b > 0$ and $a_n \ge a_m^{q_n} a_{r_n}$, so that $a_m^{q_n/n} b^{1/n} \le a_n^{1/n}$. If $n \to \infty$ then $q_n/n \to 1/m$. Hence,

$$a_m^{1/m} \le \liminf_{n\to\infty} a_n^{1/n} \le \rho(P)^{\mathrm{d}} \quad \text{for every } m \in N(x)\,.$$

This proves the first statement. If we now let $m \to \infty$, then $\limsup_m a_m^{1/m} \le \liminf_n a_n^{1/n}$, and $a_n^{1/n}$ converges. □

(1.10) Exercise. Prove the following. *If P is irreducible and aperiodic then P^k is irreducible and aperiodic for every $k \ge 1$, and $\rho(P^k) = \rho(P)^k$.*

Next, define the stopping time $\mathbf{s}^y = \min\{n \ge 0 : Z_n = y\}$ (where the minimum is ∞ when the set is empty) and the hitting probabilities plus associated generating functions

$$(1.11)\qquad f^{(n)}(x,y) = \mathbb{P}_x[\mathbf{s}^y = n] \quad \text{and} \quad F(x,y|z) = \sum_{n=0}^{\infty} f^{(n)}(x,y)\, z^n\,,$$

where $z \in \mathbb{C}$. Note that $F(x,x|z) = 1$. Finally, let

$$(1.12)\quad \mathbf{t}^x = \min\{n \ge 1 : Z_n = x\} \quad \text{and} \quad U(x,x|z) = \sum_{n=0}^{\infty} \mathbb{P}_x[\mathbf{t}^x = n]\, z^n\,.$$

The following will be useful on several occasions.

(1.13) Lemma. (a) $G(x,x|z) = \dfrac{1}{1 - U(x,x|z)}$,

(b) $G(x,y|z) = F(x,y|z)G(y,y|z)$,

(c) $U(x,x|z) = \sum_y p(x,y)z\,F(y,x|z)$ and,

(d) *if* $y \ne x$, $F(x,y|z) = \sum_w p(x,w)z\,F(w,y|z)$.

Proof. Part (a) follows from the identity

$$p^{(n)}(x,x) = \sum_{k=0}^{n} \mathbb{P}_x[\mathbf{t}^x = k]\, p^{(n-k)}(x,x)\,, \quad \text{if } n \ge 1\,,$$

while $p^{(0)}(x,x) = 1$ and $\mathbb{P}_x[\mathbf{t}^x = 0] = 0$.

Analogously, (b) is obtained by conditioning with respect to the first visit to y. Parts (c) and (d) are obtained by factoring though the first step (that is, the values of Z_1). □

We shall write $G(x,y)$ for $G(x,y|1)$. This is the expected number of visits of $(Z_n)_{n\ge0}$ to y when $Z_0 = x$. Analogously, $F(x,y)$ stands for $F(x,y|1)$, the probability of ever reaching y when starting at x, and $U(x,x) = U(x,x|1) = \mathbb{P}_x[\mathbf{t}^x < \infty]$ is the probability of ever returning after starting at x.

(1.14) Definition. The Markov chain (X,P) is called *recurrent* if $G(x,y) = \infty$ for some ($\Longleftrightarrow$ every) $x,y \in X$, or equivalently, if $U(x,x) = 1$ for some ($\Longleftrightarrow$ every) $x \in X$. Otherwise, the Markov chain is called *transient.*

If $\rho(P) < 1$ then (X,P) is transient. The converse is not true. The spectral radius will be studied in Chapter II, with sufficient transience criteria as by-products. There is a useful characterization of recurrence is terms of superharmonic functions. P acts on functions $f : X \to \mathbb{R}$ by

$$Pf(x) = \sum_y p(x,y)\, f(y)\,.$$

(We assume that $P|f|$ is finite.) We say that f is *superharmonic* if $Pf \le f$ pointwise, and *harmonic* if $Pf = f$.

(1.15) Minimum principle. *If f is superharmonic and there is $x \in X$ such that $f(x) = \min_X f$ then f is constant.*

Proof. For every n, we have $f(x) \ge \sum_y p^{(n)}(x,y)f(y)$. Hence, it cannot be that $f(y) > f(x)$ for any y such that $p^{(n)}(x,y) > 0$. Now irreducibility yields $f \equiv f(x)$. □

For harmonic functions there is an analogous *maximum principle* (the minimum principle applied to $-f$).

(1.16) Theorem. *(X,P) is recurrent if and only if all non-negative superharmonic functions are constant.*

Proof. If (X,P) is transient then for $y \in X$, the function $x \mapsto G(x,y)$ is superharmonic, non-harmonic and hence non-constant.

Conversely, assume that (X,P) is recurrent. Let $f \geq 0$ be any superharmonic function. Set $g = f - Pf$. We claim that $g \equiv 0$. Suppose $g(y) > 0$ for some y. Choose $x \in X$. For each n,

$$\sum_{k=0}^{n} p^{(n)}(x,y)g(y) \leq \sum_{k=0}^{n} P^n g(x) = f(x) - P^{n+1}f(x) \leq f(x)\,.$$

Consequently, $G(x,y) \leq f(x)/g(y)$ in contradiction with recurrence. We have shown that every non-negative superharmonic function is harmonic.

Now, for superharmonic $f \geq 0$, choose $x \in X$ and set $M = f(x)$. Then $h = f \wedge M$ (pointwise minimum) is superharmonic and hence also harmonic. It assumes its maximum M, and by the maximum principle, h is constant. Thus f is constant. □

Here are further characterizations of recurrence and transience.

(1.17) Proposition. (a) *If (X,P) is recurrent then $F(x,y) = 1$ and*

$$\mathbb{P}_x[Z_n = y \text{ for infinitely many } n] = 1 \quad \textit{for all } x,y \in X\,.$$

(b) *If (X,P) is transient then for every finite $A \subset X$,*

$$\mathbb{P}_x[Z_n \in A \text{ for infinitely many } n] = 0 \quad \textit{for all } x \in X\,.$$

Proof. First, observe that for $y \in X$, the function $x \mapsto F(x,y)$ is superharmonic (Lemma 1.13). Thus, in the recurrent case, $F(\cdot,y)$ is constant by Theorem 1.16, and equal to $F(y,y) = 1$.

Next, write $V(x,y) = \mathbb{P}_x[Z_n = y$ for infinitely many $n]$. Conditioning with respect to $\mathbf{s}^y$, one sees that $V(x,y) = F(x,y)V(y,y) \leq V(y,y)$. Factoring through the first step, one sees that $x \mapsto V(x,y)$ is harmonic. By the maximum principle, $V(x,y) = V(y,y)$ for all x,y.

Set $V_m(x,x) = \mathbb{P}_x[(Z_n)_{n\geq 0}$ visits x at least m times$]$. Then $V_1(x,x) = 1$, and conditioning with respect to $\mathbf{t}^x$, one sees that $V_m(x,x) = U(x,x)V_{m-1}(x,x)$. Hence

$$V(x,x) = \lim_{n\to\infty} V_m(x,x) = \lim_{n\to\infty} U(x,x)^{m-1}$$

is equal to 1 in the recurrent case and 0 in the transient case. This proves (a). Furthermore, as A is finite,

$$\mathbb{P}_x[Z_n \in A \text{ for infinitely many } n] \leq \sum_{y\in A} V(x,y)\,,$$

which is 0 in the transient case. □

In particular, an irreducible Markov chain on a finite state space is always recurrent. We shall be interested in the case when X is infinite.

A recurrent Markov chain (X, P) is called *positive recurrent* if $\mathbb{E}_x[\mathbf{t}^x] < \infty$, and *null recurrent* if $\mathbb{E}_x[\mathbf{t}^x] = \infty$. Noting that $\mathbb{E}_x[\mathbf{t}^x] = U'(x, x|1-)$ (derivative with respect to z), it is easy to prove (similarly to Lemma 1.7) that this does not depend on the choice of $x \in X$. Before stating a criterion, we need another definition. P acts on non-negative measures on X by

$$\nu P(y) = \sum_x \nu(x)\, p(x, y)\,.$$

(We assume that νP is finite.) We say that ν is *excessive* if $\nu P \le \nu$ pointwise, and *invariant* if $\nu P = \nu$. (Irreducibility implies $\nu(x) > 0$ for all x if this holds for some x.) We omit the proof of the following criterion.

(1.18) Theorem. (a) *(X, P) is recurrent if and only if there is an invariant measure ν such that every positive excessive measure is a multiple of ν.*

(b) *(X, P) is positive recurrent if and only if ν has finite mass.*

The recurrent Markov chains that we shall encounter in this book will usually be null recurrent.

C. Random walks on graphs

We think of a *graph* as a finite or countable set of vertices (points) X, equipped with a symmetric *neighbourhood* or *adjacency relation* $\sim$ (a subset of $X \times X$). To view X, we draw a segment (edge), sometimes denoted by $[x, y]$, between every pair of neighbours x, y (so that $[x, y] = [y, x]$). Note that we do not exclude loops. We shall also write E or $E(X)$ for the edge set. A (finite) *path* from x to y in X is a sequence $\pi = [x = x_0, x_1, \dots, x_k = y]$ such that $x_{i-1} \sim x_i$; the number $k \ge 0$ is its length. (Alternatively, we shall think of π as a sequence of edges.) We shall always assume that our graphs are *connected,* that is, every pair of vertices is joined by a path. Thus, X carries an integer-valued *metric:* $d(x, y)$ is the minimum among the lengths of all paths from x to y. A path from x to y is called *simple* if it has no repeated vertex, and *geodesic* if its length is $d(x, y)$. We denote by $\Pi(x, y)$ the set of all geodesics from x to y.

The *degree* $\deg(x)$ of a vertex x is its number of neighbours. With a few exceptions, we shall usually consider only graphs which are *locally finite,* that is, every vertex has finite degree. We say that X has *bounded geometry,* if it is connected with bounded vertex degrees, and that X is *(M-)regular,* if $\deg(\cdot) \equiv M$ is constant.

The *simple random walk* on a locally finite graph X is the Markov chain with state space X and transition probabilities

$$p(x,y) = \begin{cases} 1/\deg(x)\,, & \text{if } y \sim x\,, \\ 0\,, & \text{otherwise.} \end{cases} \tag{1.19}$$

The graph X is said to be recurrent (transient) if the simple random walk has this property. The simple random walk is the basic example of a random walk (Markov chain) adapted to the underlying structure. In the sequel, we shall consider various more general types of adaptedness properties of the transition matrix P of a Markov chain to the structure of the underlying graph X, and it is in the presence of such adaptedness properties that we speak of a random walk (instead of a Markov chain). Here is a list of some of these properties, which will be frequently used.

We say that P is of *nearest neighbour* type, if $p(x,y) > 0$ occurs only when $d(x,y) \le 1$.

The random walk is called *uniformly irreducible* if there are $\varepsilon_0 > 0$ and $K < \infty$ such that

$$x \sim y \quad \text{implies} \quad p^{(k)}(x,y) \ge \varepsilon_0 \quad \text{for some } k \le K\,. \tag{1.20}$$

Note that this implies that $\deg(x) \le (K+1)/\varepsilon_0$ for every $x \in X$. Indeed,

$$K + 1 = \sum_{y \in X} \sum_{k=0}^{K} p^{(k)}(x,y) \ge \deg(x)\,\varepsilon_0\,.$$

When $\{y : p(x,y) > 0\}$ is finite for every x, we say that P has *finite range.* In itself, finite range is not an adaptedness property. However, this is the case for *bounded range,* that is, when

$$\sup\{d(x,y) : x, y \in X\,,\ p(x,y) > 0\} < \infty\,. \tag{1.21}$$

This can be generalized by imposing conditions like tightness, uniform integrability, etc., on the family of *step length distributions* on $\mathbb{N}_0$. The latter are given for each $x \in X$ by

$$\sigma_x(n) = \mathbb{P}_x[d(Z_1, Z_0) = n] = \sum_{y : d(y,x) = n} p(x,y)\,. \tag{1.22}$$

Consider the kth moment $M_k(\sigma_x) = \sum_n n^k \sigma_x(n) = \mathbb{E}_x\big(d(Z_1,Z_0)^k\big)$. We say that P has *finite kth moment,* if $M_k(P) = \sup_X M_k(\sigma_x)$ is finite, and that P has *exponential moment of order* $c > 0$, if $\sup_X \sum_n e^{cn}\sigma_x(n) < \infty$.

Further adaptedness conditions of geometric type will be introduced later on.

D. Trees

The nearest neighbour random walk on trees, and in particular the simple random walk on homogenous trees, is the other basic example besides Pólya's walk. A *tree* is a connected graph T without loops or cycles, where by a *cycle* in a graph we mean a sequence of vertices $x_0 \sim x_1 \sim \cdots \sim x_n$, $n \geq 3$, with no repetitions besides $x_n = x_0$. One characteristic feature of a tree is that for every pair of vertices x, y there is a unique path (*geodesic arc*) $\pi(x,y)$ of length $d(x,y)$ connecting the two.

Let P be the transition matrix of an irreducible nearest neighbour random walk on T. The following is a fundamental property linking tree structure and random walk.

(1.23) Lemma. *If $w \in \pi(x,y)$ then $F(x,y|z) = F(x,w|z)F(w,y|z)$.*

Proof. By the tree structure, the random walk must pass through w on the way from x to y. Conditioning with respect to the first visit in w, this yields

$$f^{(n)}(x,y) = \sum_{k=0}^{n} f^{(k)}(x,w)f^{(n-k)}(w,y). \qquad \square$$

As another "warm up" exercise, let us now consider a particularly typical example. The *homogeneous tree* $\mathbb{T}_M$ is the tree where all vertices have degree M. ($\mathbb{T}_2$ is isomorphic with $\mathbb{Z}$. See Figure 2 for $\mathbb{T}_3$.)

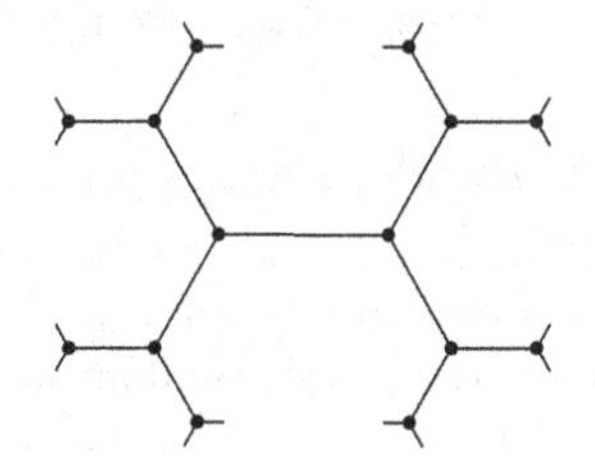

Figure 2: the homogeneous tree $\mathbb{T}_3$

(1.24) Lemma. *For the simple random walk on $\mathbb{T}_M$, one has*

$$G(x,y|z) = \frac{2(M-1)}{M-2+\sqrt{M^2-4(M-1)z^2}} \left(\frac{M-\sqrt{M^2-4(M-1)z^2}}{2(M-1)z}\right)^{d(x,y)}.$$

In particular, $\rho(P) = \dfrac{2\sqrt{M-1}}{M}$.

Proof. Obviously $F(x,y|z) = F(z)$ is the same for every pair of neighbours x, y, so that Lemma 1.23 yields $F(v,w|z) = F(z)^{d(v,w)}$. Now consider two neighbours x, y. Applying Lemma 1.13(d) we get

$$F(z) = F(x,y|z) = \sum_{w \sim x} \frac{1}{M} z\, F(z)^{d(y,w)} = \frac{1}{M} z + \frac{M-1}{M} z\, F(z)^2 \,.$$

This second order equation has two solutions. As $F(0) = 0$, the right one is (by continuity)

$$F(z) = \frac{1}{2(M-1)z} \left(M - \sqrt{M^2 - 4(M-1)z^2} \right) .$$

Using Lemma 1.13(c), (a) and (b), one now computes $U(x,x|z) = zF(z)$, $G(x,x|z)$ and the formula for $G(x,y|z)$.

The way in which $\rho(P)$ is read from this formula is typical: $G(x,x|z)$ is a power series with non-negative coefficients. By Pringsheim's theorem (see Hille [173], p. 133), the radius of convergence $\mathtt{r}(P) = 1/\rho(P)$ must be its smallest positive singularity. Thus, we have to compute the value of $z > 0$ where the term under the square root is equal to 0. □

As a consequence, the simple random walk on $\mathbb{T}_M$ is transient for every $M \geq 3$.

(1.25) Exercise. Compute $G(x,y|z)$ for the simple random walk on the bi-regular tree, that is, the tree where the vertex degrees are constant on each of the two bipartite classes. (These are the points at even or odd distance, respectively, from a given reference vertex.)

E. Random walks on finitely generated groups

Pólya's walk, besides being the simple random walk on a graph (the d-dimensional grid), can also be interpreted in terms of groups. The same is true for the simple random walk on $\mathbb{T}_M$.

Let Γ be a discrete group with unit element o (the symbol e will be used for edges), and let μ be a probability measure on Γ. The *(right) random walk* on Γ with *law* μ is the Markov chain with state space Γ and transition probabilities

$$p(x,y) = \mu(x^{-1}y) \,.$$

(Unless Γ is abelian, the group operation will be written multiplicatively.) Besides the trajectory space, in this case we may also use the product space $(\Gamma,\mu)^{\mathbb{N}}$ to obtain an equivalent model of (Z_n): the nth projections X_n of $\Gamma^{\mathbb{N}}$ onto Γ $(n \geq 1)$ constitute a sequence of independent Γ-valued random variables with common distribution μ, and the right random walk starting at $x \in \Gamma$ is

$$Z_n = x\,X_1 \cdots X_n\,, \quad n \geq 0\,.$$

This generalizes the ordinary scheme of sums of i.i.d. random variables on the integers or reals. The n-step transition probabilities are obtained by

$$p^{(n)}(x,y) = \mu^{(n)}(x^{-1}y)\,,$$

where $\mu^{(n)}$ is the n-fold convolution of μ with itself, with $\mu^{(0)} = \delta_o$, the point mass at the group identity.

Let us relate groups with graphs. Suppose that Γ is finitely generated, and let S be a symmetric set of generators of Γ. The *Cayley graph* $X(\Gamma, S)$ of Γ with respect to S has vertex set Γ, and two elements $x, y \in \Gamma$ are neighbours if $x^{-1}y \in S$. This graph is connected, locally finite and regular (all points have the same degree $|S|$). If $o \in S$ then it has a loop at each vertex. Γ acts as a group of self-isometries on the Cayley graph; the action is left multiplication.

For example, in the abelian group $\mathbb{Z}^d$, we may choose S as the set of all elements with euclidean length 1. The resulting Cayley graph is the usual grid. The homogeneous tree $\mathbb{T}_M$ is the Cayley graph of the group $\Gamma = \langle a_1, \dots, a_M \mid a_i^2 = o \rangle$ with respect to $S = \{a_1, \dots, a_M\}$. This group is the free product of M copies of the two-element group $\mathbb{Z}_2$ (see Chapter II for more on free products).

Write d_S for the graph metric on $X(\Gamma, S)$, and let S' be another finite, symmetric set of generators. Then the metrics d_S and $d_{S'}$ are *equivalent* in the sense that there are constants C, C' such that

$$d_S \leq C'\,d_{S'} \quad \text{and} \quad d_{S'} \leq C\,d_S\,. \tag{1.26}$$

Indeed, choosing $C = \max\{d_{S'}(x,o) : x \in S\}$ we get $d_{S'}(x,y) \leq C$ whenever $d_S(x,y) = 1$.

The simple random walk on $X(\Gamma, S)$ is nothing but the right random walk on Γ whose law μ is the equidistribution on S.

For arbitrary μ, we write $\operatorname{supp}\mu = \{x \in \Gamma : \mu(x) > 0\}$. Then $\operatorname{supp}\mu^{(n)} = (\operatorname{supp}\mu)^n$, and μ (that is, the random walk with law μ) is irreducible if and only if

$$\bigcup_{n=1}^{\infty} (\operatorname{supp}\mu)^n = \Gamma\,.$$

In particular, irreducibility is uniform with respect to the Cayley graph structure: for every $x \in S$, choose n_x such that $\mu^{(n_x)}(x) > 0$. Then we may set $K = \max\{n_x : x \in S\}$ and $\varepsilon_0 = \min\{\mu^{(n_x)}(x) : x \in S\}$.

Also, for random walks on groups, finite range coincides with bounded range ($\operatorname{supp} \mu$ is finite), and the step length distributions σ_x are the same for every x. Indeed, the geometric adaptedness conditions listed above for random walks on graphs can be viewed as attempts to transport typical homogeneity properties of random walks on groups to a non-algebraic setting.

As we shall see, recurrence of random walks on a group Γ is independent of the particular choice of the law μ (as long as it is irreducible and symmetric, and the step length distribution has finite second moment). Thus, one speaks of recurrent (or transient) groups. The classification of the recurrent ones among all finitely generated groups will be one of the main themes of this chapter.

F. Locally finite graphs and topological groups

Let X be a locally finite, connected graph with discrete metric d. An *automorphism* of X is a self-isometry of X with respect to d. The group $\mathrm{AUT}(X)$ of all automorphisms of X can be equipped with the topology of pointwise convergence: for a sequence $(\gamma_n)_{n\geq 1}$ in $\mathrm{AUT}(X)$, we have

$$\gamma_n \to \gamma \iff \text{for every } x \in X\,,\ \gamma_n x = \gamma x \ \text{ for all } n \geq n_x\,.$$

For an arbitrary subgroup Γ of $\mathrm{AUT}(X)$, we write

$$\Gamma_x = \{\gamma \in \Gamma : \gamma x = x\}$$

for the stabilizer of vertex x in Γ. For the topology on $\mathrm{AUT}(X)$, a subbase of the neighbourhood filter at the identity ι is given by the family of all $\mathrm{AUT}(X)_x$, $x \in X$. The stabilizers are both open and closed, and even compact. Indeed, we have the following simple lemma, whose proof is omitted. (Local finiteness is crucial here.)

(1.27) Lemma. *A subset B of $\mathrm{AUT}(X)$ is relatively compact if and only if for some ($\iff$ every) $x \in X$, the orbit $Bx = \{\gamma x : \gamma \in B\}$ is finite.*

Thus, the automorphism group of a locally finite, connected graph is a locally compact, totally disconnected Hausdorff group with countable base of the topology.

Let Γ be a closed subgroup of $\mathrm{AUT}(X)$. Then Γ carries a *left Haar measure* $d\gamma$. For a Borel subset B of Γ, we write $|B|$ for its measure. Recall the basic properties of $d\gamma$: every open set has positive measure, every compact set has finite measure, $|\gamma A| = |A|$ for every $\gamma \in \Gamma$, and as a Radon measure with these properties, $d\gamma$ is unique up to multiplication by constants. When Γ is discrete, the Haar measure is (a multiple of) the counting measure. We shall also need the *modular function* $\Delta = \Delta_\Gamma$ of Γ. Again, recall that this is a homomorphism from Γ to the multiplicative group of positive reals satisfying $|B\gamma| = \Delta(\gamma)|B|$ for every Borel set in Γ and $\int_\Gamma g(\gamma^{-1})\Delta(\gamma^{-1})\,d\gamma = \int_\Gamma g(\gamma)\,d\gamma$ for integrable functions g on Γ. The group Γ is called *unimodular* when $\Delta \equiv 1$. If $x \in X$, then the stabilizer Γ_x is open and compact, so that $0 < |\Gamma_x| < \infty$. For $\gamma \in \Gamma$, we have $\Gamma_{\gamma x} = \gamma\Gamma_x\gamma^{-1}$, whence

$$|\Gamma_{\gamma x}| = \Delta(\gamma^{-1})|\Gamma_x|. \tag{1.28}$$

(1.29) Lemma.
$$\frac{|\Gamma_x|}{|\Gamma_y|} = \frac{|\Gamma_x y|}{|\Gamma_y x|}.$$

Proof. (Of course, $|\Gamma_x y|$ and $|\Gamma_y x|$ denote the cardinalities of the respective orbits.) The open and compact subgroup $\Gamma_x \cap \Gamma_y$ is the stabilizer of y in Γ_x. Writing the left coset partition, one sees that it has finite index in Γ_x which is equal to $|\Gamma_x y|$ on the one hand and to $|\Gamma_x|/|\Gamma_x \cap \Gamma_y|$ on the other. Exchanging the roles of x and y, we also get $|\Gamma_y x| = |\Gamma_y|/|\Gamma_x \cap \Gamma_y|$. In dividing, $|\Gamma_x \cap \Gamma_y|$ cancels. □

In Chapter II we shall see examples of graphs with non-unimodular automorphism group. The following formula will be useful:

$$\int_\Gamma f(\gamma x)\,d\gamma = \sum_y \int_{\{\gamma:\gamma x = y\}} f(y)\,d\gamma = |\Gamma_x| \sum_{y \in \Gamma x} f(y), \tag{1.30}$$

whenever $f : X \to \mathbb{R}$ is a function for which the sum (integral) converges absolutely. Analogously, if $e = [x, y]$ is a (non-oriented) edge of X, then Γ_e is the set of all $\gamma \in \Gamma$ satisfying $\{\gamma x, \gamma y\} = \{x, y\}$. Again, Γ_e is open and compact ($\Gamma_x \cap \Gamma_y$ is a subgroup with index 1 or 2), formula (1.28) remains valid, and there is a formula analogous to (1.30) for functions defined on the edge set of X.

Let X_i, $i \in \mathcal{I}$, be the orbits of Γ on X. The vertex set of the *factor graph* $\Gamma\backslash X$ is $\mathcal{I}$, and $i \sim j$ in $\Gamma\backslash X$ if $u \sim v$ for some $u \in X_i$, $v \in X_j$. We obtain a connected, locally finite graph. The graph X is called *vertex-transitive* (or just *transitive*) if $\mathrm{AUT}(X)$ acts transitively on X, and it is called *quasi-transitive* if $\mathrm{AUT}(X)$ acts with finitely many orbits. Now let P be the transition matrix of a random walk on X. We define $\mathrm{AUT}(X, P)$ to be the group of all $\gamma \in \mathrm{AUT}(X)$ which leave P invariant, that is, $p(\gamma x, \gamma y) = p(x, y)$ for all $x, y \in X$. For a subgroup Γ of $\mathrm{AUT}(X, P)$, we can define the transition matrix $\tilde{P}$ of the *factor chain* on $\mathcal{I} \cong \Gamma\backslash X$ by

$$\tilde{p}(i, j) = \sum_{w \in X_j} p(x, w), \tag{1.31}$$

where $x \in X_i$ is arbitrary. It inherits irreducibility from P. We say that (X, P) is (quasi-)transitive, if $\mathrm{AUT}(X, P)$ acts transitively (with finitely many orbits) on X. Note that this in itself is a strong adaptedness property. Once more, irreducibility is automatically uniform, and there are only finitely many different step length distributions (one in the transitive case). Needless to say, not every random walk with transitive automorphism group arises as a random walk on a finitely generated discrete group. Examples will be considered in §12.C.

More generally than for finitely generated groups, in this chapter we shall give a structural classification of the recurrent ones among all locally finite, connected quasi-transitive graphs.

2. Recurrence and transience of infinite networks

In this section we shall exhibit a variety of recurrence criteria for Markov chains with a symmetry property (reversibility) and a comparison theorem for reversible and non-reversible chains. Recall that we always require irreducibility (1.5).

A. Reversible Markov chains

For the moment, let X be an infinite, countable set, not necessarily equipped with the structure of a locally finite graph. The Markov chain (X, P) is called *reversible* if there is a function (measure) $m : X \to (0, \infty)$ such that

$$m(x)p(x,y) = m(y)p(y,x) \quad \text{for all } x, y \in X. \tag{2.1}$$

In this case, $a(x,y) = m(x)p(x,y) = a(y,x)$ is called the *conductance* between x and y, and $m(x)$ is the *total conductance* at x. Conversely, if $a : X \times X \to [0, \infty)$ is a symmetric function such that $m(x) = \sum_y a(x,y)$ is positive and finite for every x, then $p(x,y) = a(x,y)/m(x)$ defines a reversible Markov chain (random walk). Note that $m(\cdot)$ is an invariant measure for P.

It will be convenient to equip X with an edge set $E = E(P)$ such that for the resulting graph, P is of nearest neighbour type: $[x,y] \in E(P)$ if and only if $a(x,y) > 0$. Note that (X, E) is connected, but not necessarily locally finite. In addition, for each – a priori non-oriented – edge $e \in E(P)$, out of its two endpoints, we specify one as its initial vertex e^- and the other as its terminal vertex e^+. Nothing of what we are going to do depends on the particular choice. (We shall think of functions on E as flows. A positive or negative sign then stands for the flow moving from e^- to e^+ or conversely.) The *resistance* of $e \in E$ is $r(e) = 1/a(e^-, e^+)$. The triple $\mathcal{N} = (X, E, r)$ is called a *network*. We may think of $\mathcal{N}$ as an infinite electrical network, where each edge e is a wire with resistance $r(e)$, and several wires are linked at each node (vertex). Or we may think of a system of tubes e with cross-section 1 and length $r(e)$, connected at the vertices. We shall say that $\mathcal{N}$ is recurrent (transient) if (X, P) has the respective property.

Later on, X will carry the structure of a locally finite graph. In this case, the simple random walk arises from the network where $a(x, y) = 1$ if $x \sim y$ (and $= 0$, otherwise). However, we stress that in general $E(P)$ does not have to be the edge set $E(X)$. The point is that we shall consider "adapted" reversible random walks on the graph X which are not necessarily nearest neighbour and may even have infinite range. For example, on groups, we shall be interested in symmetric random walks without too many restrictions on the law μ. For the moment, this is irrelevant, and we only consider the graph (X, E) of the network, locally finite or not, with $E = E(P)$.

It will be convenient to introduce a potential theoretic setup, as follows. Consider the real Hilbert spaces $\ell^2(X, m)$ and $\ell^2(E, r)$ with inner products

$$(f, g) = \sum_{x \in X} f(x)g(x)m(x) \quad \text{and} \quad \langle u, v \rangle = \sum_{e \in E} u(e)v(e)r(e),$$

respectively. We introduce the difference operator

$$\nabla : \ell^2(X, m) \to \ell^2(E, r), \quad \nabla f(e) = \frac{f(e^+) - f(e^-)}{r(e)}.$$

It is easily seen to have norm $\leq \sqrt{2}$. Its adjoint is given by

$$\nabla^* u(x) = \frac{1}{m(x)} \left(\sum_{e:e^+ = x} u(e) - \sum_{e:e^- = x} u(e) \right).$$

If we think of u as a flow in the network, then $\sum_{e:e^+ = x} u(e)$ is the amount flowing into node x, and $\sum_{e:e^- = x} u(e)$ is the amount flowing out, so that $m(x)\nabla^* u(x)$ is the "loss" at x. Below, we shall give a more precise definition of flows. The *Laplacian* is

$$\mathfrak{L} = -\nabla^* \nabla = P - I, \tag{2.2}$$

where I is the identity matrix over X and P is the transition matrix of our random walk, both viewed as operators on functions $X \to \mathbb{R}$.

Consider the space $\mathcal{D}(\mathcal{N})$ of all functions f on X (not necessarily in $\ell^2(X, m)$) such that $\nabla f \in \ell^2(E, r)$. If f is such a function, then its *Dirichlet sum* (or *Dirichlet norm*) is

$$\begin{aligned} D(f) = \langle \nabla f, \nabla f \rangle &= \sum_{e \in E} \frac{\big(f(e^+) - f(e^-)\big)^2}{r(e)} \\ &= \frac{1}{2} \sum_{x,y \in X} \big(f(x) - f(y)\big)^2 m(x)p(x, y). \end{aligned} \tag{2.3}$$

This is a quasi-norm, its kernel consisting of the constants. We shall also denote the Dirichlet sum by $D_{\mathcal{N}}(\cdot)$ or $D_P(\cdot)$. The space $\mathcal{D}(\mathcal{N})$ can be equipped with an inner product by choosing a reference point $o \in X$:

$$(f,g)_D = (f,g)_{D,o} = \langle \nabla f, \nabla g \rangle + f(o)g(o)\,.$$

Let us gather a few standard facts.

(2.4) Lemma. (a) *$\mathcal{D}(\mathcal{N})$ is a Hilbert space.*

(b) *Changing the reference point o gives rise to an equivalent Hilbert space norm.*

(c) *Convergence in $\mathcal{D}(\mathcal{N})$ implies pointwise convergence.*

(d) *If $f \in \mathcal{D}(\mathcal{N})$ then $\nabla^*(\nabla f) = -\mathfrak{L} f$.*

Proof. For (a)–(c), we need one simple inequality. Let $x \in X$, $x \neq o$. By connectedness of the graph (X,E), there are $o = x_0, x_1, \dots, x_k = x \in X$ such that $e_i = [x_{i-1}, x_i] \in E$. Let $C_1(x) = \sum_{i=1}^k r(e_i)$. Then for $f \in \mathcal{D}(\mathcal{N})$, using the Cauchy–Schwarz inequality,

$$\big(f(x) - f(o)\big)^2 = \left(\sum_{i=1}^k \frac{f(x_i) - f(x_{i-1})}{\sqrt{r(e_i)}} \sqrt{r(e_i)}\right)^2 \le C_1(x)\, D(f)\,.$$

Setting $C_2(x) = 2\max\{1, C_1(x)\}$, we get

$$f(x)^2 \le C_2(x)\,(f,f)_{D,o}\,.$$

Consequently, $(f,f)_{D,x} \le C_3(x)\,(f,f)_{D,o}$, where $C_3(x) = C_2(x) + 1$. Exchanging the roles of x and o, there is an analogous inequality: this proves (b).

Now let (f_n) be a Cauchy sequence in $\mathcal{D}(\mathcal{N})$. By the above, for every $x \in X$, $\big(f_n(x)\big)$ is Cauchy in $\mathbb{R}$ and converges to some limit $f(x)$. On the other hand, (∇f_n) is a Cauchy sequence in $\ell^2(E,r)$. Hence, there is $u \in \ell^2(E,r)$ such that $\nabla f_n \to u$ in the latter Hilbert space. It must be that $u = \nabla f$, so that $D(f) < \infty$. This proves (a).

Part (c) is now obvious.

Part (d) is a technical remark. For every $x \in X$, even when (X,P) is not locally finite, the sum $\frac{1}{m(x)} \sum_{[y,x]\in E} |f(x) - f(y)| a(x,y)$ is finite (bounded by $D(f)$). Hence, the terms in

$$\nabla^*(\nabla f)(x) = \frac{1}{m(x)} \left(\sum_{e^+ = x} \frac{f(x) - f(e^-)}{r(e)} + \sum_{e^- = x} \frac{f(x) - f(e^+)}{r(e)} \right)$$

may be reordered to give $(I - P)f(x)$. □

By $\ell_0(X)$ we denote the linear space of finitely supported functions on X, and by $\mathcal{D}_0(\mathcal{N})$ its closure in $\mathcal{D}(\mathcal{N})$. Note that $\ell^2(X,m) \subset \mathcal{D}_0(\mathcal{N})$ (as ∇ is bounded), while the converse is in general not true (see Chapter II).

Note that for every $k \geq 1$, P^k is reversible with the same invariant measure $m(\cdot)$ (but may be non–irreducible). By $D_{P^k}(\cdot)$ we denote the associated Dirichlet norm according to (2.3).

(2.5) Lemma. *$D_{P^k}(f) \leq k^2\, D_P(f)$ for every $f \in \ell_0(X)$.*

Proof. We have, using Cauchy–Schwarz and reversibility,

$$\begin{aligned} D_{P^k}(f) &= \tfrac{1}{2} \sum_{x_0,\dots,x_k \in X} \big(f(x_k) - f(x_0)\big)^2 m(x_0)p(x_0,x_1)\cdots p(x_{k-1},x_k) \\ &\leq \tfrac{1}{2} \sum_{x_0,\dots,x_k \in X} k \sum_{i=1}^{k} \big(f(x_i) - f(x_{i-1})\big)^2 m(x_0)p(x_0,x_1)\cdots p(x_{k-1},x_k) \\ &= \tfrac{k}{2} \sum_{i=1}^{k} \sum_{x_{i-1},x_i \in X} \big(f(x_i) - f(x_{i-1})\big)^2 m(x_{i-1})p(x_{i-1},x_i)\,. \end{aligned}$$

□

Before introducing and proving a first set of transience criteria, we need some preliminary material. If A is a subset of X, then we denote by P_A the restriction of P to X:

$$p_A(x,y) = p(x,y)\,, \quad \text{if } x,y \in A, \text{ and} \quad p(x,y) = 0\,, \quad \text{otherwise.} \tag{2.6}$$

Usually, we consider P_A as a matrix over the whole of X, but we shall use the same notation for the truncated matrix over A alone. In the same way, we define the restriction I_A of the identity matrix. The (x,y)-element of the matrix power P_A^n is $p_A^{(n)}(x,y) = \mathbb{P}_x[Z_n = y\,,\ Z_k \in A\ (0 \leq k \leq n)]$. In particular, $P_A^0 = I_A$. For the associated Green function we write

$$G_A(x,y|z) = \sum_{n=0}^{\infty} p_A^{(n)}(x,y)\, z^n\,, \quad G_A(x,y) = G_A(x,y|1)\,.$$

(N.B.: this is *not* the restriction of $G(\cdot,\cdot|z)$ to A.) When A is finite, then it is well known and easy to prove that G_A is finite. In matrix (or operator) notation, we have

$$(I_A - P_A)G_A = I_A\,. \tag{2.7}$$

(2.8) Lemma. *Let $A \subset X$ be finite, $x \in A$ and $f \in \ell_0(X)$ be such that* $\operatorname{supp} f \subset A$. *Then*

$$\langle \nabla f, \nabla G_A(\cdot, x) \rangle = m(x) f(x) .$$

Proof. By (2.2) and (2.7), and as the functions involved are 0 outside A,

$$\begin{aligned} \langle \nabla f, \nabla G_A(\cdot, x) \rangle &= (f, (I - P) G_A(\cdot, x)) \\ &= (f, (I_A - P_A) G_A(\cdot, x)) = (f, \delta_x) . \end{aligned}$$

□

(2.9) Lemma. *If (X, P) is transient, then $G(\cdot, x) \in \mathcal{D}_0(\mathcal{N})$ for every $x \in X$.*

Proof. Let $A \subset B$ be finite subsets of X containing x. Applying Lemma 2.8 to A, B, $f = G_B(\cdot, x)$ and $f = G_A(\cdot, x)$, respectively, we get

$$\begin{aligned} D\big(G_B(\cdot, x) - G_A(\cdot, x)\big) &= \langle \nabla G_B(\cdot, x), \nabla G_B(\cdot, x) \rangle - 2 \langle \nabla G_A(\cdot, x), \nabla G_B(\cdot, x) \rangle \\ &\quad + \langle \nabla G_A(\cdot, x), \nabla G_A(\cdot, x) \rangle \\ &= m(x) \big(G_B(x, x) - G_A(x, x)\big) . \end{aligned}$$

Now let (A_n) be an exhaustion of X (that is, an increasing sequence with union X) by finite sets containing x. By monotone convergence, $G_{A_n}(x, x)$ tends to $G(x, x)$. Hence, by the above, $\big(G_{A_n}(\cdot, x)\big)_{n \ge 1}$ is a Cauchy sequence in $\mathcal{D}(\mathcal{N})$. By Lemma 2.4, it converges to its pointwise limit, that is, $G(\cdot, x)$. Thus, $G(\cdot, x)$ can be approximated in $\mathcal{D}(\mathcal{N})$ by finitely supported functions. □

B. Flows, capacity, and Nash-Williams' criterion

Given a vertex $x_0 \in X$ and a real number i_0, a *finite energy flow from x to ∞ with input i_0* on $\mathcal{N}$ is a function $u \in \ell^2(E, r)$ such that

$$\nabla^* u(y) = -\frac{i_0}{m(x)} \delta_x(y) \quad \text{for all } y \in X .$$

Its *energy* is $\langle u, u \rangle$. Thus, we may think of the network as a system of tubes filled with (incompressible) liquid, and at the source x_0, liquid is pumped in at a constant rate of i_0 litres per second. Requiring that this be possible with finite energy $\langle u, u \rangle$ is of course absurd if the network is finite (unless $i_0 = 0$), and we shall see that the existence of such flows characterizes transient networks. In this sense, recurrent networks correspond more to our intuition of the "real world".

The *capacity* of a set $A \subset X$ is

$$\operatorname{cap}(A) = \inf\{D(f) : f \in \ell_0(X),\ f \equiv 1 \text{ on } A\} . \tag{2.10}$$

Note that

$$\text{(2.11)} \qquad \mathrm{cap}(A) = \min\{D(f) : f \in \mathcal{D}_0(\mathcal{N}),\ f \equiv 1 \text{ on } A\}.$$

Indeed, the convex set $\{f \in \mathcal{D}_0(\mathcal{N}) : f|_A \equiv 1\}$ is the closure in $\mathcal{D}(\mathcal{N})$ of $\{f \in \ell_0(X) : f|_A \equiv 1\}$. By a standard theorem in Hilbert space theory (Rudin [282], Th. 4.10), there is a unique element in the set where the norm is minimized. We write $\mathrm{cap}(x)$ for $\mathrm{cap}(\{x\})$.

With these ingredients, we can formulate the following useful collection of necessary and sufficient transience criteria.

(2.12) Theorem. *For a reversible Markov chain (X,P) and the associated network $\mathcal{N}$, the following statements are equivalent.*

(a) *The network is transient.*
(b) *For some ($\Longleftrightarrow$ every) $x \in X$, there is a finite energy flow from x to ∞ with non-zero input.*
(c) *For some ($\Longleftrightarrow$ every) $x \in X$, $\mathrm{cap}(x) > 0$.*
(d) *The constant function $\mathbf{1}$ does not belong to $\mathcal{D}_0(\mathcal{N})$.*

Proof. (a) $\Longrightarrow$ (b). If the network is transient, then $G(\cdot,x) \in \mathcal{D}_0(\mathcal{N})$. Define $u = -\frac{i_0}{m(x)}\nabla G(\cdot,x)$. Then $u \in \ell^2(E(P),r)$ and

$$\nabla^* u = \frac{i_0}{m(x)}\mathfrak{L}G(\cdot,x) = -\frac{i_0}{m(x)}\delta_x .$$

We have used Lemma 2.4(d) in the last identity.

(b) $\Longrightarrow$ (c). Suppose there is a finite energy flow u from x to ∞ with input $i_0 \neq 0$. We may assume that $i_0 = -1$. Now let $f \in \ell_0(X)$ with $f(x) = 1$. Then

$$\langle \nabla f, u\rangle = (f, \nabla^* u) = \left(f, \frac{1}{m(x)}\delta_x\right) = f(x) = 1 .$$

Hence, $1 = |\langle \nabla f, u\rangle|^2 \le D(f)\ \langle u,u\rangle$. We obtain $\mathrm{cap}(x) \ge 1/\langle u,u\rangle > 0$.

(c) $\Longleftrightarrow$ (d). This is immediate from (2.11). We have $\mathrm{cap}(x) = 0$ if and only if there is $f \in \mathcal{D}_0(\mathcal{N})$ with $f(x) = 1$ and $D(f) = 0$, that is, $f \equiv 1$.

(c) $\Longrightarrow$ (a). Let $A \subset X$ be finite and such that $x \in A$. Set $f = G_A(\cdot,x)/G_A(x,x)$. Then $f \in \mathcal{D}_0(\mathcal{N})$ and $f(x) = 1$. Therefore, applying Lemma 2.8,

$$\mathrm{cap}(x) \le D(f) = \frac{1}{G_A(x,x)^2}\langle \nabla G_A(\cdot,x), \nabla G_A(\cdot,x)\rangle = \frac{m(x)}{G_A(x,x)} .$$

We obtain $G_A(x,x) \le m(x)/\mathrm{cap}(x)$ for every finite $A \subset X$ containing x. Now take an exhaustion of X by sets of this type. Passing to the limit, monotone convergence yields $G(x,x) \le m(x)/\mathrm{cap}(x) < \infty$. □

(2.13) Exercise. Prove the following in the transient case.

The unit flow from x to ∞ with minimal energy is given by $u(x) = -\Delta G(\cdot, x)/m(x)$, its energy (the resistance between x and ∞*) is* $G(x,x)/m(x)$, and $\mathrm{cap}(x) = m(x)/G(x,x)$,

One of the useful consequences of Theorem 2.12 is that recurrence / transience depends only on the Dirichlet norm $D(\cdot)$ associated with P according to (2.3).

(2.14) Corollary. *Let P_1 and P_2 be the transition matrices of two reversible Markov chains on X with associated Dirichlet norms D_1 and D_2, respectively. Suppose that there is $\varepsilon_1 > 0$ such that*

$$D_2(f) \geq \varepsilon_1 \, D_1(f) \quad \text{for all } f \in \ell_0(X)\,.$$

Then transience of (X, P_1) implies transience of (X, P_2).

This holds, in particular, when the associated conductances satisfy $a_2(x,y) \geq \varepsilon_1 \, a_1(x,y)$ for all $x, y \in X$, where $\varepsilon_1 > 0$.

By a *subnetwork* $\mathcal{N}'$ of $\mathcal{N}$ we mean a connected subgraph of (X, E), together with the restriction of the resistance function to the remaining edges. A finite energy flow in $\mathcal{N}'$ is also a finite energy flow in $\mathcal{N}$.

(2.15) Corollary. *Transience of a subnetwork implies transience of $\mathcal{N}$.*

For example, take a locally finite graph X and a connected subgraph X' of X. Then recurrence of the simple random walk on X implies recurrence of the simple random walk on X'. More applications of Theorem 2.12 will be discussed later on, one right now as a preparation for Nash-Williams' recurrence criterion.

(2.16) Nearest neighbour random walk on $\mathbb{N}_0$. Speaking of $\mathbb{N}_0$ as a graph, we think of a one-way infinite path where neighbourhood is given by the edges $e_k = [k-1, k]$, $k \geq 1$. For an (irreducible) nearest neighbour random walk, we must specify the transition probabilities $p(k-1,k) > 0$, $p(k,k-1) > 0$ for $k \geq 1$ and $p(k,k) \geq 0$ for $k \geq 0$. We see that the random walk is reversible with

$$m(k) = \frac{p(0,1)p(1,2)\cdots p(k-1,k)}{p(k,k-1)\cdots p(2,1)p(1,0)} \quad \text{and} \quad r(e_k) = \frac{p(k-1,k-2)\cdots p(1,0)}{p(0,1)\cdots p(k-1,k)}\,.$$

The only flow from 0 to ∞ with input $i_0 = 1$ is the unit flow, and its energy is $\sum r(e_k)$. Thus, the random walk is

$$\text{transient} \iff \sum_{k=1}^{\infty} r(e_k) < \infty\,, \qquad \text{recurrent} \iff \sum_{k=1}^{\infty} r(e_k) = \infty\,,$$

and in the latter case, it is positive recurrent if and only if $\sum_{k\geq 0} m(k) < \infty$.

Note that if in an arbitrary network we can find a one-way infinite path along which the resistances have finite sum, then the network is transient. Indeed, the unit flow along that path (with input 1 at its initial point) has finite energy. □

Next, we describe the method of *shorting*. Let (X, P) be reversible with associated network $\mathcal{N}$. Suppose that we have a partition of X:

$$X = \bigcup_{i\in\mathcal{I}} X_i\,, \quad \text{such that} \quad \mathbf{1}_{X_i} \in \mathcal{D}_0(\mathcal{N}) \quad \forall i \in \mathcal{I}. \tag{2.17}$$

(The last condition is crucial. It holds in particular if $\sum_{x\in X_i} m(x) < \infty$.) Then we can define the *shorted network* $\mathcal{N}'$ with vertex set $\mathcal{I}$ and conductance function

$$a'(i,j) = \begin{cases} \displaystyle\sum_{x\in X_i, y\in X_j} a(x,y)\,, & \text{if } j \neq i\,, \\ 0\,, & \text{if } j = i\,. \end{cases} \tag{2.18}$$

Note that $m'(i) = \sum_j a'(i,j) = D(\mathbf{1}_{X_i}) < \infty$ for every i. Thus, the shorted network arises from a reversible Markov chain (I, P'), with transition probabilities $p'(i,j) = a'(i,j)/m'(i)$.

(2.19) Theorem. *Suppose that (X, P) is reversible and that (I, P') is obtained from (X, P) by shorting. If (I, P') is recurrent then so is (X, P).*

Proof. If $f \in \mathcal{D}(\mathcal{N}')$ then we set $\bar{f}(x) = f(i)$ for $x \in X_i$. Then

$$\begin{aligned} D_{\mathcal{N}}(\bar{f}) &= \frac{1}{2}\sum_{x,y\in X} \big(\bar{f}(x) - \bar{f}(y)\big)^2 a(x,y) \\ &= \frac{1}{2}\sum_{i,j\in X}\ \sum_{\substack{x\in X_i \\ y\in X_j}} \big(f(i) - f(j)\big)^2 a(x,y) \\ &= \frac{1}{2}\sum_{i,j\in X} \big(f(i) - f(j)\big)^2 a'(i,j) = D_{\mathcal{N}'}(f)\,. \end{aligned}$$

Next, we use the assumption $\mathbf{1}_{X_i} \in \mathcal{D}_0(\mathcal{N})$: if $f \in \ell_0(I)$ then

$$\bar{f} = \sum_{i\in\mathcal{I}} f(i)\,\mathbf{1}_{X_i} \in \mathcal{D}_0(\mathcal{N})\,.$$

Now, if $\mathcal{N}'$ is recurrent, then by Theorem 2.12(d), $\mathbf{1} \in \mathcal{D}_0(\mathcal{N}')$, and there is a sequence (f_n) in $\ell_0(I)$ such that $D_{\mathcal{N}'}(f_n - \mathbf{1}) \to 0$. Consequently, $(\bar{f}_n)$ is a sequence in $\mathcal{D}_0(\mathcal{N})$ satisfying $D_{\mathcal{N}}(\bar{f}_n - \mathbf{1}) \to 0$. We obtain $\mathbf{1} \in \mathcal{D}_0(\mathcal{N})$, and $\mathcal{N}$ is recurrent. □

Combining the theorem with (2.16), we get Nash-Williams' recurrence criterion as a particular instance:

(2.20) Corollary. *Under the conditions of (2.17) and (2.18), suppose that* $\mathcal{I} = \mathbb{N}_0$ *and that* $a'(i,j) = 0$ *if* $|i-j| \geq 2$. *If*

$$\sum_{i=1}^{\infty} \frac{1}{a'(i-1,i)} = \infty$$

then (X, P) *is recurrent.*

Using this criterion, there is another, instructive, way to see that the simple random walk on the square grid $\mathbb{Z}^2$ is recurrent. Let d be the graph metric of $\mathbb{Z}^2$, and set

$$X_i = \{x \in \mathbb{Z}^2 : d(x,0) = i\} = \{(k,l) \in \mathbb{Z}^2 : |k| + |l| = i\}\,,\ i \geq 0\,.$$

Recall that all edges have conductance 1. For the shorted network, we get $a'(i-1,i) = 8i - 4$, and Corollary 2.20 applies.

We conclude this subsection with another simple example, which uses the flow criterion and shows that very natural subgraphs of transient graphs may well be recurrent.

(2.21) Comb lattices in $\mathbb{Z}^d$. We write $\mathbf{C}_d$ (comb lattice) for the following spanning tree[1] of $\mathbb{Z}^d$: the vertex set is $\mathbb{Z}^d$, and neighbourhood is described by

$$(k_1, \dots, k_{j-1}, k_j, 0, \dots, 0) \sim (k_1, \dots, k_{j-1}, k_j + 1, 0, \dots, 0)\,,$$

where $j = 1, \dots, d$ and $k_i \in \mathbb{Z}$ $(1 \leq i \leq j)$. Thus, $\mathbf{C}_1 = \mathbb{Z}$, and $\mathbf{C}_d$ is obtained from $\mathbf{C}_{d-1}$ by attaching at each point a two-way infinite path (copy of $\mathbb{Z}$).

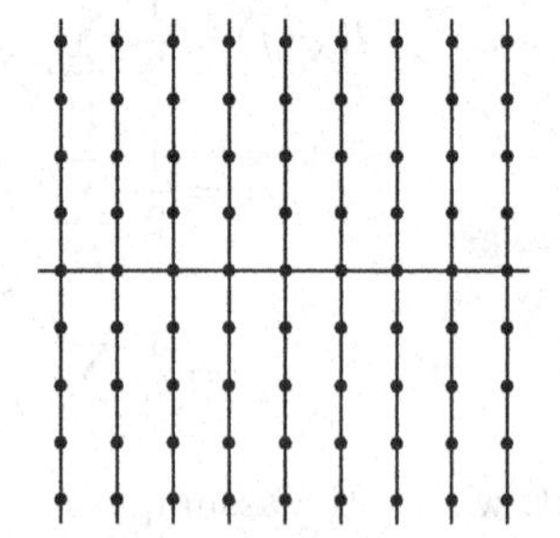

Figure 3: the comb lattice $\mathbf{C}_2$

We prove inductively that the simple random walk on $\mathbf{C}_d$ is recurrent. This is true for $\mathbf{C}_1$. Assume that $\mathbf{C}_{d-1}$ is recurrent. Suppose that $\mathbf{C}_d$ is transient. Then there is a finite energy flow u in $\mathbf{C}_d$ from the origin to infinity. But this flow must be 0 along all edges of each of the two-way infinite paths which we have attached to $\mathbf{C}_{d-1}$ for obtaining $\mathbf{C}_d$: otherwise, the energy would be infinite. Consequently, u is a finite energy flow in $\mathbf{C}_{d-1}$ from the origin to ∞. This contradicts recurrence of $\mathbf{C}_{d-1}$. □

[1] A *spanning tree* of a graph is a subgraph which is a tree and contains all vertices.

(2.22) Exercise. *Fattenings of the comb lattice* $\mathbf{C}_3$ *in* $\mathbb{Z}^3$.

(a) Take copies of the square grid $\mathbb{Z}^2 \times \{k\}$, $k \in \mathbb{Z}$, and connect each point $(0,0,k)$ with $(0,0,k+1)$ by an edge.

(b) Take $\mathbf{C}_3$ and add all those missing edges of the grid Z^3 which lie in the (x,z)-plane, that is, the edges between $(k,0,m)$ and $(k+1,0,m)$, where $k,m \in \mathbb{Z}$. The resulting graph is the square grid together with two-way infinite paths, one going through each point of the grid.

(c) Consider the Cartesian product $\mathbf{C}_2 \times \mathbb{Z}$. This is the graph with vertex set $\mathbb{Z}^3$ where $(k_1,\ell_1,m_1) \sim (k_2,\ell_2,m_2)$ if the two points are at distance 1 and either $\ell_2 = 0$ or $k_1 = k_2$.

Find the type of the simple random walk on each of these graphs. (A solution of (c) will be given in §18.)

C. Comparison with non-reversible Markov chains

Let P be an arbitrary (irreducible) transition matrix over X with excessive measure ν. Observe that P acts as a contraction on $\ell^2(X,\nu)$. Indeed, denoting by $(\cdot,\cdot)_\nu$ the inner product in this Hilbert space,

$$(2.23)\quad (Pf,f)_\nu = \sum_x \sum_y \nu(x) f(x) p(x,y) f(y)$$
$$\le \sum_x \sum_y \nu(x) p(x,y) \left(\frac{f(x)^2}{2} + \frac{f(y)^2}{2} \right) \le (f,f)_\nu\,.$$

Also, the adjoint P^* of P on $\ell^2(X,\nu)$, given by $p^*(x,y) = \nu(y)p(y,x)/\nu(x)$, is substochastic, and $\nu P^* P \le \nu$.

We want to give compatibility conditions under which the Green function of P can be compared with the Green function of a reversible Markov chain on X. The following Hilbert space lemma has become quite popular in random walk theory.

(2.24) Lemma. *Let $\mathcal{H}$ be a real Hilbert space with inner product $(\cdot,\cdot)$, and let T_1, T_2 be two invertible linear operators on $\mathcal{H}$ such that*

(i) *T_1 is self-adjoint, and*

(ii) *$(T_2 f,f) \ge (T_1 f,f) \ge 0$ for all $f \in \mathcal{H}$. Then*

$$(T_1^{-1} f,f) \ge (T_2^{-1} f,f) \quad \text{for all } f \in \mathcal{H}\,.$$

Proof. Note that $(T_i^{-1} f,f) \ge 0$ for $i = 1,2$. Applying the Cauchy–Schwarz inequality to the positive semidefinite quadratic form defined by T_1, we get

$$(T_2^{-1}f,f)^2 = (T_2^{-1}f,T_1T_1^{-1}f)^2 \le (T_2^{-1}f,T_1T_2^{-1}f)\,(T_1^{-1}f,T_1T_1^{-1}f)$$
$$\le (T_2^{-1}f,T_2T_2^{-1}f)\,(T_1^{-1}f,T_1T_1^{-1}f) = (T_2^{-1}f,f)\,(T_1^{-1}f,f)\,. \quad \square$$

(2.25) Theorem. *Let P be an irreducible transition matrix with excessive measure ν, and let Q be reversible with total conductance m. Suppose that*

(1) $\sup_X m(x)/\nu(x) < \infty$, *and*

(2) *there is $\varepsilon_0 > 0$ such that $P \geq \varepsilon_0 Q$ elementwise.*

Then recurrence of (X,P) implies recurrence of (X,Q).

Proof. Set $u(x) = m(x)/\nu(x)$. Then $u(x) \leq C$ for some constant $C > 0$. We define two new transition operators $\bar{P} = \frac{1}{2}(I+P)$ and $\bar{Q} = \left(1-\frac{1}{2C}u\right) I + \frac{1}{2C}u\,Q$, that is,

$$\bar{q}(x,y) = \left(1 - \tfrac{1}{2C}u(x)\right)\delta_x(y) + \tfrac{1}{2C}u(x)q(x,y)\,.$$

One immediately verifies that ν is an excessive measure for $\bar{P}$ and that $\bar{Q}$ is reversible with total conductance ν. Also, $\bar{P} \geq \varepsilon_1 \bar{Q}$, where $\varepsilon_1 = \min\{\varepsilon_0, 1/2\}$. We get that $\frac{1}{1-\varepsilon_1}(\bar{P} - \varepsilon_1\bar{Q})$ is also a stochastic transition operator with excessive measure ν, and by (2.23),

$$\big((\bar{P} - \varepsilon_1\bar{Q})f, f\big)_\nu \leq (1-\varepsilon_1)\,(f,f)_\nu \quad \text{for all } f \in \ell^2(X,\nu)\,.$$

Consequently, if $0 < z \leq 1$, then

$$\big((I - z\bar{P})f, f\big)_\nu \geq \varepsilon_1 \big((I - z\bar{Q})f, f\big)_\nu \geq 0$$

on $\ell^2(X,\nu)$. On this space, the operators $\bar{P}$ and $\bar{Q}$ are contractions. Thus, for $0 < z < 1$, $T_1 = \varepsilon_1(I - z\bar{Q})$ and $T_2 = I - z\bar{P}$ satisfy the assumptions of Lemma 2.24. We have

$$(I - z\bar{P})^{-1}f(x) = \sum_y G_{\bar{P}}(x,y|z)f(y)$$

(analogously for $\bar{Q}$), and setting $f = \delta_x$, we get

$$\left(1 - \tfrac{z}{2}\right)G_P\left(x,x\middle|\tfrac{z}{2-z}\right) = G_{\bar{P}}(x,x|z) \leq \frac{1}{\varepsilon_1}G_{\bar{Q}}(x,x|z) \quad \text{for every } z \in (0\,,\,1)\,.$$

Letting $z \to 1$ from below, we get $G_P(x,x) \leq G_{\bar{Q}}(x,x)/\varepsilon_1$ for every X. Thus, recurrence of (X,P) implies recurrence of $(X,\bar{Q})$.

Now let $f \in \ell_0(X)$. Then

$$\begin{aligned} D_{\bar{Q}}(f) &= \frac{1}{2}\sum_{x,y}\big(f(x) - f(y)\big)^2\nu(x)\bar{q}(x,y) \\ &= \frac{1}{2}\sum_{x,y}\big(f(x) - f(y)\big)^2\nu(x)\frac{u(x)}{2C}q(x,y) = \frac{1}{2C}D_Q(f)\,. \end{aligned}$$

By Corollary 2.14, recurrence of $(X, \bar{Q})$ implies recurrence of (X, Q). $\square$

Together with other comparison criteria for reversible chains (which will be deduced from Theorem 2.12 and its corollaries), the last theorem will be used to show that on a graph with bounded geometry (or on a group), recurrence of any "reasonable" random walk implies recurrence of the simple random walk.

3. Applications to random walks

We shall now give various applications of the results of the last section to random walks on locally finite graphs and groups. In particular, we shall determine those finitely generated groups which carry a recurrent random walk.

A. Comparison theorems for random walks on graphs

In this subsection, X will be a graph with bounded geometry. On X, we shall compare different reversible random walks (not necessarily nearest neighbour). We shall refer to the Dirichlet norm associated with the simple random walk as $D(\cdot)$ (without subscript) or $D_X(\cdot)$. In the same way, we shall write ∇ or ∇_X for the difference operator associated with the simple random walk. Thus, $\nabla f(e) = f(e^+) - f(e^-)$ for every edge of X.

(3.1) Theorem. *Let P be the transition matrix of a uniformly irreducible random walk on X with excessive measure ν satisfying $\inf_X \nu(x) > 0$. Then recurrence of (X, P) implies recurrence of the simple random walk on X.*

If, in addition, P is reversible with total conductance $m = \nu$, then there is $\varepsilon_1 > 0$ such that $D_P(f) \geq \varepsilon_1 D(f)$ for all $f \in \ell_0(X)$.

Proof. Let K and ε_0 be as in the definition of uniform irreducibility (1.20). Let $\bar{P} = \frac{1}{2}(I + P)$ and $\hat{P} = \bar{P}^K$. Then for $0 < z \leq 1$,

$$G_{\bar{P}}(x, x|z) = \tfrac{2}{2-z} G_P\left(x, x \middle| \tfrac{z}{2-z}\right),$$

so that P is recurrent if and only if $\bar{P}$ is. Next, if $\bar{P}$ is recurrent, then for $x \in X$, at least one of the series $\sum_{n=0}^{\infty} \bar{p}^{(nK+r)}(x, x)$ $(1 \leq r \leq K)$ must diverge. As $\bar{P} \geq \frac{1}{2} I$, we get

$$\sum_{n=0}^{\infty} \hat{p}^{(n)}(x, x) \geq \left(\tfrac{1}{2}\right)^{K-r} \sum_{n=0}^{\infty} \bar{p}^{(nK+r)}(x, x).$$

Consequently, $\hat{P}$ is recurrent if and only P is. The measure ν is also $\hat{P}$-excessive. We now show that $\hat{P}$ dominates a multiple of the simple random walk. If $x \sim y$, then

$$\hat{p}(x, y) \geq \frac{\varepsilon_0}{2^K} \geq \frac{\varepsilon_0}{2^K} \frac{1}{\deg(x)}.$$

Hence, Theorem 2.25 applies with $\hat{P}$ in place of P and the simple random walk in place of Q. (Recall that for the simple random walk, we have $m(x) = \deg(x)$, which is bounded above.)

We now prove the proposed inequality for the Dirichlet norms in the case when P is reversible with total conductance $m = \nu$. Then $\bar{P}$ and $\hat{P}$ are also reversible with respect to ν. By Lemma 2.5, for every $f \in \ell_0(X)$,

$$D_{\hat{P}}(f) \le K^2 D_{\bar{P}}(f) = \frac{K^2}{2} D_P(f)\,.$$

If $x \sim y$, then as above

$$\nu(x)\hat{p}(x,y) \ge \frac{\varepsilon_0}{2^K} \inf_X \nu(x) = \bar{\varepsilon} > 0\,.$$

Recall that for the simple random walk, each edge has conductance 1. Setting $\varepsilon_1 = 2\bar{\varepsilon}/K^2$, we obtain the proposed inequality. □

Next, we look for conditions under which the second statement of the last theorem can be reversed. Recall that in this subsection we are considering graphs with bounded vertex degrees.

(3.2) Theorem. *Let P be the transition matrix of a reversible bounded range random walk on X whose total conductances satisfy $\sup_X m(x) < \infty$. Then there is $\varepsilon_2 > 0$ such that $D(f) \ge \varepsilon_2 D_P(f)$ for all $f \in \ell_0(X)$.*

In particular, recurrence of the simple random walk on X implies recurrence of (X, P).

Proof. Let E be the set of edges of the graph X. Recall that $\Pi(x,y)$ is the set of geodesics from x to y, and write $\Pi_e(x,y)$ for all paths in $\Pi(x,y)$ containing edge e. Using the Cauchy–Schwarz inequality, we have

$$\big(f(x) - f(y)\big)^2 \le \frac{1}{|\Pi(x,y)|} \sum_{\pi\in\Pi(x,y)} \sum_{e\in\pi} \big(\nabla f(e)\big)^2 d(x,y)\,.$$

Hence, for $f \in \ell_0(X)$,

$$\begin{aligned} D_P(f) &\le \frac{1}{2} \sum_{x,y\in X} a(x,y) \frac{1}{|\Pi(x,y)|} \sum_{\pi\in\Pi(x,y)} \sum_{e\in\pi} \big(\nabla f(e)\big)^2 d(x,y) \\ &= \sum_{e\in E} \big(\nabla f(e)\big)^2 \phi(e)\,, \end{aligned}$$

where

$$\phi(e) = \frac{1}{2} \sum_{x,y\in X} m(x)p(x,y)d(x,y) \frac{|\Pi_e(x,y)|}{|\Pi(x,y)|}\,. \tag{3.3}$$

We get $D_P(f) \le \sup_E \phi(e)\, D(f)$. Using our assumptions, we now show that ϕ is bounded above. Let $M \ge 2$ be an upper bound for the vertex degrees in X, and let R be a bound for the range of P. Let $e \in E$. If x, y are such that $d(x,y) \le R$ and $\Pi_e(x,y) \ne \emptyset$ then both x and y must be at distance at most $M-1$ from the closer endpoint of e. There are at most $2R^{2M}$ pairs (x,y) of this type. Therefore

$$\phi(e) \le \frac{R}{2}\Bigl(\sup_X m(x)\Bigr) \sum_{x,y \in X : d(x,y) \le R} \frac{|\Pi_e(x,y)|}{|\Pi(x,y)|} \le R^{2M+1} \sup_X m(x) < \infty\,.$$

□

The last two theorems lead us to calling a reversible Markov chain (X, P) *strongly reversible* if there is a constant $C \in (0\,,\infty)$ such that

$$C^{-1} \le m(x) \le C \quad \text{for all } x \in X\,. \tag{3.4}$$

The simple random walk on a locally finite graph X is strongly reversible if and only if X has bounded geometry.

(3.5) Corollary. *If some strongly reversible, uniformly irreducible random walk with bounded range on X is recurrent, then this is true for every random walk on X with these three properties.*

For an integer $k \ge 1$ the *k-fuzz* of X is the graph $X^{(k)}$ with the same vertex set as X, where two vertices are neighbours if and only if $1 \le d(x,y) \le k$. With respect to the graph structure of X, the simple random walk on $X^{(k)}$ satisfies all requirements of Theorem 3.2.

(3.6) Corollary. *There is $\varepsilon_2 > 0$ such that*

$$D_{X^{(k)}}(f) \ge D_X(f) \ge \varepsilon_2\, D_{X^{(k)}}(f) \quad \textit{for all } f \in \ell_0(X)\,.$$

In particular, X is recurrent if and only if $X^{(k)}$ is recurrent.

Next, we want to compare different graphs. We shall use the following notion of similarity, regarding graphs as metric spaces with their natural discrete metrics.

(3.7) Definition. Let (X,d) and (X',d') be two metric spaces. A *rough isometry* is a mapping $\varphi : X \to X'$ such that

$$A^{-1} d(x,y) - A^{-1}B \le d'(\varphi x, \varphi y) \le A\, d(x,y) + B$$

for all $x, y \in X$, and

$$d'(x', \varphi X) \le B$$

for every $x' \in X'$, where $A \geq 1$ and $B \geq 0$. (We shall also refer to φ as an (A, B)-rough isometry.) In this case we say that the two spaces are *roughly isometric*, and if $B = 0$ then we say that they are *metrically equivalent*.

We can construct a *rough inverse* $\bar{\varphi}$ of φ: for $x' \in X'$, choose $x \in X$ such that $d'(x', \varphi x) \leq B$, and set $\bar{\varphi} x' = x$. Then one easily works out that $\bar{\varphi}$ is an (A', B')-rough isometry with $A' = A$ and $B' = (2A+1)B$. Furthermore, $d(\bar{\varphi}\varphi x, x) \leq (A+1)B$ for every $x \in X$, and $d'(\varphi\bar{\varphi}x', x') \leq B$ for every $x' \in X'$. It is obvious that the composition of two rough isometries is again a rough isometry. Hence, to be roughly isometric is an equivalence relation between metric spaces.

In order to get a feeling for rough isometry between graphs, we now present an example and a proposition.

(3.8) Example. Let X be a loopless graph with bounded geometry and $\deg(x) \geq 3$ for every vertex x. Then we can construct a graph X' which is 3-regular and roughly isometric with X.

We describe how to modify X locally at each vertex in order to obtain X'. Let $x \in X$, $\deg(x) = k$, and let $E(x) = \{e_0(x), \ldots, e_{k-1}(x)\}$ be an enumeration of the edges incident with x. If $k = 3$ then nothing has to be changed; we write $x^{(1)}$ for the vertex of X' corresponding to x. If $k \geq 4$ then we replace x with new vertices $x^{(1)}, \ldots, x^{(k-2)}$ and introduce new edges $e^{(i)} = [x^{(i)}, x^{(i+1)}]$, $i = 1, \ldots, k-3$. In the modified graph, (copies of) $e_0(x)$ and $e_1(x)$ are incident with $x^{(1)}$, (copies of) $e_{k-2}(x)$ and $e_{k-1}(x)$ are incident with $x^{(k-2)}$, and if $k > 4$ then (a copy of) $e_i(x)$ is incident with $x^{(i)}$, $i = 2, \ldots, k-3$; see Figure 4.

Figure 4

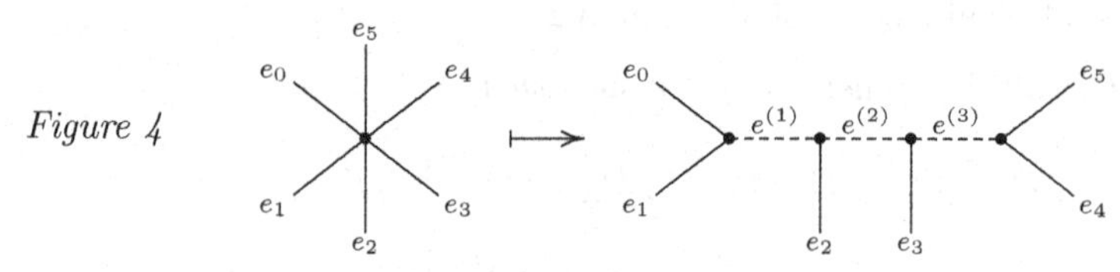

This modification is carried out at each vertex of X. Thus, we have replaced each vertex x of X with $\deg(x) > 3$ by a path of length $\deg(x) - 3$. We define $\varphi x = x^{(1)}$. It is clear that distances do not decrease under φ:

$$d(x, y) \leq d'(\varphi x, \varphi y).$$

Conversely, let $d(x, y) = n$, let $\pi = [x = x_0, x_1, \ldots, x_n = y]$ be a path in X from x to y, and let $e_i = [x_{i-1}, x_i]$ be its edges. Thinking of e_i as an edge of X', this does not necessarily connect $\varphi x_{i-1} = x_{i-1}^{(1)}$ with $\varphi x_i = x_i^{(1)}$, but it does connect $x_{i-1}^{(j)}$ with $x_i^{(k)}$ for some j, k. Now

$$d'(x^{(1)}, x^{(j)}) \leq M - 3$$

for each $x^{(j)} \in X'$, where M is an upper bound on the vertex degrees in X. Hence, $d'(\varphi x_{i-1}, \varphi x_i) \le 2(M-3)+1$, and

$$d'(\varphi x, \varphi y) \le (2M-5)\, d(x,y).$$

Consequently, φ is a $(2M-5, M-3)$-rough isometry. We remark that the natural choice for the rough inverse is given by $\bar{\varphi} x^{(j)} = x$. We also remark that this construction has the following property: when X is a tree then so is X'.

In other words, *every tree with bounded geometry and minimum degree at least* 3 *is roughly isometric with the homogeneous tree of degree* 3. □

(3.9) Proposition. *Every quasi-transitive graph is roughly isometric with a vertex-transitive graph.*

If a group $\Gamma \le \mathrm{AUT}(X)$ acts quasi-transitively and with finite vertex stabilizers then Γ is finitely generated and X is roughly isometric with a Cayley graph of Γ.

Proof. Let $\Gamma \le \mathrm{AUT}(X)$ act quasi-transitively. Consider the orbits X_i, $i \in \mathcal{I} = \{1, \dots, N\}$, and the factor graph over $\mathcal{I}$. The latter is connected on N vertices and has diameter bounded by $N-1$. Lifting back to X, we see that for every $x \in X$ there is $\bar{x} \in X_1$ such that $d(x, \bar{x}) \le N-1$ (we choose $\bar{x} = x$ when $x \in X_1$). Now consider the $(2N-1)$-fuzz $X^{(2N-1)}$. Let $X_1^{(2N-1)}$ denote the subgraph induced by X_1. We claim that it is connected. Let $x, y \in X_1$, and consider a path $\pi = [x = x_0, x_1, \dots, x_{k-1}, x_k = y]$ from x to y in X. Define $\bar{\pi} = [x = \bar{x}_0, \bar{x}_1, \dots, \bar{x}_{k-1}, \bar{x}_k = y]$. In X, we have $d(\bar{x}_{i-1}, \bar{x}_i) \le d(\bar{x}_{i-1}, x_{i-1}) + 1 + d(x_i, \bar{x}_i) \le 2N-1$. Hence, $\bar{\pi}$ is a path in $X_1^{(2N-1)}$, except that possibly $\bar{x}_{i-1} = \bar{x}_i$ for some i, in which case we simply delete $\bar{x}_i$.

It is now obvious that $\varphi x = \bar{x}$ is a rough isometry from X to $X_1^{(2N-1)}$, and the latter graph is vertex-transitive.

For the second statement, suppose that Γ has finite vertex stabilizers. By the above, we may assume that it acts transitively on X and choose a reference vertex o. We have that Γ is discrete, Haar measure is the counting measure, and all Γ_x have the same cardinality and are conjugates of each other. The (finite, symmetric) set S of all $\gamma \in \Gamma$ with $\gamma o \sim o$ generates Γ. Now consider the graph $\bar{X}$ over $X \times \Gamma_o$ where $(x, \alpha) \sim (y, \beta) \iff x \sim y$ in X. It is obvious that $\bar{X}$ and X are roughly isometric via $\varphi(x, \alpha) = x$. For each $x \in X$, choose and fix $\gamma_x \in \Gamma$ such that $\gamma_x o = x$. We define an action of Γ on $\bar{X}$:

$$\text{if} \quad \gamma x = y \quad \text{then} \quad \gamma(x, \alpha) = (y, \gamma_y^{-1} \gamma \gamma_x \alpha).$$

As the mapping $(x, \alpha) \mapsto \gamma_x \alpha$ is one-to-one from $\bar{X}$ onto Γ, one easily checks that Γ becomes a subgroup of $\mathrm{AUT}(\bar{X})$ which acts vertex-transitively. If

$\gamma(x,\alpha) = (x,\alpha)$ for a vertex of $\bar{X}$ then it must be that $\gamma_x^{-1}\gamma\gamma_x\alpha = \alpha$, whence $\gamma = \iota$. That is, the action of Γ is fixed point-free, and $\bar{X}$ must be a Cayley graph of Γ; indeed, $\bar{X} = X(\Gamma, S)$ with S as above. □

The following is the main link between rough isometries and random walks.

(3.10) Theorem. *Let X, X' be graphs with bounded geometry, and let $\varphi : X \to X'$ be a rough isometry. Then there is a constant $\varepsilon' > 0$ such that*

$$D_{X'}(f) \geq \varepsilon' D_X(f \circ \varphi) \quad \text{for all } f \in \ell_0(X').$$

In particular, X is recurrent if and only if X' is recurrent.

Proof. Let φ be an (A,B)-rough isometry. First of all, note that $f \circ \varphi \in \ell_0(X)$ for every $f \in \ell_0(X')$. Indeed, the preimage of $x' \in X'$ in X under φ has diameter bounded by B. Thus, $|\varphi^{-1}\{x'\}| \leq M^{B+1} < \infty$, where M is an upper bound on the vertex degrees in X.

Now consider the graph structure on φX induced by φ: two points $x', y' \in \varphi X$ are neighbours in φX if there are neighbours x, y in X such that $\varphi x = x'$ and $\varphi y = y'$. Let $a'(x', y')$ be the number of all such edges $[x, y]$ in X. In this situation, $d'(x', y') \leq k = A + B$, so that φX is a subgraph of $X'^{(k)}$. Applying Corollary 3.6, we get that

$$D_{\varphi X}(f) \leq D_{X'^{(k)}}(f) \leq \frac{1}{\varepsilon_2} D_{X'}(f) \quad \text{for every } f \in \ell_0(X'),$$

where $\varepsilon_2 > 0$.

Now note that φX together with the conductance function $a'(\cdot,\cdot)$ is the network obtained by shorting the network of the simple random walk on X with respect to the partition induced by φ. By the above, $a'(x', y') \leq M^{B+2}$. Hence (compare with the proof of Theorem 2.19)

$$D_X(f \circ \varphi) \leq M^{B+2} D_{\varphi X}(f) \quad \text{for every } f \in \ell_0(X').$$

This completes the proof of the inequality.

Applying the same argument to a rough inverse of φ, we get an analogous inequality in the opposite direction. Using Theorem 2.12(c) or (d), we see that recurrence of X and recurrence of X' are equivalent. □

B. Growth and the classification of recurrent groups

In this subsection we shall present the quickest possible access to the classification of those finitely generated groups which carry a recurrent (irreducible) random walk. Before this, we need some general considerations on growth of graphs.

Let X be a locally finite, connected graph. On X, we consider a reversible Markov chain (X,P) which is of *nearest neighbour type* and such that the invariant measure satisfies $\inf_X m(x) > 0$. For $x \in X$ and $n \geq 0$, the *n-ball* centred at x is $B(x,n) = \{y \in X : d(y,x) \leq n\}$. The *growth function of* (X,P) *at* x is $V_P(x,n) = m\big(B(x,n)\big)$. We set

$$(3.11) \qquad V_P(n) = \inf_X V_P(x,n) \quad \text{and} \quad \bar{V}_P(n) = \sup_X V_P(x,n)\,.$$

In the case of the simple random walk, where $m(x)$ is proportional to $\deg(x)$, we omit P or write $V_X(\cdot)$. We say that the graph X has *exponential growth,* if $V(n) \geq C\,r^n$ for some $C > 0$ and $r > 1$, and that X has *polynomial growth,* if $\bar{V}(n) \leq C\,(n+1)^d$ for some $C, d > 0$.

We shall usually use $m(x) = \deg(x)$ for the simple random walk. When X is regular, we shall often prefer the counting measure ($m(x) = 1$). In any case, polynomial and exponential growth do not depend on the chosen normalization. If X is vertex-transitive, then $V(x,n) = V(n) = \bar{V}(n)$ is independent of x.

Let us immediately exhibit a relation between growth and recurrence.

(3.12) Lemma. *If* $\liminf_n \Big(V_P(x,n)/n^2\Big) < \infty$ *for some* ($\iff$ *every*) $x \in X$ *then* (X,P) *is recurrent.*

Proof. We apply Nash-Williams' criterion (Corollary 2.20) to the distance partition, that is, we short each of the sets $S(x,n) = \{y : d(y,x) = n\}$, $n \geq 0$, to a single point. In the notation of (2.18), $a'(n,n+1)$ is the sum of the conductances of the edges between $S(x,n)$ and $S(x,n+1)$. In particular, $a'(n,n+1) \leq V_P(x,n+1) - V_P(x,n)$. By Jensen's inequality,

$$\sum_{k=n+1}^{2n} \frac{1}{a'(k-1,k)} \geq \frac{n^2}{\sum_{k=n+1}^{2n} a'(k-1,k)} \geq \frac{n^2}{V_P(x,2n)} > 0\,.$$

Therefore the series $\sum_n a'(n-1,n)^{-1}$ must diverge. □

This sheds more light on recurrence of the simple random walk on $\mathbb{Z}^2$. Note that here we did not require X to have bounded vertex degrees. The condition of the lemma cannot be replaced with $\liminf_n \big(V_P(n)/n^2\big) < \infty$ (unless X is quasi-transitive), as there are transient trees with $V(n) = 4n+2$, while $V(x,n) \geq C_x\,r^n$ with $r > 1$ (attach a half-line at the root of the binary tree). On the other hand, it is also easy to construct recurrent trees with $V(x,n) \geq 2^n$. An example will be given in §6.B below. However, we shall see that polynomial growth with degree at most 2 is not only sufficient, but also necessary for recurrence of the simple and other random walks on

Cayley graphs, and more generally on quasi-transitive graphs. Before that, we need to relate growth with the structure of groups.

We start with an observation concerning rough isometries. The following is easily obtained along the lines of Theorem 3.10 (compare also with Theorem 4.7 below); we leave the proof as an exercise.

(3.13) Lemma. *Let X and X' be two roughly isometric graphs with bounded geometry. Then X and X' have* equivalent growth functions *in the sense that there are constants c_0, k_0 such that*

$$V_X(n) \le c_0\, V_{X'}(k_0 n) \quad \text{and} \quad V_{X'}(n) \le c_0\, V_X(k_0 n) \quad \text{for all } n\,.$$

In particular, X has polynomial (exponential) growth if and only if X' has the respective property.

If Γ is a finitely generated group and S_1, S_2 are two finite, symmetric sets of generators, then the Cayley graphs $X(\Gamma, S_1)$ and $X(\Gamma, S_2)$ have equivalent metrics. Indeed, let d_i, $i = 1, 2$, be the respective Cayley graph metrics. If $A_1 = \max\{d_1(o, g) : g \in S_2\}$, then $d_1(\cdot,\cdot) \le A_1\, d_2(\cdot,\cdot)$. Consequently, the growth functions are equivalent, and when we speak of growth of finitely generated groups (polynomial, exponential, etc.), it is not necessary to specify the generating set.

From Lemma 3.13 one deduces the following.

(3.14) Lemma. *Let Γ be finitely generated and Γ_1 a subgroup with finite index. Then Γ and Γ_1 have equivalent growth functions.*

Proof. Given any Cayley graph X of Γ, we have that Γ_1 acts on X with finitely many orbits and trivial vertex stabilizers. By the second part of Proposition 3.9, X is roughly isometric with a Cayley graph of Γ_1. Now Lemma 3.13 yields the result. □

The structural classification of all groups with polynomial growth is a deep topic; proofs go beyond the scope of this book. We describe the main results. If Γ is a finitely generated abelian group then there is d such that Γ contains a finite-index subgroup isomorphic with the free abelian group $\mathbb{Z}^d$; we write $\mathrm{rk}(\Gamma) = d$, the *rank* of Γ. Next, let Γ be a finitely generated nilpotent group with lower central series

$$\Gamma = \Gamma_0 \trianglerighteq \Gamma_1 \trianglerighteq \cdots \trianglerighteq \Gamma_{r-1} \trianglerighteq \Gamma_r = \{\iota\}\,.$$

Each factor group Γ_{i-1}/Γ_i is finitely generated and abelian. Set

$$d(\Gamma) = \sum_{i=1}^{r} i\, \mathrm{rk}(\Gamma_{i-1}/\Gamma_i)\,. \tag{3.15}$$

The following was proved by Bass [27] and Guivarc'h [154].

(3.16) Theorem. *If Γ is nilpotent and $V(n)$ is its growth function with respect to some finite, symmetric set of generators, then there are constants $C_0, C_1 > 0$ such that for all $n \geq 1$*

$$C_0\, n^{d(\Gamma)} \leq V(n) \leq C_1\, n^{d(\Gamma)}\,.$$

Thus (Lemma 3.14), finitely generated groups which have a finite-index nilpotent subgroup have polynomial growth. The main theorem of the theory is the following famous result of Gromov [149].

(3.17) Theorem. *Let Γ be finitely generated with growth function $V(n)$ with respect to some generating set. If there are C, d such that $V(n) \leq C\, n^d$ for infinitely many n, then Γ has a nilpotent subgroup with finite index.*

(3.18) Corollary. *If Γ is finitely generated then either $V(n) \geq C\, n^3$ for all n or $V(n) \leq C\, n^2$ for all n. In the second case, Γ contains a finite-index subgroup isomorphic with $\mathbb{Z}$ or $\mathbb{Z}^2$.*

Proof. The first part follows from Theorem 3.17. Assume that $V(n) \leq C\, n^2$ for all n. Theorem 3.17 implies that Γ has a nilpotent subgroup $\mathfrak{N}$ with finite index. It is well known that every finitely generated nilpotent group has a torsion-free subgroup with finite index. Hence we may assume that $\mathfrak{N}$ itself is torsion-free. From Lemma 3.14 we get $d(\mathfrak{N}) \in \{1, 2\}$.

We now look at the lower central series $\mathfrak{N} = \mathfrak{N}_0 \trianglerighteq \ldots \trianglerighteq \mathfrak{N}_r = \{o\}$. We see from formula (3.15) that $\mathrm{rk}(\mathfrak{N}_{i-1}/\mathfrak{N}_i) = 0$ for all $i > 2$. In particular, $\mathfrak{N}_2$ is finite, whence trivial, and $r \leq 2$.

If $d(\mathfrak{N}) = 1$ then it must also be that $\mathrm{rk}(\mathfrak{N}_1/\mathfrak{N}_2) = 0$, so that $\mathfrak{N}_2$ is trivial and $r = 1$. Thus, $\mathfrak{N}$ is abelian and $\mathrm{rk}(\mathfrak{N}) = 1$. Being torsion-free, $\mathfrak{N} \cong \mathbb{Z}$.

If $d(\mathfrak{N}) = 2$ then it cannot be that $\mathrm{rk}(\mathfrak{N}_0/\mathfrak{N}_1) = 0$ and $\mathrm{rk}(\mathfrak{N}_1) = 1$, as otherwise $\mathfrak{N}$ would have a finite-index cyclic subgroup. Now the only possibility is $\mathrm{rk}(\mathfrak{N}_0/\mathfrak{N}_1) = 2$ and $\mathrm{rk}(\mathfrak{N}_1/\mathfrak{N}_2) = 0$, so that $\mathfrak{N}_1$ is trivial, $\mathfrak{N}$ is abelian and $\mathrm{rk}(\mathfrak{N}) = 2$, whence (being torsion-free) $\mathfrak{N} \cong \mathbb{Z}^2$. □

The simplest example of an infinite nilpotent group which is not abelian is the *Heisenberg group* over the integers. It consists of all 3×3 upper diagonal matrices with ones in the diagonal and integer entries. It is "three-dimensional" in the sense that its elements are parametrized by three integers, but it has polynomial growth with degree 4.

(3.19) Exercise. Draw the Cayley graph of the Heisenberg group with respect to a suitable set of generators.

To complete the picture concerning growth of finitely generated groups, we mention that all non-amenable groups have exponential growth (see Chapter II), and that a solvable group has exponential growth unless it contains a nilpotent subgroup with finite index. Finally, there also are examples of groups with *intermediate* growth: faster than any polynomial in n, but slower than exponential; see Grigorchuk [147].

We now return to random walks. Again, let Γ be a finitely generated group, and let $X = X(\Gamma, S)$ be its Cayley graph with respect to some finite, symmetric set of generators S. The simple random walk on X is symmetric, and hence reversible with respect to the counting measure ($m(x) = 1$). Indeed, if μ is any (irreducible) probability measure on Γ, then m is an invariant measure for the associated random walk. From this and Theorem 3.1 we get that recurrence of the random walk with law μ implies recurrence of the simple random walk on X.

We can generalize Theorem 3.2. Recall the definition of moments (§1.C). We write $M_k(\mu)$ for the kth moment of the random walk with law μ, where the distance $d = d_S$ is that of our Cayley graph X. It is easy to see that finiteness of $M_k(\mu)$ does not depend on the particular choice of S and the resulting Cayley graph.

(3.20) Proposition. *The group Γ carries a recurrent random walk if and only if some ($\Longleftrightarrow$ every) symmetric irreducible probability μ on Γ with $M_2(\mu) < \infty$ induces a recurrent random walk.*

Proof. In view of what we have just said, what is left is to show that recurrence of the simple random walk on X implies recurrence of the random walk with law μ. As μ is symmetric, it is reversible with respect to the counting measure just like the simple random walk. Thus, we prove that the conclusion of Theorem 3.2 still holds under our moment condition. We show that the function ϕ of (3.3) is bounded. Let $e_0 \in E(X)$. Then

$$\phi(e_0) = \frac{1}{2} \sum_{x,y \in \Gamma} \mu(x^{-1}y)\, d(o, x^{-1}y)\, \frac{|\Pi_{x^{-1}e_0}(o, x^{-1}y)|}{|\Pi(o, x^{-1}y)|}$$

(setting $w = x^{-1}y$)

$$= \frac{1}{2} \sum_{w \in \Gamma} \mu(w)\, d(o,w) \sum_{x \in \Gamma} \frac{|\Pi_{x^{-1}e_0}(o,w)|}{|\Pi(o,w)|}$$

$$\leq \frac{1}{2} \sum_{w \in \Gamma} \mu(w)\, d(o,w) \sum_{e \in E(X)} \frac{|\Pi_e(o,w)|}{|\Pi(o,w)|} = M_2(\mu)\,.$$

The last identity holds because

$$\sum_{e \in E(X)} \frac{|\Pi_e(o,w)|}{|\Pi(o,w)|} = \frac{1}{|\Pi(o,w)|} \sum_{\pi \in \Pi(o,w)} |\{e \in E(X) : e \in \pi\}| = d(o,w)\,. \quad \square$$

We know (Lemma 3.12) that the conclusion of the last proposition holds when the group has at most quadratic growth. In order to show that otherwise the group is transient, we shall construct a symmetric probability with finite second moment which induces a transient random walk. We need the following preparatory lemma.

(3.21) Lemma. *Let μ be a probability measure on Γ, and decompose $\mu = \mu_1 + \mu_2$, where the μ_i are non-negative measures. Then for every $x \in \Gamma$ and $n \in \mathbb{N}$*

$$\mu^{(n)}(x) \le \mu_1(\Gamma)^n + n\,\|\mu_2\|_\infty\,,$$

where $\|\mu_2\|_\infty = \sup_{y\in\Gamma} \mu_2(y)$.

Proof. The statement is true for $n = 1$. Suppose it is true for n. Then

$$\begin{aligned}\mu^{(n+1)}(x) &= \sum_{y\in\Gamma} \big(\mu_1(y) + \mu_2(y)\big)\mu^{(n)}(y^{-1}x)\\ &\le \mu_1(\Gamma)\big(\mu_1(\Gamma)^n + n\,\|\mu_2\|_\infty\big) + \|\mu_2\|_\infty\,,\end{aligned}$$

which is smaller than the proposed bound for $n+1$. □

Now let $B(k) = B(n,k)$ be the k-balls in our Cayley graph ($k \ge 1$), and define a symmetric probability on Γ by

$$\mu(x) = \sum_{k=1}^{\infty} \frac{\lambda_k}{V(k)} \mathbf{1}_{B(k)}(x)\,, \tag{3.22}$$

where $\lambda_k = 1/k^3 \log^2 k$ for $k \ge 2$ and $\lambda_1 = 1 - \sum_{k\ge 2} \lambda_k$.

(3.23) Proposition. *$M_2(\mu) < \infty$, and if the growth function satisfies $V(n) \ge C\,n^3$ then μ induces a transient random walk.*

Proof. We have

$$M_2(\mu) = \sum_{k=1}^{\infty} \frac{\lambda_k}{V(k)} \sum_{x\in\Gamma} d(o,x)^2\,\mathbf{1}_{B(k)}(x) \le \sum_{k=1}^{\infty} \lambda_k\,k^2 < \infty\,.$$

Now let $m \ge 2$ be arbitrary, and decompose $\mu = \mu_1 + \mu_2$, where $\mu_1(x) = \sum_{k=1}^{m-1} \frac{1}{V(k)} \lambda_k\,\mathbf{1}_{B(k)}(x)$. Putting $s_m = \sum_{k\ge m} \lambda_k$, we get $\mu_1(\Gamma) = 1 - s_m$ and $\|\mu_2\| = \sum_{k\ge m} \big(\lambda_k / V(k)\big) \le s_m / V(m)$. Lemma 3.21 gives

$$\mu^{(n)}(o) \le (1 - s_m)^n + n\frac{s_m}{V(m)}\,.$$

As $m \to \infty$, we have $s_m\, m^2 \log^2 m \to 1/2$. Using in addition that $V(m) \geq C\, m^3$, we find $c_1, c_2 > 0$ such that

$$\mu^{(n)}(o) \leq \exp\Big(-\frac{c_1\, n}{m^2 \log^2 m}\Big) + \frac{c_2\, n}{m^5 \log^2 m}$$

for all n and sufficiently large m. Setting $m = m(n) = n^{2/5}$, we see that $\sum_n \mu^{(n)}(o)$ converges. □

Hence we conclude:

(3.24) Theorem. *Let Γ be a finitely generated, infinite group. Then Γ carries a recurrent random walk if and only if it contains a finite-index subgroup isomorphic with $\mathbb{Z}$ or $\mathbb{Z}^2$. In this case, every symmetric random walk with finite second moment is recurrent.*

In §6.A we shall also give conditions for recurrence of non-symmetric random walks on groups with linear or square growth. Also, we shall extend Theorem 3.24 to quasi-transitive random walks, and we shall show in Section 5 that every transient quasi-transitive graph has a transient subtree. We next need some preparatory results.

C. Random walks on quasi-transitive graphs

Again, X will be an infinite graph with bounded geometry and P will be the transition matrix of an irreducible random walk on X. We shall study the case when (X, P) is (quasi-)transitive, including in particular the case of random walks on finitely generated groups. All subgroups Γ of $\mathrm{AUT}(X)$ considered in the sequel are assumed to be closed with respect to pointwise convergence. This is not a restriction, but a natural assumption, as those properties that we shall study pass from an arbitrary subgroup to its closure.

(3.25) Lemma. *Let Γ be a subgroup of $\mathrm{AUT}(X, P)$. Suppose that the factor chain $(I, \tilde{P})$, as defined by (1.31), has an excessive distribution $\tilde{\nu}$. Then*

$$\nu(x) = \tilde{\nu}(i)\,\big|\Gamma_x\big|, \quad x \in X_i\,,$$

defines an excessive measure for P on X. If $\tilde{\nu}$ is invariant then so is ν.

In particular, when Γ is unimodular then ν is constant on each orbit. If, besides being unimodular, Γ also acts transitively, then P is doubly stochastic, that is, $\sum_x p(x, y) = 1$ for all y.

Proof. Choose a reference point o_i in each orbit X_i. Let $y \in X_j$. Using

(1.28) and (1.30), we compute

$$\begin{aligned}\sum_{x\in X}\nu(x)p(x,y) &= \sum_{i\in\mathcal{I}}\tilde{\nu}(i)\sum_{x\in X_i}|\Gamma_x|p(x,y) = \sum_{i\in\mathcal{I}}\tilde{\nu}(i)\int_\Gamma \frac{|\Gamma_{\gamma o_i}|}{|\Gamma_{o_i}|}p(\gamma o_i,y)\,d\gamma \\ &= \sum_{i\in\mathcal{I}}\tilde{\nu}(i)\int_\Gamma \Delta(\gamma^{-1})p(o_i,\gamma^{-1}y)\,d\gamma = \sum_{i\in\mathcal{I}}\tilde{\nu}(i)\int_\Gamma p(o_i,\gamma y)\,d\gamma \\ &= \sum_{i\in\mathcal{I}}\tilde{\nu}(i)\sum_{w\in X_j}|\Gamma_y|p(o_i,w) = |\Gamma_y|\sum_{i\in\mathcal{I}}\tilde{\nu}(i)\tilde{p}(i,j) \\ &\le |\Gamma_y|\tilde{\nu}(j) = \nu(y)\,.\end{aligned}$$

If $\tilde{\nu}$ is invariant, then the last inequality is an equality. Furthermore, if Γ is unimodular, then $|\Gamma_x| = |\Gamma_{o_i}|$ for every $x \in X_i$. □

Note that when P is reversible with total conductance m then it is not necessarily true that $\nu = m$ for the measure of Lemma 3.25. For example, the random walk on $\mathbb{Z}$ with $p(k,k+1) = p$, $p(k,k-1) = q$ $(p+q = 1)$ has $m(k) = (p/q)^k$, while $\nu(k) = 1$. (See also Lemma 3.30 below.) The next theorem shows that we can eliminate non-unimodularity from all considerations concerning recurrence.

(3.26) Theorem. *Suppose that* AUT(X,P) *contains a closed subgroup* Γ *which is non-unimodular. Then* (X,P) *is transient.*

Proof. *Step 1.* We suppose that Γ acts transitively. In the following, it will not be used that Γ arises as a group of automorphisms of a locally finite graph. The relevant facts are that Γ is a transitive group of permutations of X, acting continuously with respect to pointwise convergence, locally compact with compact point stabilizers, and leaving P invariant.

First observe that for every positive $z < \rho(P)$ (in particular for $z < 1$) and every $x \in X$, the measure $\nu(y) = G(x,y|z)/G(x,x|z)$ satisfies $\nu P \le \frac{1}{z}\nu$. Now, by transitivity, $G(x,x|z) = G(y,y|z)$ for all x,y. Using Lemma 1.13 we get, in the recurrent as well as in the transient case,

$$F(x,y) = \lim_{z\to 1-}\frac{G(x,y|z)}{G(y,y|z)} = \lim_{z\to 1-}\frac{G(x,y|z)}{G(x,x|z)}\,.$$

From the above and Fatou's lemma we infer that for every x, the measure $F(x,\cdot)$ is excessive. On the other hand, by Lemma 3.25, $y \mapsto |\Gamma_y|$ defines an invariant measure.

Now suppose that (X,P) is recurrent. Then, by Proposition 1.17, $F(x,y) = 1$ for all x,y. Also, by Theorem 1.18, up to multiplication by a constant there is a unique excessive measure, which has to be invariant. Consequently $|\Gamma_y| = c\,F(x,y) = c > 0$ for all y. This yields $\Delta(\gamma) = |\Gamma_y|/|\Gamma_{\gamma y}| = 1$ for all $\gamma \in \Gamma$.

Step 2. Coming to the general case, suppose that (X,P) is recurrent. As Γ is non-unimodular, all its orbits must be infinite: otherwise, Γ would be compact by Lemma 1.27, whence also unimodular. Let $Y \subset X$ be one of the orbits. Define the stopping time

$$\mathbf{t}^Y = \min\{n \geq 1 : Z_n \in Y\}. \tag{3.27}$$

By recurrence, it is $\mathbb{P}_x$-almost surely finite for every x. The *induced random walk* (Y, P^Y) has transition probabilities

$$p^Y(x,y) = \mathbb{P}_x[Z_{\mathbf{t}^Y} = y], \quad x,y \in Y. \tag{3.28}$$

Recurrence of (X,P) implies recurrence of the induced walk. Indeed, if $x,y \in Y$, then $F(x,y)$ and $U(x,x)$ are the same for the original and for the induced random walk. Now, Γ acts transitively on Y, and we are in the situation of the first step. Therefore Γ cannot be recurrent. $\square$

If Γ in Lemma 3.25 acts quasi-transitively, then the factor chain is an irreducible Markov chain on a finite state space. As such, it is recurrent, and by Theorem 1.18 it must be positive recurrent and admit an invariant probability measure $\tilde{\nu}$. We obtain an invariant measure ν for P which in the unimodular case is constant on each orbit, whence $\inf_X \nu(x) > 0$. Recalling that in the quasi-transitive case irreducibility is automatically uniform, we obtain the following from Theorem 3.1.

(3.29) Corollary. *Suppose that* $\mathrm{AUT}(X,P)$ *acts quasi-transitively. Then recurrence of* (X,P) *implies recurrence of the simple random walk on* X.

Now suppose that P is reversible with respect to m. Let $x,y \in X$, and choose n such that $p^{(n)}(x,y) > 0$. If $\gamma \in \mathrm{AUT}(X,P)$ then

$$m(x)p^{(n)}(x,y) = m(y)p^{(n)}(y,x) \quad \text{and} \quad m(\gamma x)p^{(n)}(x,y) = m(\gamma y)p^{(n)}(y,x).$$

Dividing, we get $m(\gamma x)/m(x) = m(\gamma y)/m(y)$, that is, the function $g(\gamma) = m(\gamma x)/m(x)$ does not depend on $x \in X$. Consequently, g is an exponential on Γ: it satisfies $g(\beta\gamma) = g(\beta)g(\gamma)$. We have two possibilities: either $g \equiv 1$, in which case m is constant on each orbit of $\mathrm{AUT}(X,P)$, or g and m are unbounded. In particular, when (X,P) is transitive, then strong reversibility coincides with symmetry: $p(x,y) = p(y,x)$.

(3.30) Lemma. *Let* (X,P) *be reversible and quasi-transitive. If* P *is not strongly reversible, then it is transient.*

Proof. If $\Gamma = \mathrm{AUT}(X,P)$ is non-unimodular then P is transient in any case. Otherwise, by Lemma 3.25, P admits an invariant measure ν which is constant on each orbit and hence bounded. If P is recurrent, then m must be a multiple of ν by Theorem 1.18 and cannot be unbounded. $\square$

Next, recall the definition (1.22) of the step length distributions σ_x on $\mathbb{N}$ and associated moments.

(3.31) Theorem. *Let P be the transition matrix of a strongly reversible random walk on X with finite second moment. Suppose that there is a quasi-transitive subgroup Γ of $\mathrm{AUT}(X,P)$ which is unimodular. Then there are $\varepsilon_1, \varepsilon_2 > 0$ such that $\varepsilon_1^{-1} D_P(f) \geq D(f) \geq \varepsilon_2 D_P(f)$ for all $f \in \ell_0(X)$.*

In particular, (X,P) is recurrent if and only if the simple random walk on X is recurrent.

The first inequality is immediate from Theorem 3.1. Also, for P having bounded range, the second inequality follows from Theorem 3.2. None of this requires unimodularity of Γ, which is only needed when extending to finite second moment. We omit the proof, which generalizes that of Proposition 3.20. (In §6.A we shall give a criterion which does not require reversibility.)

Let us recapitulate what we now know about recurrence in the case of vertex-transitive graphs. Let (X,P) be transitive.

- If P is recurrent then the simple random walk on X is recurrent.
- If the simple random walk on X is recurrent and P is symmetric with finite second moment, then P is recurrent.
- If P is reversible but not symmetric, then P is transient.

Furthermore, if (X,P) is quasi-transitive and recurrent then there is a vertex-transitive graph which is roughly isometric with X (Proposition 3.9) and such that the simple random walk is recurrent.

4. Isoperimetric inequalities

This section will consist only of structure theoretical considerations with an analytical flavour. Their significance for recurrence will become clear in the main theorems of the next section. We shall also need this material in Chapters II and III.

A. Isoperimetric and Sobolev inequalities

Let (X,P) be reversible with associated total conductance m, edge conductance $a(\cdot,\cdot)$ and network $\mathcal{N} = (X, E(P), r)$. Thinking of m as a measure on X, $m(A) = \sum_{x \in A} m(x)$, if $A \subset X$. Similarly, if $D \subset E(P)$, we write $a(D) = \sum_{e \in D} a(e^-, e^+)$. For $A \subset X$, we define ∂A as the set of all edges in $E(P)$ having one endpoint in A and the other in $X \setminus A$. Then $a(\partial A)$ is a discrete analogue for the surface area.

Let $\mathfrak{F} : \mathbb{R}^+ \to \mathbb{R}^+$ be a non-decreasing function.

(4.1) Definition. We say that $\mathcal{N}$ (or (X,P)) satisfies an $\mathfrak{F}$-isoperimetric inequality $IS_{\mathfrak{F}}$, if there is a constant $\kappa > 0$ such that

$$\mathfrak{F}\big(m(A)\big) \leq \kappa\, a(\partial A)$$

for every finite $A \subset X$.

If this holds for the simple random walk then we say that the graph X itself satisfies an $\mathfrak{F}$-isoperimetric inequality.

If, in particular, $\mathfrak{F}(t) = t^{1-1/d}$ (with $1 \le d \le \infty$) then we speak of a *d-dimensional isoperimetric inequality,* for short IS_d. Note that if $\mathcal{N}$ has bounded resistances on $E(P)$ and $\inf_X m(x) > 0$ (which holds, in particular, for the simple random walk on a graph with bounded geometry), then IS_1 is equivalent to X being infinite. When $d = \infty$, we intend $1/d = 0$ and $d/(d-1) = 1$ and speak of a *strong isoperimetric inequality,* usually denoted by *IS* (without subscript). The latter will become important in the next chapter. We shall see that isoperimetric inequalities serve as an important tool for linking geometrical properties with transience and other probabilistic features of (X, P).

The isoperimetric inequality IS_d is equivalent to a *d-dimensional Sobolev inequality.* For a function $f : X \to \mathbb{R}$, define its *Sobolev norm*

$$(4.2) \qquad S_P(f) = \sum_{e \in E(P)} |\nabla f(e)| = \frac{1}{2} \sum_{x,y \in X} |f(x) - f(y)| a(x,y)$$

and its norm in $\ell^p(X, m)$

$$\|f\|_p = \left(\sum_{x \in X} |f(x)|^p m(x) \right)^{1/p},$$

whenever these sums converge. (If $p = \infty$ then we mean the sup-norm.) Here, $\nabla = \nabla_{\mathcal{N}}$. We shall also write $S_{\mathcal{N}}(f)$ instead of $S_P(f)$, while we shall reserve the notation $S(f)$ or $S_X(f)$ for the Sobolev norm associated with the simple random walk.

(4.3) Proposition. *(X, P) satisfies IS_d $(1 \le d \le \infty)$ if and only if*

$$\|f\|_{\frac{d}{d-1}} \le \kappa\, S_P(f) \quad \textit{for every } f \in \ell_0(X).$$

(The constant κ is the same as in IS_d.)

Proof. The isoperimetric inequality is obtained from the Sobolev inequality by setting $f = \mathbf{1}_A$.

We now show that IS_d implies the Sobolev inequality. As $S_P(f) \ge S_P(|f|)$, it is enough to prove this for $f \ge 0$. For $t > 0$, we write $[f > t]$ for the (finite) set $\{x : f(x) > t\}$. First, we have

$$
\begin{aligned}
S_P(f) &= \sum_x \sum_{y:f(y)>f(x)} \big(f(y) - f(x)\big)a(x,y) \\
&= \sum_x \sum_{y:f(y)>f(x)} a(x,y) \int_0^\infty \mathbf{1}_{[f(x)\,,\,f(y))}(t)\,dt \\
&= \int_0^\infty \Bigg(\sum_{x,y:f(x)\le t<f(y)} a(x,y) \Bigg)\,dt = \int_0^\infty a(\partial[f>t])\,dt\,.
\end{aligned}
$$

If $d=1$ then $\partial[f>t] \ne \emptyset \iff 0 \le t < \|f\|_\infty$, and the last term becomes

$$
= \int_0^{\|f\|_\infty} a(\partial[f>t])\,dt \ge \frac{1}{\kappa}\,\|f\|_\infty\,.
$$

Otherwise, we set $p = \frac{d}{d-1}$ and get that the above is

$$
\ge \frac{1}{\kappa} \int_0^\infty m[f>t]^{1/p}\,dt\,.
$$

Next, the function $F(t) = m[f>t]^{1/p}$ is non-negative and decreasing for $t \ge 0$. Therefore

$$
p\big(tF(t)\big)^{p-1} F(t) \le p \left(\int_0^t F(z)\,dz\right)^{p-1} F(t) = \frac{d}{dt}\left(\int_0^t F(z)\,dz\right)^p .
$$

Integrating and letting $t \to \infty$, we obtain

$$
\int_0^\infty p\,t^{p-1} F(t)^p\,dt \le \left(\int_0^\infty F(t)\,dt\right)^p
$$

(Hardy–Littlewood–Pólya inequality). Thus

$$
\kappa\, S_P(f) \ge \left(\int_0^\infty p\,t^{p-1} m[f>t]\,dt\right)^{1/p} .
$$

But the last term is well known to be equal to $\|f\|_p$. Indeed, let $0 = t_0 < t_1 < \cdots < t_n$ be the values assumed by f. Then

$$
\|f\|_p^p = \sum_{i=1}^{n} t_i^p \big(m[f>t_{i-1}] - m[f>t_i]\big) = \sum_{i=0}^{n-1} (t_{i+1}^p - t_i^p)\, m[f>t_i]\,.
$$

Observing that $m[f>t]$ is constant in $[t_i\,,\,t_{i+1})$, this becomes

$$
= \sum_{i=0}^{n-1} \int_{t_i}^{t_{i+1}} p\,t^{p-1} m[f>t]\,dt = \int_0^\infty p\,t^{p-1} m[f>t]\,dt\,,
$$

and the proof is completed. □

We have a series of comparison results which are completely analogous with those proved for the Dirichlet norm in the preceding sections. We suppose that X is a graph with bounded geometry (although several of the results hold without this assumption). The following is the analogue to Lemma 2.5.

(4.4) Lemma. $S_{P^k}(f) \le k\, S_P(f)$.

(4.5) Proposition. *Let P be reversible with total conductance $m(\cdot)$.*

(a) *Suppose that P is uniformly irreducible and $\inf_X m(x) > 0$. Then there is $c_1 > 0$ such that $S_P(f) \ge c_1 S(f)$ for all $f \in \ell_0(X)$.*

(b) *In general, $S_P(f) \le \big(\sup_{E(X)} \psi(e)\big) S(f)$ for all $f \in \ell_0(X)$, where*

$$\psi(e) = \frac{1}{2} \sum_{x,y \in X} m(x)p(x,y) \frac{|\Pi_e(x,y)|}{|\Pi(x,y)|}\,.$$

(c) *The following conditions are sufficient for ψ being bounded above: $\sup_X m(x) < \infty$ and (1) P has bounded range, or (2) P has finite first moment and there is a unimodular, quasi-transitive subgroup of* $\mathrm{AUT}(X,P)$.

For (a), compare with the second part of Theorem 3.1. For (b) and (c), see Theorem 3.2, Proposition 3.20 and also Theorem 3.31.

(4.6) Corollary. *Let (X,P) be strongly reversible and uniformly irreducible. Suppose that P has bounded range, or that P has finite first moment and there is a unimodular, quasi-transitive subgroup of* $\mathrm{AUT}(X,P)$. *Then (X,P) satisfies IS_d $(1 \le d \le \infty)$ if and only if the graph X satisfies IS_d.*

Finally, different graphs can be compared as follows.

(4.7) Theorem. *Let X, X' be graphs with bounded geometry, and let $\varphi : X \to X'$ be a rough isometry. Then there is a constant $c_0' > 0$ such that*

$$S_{X'}(f) \ge c_0'\, S_X(f \circ \varphi) \quad \text{for all } f \in \ell_0(X')\,.$$

In particular, X satisfies IS_d $(1 \le d \le \infty)$ if and only if X' satisfies IS_d.

Proof. The inequality between Sobolev norms is proved like Theorem 3.10. Using a rough inverse $\bar{\varphi}$ of φ, we get an analogous inequality in the other direction.

Next, note that with respect to the simple random walk, $m(A) = \sum_{x \in A} \deg(x)$. In particular, if M is a bound for the vertex degrees in our two graphs, we have $|A| \le m(A) \le M\,|A|$.

Now suppose that X satisfies IS_d. For finite $A' \subset X'$, consider the enlarged set $A'' = \{x' \in X' : d'(x', A') \le B\}$. If $[x', y'] \in \partial A''$ with $x' \in A''$

then there is $v' \in A'$ with $d'(v', x') = B$ (it cannot be that $d'(v', x') < B$ as then $y' \in A'$), and if w' is the neighbour of v' on some shortest path from v' to y', then $[v', w'] \in \partial A'$. Therefore

$$M^B \, a'(\partial A') \geq a'(\partial A'') = S_{X'}(\mathbf{1}_{A''}) \geq c_0' \, S_X(\mathbf{1}_{A''} \circ \varphi) = c_0' \, a(\partial \varphi^{-1} A'')$$
$$\geq c_0' \, \kappa \, m(\varphi^{-1} A'')^{1-1/d} \geq c_0' \, \kappa \, |\varphi\varphi^{-1} A''|^{1-1/d} .$$

Every point in A' is at distance at most B from some point in $\varphi\varphi^{-1}A''$, whence $|\varphi\varphi^{-1}A''| \geq M^{-B-1}\,|A'| \geq M^{-B-2}\, m'(A')$. □

B. Cartesian products

If X_1 and X_2 are two graphs then their *direct product* $X_1 \otimes X_2$ is the graph with vertex set consisting of all pairs x_1x_2 where $x_i \in X_i$, and neighbourhood given by

$$x_1x_2 \sim y_1y_2 \iff x_1 \sim y_1 \text{ and } x_2 \sim y_2 .$$

The *Cartesian product* $X_1 \times X_2$ has the same vertices, while

$$x_1x_2 \sim y_1y_2 \iff x_1 \sim y_1 \text{ and } x_2 = y_2 \quad \text{or} \quad x_1 = y_1 \text{ and } x_2 \sim y_2 .$$

Note that if both factors are bipartite graphs (that is, they have no closed paths with odd length), then the direct product is disconnected. Also, if Γ_1 and Γ_2 are finitely generated groups with Cayley graphs $X(\Gamma_i, S_i)$, then the direct product is not necessarily a Cayley graph of the group $\Gamma_1 \times \Gamma_2$, while this *is* the case for the Cartesian product.

Analogously, we define the direct and Cartesian products of two networks $\mathcal{N}_1$, $\mathcal{N}_2$: if a_i, $i = 1, 2$, are the conductance functions of the two, then $\mathcal{N}_1 \otimes \mathcal{N}_2$ has conductances $a(x_1x_2, y_1y_2) = a_1(x_1, y_1)a_2(x_2, y_2)$ (again, this is not necessarily connected), while for the Cartesian product $\mathcal{N}_1 \times \mathcal{N}_2$ we define

$$a(x_1x_2, y_1y_2) = a_1(x_1, y_1)\delta_{x_2}(y_2) + \delta_{x_1}(y_1)a_2(x_2, y_2) . \tag{4.8}$$

Finally, for two transition matrices P_1, P_2 over X_1 and X_2, respectively, their direct or *tensor* product $P_1 \otimes P_2$ over $X_1 \times X_2$ is given by $p(x_1x_2, y_1y_2) = p_1(x_1, y_1)p_2(x_2, y_2)$. This definition is "natural" in the sense that in the reversible case, $P_1 \otimes P_2$ is the reversible chain associated with the direct product of the corresponding networks. On the other hand, the tensor product does not in general preserve irreducibility. Also, it is only for dimension 2 that the direct product of d simple random walks on $\mathbb{Z}$ is isomorphic with the simple random walk on $\mathbb{Z}^d$.

A Cartesian product of P_1 and P_2 is a transition operator of the form

$$P = c\,P_1 \otimes I_2 + (1-c)I_1 \otimes P_2\,, \tag{4.9}$$

where $0 < c < 1$. The simple random walk on the Cartesian product of two *regular* graphs is a Cartesian product of the simple random walks on the factors. In general, the reversible Markov chain associated with the Cartesian product of two networks is not a Cartesian product of the reversible Markov chains on the factors. Still, under our usual adaptedness conditions the two are comparable. Throughout this book, we shall prefer Cartesian products to direct products.

(4.10) Theorem. *Let $\mathcal{N}_i$, $i = 1,2$, be two networks with associated invariant measures m_i such that $\sup_{X_i} m_i(x) < \infty$. If $\mathcal{N}_1$ satisfies IS_{d_1} and $\mathcal{N}_2$ satisfies IS_{d_2}, then $\mathcal{N} = \mathcal{N}_1 \times \mathcal{N}_2$ satisfies $IS_{d_1+d_2}$.*

Proof. Recall the definition of mixed norms: let $1 \le p, q \le \infty$ and f be a real function on $X_1 \times X_2$. Define $g(x_2) = \|f(\cdot, x_2)\|_p$, taken in $\ell_p(X_1, m_1)$. Then $\|f\|_{p,q} = \|g\|_q$, the latter taken in $\ell_q(X_2, m_2)$.

Now set $p_i = \frac{d_i}{d_i - 1}$ (so that $1 < p_i \le \infty$). Let $f \in \ell_0(X_1 \times X_2)$, and write

$$\bar{f}(x_2) = \sum_{x_1 \in X_1} |f(x_1, x_2)|\, m_1(x_1)\,.$$

Then we have

$$\begin{aligned}
\|f\|_{(1,p_2)} = \|\bar{f}\|_{p_2} &\le \kappa_2\, S_{\mathcal{N}_2}(\bar{f}) \\
&\le \frac{\kappa_2}{2} \sum_{x_2, y_2 \in X_2} \sum_{x_1 \in X_1} |f(x_1, x_2) - f(x_1, y_2)|\, m_1(x_1)\, a_2(x_2, y_2) \\
&\le \kappa_2 \left(\sup\nolimits_{X_1} m_1(x_1)\right) S_{\mathcal{N}}(f)\,.
\end{aligned}$$

On the other hand,

$$\begin{aligned}
\|f\|_{(p_1,1)} &= \sum_{x_2 \in X_2} \|f(\cdot, x_2)\|_{p_1}\, m_2(x_2) \\
&\le \sum_{x_2 \in X_2} \frac{\kappa_1}{2} \sum_{x_1, y_1 \in X_1} |f(x_1, x_2) - f(y_1, x_2)|\, a_1(x_1, y_1)\, m_2(x_2) \\
&\le \kappa_1 \left(\sup\nolimits_{X_2} m_2(x_2)\right) S_{\mathcal{N}}(f)\,.
\end{aligned}$$

By the interpolation theorem for mixed norms, there is $\kappa > 0$ such that for every choice of $s, t \ge 0$, $s + t = 1$, setting $(\frac{1}{p}, \frac{1}{q}) = s\,(1, \frac{1}{p_1}) + t\,(\frac{1}{p_2}, 1)$ we have

$$\|f\|_{(p,q)} \le \kappa\, S_{\mathcal{N}}(f) \quad \text{for all } f \in \ell_0(X_1 \times X_2)\,.$$

Choosing $s = \frac{d_1}{d_1+d_2}$, we get $p = q = \frac{d_1+d_2}{d_1+d_2-1}$, which proves that $\mathcal{N}$ satisfies $IS_{d_1+d_2}$. □

(4.11) Remarks. (1) Similarly, one proves the following. Suppose that for $i = 1, 2$, the $\mathcal{N}_i$ are networks satisfying IS_{d_i} and that for the corresponding transition matrices, one has $P_i \geq c_i\, I_i$ elementwise (where $c_i > 0$ and I_i is the identity matrix over X_i). Then the direct product $\mathcal{N}_1 \otimes \mathcal{N}_2$ satisfies $IS_{d_1+d_2}$.

(2) Usually, the interpolation theorem for mixed norms is formulated for a linear operator, say T, taking values in a product space, such that $\|T(f)\|_{(p,q)} \leq \kappa_{p,q}\|f\|$ for different pairs (p, q).

To see that the above proof is correct in these terms, let $\mathcal{N} = (X, E, r)$ be an arbitrary infinite network associated with a reversible Markov chain. Let $\ell^1(E)$ be the space of all $u : E \to \mathbb{R}$ such that $\|u\| = \sum_E |u(e)|$ is finite. For a reference vertex o, consider the subspace $\ell^1_o(\mathcal{N}) = \{\nabla_{\mathcal{N}} f : f \in \ell_0(X)\,,\ f(o) = 0\}^-$. For each $u \in \ell^1_o(\mathcal{N})$ there is precisely one $f : X \to \mathbb{R}$ such that $f(o) = 0$ and $u = \nabla_{\mathcal{N}} f$. Define $T(u) = f$. Then IS_d implies $\|T(u)\|_p \leq \kappa\, \|u\|$ for all $u \in \ell^1_o(\mathcal{N})$, where $p = \frac{d}{d-1}$. Conversely, if this holds for any choice of the base point, then we get IS_d : indeed, every $f \in \ell_0(X)$ is 0 in some $o \in X$. Transporting this to the context of products of networks, we see that the above proof works in the correct setting.

(4.12) Corollary. *Let X_1 and X_2 be two infinite graphs with bounded geometry. If X_1 satisfies IS_{d_1} and X_2 satisfies IS_{d_2} then $X_1 \times X_2$ satisfies $IS_{d_1+d_2}$. In every case, $X_1 \times X_2$ satisfies IS_2.*

In this corollary, the simple random walk on $X_1 \times X_2$ may be replaced by any Cartesian product of the simple random walks on the factors (by virtue of Proposition 4.5).

C. Isoperimetric inequalities and growth

From Corollary 4.12, we obtain the well-known fact that the grid $\mathbb{Z}^d$ satisfies IS_d. However, in general it is not a very simple task to prove an isoperimetric inequality for a general graph. The goal of this subsection is to relate isoperimetric inequalities and growth.

Recall the definition (§3.B) of the growth function of a reversible nearest neighbour random walk (X, P) on a locally finite, connected graph X. Also recall our requirement that the invariant measure (total conductance) satisfies $\inf_X m(x) > 0$.

(4.13) Lemma. *If (X, P) satisfies IS_d then $V_P(n) \geq C\, n^d$ for some $C > 0$.*

Proof. We use induction on n. If $n = 1$, then the statement is true for any $C \leq 2 \min_X V_P(x, 1)$. (As X is infinite and connected, $V_P(x, 1) \geq 2 \inf_X m(\cdot) > 0$ for every x.)

Suppose it holds for $n-1 \geq 1$. Observe that

$$n^d - (n-1)^d \leq d\,n^{d-1} \leq d\,\big(2(n-1)\big)^{d-1}.$$

Let $x \in X$. Then (using that P is nearest neighbour)

$$\begin{aligned} V_P(x,n) &\geq V_P(x,n-1) + a\big(\partial B(x,n-1)\big) \\ &\geq V_P(x,n-1) + \frac{1}{\kappa} V_P(x,n-1)^{1-1/d} \\ &\geq C\,(n-1)^d + \frac{1}{\kappa} C^{1-1/d}(n-1)^{d-1} \\ &\geq C\,n^d + \Big(\frac{1}{\kappa} C^{1-1/d} - 2^{d-1}Cd\Big)(n-1)^{d-1}. \end{aligned}$$

Hence, the statement holds for n if C is chosen such that $C^{1-1/d}/\kappa \geq 2^{d-1}Cd$, that is, $C \leq 1/(2^{d-1}\kappa d)^d$. □

Under an additional condition, there is a converse to the last lemma. For $f : X \to \mathbb{R}$, define

$$P_n f(x) = \frac{1}{V_P(x,n)} \sum_{y \in B_P(x,n)} f(y) m(y).$$

This is the transition operator of a nearest neighbour random walk on the n-fuzz $X^{(n)}$. We say that (X,P) (or, in the case of the simple random walk, X itself) is *quasi-homogeneous* if there is a constant $\eta < \infty$ such that

$$\|f - P_n f\|_1 \leq \eta\, n\, S_P(f) \quad \text{for all } f \in \ell_0(X). \tag{4.14}$$

With the growth function, we associate the right semicontinuous functions

$$\mathfrak{f}(t) = \min\{n : V_P(n) > t\} \quad \text{and} \quad \mathfrak{F}(t) = \mathfrak{F}_P(t) = t/\mathfrak{f}(2t), \quad t \geq \tfrac{1}{2} \inf_X m(x).$$

(4.15) Proposition. *If (X,P) is quasi-homogeneous (with constant η) then it satisfies $IS_{\mathfrak{F}_P}$ with constant $\kappa = 2\eta$.*

Proof. Let $A \subset X$ be finite. Then, for arbitrary n,

$$m(A) = m[\mathbf{1}_A \geq 1] \leq m[|\mathbf{1}_A - P_n \mathbf{1}_A| \geq 1/2] + m[P_n \mathbf{1}_A \geq 1/2].$$

By Markov's inequality, $m[|\mathbf{1}_A - P_n\mathbf{1}_A| \geq 1/2] \leq 2\|\mathbf{1}_A - P_n\mathbf{1}_A\|_1$, which by quasi-homogeneity is $\leq 2n\eta a(\partial A)$.

Now choose $n = \mathfrak{f}\big(2m(A)\big)$. Then $V_P(x,n) > 2m(A)$ and consequently $P_n\mathbf{1}_A(x) \leq m(A)/V_P(x,n) < 1/2$ for every $x \in X$. We get

$$m(A) \leq 2\eta\, \mathfrak{f}\big(2m(A)\big)\, a(\partial A). \qquad \square$$

(4.16) Corollary. *If (X,P) is quasi-homogeneous and $V_P(n) \geq C\,n^d$ ($C > 0$) then (X,P) satisfies IS_d.*

In Proposition 4.15 we did not use the assumption that P is of nearest neighbour type. However, that will be used in the next steps, where we study quasi-homogeneity.

(4.17) Lemma. *If $f \in \ell_0(X)$ then*

$$\|f - P_n f\|_1 \leq \Big(\sup_E \eta(e,n)\Big)\, S_P(f)\,,$$

where

$$\eta(e,n) = r(e) \sum_{x\in X} \frac{m(x)}{V_P(x,n)} \sum_{y\in B(x,n)} \frac{m(y)|\Pi_e(x,y)|}{|\Pi(x,y)|}\,.$$

Proof. Let ∇ be the difference operator associated with our network. Observe that for $x,y \in X$ and $\pi \in \Pi(x,y)$ we have $|f(y) - f(x)| \leq \sum_{e\in\pi} |\nabla f(e)|\, r(e)$. We compute

$$\begin{aligned}
\|f - P_n f\|_1 &= \sum_{x\in X} \left| \frac{1}{V_P(x,n)} \sum_{y\in B(x,n)} \big(f(x) - f(y)\big) m(y) \right| m(x) \\
&\leq \sum_{x\in X} \frac{m(x)}{V_P(x,n)} \sum_{y\in B(x,n)} \frac{m(y)}{|\Pi(x,y)|} \sum_{\pi\in\Pi(x,y)} \sum_{e\in\pi} |\nabla f(e)| r(e) \\
&= \sum_{e\in E} |\nabla f(e)| \left(r(e) \sum_{x\in X} \frac{m(x)}{V_P(x,n)} \sum_{y\in B(x,n)} \frac{m(y)|\Pi_e(x,y)|}{|\Pi(x,y)|} \right). \quad \square
\end{aligned}$$

For Cayley graphs and, more generally, for quasi-transitive random walks, the situation is as follows.

(4.18) Theorem. *Let P be the transition matrix of a strongly reversible nearest neighbour random walk on X. Suppose that (X,P) is quasi-transitive. Then (X,P) is quasi-homogeneous and satisfies $IS_{\mathfrak{F}_P}$.*

Proof. We first consider the (easier) case when (X,P) satisfies a strong isoperimetric inequality. Actually, then we do not really need quasi-homogeneity which serves here only as a tool for obtaining isoperimetric inequalities. However, it is easy to see that (X,P) is quasi-homogeneous. First, use quasi-transitivity and the fact that m is constant on each orbit of $\mathrm{AUT}(X,P)$ to work out that there is a constant $C > 0$ such that $V(y,n) \leq C\,m(y)\,V_P(x,n)$ for all $x,y \in X$ and $n \in \mathbb{N}$. This yields $\|P_n f\|_1 \leq C\,\|f\|_1$, and by *IS*,

$$\|f - P_n f\|_1 \leq (1+C)\|f\|_1 \leq (1+C)\kappa\, S_P(f)\,.$$

So now suppose that (X, P) does not satisfy *IS*. Then there must be a sequence of finite subsets A_k of X such that $a(\partial A_k) \le m(A_k)/k$. In view of Lemma 4.17, we show that $\eta(e,n) \le n\,\eta$ for some finite η.

Let X_i $(i \in \mathcal{I})$ and E_j $(j \in J)$ be the finitely many orbits of $\Gamma = \mathrm{AUT}(X,P)$ on X and E, respectively. For $e \in E$ and $x \in X$, we let $d(x,e) = d(e,x) = \min\{d(e^-,x), d(e^+,x)\}$ and define $B(e,n) = \{x \in X : d(x,e) \le n\}$ and $V_P(e,n)$ accordingly. We introduce

$$s = \max\{d(x,E_j) : x \in X\,,\ j \in J\} \quad \text{and} \quad \eta = \max\{r(e)V_P(e,s) : e \in E\}\,;$$

s and η are finite by quasi-transitivity and Γ-invariance of $m(\cdot)$ and $r(\cdot)$. Now set $E_{j,k} = \{e \in E_j : d(e,A_k) \le s\}$. For each $x \in A_k$ we choose $e = e(x) \in E_{j,k}$ such that $d(e,x) \le s$. Then

$$m(A_k) = \sum_{e \in E_{j,k}} m(\{x \in A_k : e(x) = e\}) \le \sum_{e \in E_{j,k}} m\big(B(e,s)\big) \le \eta\, a(E_{j,k})\,.$$

Observe that $\eta(e,n) = \eta_j(n)$ is constant on each E_j. Therefore

$$\begin{aligned}\eta_j(n) &\le \frac{1}{a(E_{j,k})} \sum_{e \in E_{j,k}} \frac{1}{r(e)} \eta(e,n) \\ &= \frac{1}{a(E_{j,k})} \sum_{x,y \in X : d(x,y) \le n} \frac{m(x)m(y)}{V_P(x,n)} \sum_{e \in E_{j,k}} \frac{|\Pi_e(x,y)|}{|\Pi(x,y)|}\,.\end{aligned}$$

In order that $\Pi_e(x,y)$ be non-empty for some $e \in E_{j,k}$, it must be that $x \in A_k^{(n+s)} = \{w \in X : d(w,A_k) \le n+s\}$. Also, we have the identity $\sum_{e \in E} |\Pi_e(x,y)| = d(x,y)\,|\Pi(x,y)|$. Hence,

$$\eta_j(n) \le \frac{1}{a(E_{j,k})} \sum_{x \in A_k^{(n+s)}} \frac{m(x)}{V_P(x,n)} \sum_{y \in B(x,n)} m(y) d(x,y) \le n\,\eta\, \frac{m(A_k^{(n+s)})}{m(A_k)}\,.$$

This holds for every $n, k \in \mathbb{N}$ and $j \in J$. Now

$$m(A_k^{(n+s)} \setminus A_k) \le |\partial A_k|\, \bar{V}_P(n+s) \le a(\partial A_k) \sup_E r(e)\, \bar{V}_P(n+s)\,.$$

By the choice of the A_k, we get $\lim_k \big(m(A_k^{(n+s)})/m(A_k)\big) = 1$. Hence $\eta_j(n) \le \eta\, n$. □

5. Transient subtrees, and the classification of the recurrent quasi-transitive graphs

We shall now present a purely graph theoretical theorem which, combined with the results of the preceding sections, can be used to give a structural classification of those graphs which carry a recurrent quasi-transitive random walk, thereby extending the corresponding result on groups (Theorem 3.24).

A. Transient subtrees

In the sequel, $m(\cdot)$ and $a(\cdot)$ will be the measures associated with the simple random walk on the graph X. (In particular, $m(\cdot)$ is integer-valued, and $a(\cdot)$ is the counting measure on $E(X)$.) Let $\mathfrak{F} : \mathbb{N} \to \mathbb{R}^+$ be a non-decreasing function such that $\mathfrak{F}(n) \to \infty$. Consider the following weaker version of the $\mathfrak{F}$-isoperimetric inequality.

(5.1) Definition. Given a "root" $o \in X$, we say that X satisfies $IS_{\mathfrak{F},o}$, if there is a constant $\kappa > 0$ such that

$$\mathfrak{F}\big(m(A)\big) \le \kappa\, a(\partial A)$$

for every finite $A \subset X$ which is connected and contains o.

(5.2) Theorem. *Let X be a graph with bounded geometry satisfying $IS_{\mathfrak{F},o}$ with respect to some root $o \in X$. If $\sum_n \mathfrak{F}(n)^{-2} < \infty$ then X contains a transient tree with maximum degree 3.*

Besides its application to the above classification problem, this theorem is of interest by itself: trees are a well-understood class of graphs with a particularly simple structure, and so it is of considerable interest to know which graphs have transient subtrees. The proof of Theorem 5.2 is going to be "elementary", but long. As a basic graph theoretical ingredient, we need Menger's theorem, which can be derived from the max flow–min cut theorem (Ford and Fulkerson [119]) and is proved in most books on graph theory.

(5.3) Theorem. *Let X be a graph and $U, V \subset X$ such that $|U| = |V| = k$. Suppose that for every $A \subset X$ with $|A| < k$, there is a path in X from U to V which does not meet A.*

Then there are k pairwise disjoint paths in X from U to V.

We now set up some notation for the proof of Theorem 5.2. First of all, as usual, we write M for an upper bound on the vertex degrees in X.

Components. Let X be a connected, locally finite graph with edge set $E = E(X)$. Let $A \subset X \cup E$ (usually finite), and let Y be a subgraph of X.

Then $Y \setminus A$ is defined as the graph obtained from Y by deleting A and all edges incident with vertices in A. If $U \subset X \setminus A$ is connected, then we write $\mathcal{C}(U, A)$ for the connected component of $X \setminus A$ which contains U. The *finite (infinite) part* of $X \setminus A$ is the union of all its finite (infinite) components; if A is finite then the finite part is a finite graph by local finiteness.

We shall need that the root o has the property that $\mathcal{C}(o, y)$ is infinite for every $y \in X$, $y \neq o$.

(5.4) Lemma. *If X has bounded geometry and satisfies $IS_{\mathfrak{F},o}$, where $\mathfrak{F}(k) \to \infty$, then there is $x \in X$ such that $IS_{\mathfrak{F},x}$ holds and $\mathcal{C}(x, y)$ is infinite for all $y \neq x$.*

Proof. Define

$$\mathfrak{g}(t) = \max\{k : \mathfrak{F}(k) \leq t\}, \quad \text{where } t \geq \mathfrak{F}(0). \tag{5.5}$$

If $\mathcal{C}(o, y)$ is infinite for all $y \neq o$, then we set $x = o$. So suppose that there is x such that $C = \mathcal{C}(o, x)$ is finite. Then $\mathfrak{F}\big(m(C)\big) \leq \kappa\, a(\partial C) \leq \kappa(M-1)$, so that $d(o, x) \leq m(C) \leq \mathfrak{g}\big(\kappa(M-1)\big)$. Hence we may choose x such that $d(o, x)$ is maximal. We claim that x has the desired properties.

Let $y \neq x$. If $y \in \mathcal{C}(o, x)$ then $\mathcal{C}(x, y)$ contains the infinite set $X \setminus \mathcal{C}(o, x)$. If $y \notin \mathcal{C}(o, x)$ then $d(y, o) > d(x, o)$ and $\mathcal{C}(x, y) = \mathcal{C}(o, y)$, which is infinite by the choice of x. Now let $A \subset X$ be finite, connected and containing x. Then, by monotonicity of $\mathfrak{F}$,

$$\mathfrak{F}\big(m(A)\big) \leq \mathfrak{F}\big(m(A \cup C)\big) \leq \kappa\, a\big(\partial(A \cup C)\big) \leq \kappa\, a(\partial A). \qquad \square$$

By virtue of this lemma, we may assume for the rest of this subsection that o itself has the desired property.

Rooted trees. We shall only consider finite or infinite rooted trees T with maximum degree 3. Unless $T = \{o\}$, the root o will have degree $\deg_T(o) = 2$. We define the *levels* $S_k = S_k(T)$ as follows. $S_0(T) = \{o\}$, and for $k \geq 1$, a vertex $x \in T$ lies in S_k if $\deg_T(x) = 3$ and on $\pi(o, x)$ there are precisely k vertices with degree 3. We write $S(T) = \bigcup_{k \geq 0} S_k(T)$.

The infinite subtree of X that we shall construct will be a *subdivision of the binary tree with root o,* that is, the minimum degree is 2, and $|S_k| = 2^k$ for all $k \geq 0$. This tree will be constructed as the limit of an increasing sequence of finite rooted trees.

If T is finite, then the *leaves* of T are the elements of $L = L(T) = \{x \in T : \deg_T(x) = 1\}$. The *level height* of a leaf $x \in L(T)$ is

$$l(x) = \max\{k : S_k(T) \cap \pi(o, x) \neq \emptyset\},$$

and the level height of T is $l(T) = \min_{L(T)} l(x)$. We shall only consider finite trees with the above features such that

$$(5.6) \qquad l(T) \le l(x) \le l(T) + 1 \quad \text{for all } x \in L(T),$$

and we write $L'(T) = \{x \in L(T) : l(x) = l(T)\}$.

Now suppose that T is a finite subtree of our graph X, with root o, and satisfying the above requirements, in particular (5.6). We write $D(T)$ for the set of all edges in $E(X) \setminus E(T)$ which are incident with some vertex in $S(T)$, unless $T = \{o\}$, in which case we set $D(T) = \emptyset$. ($D(T)$ is also defined when T is an infinite subtree of X.) We say that T is *expandable* if either $T = \{o\}$ or the following technical conditions hold.

(a) The component $\mathcal{C}\big(T \setminus L(T), D(T) \cup L(T)\big)$ in X is finite, and
(b) if $A \subset X$ is such that $A \cap T \subset L(T)$ and $|A| < |L(T)|$ then there is $x \in L(T) \setminus A$ such that $\mathcal{C}\big(x, D(T) \cup A\big)$ is infinite.

(5.7) Lemma. *If T is an expandable subtree of X and $l(T) = k$ then the boundary of $C = \mathcal{C}\big(T \setminus L(T), D(T) \cup L(T)\big)$ in X satisfies*

$$a(\partial C) \le (M-1)\, 2^{k+2}.$$

In particular, every $x \in L(T)$ satisfies $d(x,o) \le R_k$, where $R_k = \mathfrak{g}\big(\kappa\,(M-1)\,2^{k+2}\big)$.

Proof. ∂C consists of $D(T)$ and the edges in T incident with vertices in $L(T)$. This proves the first part. For the second part, observe that $\mathfrak{F}\big(d(x,o)\big) \le \mathfrak{F}\big(m(C)\big) \le \kappa\, a(\partial C)$. □

Starting with $T_0 = \{o\}$, we shall now construct an increasing sequence of finite, expandable trees T_n in X, $n \ge 0$, such that each T_n is a proper subtree of T_{n+1}. We describe the expansion algorithm.

Suppose that we have $T = T_n$. Let Y be the infinite part of $X \setminus (D \cup L)$, where $D = D(T)$ and $L = L(T)$.

Case 1. $T \ne \{o\}$, and there is $B \subset Y \cup L$ such that (i) $|B| = |L|$, (ii) $B \ne L$, and (iii) $\mathcal{C}(T \setminus B, D \cup B)$ is finite.

We consider the subgraph W of X induced by all paths in $X \setminus D$ from B to L which meet $B \cup L$ only in the endpoints. Then W is finite and intersects neither $T \setminus L$ nor the infinite part of $X \setminus (D \cup B)$. Also, if $A \subset W$ with $|A| < k = |B| = |L|$, then by property (2) of T, it must be that $|\mathcal{C}(x, D \cup A)| = \infty$ for some $x \in L \setminus A$. Now either $x \in B$ or $|\mathcal{C}(x, D \cup B)| < \infty$ by (iii), and in both cases there is a path in W from x to B. Theorem 5.3 applies, and there are k disjoint paths in W from L to B. We now construct T_{n+1} by attaching these paths to L.

We get a larger subtree of X with $L(T_{n+1}) = B$, $S(T_{n+1}) = S(T_n)$ and the same level heights as T_n. Property (a) holds for T_{n+1} by (iii), and property (b) is inherited from T_n. Thus, T_{n+1} is again expandable.

Case 2. $T = \{o\}$, or there is no set B as in Case 1.

Let $x \in L' = L'(T)$, and choose a neighbour v of x in the infinite part of $X \setminus (D \cup L)$. If $x = o$, then the existence of such a v is clear. Otherwise, existence follows from property (b), setting $A = L \setminus \{x\}$.

Case 2.1 There is a set $B' \subset (L \cup Y) \setminus \{x\}$ such that (i) $|B'| = |L|$, and (ii) $\{v\} \cup L \setminus (B' \cup \{x\})$ is non-empty and belongs to the finite part of $X \setminus (D \cup B' \cup \{x\})$.

Then either $L \setminus B'$ contains an element different from x, or $B' \cup \{x\} \supset L$; in both cases, $T \setminus (B' \cup \{x\})$ also belongs to the finite part of $X \setminus (D \cup B' \cup \{x\})$. By Lemma 5.7, we may choose B' such that the measure of $\mathcal{C}\big(T \setminus L, (D \cup B' \cup \{x\})\big)$ is maximal.

Now consider the subgraph W' of X induced by all paths in $X \setminus (D \cup \{x\})$ from B' to $\{v\} \cup L \setminus \{x\}$ which meet these two sets only in the endpoints. Then W' is finite and intersects neither $T \setminus (L \setminus \{x\})$ nor the infinite part of $X \setminus (D \cup B' \cup \{x\})$. Now let $A \subset W'$ with $|A| < k = |B'| = |\{v\} \cup L \setminus \{x\}|$. Suppose there is no path in W' from B' to $\{v\} \cup L \setminus \{x\}$. Then it cannot be that $A = L \setminus \{x\}$ by the choice of v. But otherwise, $B = A \cup \{x\}$ is as in Case 1, a contradiction. (When $T = \{o\}$, we have $x = o$, $|B'| = 1$ and $A = \emptyset = L \setminus \{o\}$.)

Again, we may apply Menger's theorem to obtain k disjoint paths in W' from $\{v\} \cup L \setminus \{x\}$ to B'. Together with the edge $[x, v]$, we attach these paths to T.

Next, x must have a neighbour w in the infinite part of $X \setminus (D \cup B' \cup \{x\})$. Otherwise, B' would have the properties of B in Case 1. (If $T = \{o\} = \{x\}$, then $B' = \{v'\}$ is such that $\mathcal{C}(v, \{o, v'\})$ is finite. On the other hand, $\mathcal{C}(o, v')$ is infinite by assumption. In this situation there must also be a neighbour w of x in the infinite part of $X \setminus \{o, v'\}$.) We also attach w and the edge $[x, w]$ to T.

This completes the construction of T_{n+1} in Case 2.1. We have that $D(T_{n+1})$ is the union of D and the set of all edges incident with x besides the three lying in T_{k+1}. Also, $L(T_{n+1}) = B' \cup \{w\}$. The level heights of the points in B' coincide with those of the corresponding points in L, with the exception of the endpoint in B' of the path in W' starting in v: this point, as well as w, has level height $l(x) + 1$. As $x \in L'(T)$, we see that T_{n+1} has property (5.6). We show that T_{n+1} is expandable: it has property (a) by construction. To see that (b) holds, let $|A| \le |L(T_{n+1})| - 1 = |B'|$ and $A \cap T_{n+1} \subset L(T_{n+1})$, without loss of generality $|A| = |B'|$. Either we have $A = B'$, in which case $w \in L(T_{n+1}) \setminus A$ satisfies the requirement of (b), or else A does not have the defining property of B' by the choice of

B' (maximality). In that case, there is $u \in \{v\} \cup L \setminus (A \cup \{x\})$ such that $\mathcal{C}(u, D \cup A \cup \{x\})$, and hence also $\mathcal{C}\big(u, D(T_{n+1}) \cup A\big)$, is infinite.

Case 2.2 If B' as in Case 2.1 does not exist, then x must have a neighbour $w \neq v$ in the infinite part of $X \setminus (D \cup L)$. Otherwise, $B = (L \setminus \{x\}) \cup \{v\}$ would be as in Case 1. (If $T = \{o\}$, then existence of such a w follows from the assumption that $\mathcal{C}(o, v)$ is infinite; recall Lemma 5.4.) We construct T_{k+1} by attaching v, w and the edges $[x, v]$ and $[x, w]$ to T.

Again, T_{n+1} satisfies (5.6). We have $L(T_{n+1}) = (L \setminus \{x\}) \cup \{v, w\}$, and $D(T_{n+1})$ is D plus all edges outside T_{n+1} which are incident with x. Once more, T_{n+1} has property (a) by construction. If $A \subset X$ does not meet interior points (non-leaves) of T_{n+1} and $|A| = |L(T_{n+1})| - 1 = |L|$, then there are two possibilities: either $A = (L \setminus \{x\}) \cup \{v\}$, in which case $\mathcal{C}(w, D(T_{n+1}) \cup A)$ is infinite, or else A cannot be of type B' as in Case 2.1, whence $\mathcal{C}(u, D \cup A \cup \{x\})$ is infinite for some $u \in (L \setminus \{x\}) \cup \{v\}$. But then also $\mathcal{C}\big(u, D(T_{n+1}) \cup A\big)$ is infinite, and property (b) is satisfied.

This describes the algorithm for constructing the sequence of trees T_n. By Lemma 5.7, the expansion of Case 1 (which changes no level height) cannot be carried out infinitely often in succession. Hence (1) we have $l(T_n) \to \infty$, and (2) for each $k \geq 0$ there is $n(k)$ such that $l(x) = k$ for each leaf x of $T_{n(k)}$. Write L_k for $L(T_{n(k)})$, and consider the union (limit) $\bar{T}$ of the T_n. Then we can summarize the essence of our construction as follows.

(5.8) Proposition. *Suppose that X is a graph with bounded geometry satisfying $IS_{\mathfrak{F},o}$, where $\mathfrak{F}$ is non-decreasing and $\mathfrak{F}(k) \to \infty$. Then X has a subtree $\bar{T}$ which is a subdivision of the binary tree and such that for each k there is a set L_k consisting of 2^{k+1} vertices satisfying*

(1) *$L_k \cap S_k(\bar{T}) = \emptyset$, and there is precisely one element of L_k on each of the 2^{k+1} paths in $\bar{T}$ which go from $S_k(\bar{T})$ to $S_{k+1}(\bar{T})$;*
(2) *if o' is the root of $\bar{T}$, then the component $\mathcal{C}\big(o', D(\bar{T}) \cup L_k\big)$ in X is finite for each k.*

Note that it is not necessarily true that $o' = o$; compare with Lemma 5.4. The component in (2) is that of $T_{n(k)} \setminus L_k$.

Proof of Theorem 5.2. We show that $\bar{T}$ has a finite energy flow with input 1 from the root to ∞. This implies transience of the simple random walk on $\bar{T}$ and hence also of the simple random walk on X.

Given $e \in E(\bar{T})$, we choose e^- as the endpoint closer to the root. We use the "simple" flow u on $\bar{T}$: at each branching (vertex in $S(\bar{T})$), it subdivides equally between the two outgoing edges. Thus, $u(e) = 1/2^k$ on each edge between $S_{k-1}(T)$ and $S_k(T)$. Let s_k denote the number of all these edges ($k \geq 1$). The energy of the flow is

$$\langle u, u \rangle = \sum_{e \in E(T)} u(e)^2 = \sum_{k=1}^{\infty} s_k / 4^k \,.$$

Now, by monotonicity of $\mathfrak{F}$ and the isoperimetric inequality (recall Lemma 5.4),

$$\mathfrak{F}(s_1 + \cdots + s_k) \le \mathfrak{F}\Big(m\Big(C\big(o', D(\bar{T}) \cup L_k\big)\Big)\Big) \le \kappa\,(M-1)\,2^{k+2} \,.$$

Setting $c = \big(4\,\kappa\,(M-1)\big)^2$, we get that $\langle u, u \rangle / c$ is bounded above by

$$\sum_{k=1}^{\infty} s_k\, \mathfrak{F}(s_1 + \cdots + s_k)^{-2} \le \sum_{k=1}^{\infty} \sum_{i=1}^{s_k} \mathfrak{F}(s_1 + \cdots + s_{k-1} + i)^{-2} = \sum_{n=1}^{\infty} \mathfrak{F}(n)^{-2} < \infty \,.$$

□

B. Transient subtrees in quasi-transitive graphs

We shall now complete the structural classification of all recurrent quasi-transitive graphs. We start with the following observation, whose (easy) proof is omitted.

(5.9) Lemma. *Suppose $\Gamma \le \mathrm{AUT}(X)$ acts quasi-transitively and that $\mathfrak{K}$ is a compact normal subgroup of Γ. Then X is roughly isometric with the factor graph $\mathfrak{K}\backslash X$, and the two graphs have equivalent growth functions.*

Losert [216] has extended the main result on groups with polynomial growth (Theorem 3.17) to locally compact topological groups. Let Γ be such a group, generated by a compact, symmetric neighbourhood U of the identity, and with Haar measure $|\cdot|$. The growth function of Γ with respect to U is $V(n) = |U^n|$. Change to another generating neighbourhood gives rise to an equivalent growth function. For our purpose, the following result of Losert [216] is important.

(5.10) Theorem. *Let Γ be locally compact, generated by some compact, symmetric neighbourhood of the identity. If there are C, d such that the associated growth function satisfies $V(n) \le C\,n^d$ for infinitely many n, then Γ has a compact normal subgroup $\mathfrak{K}$ such that $\Gamma/\mathfrak{K}$ is a (possibly zero-dimensional) Lie group.*

From this, we can deduce the following.

(5.11) Theorem. *Let X be a quasi-transitive graph whose growth function satisfies $V(n) \le C\,n^d$ for infinitely many n. Then X is roughly isometric with a Cayley graph of some finitely generated nilpotent group. In particular, there are an integer $d(X)$ and constants $C_0, C_1 > 0$ such that*

$$C_0\, n^{d(X)} \le V(n) \le C_1\,(n+1)^{d(X)} \quad \textit{for all } n \,.$$

Proof. Let $\Gamma = \mathrm{AUT}(X)$. Given an orbit X_1, consider the graph $Y = X_1^{(2N-1)}$ as constructed in the proof of Proposition 3.9. It is roughly isometric with X, so that $V_Y(n) \le C' n^d$ for infinitely many n by Lemma 3.13. Choose $o \in Y$ and let $U = \{\gamma \in \Gamma : d_Y(\gamma o, o) \le 1\}$. This is a compact, symmetric neighbourhood of the identity which generates Γ, and $U^n o = B_Y(n, o)$. We get that

$$|U^n| = \sum_{x \in B_Y(n,o)} |\{\gamma \in \Gamma : \gamma o = x\}| = |\Gamma_o| \, V_Y(n)\,.$$

The assumption and Theorem 5.10 now yield that Γ has a compact normal subgroup such that $\Gamma/\mathfrak{K}$ is Lie. As Γ is totally disconnected and compactly generated, $\Gamma/\mathfrak{K}$ must be discrete (zero-dimensional) and finitely generated.

Consider the factor graph $X' = \mathfrak{K}\backslash X$. It is roughly isometric with X (Lemma 5.9), and $\Gamma/\mathfrak{K}$ acts quasi-transitively on X' as a closed subgroup of $\mathrm{AUT}(X')$. Vertex stabilizers in $\Gamma/\mathfrak{K}$ are open and compact, whence finite. By Proposition 3.9, X' is roughly isometric with a Cayley graph of $\Gamma/\mathfrak{K}$. Applying Theorems 3.17 and 3.16 and Lemma 3.14 now yields the result. □

For $d \le 2$, more is known.

(5.12) Theorem. *If X is quasi-transitive with polynomial growth of degree $d \in \{1, 2\}$ then every quasi-transitive subgroup of* $\mathrm{AUT}(X)$ *contains a discrete subgroup isomorphic with $\mathbb{Z}^d$.*

In the case of linear growth ($d = 1$) this is obvious: in the above notation, there is an element with infinite order in $\Gamma/\mathfrak{K}$ acting with finitely many orbits on the factor graph. Any of its preimages in Γ under the factor map has infinite order and acts with finitely many orbits. On the other hand, for square growth, the theorem is not easy and has been proved very recently by Seifter and Trofimov [296]. We will not give the proof here.

The last theorems combined with Theorem 5.2 yield the result that we have been looking for.

(5.13) Theorem. *Let X be a quasi-transitive infinite graph. If some quasi-transitive random walk (X, P) is recurrent, then X is roughly isometric with the one- or two-dimensional grid:* $\mathrm{AUT}(X, P)$ *has a discrete subgroup isomorphic with $\mathbb{Z}$ or $\mathbb{Z}^2$ which acts quasi-transitively and fixed-point-freely.*

In this case, every strongly reversible, quasi-transitive random walk on X with finite second moment is recurrent.

Otherwise, X contains a transient subtree.

Proof. If (X, P) as given is recurrent, then Γ must be unimodular (Theorem 3.26) and the simple random walk on X recurrent (Corollary 3.29). This in turn yields unimodularity of every quasi-transitive subgroup of $\mathrm{AUT}(X)$, and Theorem 3.31 now yields recurrence of every strongly reversible, quasi-transitive random walk with finite second moment.

Let $V(n)$ be the growth function of X. If $V(n) \le C\,n^2$ then the simple random walk on X is indeed recurrent (Lemma 3.12), and from Theorems 5.11 (and its proof) and 5.12 we see that $\mathrm{AUT}(X, P)$ must have the stated properties. (The fact that the action of $\mathbb{Z}$ or $\mathbb{Z}^2$, respectively, is fixed-point-free follows from commutativity of these groups.)

Otherwise, Theorem 5.11 tells us that $V(n) \ge C\,n^3$. From Theorem 4.18 we get that X satisfies IS_3. We can now apply Theorem 5.2 to get a transient subtree. □

6. More on recurrence

In this final section, we present further results on recurrent random walks on quasi-transitive graphs, trees, and planar graphs associated with tilings and circle packings.

A. Generalized lattices

By a *d-dimensional generalized lattice* we mean a locally finite graph whose automorphism group contains the free abelian group $\mathbb{Z}^d$ as a quasi-transitive subgroup. From Proposition 3.9 we know that such a graph is roughly isometric with the d-dimensional grid and that it has polynomial growth with degree d. If X is a recurrent quasi-transitive graph, then we know from Theorems 5.12 and 5.13 that X is a generalized lattice with dimension $d = 1$ or $d = 2$.

The purpose of this subsection is to determine further recurrence criteria (besides strong reversibility) for quasi-transitive random walks on generalized lattices with dimension 1 or 2.

First, we recall what is known for random walks on the abelian groups $\mathbb{Z}$ and $\mathbb{Z}^2$ themselves.

(6.1) Theorem. *Let μ be an irreducible probability measure on $\mathbb{Z}^d$ with finite first moment $M_1(\mu)$, and write*

$$\mathfrak{m}(\mu) = \sum_{\mathbf{k} \in \mathbb{Z}^d} \mathbf{k}\,\mu(\mathbf{k})$$

for the mean vector *or* drift *of the associated random walk.*

(a) *If $\mathfrak{m}(\mu) \ne \mathbf{0}$ then the random walk is transient.*

(b) *If $d = 1$ and $\mathfrak{m}(\mu) = 0$ then the random walk is recurrent.*

(c) *If $d = 2$, $M_2(\mu) < \infty$ and $\mathfrak{m}(\mu) = \mathbf{0}$ then the random walk is recurrent.*

Proof. Part (a) follows immediately from the strong law of large numbers.

(b) We use the following simple inequality, valid for any Markov chain (X, P): for all $x, y \in X$ and $N \in \mathbb{N}$,

$$\sum_{n=0}^{N} p^{(n)}(y,y) = \sum_{k=0}^{N} p^{(k)}(x,y) \sum_{j=0}^{N-k} f^{(j)}(x,y) \leq F(x,y) \sum_{n=0}^{N} p^{(n)}(y,y).$$

Now we consider the translation-invariant random walk $P = P_\mu$ on $\mathbb{Z}$. Fix $M \in \mathbb{N}$ and set $\varepsilon = 1/M$. Then, for any $k \in \mathbb{Z}$ and $N \in \mathbb{N}$,

$$\sum_{n=0}^{MN} p^{(n)}(0,k) \leq \sum_{n=0}^{MN} p^{(n)}(k,k) = \sum_{n=0}^{MN} p^{(n)}(0,0).$$

Consequently

$$\begin{aligned} \sum_{n=0}^{MN} p^{(n)}(0,0) &\geq \frac{1}{2N+1} \sum_{k:|k|\leq N} \sum_{n=0}^{MN} p^{(n)}(0,k) \\ &\geq \frac{MN}{2N+1} \frac{1}{MN} \sum_{n=0}^{MN} \sum_{k:|k|\leq n\varepsilon} p^{(n)}(0,k). \end{aligned}$$

Now, by the weak law of large numbers and the fact that $\mathfrak{m}(\mu) = 0$,

$$\lim_{n\to\infty} \sum_{k:|k|\leq n\varepsilon} p^{(n)}(0,k) = 1.$$

Hence, as $N \to \infty$, the right hand side of the last inequality tends to $M/2$, and $G(0,0) \geq M/2$ for every $M \in \mathbb{N}$.

Part (c) is obtained by using Fourier transformation. We postpone the proof to Chapter III, where we shall derive the asymptotic behaviour of $p^{(n)}(x,y)$. □

We now consider an irreducible random walk (X, P), where X is a generalized lattice and $\mathrm{AUT}(X, P)$ contains a quasi-transitive subgroup isomorphic with $\mathbb{Z}^d$. For $\mathbf{k} \in \mathbb{Z}^d$, let $\gamma_{\mathbf{k}}$ be the corresponding automorphism of (X, P). Let X_i, $i \in \mathcal{I}$, be the finite family of orbits of the latter group. As mentioned in the proof of Theorem 5.13, the action of $\mathbb{Z}^d$ must be fixed-point-free. (Indeed, if some $\gamma_{\mathbf{k}}$ fixes a vertex in some X_i, then by commutativity it must fix every point in X_i. If it is different from the identity, then it has infinite order and hence must have infinite orbit in some X_j, a contradiction. Note that finiteness of $\mathcal{I}$ is crucial in this argument.) Therefore we may choose a "root" o_i in each of the X_i and we can identify $\mathcal{I} \times \mathbb{Z}^d$ with X via the one-to-one correspondence

$$i\mathbf{k} \leftrightarrow \gamma_{\mathbf{k}} o_i \,.$$

In this notation, we have

$$(6.2) \qquad p\big(i\mathbf{k}, j(\mathbf{k}+\mathbf{l})\big) = p(i\mathbf{0}, j\mathbf{l}) \quad \text{for all } i,j \in \mathcal{I} \text{ and } \mathbf{k},\mathbf{l} \in \mathbb{Z}^d \,.$$

The factor chain $(I, \widetilde{P})$ has transition probabilities

$$\widetilde{p}(i,j) = \sum_{\mathbf{k}} p(i\mathbf{0}, j\mathbf{k}) \,.$$

As $\mathcal{I}$ is finite, it is positive recurrent and has a unique invariant probability measure $\widetilde{\nu}$; see Theorem 1.18. The random walk on X can be decomposed as a pair

$$Z_n = \widetilde{Z}_n Y_n \,,$$

where $(\widetilde{Z}_n)_n$ is the factor chain and $(Y_n)_n$ is a sequence of $\mathbb{Z}^d$-valued random variables. For each $i \in \mathcal{I}$, define a sequence of stopping times by

$$\mathbf{t}_1^i = \mathbf{t}^i = \min\{k > 0 : \widetilde{Z}_k = i\} \quad \text{and} \quad \mathbf{t}_{n+1}^i = \min\{k > \mathbf{t}_n^i : \widetilde{Z}_k = i\} \,;$$

compare with (1.12). Here $\mathbf{t}_n^i$ is the time of the nth visit of $(Z_k)_{k\ge 1}$ to the orbit X_i. By recurrence of $\widetilde{P}$, this is a.s. finite. From the general theory of finite-state Markov chains (e.g. Chung [75]) it is known that $\mathbf{t}_1$ has moments of all orders (indeed, $\widetilde{\mathbb{E}}_j\big(\exp(c\,\mathbf{t}_1^i)\big) < \infty$ for some $c > 0$), and $\widetilde{\nu}(i) = 1/\widetilde{\mathbb{E}}_i(\mathbf{t}^i)$.

Furthermore, $(\mathbf{t}_n^i - \mathbf{t}_{n-1}^i)_{n\ge 2}$ is a sequence of i.i.d random variables. They have the same distribution as $\mathbf{t}_1^i$, when $Z_0 \in X_i$. In this case we set $\mathbf{t}_0^i = 0$, and obtain the induced random walk $(Z_{\mathbf{t}_n^i})_{n\ge 0}$ on X_i. The following is obvious.

(6.3) Lemma. *(X,P) is recurrent if and only if the induced random walk on some ($\Longleftrightarrow$ every) X_i is recurrent.*

Now, the induced random walk on X_i may be identified with the Markov chain

$$(Y_{\mathbf{t}_n^i})_{n\ge 0}$$

on $\mathbb{Z}^d$. Its transition probabilities inherit from P the translation invariance under the action of $\mathbb{Z}^d$. Thus, it is a random walk on $\mathbb{Z}^d$ whose law is given by

$$\mu_i(\mathbf{k}) = \mathbb{P}_{i\mathbf{0}}[\mathbf{t}^i < \infty \,,\; Z_{\mathbf{t}^i} = i\mathbf{k}] \,.$$

(6.4) Proposition. *If $M_r(P) < \infty$ $(r \geq 1)$ then the rth absolute moment $M_r(\mu_i)$ is also finite.*

Proof. Consider the usual Cayley graph of $\mathbb{Z}^d$, that is, the grid. We know from Proposition 3.9 and its proof that the mapping $\varphi : X \to \mathbb{Z}^d$, $\varphi(i\mathbf{k}) = \mathbf{k}$ is a rough isometry. Hence,

$$\sum_{\mathbf{k}\in\mathbb{Z}^d} |\mathbf{k}|^r \, \mu_i(\mathbf{k}) < \infty \iff \mathbb{E}_{o_i}\big(d(Z_0, Z_{\mathbf{t}^i})^r\big) < \infty \,.$$

We know that $\mathbf{t}^i$ has moments of all orders. Here, we shall need finiteness of the following three quantities:

$$A = \mathbb{E}_{o_i}(\mathbf{t}^i)\,, \quad B = \sup_{x\in X} \mathbb{E}_x\big((\mathbf{t}^i)^{r-1}\big)\,, \quad C = \sum_{k=1}^{\infty} k^{r-1}\, \mathbb{P}_{o_i}[\mathbf{t}^i \geq k]\,.$$

Using the inequality $(a_1 + \cdots + a_n)^r \leq n^{r-1}(a_1^r + \cdots + a_n^r)$ (where $a_k \geq 0$), we obtain

$$\mathbb{E}_{o_i}\big(d(Z_0, Z_{\mathbf{t}^i})^r\big) \leq \sum_{n=1}^{\infty}\sum_{k=1}^{n} n^{r-1}\mathbb{E}_{o_i}\big(d(Z_{k-1}, Z_k)^r\, \mathbf{1}_{[\mathbf{t}^i=n]}\big) = (I) + (II)\,,$$

where

$$(I) = \sum_{k=1}^{\infty} k^{r-1}\, \mathbb{E}_{o_i}\big(d(Z_{k-1}, Z_k)^r\, \mathbf{1}_{[\mathbf{t}^i=k]}\big)$$

and

$$(II) = \sum_{k=1}^{\infty}\sum_{n=k+1}^{\infty} n^{r-1}\, \mathbb{E}_{o_i}\big(d(Z_{k-1}, Z_k)^r\, \mathbf{1}_{[\mathbf{t}^i=n]}\big)\,.$$

In order to bound (I), we write

$$\begin{aligned}&\mathbb{E}_{o_i}\big(d(Z_{k-1}, Z_k)^r\, \mathbf{1}_{[\mathbf{t}^i=n]}\big)\\ &= \sum_{x\in X\setminus X_i} \mathbb{P}_{o_i}[Z_{k-1} = x\,,\ \mathbf{t}^i \geq k] \sum_{y\in X_i} p(x,y)\, d(x,y)^r \leq M_r(P)\, \mathbb{P}_{o_i}[\mathbf{t}^i \geq k]\,.\end{aligned}$$

We get $(I) \leq C\, M_r(P)$. Next, for $n > k$ we write

$$\begin{aligned}&\mathbb{E}_{o_i}\big(d(Z_{k-1}, Z_k)^r\, \mathbf{1}_{[\mathbf{t}^i=k]}\big)\\ &= \sum_{x\in X\setminus X_i} \mathbb{P}_{o_i}[Z_{k-1} = x\,,\ \mathbf{t}^i \geq k] \sum_{y\in X\setminus X_i} p(x,y)\, d(x,y)^r\, \mathbb{P}_y[\mathbf{t}^i = n-k]\,,\end{aligned}$$

substitute $n-k$ with n in (II) and use $(n+k)^{r-1} \le 2^{r-2}(n^{r-1}+k^{r-1})$ to obtain

$$(II) \le 2^{r-2} \sum_{k=1}^{\infty} \sum_{x \in X \setminus X_i} \mathbb{P}_{o_i}[Z_{k-1}=x\,,\; \mathbf{t}^i \ge k] \sum_{y \in X \setminus X_i} p(x,y)\, d(x,y)^r \times \sum_{n=1}^{\infty} (n^{r-1}+k^{r-1})\, \mathbb{P}_y[\mathbf{t}^i = n]$$

$$\le 2^{r-2} \sum_{k=1}^{\infty} \sum_{x \in X \setminus X_i} \mathbb{P}_{o_i}[Z_{k-1}=x\,,\; \mathbf{t}^i \ge k] \sum_{y \in X \setminus X_i} p(x,y)\, d(x,y)^r\, (B+k^{r-1})$$

$$\le 2^{r-2} \sum_{k=1}^{\infty} \mathbb{P}_{o_i}[\mathbf{t}^i \ge k]\, M_r(P)\, (B+k^{r-1}) \le 2^{r-2}\, M_r(P)\, (A\,B + C)\,.$$

Thus, both (I) and (II) are finite. □

In view of Theorem 6.1, we are now looking for a condition which guarantees $\mathfrak{m}(\mu_i) = \mathbf{0}$. Let us define a new transition matrix $\bar{P}$ on X:

$$\bar{p}(i\mathbf{k}, j\mathbf{l}) = p(i\mathbf{0}, j\mathbf{l})\,.$$

The resulting Markov chain will in general not be irreducible, as it is absorbed after one step in the set

$$\bar{X} = \{j\mathbf{l} : p(i\mathbf{0}, j\mathbf{l}) > 0 \text{ for some } i \in \mathcal{I}\}\,.$$

The significance of $(X, \bar{P})$ is that it keeps track of the increments in the $\mathbb{Z}^d$-component of the original chain. More precisely, if we have the original sequence of random variables $Z_n = \widetilde{Z}_n Y_n$, then we define

$$\text{(6.5)} \qquad \bar{Z}_n = \widetilde{Z}_n \bar{Y}_n\,, \quad \text{where} \quad \bar{Y}_0 = Y_0\,,\; \bar{Y}_n = Y_n - Y_{n-1}\; (n \ge 1)\,.$$

We get that $(\bar{Z}_n)_n$ is (a copy of) the Markov chain with transition matrix $\bar{P}$ and starting with Z_0.

(6.6) Lemma. *Restricted to $\bar{X}$, the transition matrix $\bar{P}$ is irreducible and has an invariant probability distribution $\bar{\nu}$, given by*

$$\bar{\nu}(j\mathbf{k}) = \sum_{i \in \mathcal{I}} \widetilde{\nu}(i)\, p(i\mathbf{0}, j\mathbf{k})\,.$$

Proof. Using that $\widetilde{\nu}$ is the invariant probability distribution of the factor chain $(I, \widetilde{P})$, it is straightforward to show that $\bar{\nu}$ has total mass 1 and is invariant for $\bar{P}$.

To see irreducibility, let $j_1\mathbf{l}_1, j_2\mathbf{l}_2 \in \bar{X}$. Then there is $i \in \mathcal{I}$ such that $\bar{p}(i\mathbf{k}, j_2\mathbf{l}_2) = p(i\mathbf{0}, j_2\mathbf{l}_2) > 0$ for all $\mathbf{k} \in \mathbb{Z}^d$. Using irreducibility of (X, P), we can find n such that $p^{(n)}(j_1\mathbf{l}_1, i\mathbf{0}) > 0$. Translating this into $\bar{P}$, we get that there is some $\mathbf{k} \in \mathbb{Z}^d$ (the increment of the last step) such that $\bar{p}^{(n)}(j_1\mathbf{l}_1, i\mathbf{k}) > 0$. In combination, we get $\bar{p}^{(n+1)}(j_1\mathbf{l}_1, j_2\mathbf{l}_2) > 0$. □

Now we can deduce a law of large numbers for random walks on generalized lattices.

(6.7) Theorem. *In the above setting, suppose that $M_1(P) < \infty$ and define the drift of P by*

$$\mathfrak{m}(P) = \sum_{\mathbf{k} \in \mathbb{Z}^d} \sum_{i,j \in \mathcal{I}} \mathbf{k}\,\widetilde{\nu}(i)\,p(i\mathbf{0}, j\mathbf{k})\,.$$

Then

$$\lim_{n\to\infty} \tfrac{1}{n} Y_n = \mathfrak{m}(P) \quad \mathbb{P}_x\text{-almost surely for every } x \in X\,, \quad \text{and}$$

$$\lim_{n\to\infty} \tfrac{1}{n} d(Z_0, Z_n) = 0 \quad \text{almost surely} \iff \mathfrak{m}(P) = \mathbf{0}\,.$$

Proof. We consider $(\bar{Z}_n)_{n\ge 1}$ as defined in (6.5). This positive recurrent Markov chain lives on $\bar{X}$ with invariant probability distribution $\bar{\nu}$ (Lemma 6.6). If $f : \bar{X} \to \mathbb{R}^d$ is any function such that $\int_{\bar{X}} |f|\,d\bar{\nu} < \infty$, then the law of large numbers for functionals of a recurrent Markov chain (see Revuz [276], Thm. 4.3 and remark, p. 140) tells us that

$$\lim_{n\to\infty} \frac{1}{n} \sum_{k=1}^{n} f(\bar{Z}_n) = \int_{\bar{X}} f\,d\bar{\nu} \quad \text{almost surely}$$

(for every starting distribution). Setting $f(i\mathbf{k}) = \mathbf{k}$, finiteness of $M_1(P)$ implies $\bar{\nu}$-integrability of $|f|$. Furthermore, $\int_{\bar{X}} f\,d\bar{\nu} = \mathfrak{m}(P)$, and we get that

$$\frac{1}{n} \sum_{k=1}^{n} \bar{Y}_k = \frac{1}{n}(Y_n - Y_0) \to \mathfrak{m}(P) \quad \text{almost surely.}$$

The statement on convergence to 0 follows from the fact that $\varphi(i\mathbf{k}) = \mathbf{k}$ defines a rough isometry. □

Now we can finally state the recurrence criterion that we have been looking for.

(6.8) Corollary. *Suppose that* AUT(X, P) *contains a quasi-transitive subgroup isomorphic with $\mathbb{Z}^d$, and that $M_1(P) < \infty$.*

(a) *If $\mathfrak{m}(P) \ne \mathbf{0}$, or if $d \ge 3$, then the random walk is transient.*

(b) *If $d = 1$, or if $d = 2$ and $M_2(P) < \infty$, then the random walk is recurrent if and only if $\mathfrak{m}(P) = \mathbf{0}$.*

Proof. Transience for $d \ge 3$ is clear. Let μ_i be the law of $(Y_{\mathbf{t}_n^i})_{n\ge 0}$, which, as well as $\mathbf{t}_n^i$, is a sequence of sums of i.i.d. random variables. By Proposition 6.4, μ_i inherits finiteness of the first moment from P. By the law of large numbers,

$$\frac{Y_{\mathbf{t}_n^i}}{n} \to \mathfrak{m}(\mu_i) \quad \text{and} \quad \frac{\mathbf{t}_n^i}{n} \to \widetilde{\mathbb{E}}_i(\mathbf{t}_1^i) = \frac{1}{\widetilde{\nu}(i)} \quad \text{almost surely.}$$

From Theorem 6.7 we get that

$$\frac{Y_{\mathbf{t}_n^i}}{n} = \frac{Y_{\mathbf{t}_n^i}}{\mathbf{t}_n^i}\frac{\mathbf{t}_n^i}{n} \to \frac{\mathfrak{m}(P)}{\widetilde{\nu}(i)} \quad \text{almost surely.}$$

Hence, $\mathfrak{m}(\mu_i) = \mathfrak{m}(P)/\widetilde{\nu}(i)$ is equal to 0 if and only if $\mathfrak{m}(P) = \mathbf{0}$. Using Proposition 6.4, the results now follow from Theorem 6.1. □

Consequently, in view of Theorem 5.12, if X is a graph with linear or square growth, and if (X, P) is a quasi-transitive random walk with finite first moment, then we can find a quasi-transitive subgroup of $\mathrm{AUT}(X, P)$ isomorphic with $\mathbb{Z}$ or $\mathbb{Z}^2$, so that we can compute $\mathfrak{m}(P)$ and decide on recurrence or transience. (If X has square growth then we also need P to have finite second moment.)

B. More on trees

Here, we shall present a recurrence criterion for nearest neighbour random walks on trees, which involves the boundary at infinity and the notion of logarithmic capacity.

(6.9) Definition. Let (M, θ) be a compact metric space, and let ν be a Borel measure on M. The *logarithmic potential* of ν is the function

$$\phi_\theta(x|\nu) = \int_M -\log\theta(x,y)\,d\nu(y)$$

on M. The *logarithmic energy* of ν is

$$I_\theta(\nu) = \int_M \phi_\theta(x|\nu)\,d\nu(x)\,.$$

The *logarithmic capacity* of a Borel set $B \subset M$ is

$$\mathrm{Cap}_\theta(B) = \sup\{\nu(B) : \nu \text{ a non-negative Borel measure with } I_\theta(\nu) \le 1\}\,.$$

Note that $\mathrm{Cap}_\theta(B) > 0$ if and only if there is a probability measure with finite logarithmic energy such that $\nu(M \setminus B) = 0$.

Now let T be an infinite, locally finite tree; compare with §1.D. A (*geodesic*) *ray* in T is an infinite path $\pi = [x_0, x_1, x_2, \dots]$ without repeated vertices. (We also think of π as a sequence of edges.) Two rays π, π' are said to be equivalent if their symmetric difference has finitely many vertices. An *end* of T is an equivalence class of geodesic rays. The *boundary* of T is the set ϑT of all ends of T. We set $\widehat{T} = T \cup \vartheta T$. If $x \in T$ and $\xi \in \vartheta T$, then ξ (as an equivalence class) has a unique representative which is a ray starting at x, denoted by $\pi(x,\xi)$. If x, y are two distinct vertices of T, then we define the "branch" $T_{x,y}$, its closure $\widehat{T}_{x,y}$ and its boundary by

$$\widehat{T}_{x,y} = \{w \in \widehat{T} : y \in \pi(x,w)\}, \quad T_{x,y} = T \cap \widehat{T}_{x,y} \quad \text{and} \quad \vartheta T_{x,y} = \vartheta T \cap \widehat{T}_{x,y}.$$

T becomes a compact, totally disconnected Hausdorff space: a subbasis of the topology is given by the family of all augmented branches $\widehat{T}_{x,y}$; each of these is compact and open. Fix a root o. For $v \in \widehat{T}$, write $|v| = d(v,o)$. For $v, w \in \widehat{T}$, their *confluent* $v \wedge w$ is the last common vertex on the geodesics $\pi(o,v)$ and $\pi(o,w)$. This is a vertex of T, unless $v = w \in \vartheta T$, in which case $v \wedge w = v$. Now

$$\theta(v,w) = \begin{cases} 0, & v = w, \\ \exp(-|v \wedge w|), & v \neq w, \end{cases} \tag{6.10}$$

defines an ultrametric which induces the topology of $\widehat{T}$. In order to define a Borel measure ν on ϑT, it is enough to specify the values $\nu(\vartheta T_{o,x})$, $x \neq o$, consistently. As all these sets are open and compact, it is enough to check finite additivity. The usual extension machinery (Carathéodory's theorem) does the rest for us.

Choose the orientation of edges such that for each $e \in E(T)$, the endpoint closer to o is e^-. Measures ν on ϑT are in one-to-one correspondence with flows u from o to ∞ via

$$u(e) = \nu(\vartheta T_{o,e^+}). \tag{6.11}$$

The input at o is $\nu(\vartheta T)$. No energy is yet involved; u being a flow means $\sum_{e^+=x} u(e) - \sum_{e^-=x} u(e) = -\nu(\vartheta T)\,\delta_o(x)$. If we think of T being grounded at ϑT, then for $B \subset \vartheta T$, the amount flowing out of T through B is $\nu(B)$.

Now let P be the transition matrix of a nearest neighbour random walk on T. Then P is reversible with respect to the measure

$$m(x) = \begin{cases} 1, & x = o, \\ \prod_{e \in \pi(o,x)} \frac{p(e^-,e^+)}{p(e^+,e^-)}, & x \neq o. \end{cases} \tag{6.12}$$

Let $r : E(T) \to (0,\infty)$ be the associated resistance function. Think of r as a length element, inducing a new distance d_r on T by $d_r(x,y) = \sum_{e \in \pi(x,y)} r(e)$ for $x \neq y$. We write $|v|_r = d_r(v,o)$. As in (6.10), one can use this metric to define a new ultrametric θ_r on $\widehat{T}$.

(6.13) Theorem. *The random walk on T is transient if and only if $\mathrm{Cap}_{\theta_r}(\vartheta T) > 0$.*

Proof. Let ν be a Borel measure on ϑT and u the associated flow according to (6.11). We write $\phi_r = \phi_{\theta_r}$ and claim that for every $\xi \in \vartheta T$,

$$\phi_r(\xi|\nu) = \sum_{e \in E(T)} r(e) u(e)\, \mathbf{1}_{\vartheta T_{o,e^+}}(\xi). \tag{6.14}$$

Let e_n, $n \geq 1$, be the successive edges on the ray $\pi(o,\xi)$, such that $e_1^- = o$ and $e_n^+ = e_{n+1}^- = x_n$. Then (with $0 \cdot \infty = 0$ as usual)

$$\begin{aligned}
\phi_r(\xi|\nu) &= \int_{\vartheta T} |\xi \wedge \eta|_r \, d\nu(\eta) \\
&= |\xi|_r \, \nu(\{\xi\}) + \sum_{n=1}^{\infty} |x_n|_r \, \nu(\vartheta T_{o,x_n} \setminus \vartheta T_{o,x_{n+1}}) \\
&= |\xi|_r \, \nu(\{\xi\}) + \lim_{N\to\infty} \left(\sum_{n=1}^{N} r(e_n)\, \nu(\vartheta T_{o,x_n}) - |x_N|_r \, \nu(\vartheta T_{o,x_{N+1}}) \right) \\
&= |\xi|_r \, \nu(\{\xi\}) + \lim_{N\to\infty} \sum_{n=1}^{N} r(e_n) \big(\nu(\vartheta T_{o,x_n}) - \nu(\vartheta T_{o,x_{N+1}}) \big) .
\end{aligned}$$

Set $f_N(n) = r(e_n) \max\{0, \nu(\vartheta T_{o,x_n}) - \nu(\vartheta T_{o,x_{N+1}})\}$. This is non-decreasing in N, and as $N \to \infty$, the pointwise limit is $f(n) = r(e_n) \big(\nu(\vartheta T_{o,x_n}) - \nu(\{\xi\}) \big)$. By monotone convergence, we get

$$\phi_r(\xi|\nu) = |\xi|_r \, \nu(\{\xi\}) + \sum_{n=1}^{\infty} r(e_n) \big(u(e_n) - \nu(\{\xi\}) \big) .$$

If $\nu(\{\xi\}) = 0$ then we see that (6.14) is true. If $\nu(\{\xi\}) > 0$ and $|\xi|_r < \infty$ then $\sum_n r(e_n)\, \nu(\{\xi\})$ cancels with $|\xi|_r \, \nu(\{\xi\})$, and (6.14) is again true. Finally, if $\nu(\{\xi\}) > 0$ and $|\xi|_r = \infty$ then each side of (6.14) is equal to $+\infty$.

Integrating both sides of (6.14), we see that $I_{\theta_r}(\nu) = \langle u, u \rangle$, the energy of the flow with respect to $\ell^2\big(E(T), r\big)$. Consequently, there is a finite energy flow with input $i_0 = 1$ from o to ∞ if and only if there is a probability measure ν on ϑT with $I_\theta(\nu) < \infty$, that is, if and only if $\mathrm{Cap}_{\theta_r}(\vartheta T) > 0$. The result follows from Theorem 2.12. $\square$

As a simple application, we can now explain recurrence of the comb lattices in (2.21) from a more general viewpoint.

(6.15) Corollary. *If ϑT is countable and $|\xi|_r = \infty$ for all $\xi \in \vartheta T$, then the random walk is recurrent.*

Proof. If ν is a probability measure on ϑT then there must be $\xi \in \vartheta T$ with $\nu(\{\xi\}) > 0$. But then $I_{\theta_r}(\nu) \geq |\xi|_r \, \nu(\{\xi\})^2 = \infty$. $\square$

This applies, in particular, to the simple random walk on a tree with countably many ends and also to all nearest neighbour random walks on such trees with $\inf_{E(T)} r(e) > 0$.

(6.16) A recurrent tree with exponential growth around each point. Let T be the tree with root o constructed as follows: o has two "sons". For $n \geq 1$, the sphere $S(o,n) = \{x \in T : d(x,o) = n\}$ has 2^n elements $x_{i,n}$, $i = 1, \ldots, 2^n$. Each of these has its "father" in $S(o, n-1)$; for $i = 1, \ldots, 2^{n-1}$, the vertex $x_{i,n}$ has the three "sons" $x_{3i-2,n+1}$, $x_{3i-1,n+1}$ and $x_{3i,n+1}$, while $x_{2^{n-1}+i,n}$ has only one "son" $x_{3\cdot 2^{n-1}+i,n}$. Thus, $V(o,n) = 2^{n+1} - 1$. Besides the "leftmost" ray $[o, x_{1,1}, x_{1,2}, x_{1,3}, \ldots]$, every geodesic ray starting from o has only finitely many vertices with three "sons". In particular, ϑT is countable and the simple random walk on T is recurrent.

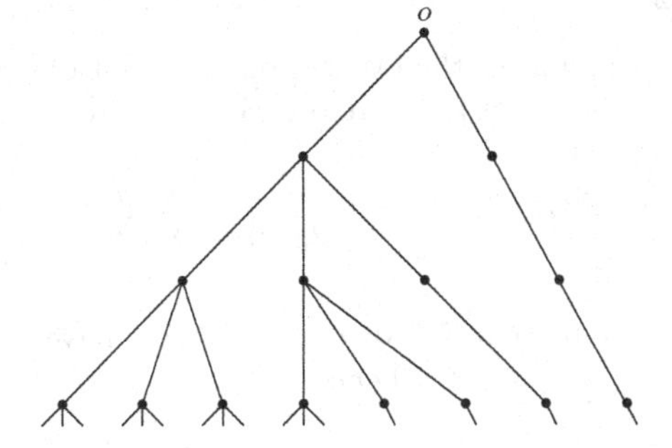

Figure 5: a recurrent tree

(6.17) Exercise. Prove recurrence of the tree of Figure 5 by direct use of the flow criterion (Theorem 2.12(b)).

(6.18) Trees associated with subsets of $[0\,,\,1]$**.** Let Λ be a closed subset of the unit interval $[0\,,\,1]$, and let $q \geq 2$ be an integer. We associate with Λ a tree $T = T(\Lambda, q)$ as follows. Set $J_{j,n} = [(j-1)/q^n\,,\, j/q^n]$. Then the vertices of T are points $x_{j,n}$ ($n \geq 0$, $j \in \{1, \ldots, q^n\}$), such that $J_{j,n}$ has non-empty intersection with Λ. Two vertices are neighbours if they are of the form $x_{j,n}$ and $x_{k,n+1}$, and $J_{k,n+1} \subset J_{j,n}$.

We ask when the simple random walk on $T(\Lambda, q)$ is recurrent. For example, if Λ is the "middle third removal" Cantor set, then $T(\Lambda, 3)$ is the binary tree, which is transient.

Take the vertex $o = x_{1,0}$ corresponding to $[0\,,\,1]$ as the origin. Let $\xi \in \vartheta T$. The geodesic ray $\pi(o,\xi)$ corresponds to a decreasing sequence of intervals whose intersection is a point $z \in \Lambda$. We define a mapping $\varphi : \vartheta T \to \Lambda$ by $\varphi(\xi) = z$. It is continuous, indeed

$$|\varphi(\xi) - \varphi(\eta)| \leq \theta(\xi,\eta)^{\log q}\,, \tag{6.19}$$

where θ is the metric defined in (6.10). Also, φ is one-to-one except possibly at countably many points which are mapped to q-adic rationals and where φ is two-to-one.

(6.20) Theorem. *$T(\Lambda, q)$ is transient if and only if Λ has positive logarithmic capacity with respect to the metric of $\mathbb{R}$.*

Proof. Let ν be a probability measure on ϑT, and let $\bar{\nu}$ denote its image under φ, that is, $\bar{\nu}(B) = \nu(\varphi^{-1}(B))$. Then (6.19) implies that for every $\xi \in \vartheta T$

$$\phi_\theta(\xi, \nu) = \int_{\vartheta T} -\log \theta(\xi, \eta)\, d\nu(\eta) \leq \frac{1}{\log q} \int_0^1 -\log |\varphi(\xi) - y|\, d\bar{\nu}(y)$$

and hence

$$I(\bar{\nu}) \geq I_\theta(\nu) \log q\,. \tag{6.21}$$

For a bound in the other direction, first observe that $\vartheta T_{o,x_{j,n}} = \varphi^{-1}(J_{j,n})$ for $x_{j,n} \in T \setminus \{o\}$. Therefore we can write, using (6.14) and integrating with respect to ν,

$$I_\theta(\nu) = \sum_{n=1}^{\infty} \sum_{j=1}^{q^n} \bar{\nu}(J_{j,n})^2\,.$$

Next note that for $0 < a < 1$, we have $q^{k-1} < a^{-1} \leq q^k$ with $k = \sum_{n \geq 0} \mathbf{1}_{[0\,,\,q^{-n}]}(a)$, whence

$$-\log |x - y| \leq \log q \sum_{n=0}^{\infty} \mathbf{1}_{[0\,,\,b^{-n}]}(|x - y|)$$

for $x, y \in [0\,,\,1]$. Now, if $|x - y| \leq q^{-n}$, $n \geq 2$, then

$$(x, y) \in J_{j,n-1} \times J_{j,n-1} \cup J_{jq,n} \times J_{jq+1,n} \cup J_{jq+1,n} \times J_{jq,n}$$

for some $j \in \{1, \ldots, q^{n-1}\}$. Hence (using $2ab \leq a^2 + b^2$)

$$\begin{aligned}
&\bar{\nu} \times \bar{\nu}\big(\{(x, y) \in [0\,,\,1]^2 : |x - y| \leq q^{-n}\}\big) \\
&\qquad \leq \sum_{j=1}^{q^{n-1}} \Big(\bar{\nu}(J_{j,n-1})^2 + \bar{\nu}(J_{jq,n})^2 + \bar{\nu}(J_{jq+1,n})^2\Big) \leq 2 \sum_{j=1}^{q^{n-1}} \bar{\nu}(J_{j,n-1})^2
\end{aligned}$$

for $n \geq 2$. Now we compute the logarithmic energy of $\bar{\nu}$ on $[0\,,\,1]$:

$$\begin{aligned}
I(\bar{\nu}) &= \int_0^1 \int_0^1 -\log |x - y|\, d\bar{\nu}(x) d\bar{\nu}(y) \\
&\leq \log q \sum_{n=0}^{\infty} \bar{\nu} \times \bar{\nu}\big(\{(x, y) : |x - y| \leq q^{-n}\}\big) \\
&\leq \log q \Big(2 + 2 \sum_{n=2}^{\infty} \sum_{j=1}^{q^{n-1}} \bar{\nu}(J_{j,n-1})^2\Big)\,.
\end{aligned}$$

Hence

$$I(\bar{\nu}) \le 2\big(1 + I_\theta(\nu)\big)\log q\,. \tag{6.22}$$

We can now conclude the proof. We apply Theorem 6.13 (with $r \equiv 1$). If $\mathrm{Cap}_\theta(\vartheta T) > 0$ then there is a probability ν on ϑT with $I_\theta(\nu) < \infty$. Defining $\bar{\nu}$ on $[0\,,\,1]$ as above, we get from (6.22) that $I(\bar{\nu}) < \infty$, and $\mathrm{Cap}(\Lambda) > 0$. Conversely, if $I(\bar{\nu}) < \infty$ for some probability measure supported by Λ, then $\bar{\nu}$ carries no point mass. Hence, using the fact that φ is two-to-one at at most countably many points, we can define a probability measure ν on ϑT by $\nu(B) = \bar{\nu}\big(\varphi(B)\big)$, and the image of ν under φ is $\bar{\nu}$. Thus, by (6.21), $I_\theta(\nu) < \infty$. □

In particular, we see that recurrence (or transience) of $T(\Lambda, q)$ does not depend on the choice of q. Also, if Λ has positive Hausdorff dimension then $T(\Lambda, q)$ is transient.

C. Extremal length and plane tilings

In this section we present another recurrence criterion for reversible Markov chains with *finite range,* that is, giving rise to a locally finite network. The criterion will be applied to the edge graphs of a class of tilings of the plane. Let (X, P) be reversible with $E = E(P)$, resistance $r(\cdot)$, total conductance $m(\cdot)$ and associated network $\mathcal{N}$.

(6.23) Definition. Let Π be a set of non-empty simple paths in $\mathcal{N}$. The *extremal length* $EL(\Pi)$ of Π is given by

$$\begin{aligned} EL(\Pi) &= \Big(\inf\{\langle u, u\rangle : u \in \ell^2_+(E, r)\,,\ L(\Pi|u) \ge 1\}\Big)^{-1} \\ &= \sup\{L(\Pi|u)^2 : u \in \ell^2_+(E, r)\,,\ \langle u, u\rangle = 1\}\,, \end{aligned}$$

where $L(\Pi|u) = \inf\{L(\pi|u) : \pi \in \Pi\}$, with $L(\pi|u) = \sum_{e\in\pi} u(e)r(e)$.

(It is a straightforward exercise to show that the two expressions for $EL(\Pi)$ coincide.) For $x \in X$, let Π_x denote the set of all one-sided infinite simple paths starting in x.

(6.24) Theorem. *If (X, E) is locally finite then $\mathrm{cap}(x) = EL(\Pi_x)^{-1}$ for every $x \in X$. In particular, (X, P) is recurrent if and only if $EL(\Pi_x) = \infty$ for some ($\iff$ every) $x \in X$.*

Proof. The capacity used here is of course the one of (2.10) and Theorem 2.12.

First, let $f \in \ell_0(X)$ be such that $f(x) = 1$. Define $u(e) = |\nabla f(e)|$. If $\pi = [x = x_0, x_1, x_2, \dots] \in \Pi_x$ then $f(x_n) = 0$ for all but finitely many n. Therefore

$$L(\pi|u) \ge \sum_{n=0}^{\infty}\big(f(x_n) - f(x_{n+1})\big) = 1\,.$$

Consequently $D(f) = \langle u,u\rangle \geq EL(\Pi_x)^{-1}$, and thus $\text{cap}(x) \geq EL(\Pi_x)^{-1}$.

The reverse inequality needs more work. Let $B \subset X$ be finite and connected, with $x \in B \setminus dB$, where $dB = \{y \in B : d(y, X \setminus B) = 1\}$. Write $\Pi_{x,y}(B)$ for the set of all simple paths in B from x to y, and $\Pi_x(B) = \bigcup_{y\in dB} \Pi_{x,y}(B)$. We show that

$$\text{(6.25)} \qquad \text{cap}(x) \leq EL\big(\Pi_x(B)\big)^{-1}.$$

Let $\varepsilon > 0$. Then there is $u \in \ell^2_+(E,r)$ such that $L(\pi|u) \geq 1$ for all $\pi \in \Pi_x(B)$ and $\langle u,u\rangle < EL\big(\Pi_x(B)\big)^{-1} + \varepsilon$. We define a function $g : B \to [0,\infty)$ as follows.

$$g(x) = 0, \quad \text{and} \quad g(y) = \inf\{L(\pi|u) : \pi \in \Pi_{x,y}(B)\}, \text{ if } y \neq x.$$

If $y \in dB$ then $g(y) \geq 1$. If $y, w \in B$ are neighbours and $\pi = [x = x_0, x_1, \dots, x_n = y] \in \Pi_{x,y}(B)$, then the path $[x_0, x_1, \dots, x_n, w]$ is not necessarily simple but can be made simple by deleting the piece after the first appearance of w. Hence

$$g(w) \leq L(\pi|u) + u([y,w])r([y,w]),$$

and passing to the infimum over all $\pi \in \Pi_{x,y}(B)$, we obtain $g(w) \leq g(y) + u([y,w])r([y,w])$. Exchanging the roles of y and w, we see that $g(e^+) - g(e^-) \leq u(e)r(e)$ for all edges with endpoints in B. Now define $f \in \ell_0(X)$ by

$$f(x) = \max\{1 - g(x), 0\} \text{ for } x \in B \quad \text{and} \quad f(x) = 0 \text{ for } x \in X \setminus B.$$

We get $f(x) = 1$, $\text{supp} f \subset B$ and $|\nabla f| \leq u$. Therefore $\text{cap}(B) \leq D(f) \leq \langle u,u\rangle$. This proves (6.25).

Now replace B with $B_n = B(x,n)$, which is finite by assumption, and let $n \to \infty$. The theorem will be proved when we show that $EL\big(\Pi_x(B_n)\big) \to EL(\Pi_x)$. To this end, start with $u \in \ell^2_+(E,r)$ having norm 1, and consider $L_n(u) = L\big(\Pi_x(B_n)|u\big)$. By local finiteness, each $\Pi_x(B_n)$ is finite and there is a path $\pi_n \in \Pi_x(B_n)$ such that $L_n(u) = L(\pi_n|u)$. By truncating π_{n+1} at the end, we obtain a path $\pi'_{n+1} \in \Pi_x(B_n)$, and we get

$$L_n(u) \leq L(\pi'_{n+1}|u) \leq L(\pi_{n+1}|u) = L_{n+1}(u).$$

In a similar way one sees that $L_n(u) \leq L(\Pi_x|u)$.

Again using local finiteness, there must be an (infinite) path $\pi = [x = x_0, x_1, \dots] \in \Pi_x$ such that $\pi_n \to \pi$ pointwise. That is, for every m, all but finitely many of the π_n start with $[x_0, \dots, x_m]$, so that

$$L([x_0, \dots, x_m]|u) \leq L(\pi_n|u) = L_n(u) \leq \lim_{n\to\infty} L_n(u).$$

Letting $m \to \infty$, we see that $L(u) \le L(\pi|u) \le \lim_n L_n(u)$. Therefore

$$\lim_{n\to\infty} L_n(u) = L(\Pi_x|u).$$

We now get, using monotonicity of $EL(\Pi_x(B_n))$ in n,

$$L(\Pi_x|u) \le \lim_{n\to\infty} EL(\Pi_x(B_n)) \le EL(\Pi_x).$$

Taking the supremum over all u, we obtain the required limit. □

Next, we study limits along paths.

(6.26) Lemma. *Let $f \in \mathcal{D}(\mathcal{N})$. For $x \in X$, set*

$$\Pi'_x = \left\{ \pi \in \Pi_x : \sum_{e\in\pi} |\nabla f(e)| r(e) < \infty \right\}.$$

Then $EL(\Pi_x \setminus \Pi'_x) = \infty$.

Proof. Set $u = |\nabla f|$. Then $u \in \ell^2_+(E, r)$ and $L(\pi|u) = \infty$ for all $\pi \in \Pi_x \setminus \Pi'_x$. Now use the second of the two definitions of extremal length. □

In particular, f converges along each path $\pi = [x_0, x_1, \dots] \in \Pi'_x$ in the sense that $\lim_n f(x_n)$ exists. Combining the lemma with Theorem 6.24, we get another sufficient criterion for recurrence.

(6.27) Corollary. *If there is $f \in \mathcal{D}(\mathcal{N})$ such that f does not converge along any path in Π_x, then the random walk is recurrent.*

We shall now apply this criterion to a class of planar graphs arising from tilings with certain properties.

Let $\mathcal{O}$ be an open, simply connected subset of the Euclidean plane $\mathbb{R}^2$. A *tiling* of $\mathcal{O}$ is a family $\mathcal{T}$ of closed topological disks $T \subset \mathcal{O}$ (the tiles) with pairwise disjoint interiors and such that $\bigcup_{T\in\mathcal{T}} T = \mathcal{O}$. Here we shall always assume that $\mathcal{T}$ is *locally finite,* that is, every compact subset of $\mathcal{O}$ intersects only finitely many tiles. A connected component of the intersection of two or more tiles is called a *vertex* if it is a point and an *edge* if it is an arc, in which case it arises from the intersection of two tiles. Local finiteness implies that the boundary of each tile is the union of a finite number of edges. With this definition of vertices and edges, we obtain the *edge graph* $X(\mathcal{T})$ of the tiling. In order to avoid multiple edges, we always require that the boundary of each tile has at least three edges. Thus, $X(\mathcal{T})$ is a planar, locally finite, connected graph. The reader is referred to the book of Grünbaum and Shephard [152] for many figures and all basic facts concerning tilings.

Given any tile T, we denote by $\mathfrak{d}_T$ the maximal distance from the origin to a point in T.

(6.28) Definition. Given $m \in \mathbb{N}$, $0 < \lambda < 1$ and $\kappa > 0$, a tile T is called (m, λ, κ)-regular if

(1) the boundary of T consists of no more than m edges,
(2) $\operatorname{diam}(T) \le \lambda\, \eth_T$, and
(3) $\operatorname{diam}(T)^2 \le \kappa \operatorname{area}(T)$.

The tiling $\mathcal{T}$ of $\mathcal{O}$ is called *quasi-regular,* if there are m, λ, κ such that all but finitely many edges lie on an (m, λ, κ)-regular tile of $\mathcal{T}$.

Quasi-regularity of tilings is scale- and translation-invariant. Soardi [304] calls a tile *quasi-normal,* if (1) holds, and instead of (2) and (3) one has the more restrictive conditions $\operatorname{diam}(T) \le \delta < \infty$ and $\operatorname{diam}(T) \le \kappa \operatorname{inradius}(T)$ for all tiles. A picture of a quasi-normal tiling of $\mathbb{R}^2$ can be found on the cover of Soardi's book [304]; its edge graph has exponential growth at each vertex. In particular, the tiles in a quasi-normal tiling may be arbitrarily small, and in a quasi-regular tiling they may even be arbitrarily big, as long as (2) holds.

(6.29) Theorem. *If $\mathcal{T}$ is a quasi-regular tiling of the plane $\mathbb{R}^2$, then the simple random walk on $X(\mathcal{T})$ is recurrent.*

Proof. Define the radial function

$$F(z) = \sin\Big(\log\big(1 + \log(1 + |z|^2)\big)\Big), \quad z \in \mathbb{R}^2$$

($|\cdot|$ denotes the Euclidean norm). Let f be the restriction of F to the vertices of the tiling. We claim that f has finite Dirichlet sum.

First of all calculate $|\nabla F(z)|^2 \le h(|z|)$, where ∇ is the ordinary gradient, and

$$h(r) = 4\Big((1 + r^2)\big(1 + \log(1 + r^2)\big)^2\Big)^{-1}.$$

Let m, λ, κ be the constants of Definition 6.28. Call an edge good if it lies on a regular tile (and bad, otherwise). Now consider a good edge $[x, y]$ and a regular tile T containing it. Then $|x - y| \le \operatorname{diam}(T)$, and

$$\begin{aligned} |f(x) - f(y)|^2 &\le |x - y|^2 \int_0^1 |\nabla F\big((1-t)x + t\,y\big)|^2\, dt \\ &\le \operatorname{diam}(T)^2 \int_0^1 h\big(|(1-t)x + t\,y|\big)\, dt. \end{aligned}$$

Choose a point $z_T \in T$ such that $|z_T| = \eth_T$. Every point w on the line segment between x and y is at distance at most $\operatorname{diam}(T)$ from z_T. Hence $|w| \ge \eth_T - \operatorname{diam}(T)$. Using the monotonicity of h and condition (3), we continue:

$$|f(x) - f(y)|^2 \le \operatorname{diam}(T)^2 h\big(\eth_T - \operatorname{diam}(T)\big) \le \kappa \operatorname{area}(T)\, h(\eth_T) \frac{h\big((1-\lambda)\eth_T\big)}{h(\eth_T)}.$$

Now $h(\mathfrak{d}_T)$ is the minimum of $h(|z|)$ on T, while the function which maps r to $h\big((1-\lambda)r\big)/h(r)$ is bounded for $r \in [0\,,\infty)$. We obtain

$$|f(x)-f(y)|^2 \le C\,\text{area}(T)\,h(\mathfrak{d}_T) \le C\int_T h(|z|)dz\,,$$

where C depends only on κ and λ. The tile T has at most m edges, so that summing $|f(x)-f(y)|^2$ over all edges of T yields the upper bound $m\,C\int_T h(|z|)dz$. Each good edge is common to one or two regular tiles. Therefore, writing A for the (finite) sum of $|f(x)-f(y)|^2$ over all bad edges,

$$D(f) \le A + m\,C \sum_{T\in\mathcal{T}} h(|z_T|)\,\text{area}(T) \le A + m\,C\int_{\mathbb{R}^2} h(|z|)\,dz < \infty\,.$$

To conclude the proof, in view of Corollary 6.27 we now show that f does not converge along any infinite simple path in $X(\mathcal{T})$. A (formal) simple path $[x_0, x_1, \dots]$ corresponds to a curve in $\mathbb{R}^2$ without self-intersections; it is the union of all edges $[x_n, x_{n+1}]$ (arcs on the boundaries of tiles). As the tiling is locally finite, this curve is infinite and not contained in any compact subset of $\mathbb{R}^2$. As the function F is radial and oscillates between -1 and 1, there is a sequence of points w_k on the curve such that $|w_k| \to \infty$ and $F(w_k) = (-1)^k$. Each w_k lies on some edge $[x_{n(k)}, x_{n(k)+1}]$, which in turn lies on some tile denoted by T_k. Suppose without loss of generality that T_k is regular, and write $\mathfrak{d}_k = \mathfrak{d}_{T_k}$. As above, we use the fact that each $z \in T_k$ satisfies $(1-\lambda)\mathfrak{d}_k \le \mathfrak{d}_k - \text{diam}(T_k) \le |z| \le \mathfrak{d}_k$. Therefore, for each pair of points $w, z \in T_k$,

$$\begin{aligned}|F(w)-F(z)| &\le \big|\log\big(1+\log(1+|w|^2)\big) - \log\big(1+\log(1+|z|^2)\big)\big| \\ &\le \log\frac{1+\log(1+\mathfrak{d}_k^2)}{1+\log\big(1+(1-\lambda)^2\mathfrak{d}_k^2\big)}\,.\end{aligned}$$

As $k\to\infty$, also $\mathfrak{d}_k \to \infty$, and the last expression tends to 0. Consequently

$$|f(x_{n(k)}) - (-1)^k| = |F(x_{n(k)}) - F(w_k)| < 1/2$$

for all sufficiently large k. As $k\to\infty$, it must be that $n(k)\to\infty$, so that f does not converge along the given path. □

D. Circle packings and random walks

A *packing* in the plane consists of a collection $\mathcal{P}$ of closed topological disks C_x ($x \in X$, where X is an an index set) with pairwise disjoint interiors. The

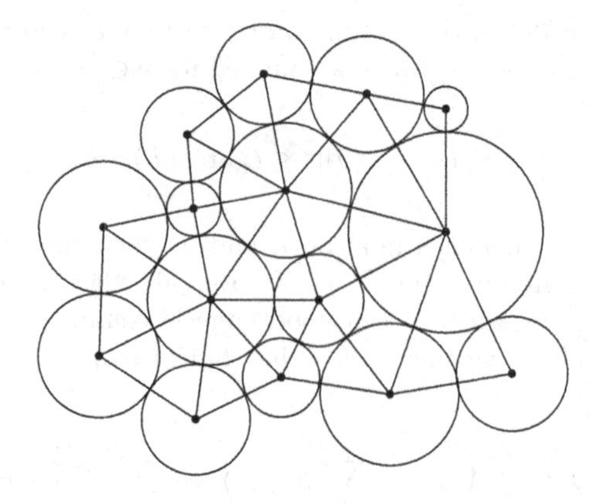

Figure 6:
a (finite) circle packing and its contacts graph

contacts graph or *nerve* of the packing has vertex set X, and two vertices x, y are neighbours if $C_x \cap C_y \neq \emptyset$. Here we shall only consider *circle packings,* that is, packings where each C_x is a geometrical disk; see Figure 6. (The terminology "disk packing" might be more appropriate.)

A *disk triangulation graph* X is defined as the edge graph of a *triangulation* of an open topological disk A, that is, a locally finite tiling of A all of whose tiles are triangles. In this context, the circle packing theorem (originally stated and proved by Koebe [207] for finite planar graphs) says that there is a circle packing in $\mathbb{R}^2 = \mathbb{C}$ whose contacts graph is isomorphic with X. We can realize the contacts graph by taking the centres of all circles and connecting two of them by an edge whenever the corresponding circles (disks) touch each other. The new graph thus obtained is isomorphic with X, and the union of all its triangles is called the *carrier* $\mathrm{carr}(\mathcal{P})$ of $\mathcal{P}$. The packing is locally finite in $\mathrm{carr}(\mathcal{P})$ (each compact subset of $\mathrm{carr}(P)$ intersects only finitely many circles). For more details, see Beardon and Stephenson [30] (for the case of bounded vertex degrees) and He and Schramm [168] (general case), who have proved the following fundamental result.

(6.30) Theorem. *Let X be a disk triangulation graph as above. Then there is either a circle packing $\mathcal{P}$ with contacts graph X whose carrier is the whole plane $\mathbb{C}$, or one such that its carrier is the open unit disk $\mathbb{D}$, but not both.*

In the first case ($\mathrm{carr}(\mathcal{P}) = \mathbb{C}$), the disk triangulation graph X is called *CP-parabolic* and in the second case *CP-hyperbolic.* We would like to link these properties with the behaviour of the simple random walk on X. The latter depends only on the "abstract" graph structure and not on the particular realization of the graph, so that the study of the simple random walk will lead to *a priori* criteria for CP-parabolicity or -hyperbolicity.

Let $\mathcal{P}$ be a circle packing whose tangency graph is X. Instead of the

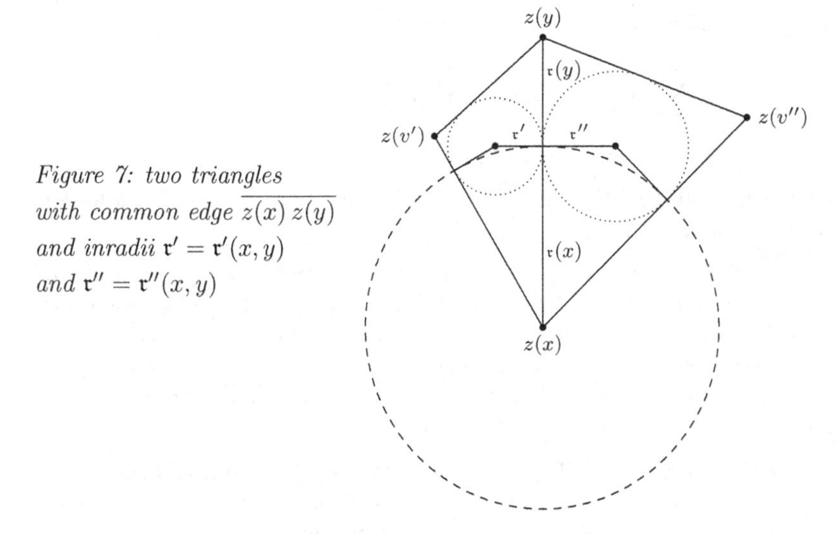

Figure 7: two triangles with common edge $\overline{z(x)\,z(y)}$ and inradii $\mathfrak{r}' = \mathfrak{r}'(x,y)$ and $\mathfrak{r}'' = \mathfrak{r}''(x,y)$

simple random walk we consider a reversible nearest neighbour random walk with conductances $a(x,y) = a_{\mathcal{P}}(x,y)$ depending on $\mathcal{P}$. Given $x \in X$, write $z(x) = z_{\mathcal{P}}(x)$ for the centre of the circle C_x in $\mathcal{P}$, and $\mathfrak{r}(x) = \mathfrak{r}_{\mathcal{P}}(x)$ for the radius of C_x. If $x \sim y$, then x and y have two common neighbours v' and v'' in X. Let $\mathfrak{r}'(x,y) = \mathfrak{r}'_{\mathcal{P}}(x,y)$ and $\mathfrak{r}''(x,y) = \mathfrak{r}''_{\mathcal{P}}(x,y)$ be the inradii of the triangles $z(x)z(y)z(v')$ and $z(x)z(y)z(v'')$, respectively; See Figure 7. Then we set

$$a(x,y) = \frac{\mathfrak{r}'(x,y)+\mathfrak{r}''(x,y)}{\mathfrak{r}(x)+\mathfrak{r}(y)} = \frac{1}{\mathfrak{r}(x)+\mathfrak{r}(y)}\left(\sqrt{\frac{\mathfrak{r}(x)\mathfrak{r}(y)\mathfrak{r}(v')}{\mathfrak{r}(x)+\mathfrak{r}(y)+\mathfrak{r}(v')}} + \sqrt{\frac{\mathfrak{r}(x)\mathfrak{r}(y)\mathfrak{r}(v'')}{\mathfrak{r}(x)+\mathfrak{r}(y)+\mathfrak{r}(v'')}}\right).$$

Let $P = P_{\mathcal{P}}$ be the transition matrix of the resulting random walk.

(6.31) Proposition. *The function $x \mapsto z(x)$ is a complex-valued harmonic function for P.*

Proof. Let $x \in X$, and enumerate the neighbours $y_0, \dots y_{m-1}, y_m = y_0$ of x in circular order. Let w_j (j modulo m) be the incentre of the triangle $z(x)z(y_j)z(y_{j+1})$, and look at the closed polygon $[w_0, w_1, \dots, w_m = w_0]$. We have $|w_j - w_{j-1}| = \mathfrak{r}'(x,y_j) + \mathfrak{r}''(x,y_j)$. Also, $\eta_j = \frac{z(y_j)-z(x)}{\mathfrak{r}(y_j)+\mathfrak{r}(x)}$ is the outward normal unit vector to the segment $\overline{w_{j-1}\,w_j}$. Therefore

$$\sum_{y\sim x} a(x,y)\big(z(y)-z(x)\big) = \sum_{j=1}^{m} \frac{|w_j - w_{j-1}|}{\mathfrak{r}(y_j)+\mathfrak{r}(x)}\big(z(y_j)-z(x)\big) = \sum_{j=1}^{m} |w_j - w_{j-1}|\,\eta_j\,.$$

Now write $w_j - w_{j-1} = |w_j - w_{j-1}|\, e^{i\varphi_j}$, where $\varphi_j \in [0\,, 2\pi)$, so that $\eta_j = e^{i(\varphi_j - \pi/2)}$. Then

$$\sum_{j=1}^{m} |w_j - w_{j-1}|\, \eta_j = \sum_{j=1}^{m} (w_j - w_{j-1})\, e^{-i\pi/2} = 0\,. \qquad \square$$

(6.32) Corollary. *If X is CP-hyperbolic then the simple random walk on X is transient.*

Proof. There is a circle packing $\mathcal{P}$ with carrier $\mathbb{D}$ and tangency graph X. Let $P = P_{\mathcal{P}}$ be as above. Then $h(x) = \text{Re}\big(z(x)\big)$ is a bounded, non-constant harmonic function for (X, P). Hence there is a non-constant positive harmonic function, and by Theorem 1.16, (X, P) must be transient. The conductances associated with P satisfy $a(x, y) \le 1$. Now Corollary 2.14 implies transience of the simple random walk. $\square$

Next, we study the CP-parabolic case under the additional assumption that X has bounded geometry. We shall need the following simple observation regarding any circle packing $\mathcal{P}$ with tangency graph X.

(6.33) Ring Lemma. *Suppose that X has vertex degrees bounded by M. Then there is a constant $a = a_M \ge 1$ such that $a^{-1} \le \mathfrak{r}(x)/\mathfrak{r}(y) \le a$ for any pair of neighbours x, y in X.*

The point is that one cannot surround a given circle by M circles, if one of the latter is too small. For more details and an illustration, see Rodin and Sullivan [278]. We can now prove the main theorem of this subsection.

(6.34) Theorem. *Let X be a disk triangulation graph with bounded geometry. Then X is CP-parabolic if and only if the simple random walk on X is recurrent.*

Proof. The "if" is Corollary 6.32. To prove the "only if", let $\mathcal{P}$ be a circle packing of the plane with tangency graph X. Then $\mathcal{P}$ itself gives rise to a tiling of the plane, still denoted by $\mathcal{P}$, whose tiles are the disks of the packing and the "triangular" interstices between the circles. Using Theorem 6.29, we first show that the edge graph $X^* = X(\mathcal{P})$ of this tiling is recurrent. Each circle C_x $(x \in X)$ is subdivided into $\deg(x) \le M$ arcs between the consecutive points where C_x touches an adjacent circle. The vertices and edges of X^* are precisely these points and arcs. We now show that all but finitely many C_x are $\big(M, 2a/(1+2a), \pi^{-1}\big)$-regular in the sense of Definition 6.28, where a is the constant of Lemma 6.33. Conditions (1) and (3) are obviously satisfied.

We now assume without loss of generality that the origin is the centre of a circle C_o in $\mathcal{P}$. Let $x \ne o$, and let $\mathfrak{s}(x)$ be the smallest radius of a circle tangent with x. Then $\mathfrak{s}(x) \ge \mathfrak{r}(x)/a$ by the ring lemma. Also, the open disk centred at $z(x)$ with radius $\mathfrak{r}(x) + \mathfrak{s}(x)$ contains no centre of any other circle in the packing. In particular, it does not contain $z(o) = 0$. Therefore $|z(x)| \ge \mathfrak{r}(x) + \mathfrak{s}(x) \ge \mathfrak{r}(x)\big(1 + \frac{1}{a}\big)$. For the maximal distance from a point in C_x to 0 we now get

$$\mathfrak{d}_{C_x} = \mathfrak{r}(x) + |z(x)| \ge 2\mathfrak{r}(x)\big(1 + \tfrac{1}{2a}\big). \tag{6.35}$$

This proves (2), and all but finitely many edges of X^* lie on such a C_x. By Theorem 6.29, the simple random walk on X^* is recurrent.

It is rather easy to see that this implies recurrence of X: first, observe that X^* has bounded geometry (all vertices have degree 4). Second, construct a new graph Y^* out of X^* by drawing edges between *all* pairs of distinct vertices x^*, y^* of X^* lying on the same circle. With respect to the graph structure of X^*, the simple random walk on Y^* satisfies all requirements of Theorem 3.2. (Bounded geometry of X is used here!) Hence Y^* is recurrent. Now let $f^* \in \ell_0(Y^*) = \ell_0(X^*)$. We define a function $f = \Phi(f^*) \in \ell_0(X)$ by

$$f(x) = \frac{1}{\deg(x)} \sum_{v^* \in C_x} f^*(v^*)$$

(summation over vertices of Y^*). Let $y \sim x$ in X, and let x^* be the point where C_x and C_y meet. Then (as $\deg(x) \ge 3$)

$$\begin{aligned}
&\big(f(x) - f(y)\big)^2 \le 2\big(f(x) - f^*(x^*)\big)^2 + 2\big(f^*(x^*) - f(y)\big)^2 \\
&= 2\Big(\tfrac{1}{\deg(x)} \sum_{v^* \in C_x} \big(f^*(v^*) - f^*(x^*)\big)\Big)^2 + 2\Big(\tfrac{1}{\deg(y)} \sum_{w^* \in C_y} \big(f^*(x^*) - f^*(w^*)\big)\Big)^2 \\
&\le \tfrac{2}{\deg(x)} \sum_{v^* \in C_x} \big(f^*(v^*) - f^*(x^*)\big)^2 + \tfrac{2}{\deg(y)} \sum_{w^* \in C_y} \big(f^*(w^*) - f^*(x^*)\big)^2 . \\
&\le \sum_{v^* \sim x^*} \big(f^*(v^*) - f^*(x^*)\big)^2 ,
\end{aligned}$$

where $\sim$ is neighbourhood in Y^*. Summing over all edges of X is the same as summing over all vertices x^* of Y^*. Therefore $D_X(f) \le 2D_{Y^*}(f^*)$. Also, constant functions are preserved by Φ. Using Theorem 2.12, we see that the simple random walk on X is recurrent. □

We now give a criterion for CP-parabolicity and recurrence in terms of the vertex degrees.

(6.36) Proposition. *If* $\deg(x) \le 6$ *for all but finitely many vertices then the simple random walk on* X *is recurrent, and* X *is CP-parabolic.*

Proof. Let A_0 be a finite, connected subgraph of X containing all vertices with degree > 6. We define inductively an increasing sequence of subgraphs: $A_{n+1} = A_n \cup B_n$, where $B_n = \{x \in X \setminus A_n : d(x, A_n) = 1\}$. We decompose $B_n = C_n \cup D_n$, were C_n consists of those vertices having precisely one neighbour in A_n, and D_n is the rest.

Let $x \in B_n$ be a vertex having some ("forward") neighbour in $X \setminus A_{n+1}$. This x has a neighbour y_0 in A_n. Let $y_0, y_1, \ldots, y_{m-1}, y_m = y_0$ denote all the neighbours of x in cyclic order, with $m \le 6$. Let $j(1)$ and $j(2)$ be the minimal and the maximal index such that $y_j \notin A_{n+1}$. Then $y_{j(1)-1}, y_{j(2)+1} \in B_n$ must be different from y_0, so that $2 \le j(1) \le j(2) \le m-2$. Also, $y_{j(1)}$ and $y_{j(2)}$ lie in D_{n+1}.

Case 1. $x \in D_n$. If $j(1) < j(2)$ then x has two neighbours in A_n, two neighbours in B_n and two in D_{n+1}, and there can be no further neighbours of x. If $j(1) = j(2)$ then $y_{j(1)}$ is the only neighbour of x outside of A_{n+1}, and it lies in D_{n+1}. Thus, x has one or two neighbours in D_{n+1} and none in C_{n+1}.

Case 2. $x \in C_n$. In the same way as above, we conclude that x has one, two or three neighbours in D_{n+1} and one or no neighbour in C_{n+1}.

Combining the two cases, we infer that the number of edges between A_{n+1} and B_{n+1} satisfies $|\partial A_{n+1}| \le 2|D_n| + 3|C_n|$ and that

$$|C_{n+1}| \le |C_n| \quad \text{and} \quad 2|D_{n+1}| \le 2|D_n| + 3|C_n| .$$

We obtain

$$|\partial A_{n+1}| \le 2|D_0| + 3n|C_0| .$$

Shorting each of the finite sets A_0 and B_n ($n \ge 0$) to a single point, we can now apply Nash-Williams' criterion (Corollary 2.20) to get recurrence of the simple random walk on X. □

We remark that X being infinite, it cannot be that $\deg(x) \le 5$ for all but finitely many n. This is obtained by modifying the above counting argument in the obvious way. In the next chapter, we shall also provide conditions on the vertex degrees which imply $\rho(P) < 1$ for the spectral radius of the simple random walk, and hence transience.

Notes and remarks

1. Basic facts

§**A.** In his original article, Pólya [269] uses characteristic functions to deduce (1.4).

§**B.** For the general theory of Markov chains, see e.g. the monographs of Chung [75], Kemeny et al. [197] and Revuz [276]. Lemma 1.9 is due to Kingman [205].

It has been pointed out to me that the non-probabilist reader might appreciate the following clarification regarding the measures $\mathbb{P}_x$. The transition matrix contains no information on the distribution of Z_0. Hence the latter has to be specified when one constructs the probability measure on the trajectory space. If it is δ_x, i.e., $Z_0 = x$, then one obtains $\mathbb{P}_x$. If one wants to avoid the introduction of an infinity of different $\mathbb{P}_x$ (one for each x), then one may also use a starting distribution ν supported by all of X. For the resulting $\mathbb{P} = \mathbb{P}_\nu$ one gets $\mathbb{P}_x = \mathbb{P}[\,\cdot\,|Z_0 = x]$.

§**C.** Uniform irreducibility first appeared in Picardello and Woess [258] and Ancona [5]. In the mid 80s, it appeared to be a new idea to replace algebraic with geometric adaptedness conditions.

§**D.** Lemma 1.24 and related computations are recurrent: they have been repeated by many authors in the illusion of making a new discovery. The first traces can be found in Kesten [198] and Dynkin and Malyutov [111]. To my knowledge, it first appears explicitly in the context of free groups in Gerl [130] (for $M = 4$) and in Levit and Molchanov [215].

§**E.** The theory of random walks on groups was founded by Kesten [198].

§**F.** I first learned the nice formula of (1.28) and Lemma 1.29 for the modular function from Trofimov [319]; previously, it had appeared in Schlichting [293]. Besides [319], Woess [346] also contains an exposition of some basic facts concerning the topology of the automorphism group of a graph.

2. Recurrence and transience of infinite networks

§**A.** This potential theoretic framework is a special case of the theory of Dirichlet spaces, see Fukushima [123].

§**B.** Theorem 2.12 is the discrete analogue of a set of "hyperbolicity" ($\equiv$ transience) criteria for Riemannian manifolds equipped with the Laplace–Beltrami operator ($\equiv$ Brownian motion). For example, the flow criterion (b) corresponds to the Kelvin–Nevanlinna–Royden criterion. In the context of reversible Markov chains, it became popular through the paper of T. Lyons [222]. The capacity criterion appears in Varopoulos [323], see also Gerl [136]. Before that, these criteria had been proved for locally finite networks by Yamasaki [354], [355] without any use of probabilistic notions. (Yamasaki's work seems to have been unknown to random walk people until 1988, when I "discovered" one of his papers.) For the non-locally-finite case, see also Soardi and Yamasaki [306].

First variants of Corollary 2.14 appeared in Griffeath and Liggett [146] and Doyle and Snell [103]. The first to use shorting for recurrence was Nash-Williams [245], who proved Corollary 2.20. Extensions were given by Griffeath and Liggett [146], Doyle and Snell [103], McGuinness [230] and Woess [348].

In the literature there are many concrete examples of applications of the above criteria; not all are reported in this book. See e.g. Doyle and Snell [103] (among others for certain radial trees which can be embedded in $\mathbb{Z}^2$ or $\mathbb{Z}^3$, respectively), T. Lyons [222] (a class of sublattices of $\mathbb{Z}^d$, also explained in Soardi [304]) or Markvorsen et al. [226] (in connection with the classification of Riemannian manifolds). The comb lattices were considered by Gerl [138] and Cassi and Regina [69].

Concerning the material of this subsection, until the late 70s the only source had remained Nash-Williams [245], but since then there has been a growing number of contributors. I do not claim that the references given here cover every single step of the development.

§**C.** This technique goes back to Baldi et al. [18], who proved Lemma 2.24 and used it for the classification of recurrent (connected) Lie groups. Generalizations are due to Varopoulos [321], [322] and Chen [71]. Theorem 2.25 is taken from Woess [350] and extends the corresponding result of Chen [71]. (In the proof of the latter, the final inequality is unclear.)

3. Applications to random walks

§**A.** As stated here, Theorems 3.1 and 3.2 may appear new, but I do not deserve any credit. I am unable to trace back all references which should be considered as forerunners. One of them is Ancona [6], Thm. 3.1, whose part a) is our Corollary 3.5. For the k-fuzz, see Doyle and Snell [103].

The definition of rough isometry goes back to Gromov [150], [151] (using the terminology "quasi-isometry") and Kanai [192], [193], [194]. Example 3.8 modifies a construction explained to me by C. Thomassen; compare with Thomassen [315], Thm. 3.3. For the first part of Proposition 3.9, see e.g. Salvatori [288]. The second part is taken from Sabidussi [284]. For Theorem 3.10, see Kanai [193] and Markvorsen et al. [226].

§**B.** Growth was first introduced for finitely generated groups by Efremovic [112] and Schwarcz [294], and later on independently by Milnor [236] in relation to Riemannian manifolds. Since then, growth of groups and graphs has been attracting considerable interest. I will not give an exhaustive bibliography here. For groups, see e.g. Bass [27], Gromov [149], Grigorchuk [147] and the references therein. For graphs, see the survey by Imrich and Seifter [179], and for groups, the more recent one by Grigorchuk and de la Harpe [148].

I learned Lemma 3.12 from the anonymous referee of the paper by Rigoli et al. [277]. The present simple proof was communicated to me by N. Pintacuda. Theorem 3.16 is usually attributed to Bass [27], but Guivarc'h [154] proved it independently at about the same time; see also his earlier note [153]. Regarding Theorem 3.17, the version stated here is in fact due to Van den Dries and Wilkie [104]; the original result of Gromov is stated with "for all n" in place of "for infinitely many n". Proposition 3.20 was first noted by Varopoulos [325]. Lemma 3.21 and Proposition 3.23 are also due to Varopoulos [325] and constitute the final and surprisingly simple step in the solution of Kesten's problem. A nice exposition is also contained in Ancona [6].

§**C.** Most of these results were "born" in the preparatory work for this book, see Woess [350], which also contains a proof of Theorem 3.31. For (3.30) and part of (3.31), see Salvatori [289].

4. Isoperimetric inequalities

§**A.** It was the merit of Varopoulos [323] to realize the importance of isoperimetric inequalities in the study of reversible Markov chains, not only for the question of recurrence. The paper by Dodziuk [98] appeared at about the same time and introduced the strong isoperimetric inequality ($\mathfrak{F}(t) = t$) in connection with the spectral radius; see Chapter II. Many references can be found in Varopoulos et al. [326], some more in my survey [348]. For the purpose of §A and §C, I found that the most useful and readable source was Saloff-Coste [285]. Theorem 4.7 is due to Kanai [192], [194].

§**B.** The version of Theorem 4.10 for direct products (see Remarks 4.11) is due to Varopoulos [323]. For clarifications and the adaptation to Cartesian products, I am most grateful to Vadim Kaimanovich.

§**C.** To my knowledge, the Poincaré type property (4.14) that I call "quasi-homogeneity" here was first introduced by Coulhon and Saloff-Coste [86], who also proved (4.15) and (4.16). The principal part of Theorem 4.18 (when IS does not hold) is due to Saloff-Coste [285]. For another proof, see Woess [350].

5. Transient subtrees, and the classification of the recurrent quasi-transitive graphs

§**A.** These "elementary" results, which left me very impressed, are due to Thomassen [315] (for graphs with maximum vertex degree 3) and [316] (general case). With few modifications, they also work for unbounded (finite) vertex degrees; see Thomassen [316]. Note that it is not even simple to find a transient subtree in the grid $\mathbb{Z}^3$; see Doyle and Snell [103] and Gerl [135] and the figures there.

Hyperbolic graphs and their boundaries will be studied in Chapter IV. There is an interesting result regarding transient subtrees, due to Bowers [47]: if X is a hyperbolic graph whose boundary contains a non-trivial continuum, then X admits a roughly isometric embedding of the binary tree. Thus, X contains a subtree that satisfies IS. This applies, in particular, to one-ended hyperbolic graphs with non-trivial hyperbolic boundary. (One-ended means that $X \setminus A$ has only one infinite component for every finite $A \subset X$; see §21 for more on ends.)

§**B.** Theorem 5.11 is originally due to Trofimov [318]. The present proof, based on Losert [216], is adapted from Woess [346]. Theorem 5.13 was outlined in Woess [348] and proved by Saloff-Coste [285]. Five different proofs are outlined in Woess [350].

6. More on recurrence

§**A.** For Theorem 6.1, see Chung and Fuchs [76], Chung and Ornstein [77] and Spitzer [307]. Statement (c) can be directly deduced from the recurrence criterion of Kesten and Spitzer [203]: the random walk with law μ on the abelian group Γ is recurrent if and only if

$$\int_{\hat{\Gamma}} \frac{1}{1-\hat{\mu}(z)}\, dz = \infty\,,$$

where $\hat{\Gamma}$ is the dual of Γ, $\hat{\mu}$ is the Fourier transform of μ and dz is Haar measure on $\hat{\Gamma}$. Many results concerning random walks on abelian groups ans, in particular, integer lattices, are very well documented in Spitzer's book [307] and in the fundamental paper of Kesten and Spitzer [203]. For this reason, and in order to save space, we "sacrifice" this material here.

The remaining results of §A are taken from Salvatori [289] (some previous results are due to Hammersley [165], Bender and Richmond [32], McGuinness [229] and Guivarc'h [157]). For Proposition 6.4, see Kaimanovich [185]. Salvatori's proof of Theorem 6.7 has been simplified thanks to Vadim Kaimanovich who suggested the use of $\bar{P}$ and the law of large numbers for functionals.

§**B.** These results are due to R. Lyons [219] and Benjamini and Peres [36], with Furstenberg [125] as a common root.

§**C.** Theorem 6.24 was proved by Yamasaki [354]; extremal length was introduced into the context of (finite) networks by Duffin [106]. (Note that in my survey [348], I made two mistakes regarding extremal length: first, in the definition I forgot the edge resistances, and second, criterion (f) in [348], Thm. 4.8, is known only for locally finite networks.)

Theorem 6.29, one of my favourites in Chapter I, is a slightly simplified and extended version of a theorem of Soardi [303] (see also [304]), one of the first significant results concerning random walks on planar graphs.

§**D.** It was a "last minute" decision (in 1996, when this chapter was completed) to include some aspects of circle packings, an area rather new to random walks. There are several proofs of Corollary 6.32 and Theorem 6.34. A forerunner of these results which might have deserved more attention was DeBaun [89]. The most general version is due to He and Schramm [169], considering packings of more general sets than ordinary disks. Another variant is due to McCaughan [228]. The ingeniously simple proof of (6.32) via Proposition 6.31 is due to Dubejko [105]. The proof of recurrence in Theorem 6.34 is a modest contribution of my own with the aim of integrating packings with tilings; I am grateful to O. Schramm for pointing out a simplification. Dubejko's slightly less general proof in [105] is from about the same time. Both are of course preceded by the stronger result of [169].

Still more on recurrence

There are various further interesting results which have not been expounded here only because of lack of space. Rather complete references up to 1994 can be found in Woess [348]. Let me outline two recent topics which are particularly appealing.

R. Lyons [219] gives an analytic definition of the *branching number* of a locally finite, infinite tree in terms of cuts between a root o and infinity. It does not depend on the choice of o. Put resistance λ^n on all edges at distance n from o, where $\lambda > 0$. For the resulting reversible random walk, the one-step probability of moving backwards towards the origin increases with λ (the random walk becomes more and more "homesick"). The supremum over all λ for which the random walk is transient is called the *branching number* of the tree. For larger values, the random walk is recurrent. This is directly related to the Hausdorff dimension of the space of ends (compare also with Furstenberg [125]), and also to the critical probability of bond percolation on the tree. In R. Lyons [220] this is applied to "homesick" random walks on Cayley graphs of groups via natural spanning subtrees, thus relating the critical value of λ to the growth rate of the group. These random walks are not group invariant but capture much of the group structure. Homesick random walks are studied in detail for a particularly interesting group by R. Lyons et al. [221].

The other topic is concerned with an application of recurrence in statistical mechanics. Regarding the Ising model on grids and other graphs, the notes of Kindermann and Snell [204] provide an excellent introduction for mathematicians. Given the grid $\mathbb{Z}^d$ with homogeneous nearest neighbour interactions, it is known that no phase transition occurs in one dimension, while it may occur for $d = 2$. The situation is different for the Heisenberg model, where no phase transition occurs in dimensions 1 and 2, while it may occur for $d \geq 3$. The same holds, more generally, when the spins, instead of values ± 1 (Ising model), take their values in a connected, compact Lie group (or homogeneous space thereof); see Mermin and Wagner [235] and Mermin [234]. Now consider the latter class of models on a general graph X, with interactions induced by the conductances of a reversible nearest neighbour random walk (X, P). Then Cassi [68] and (generalizing his result) Merkl and Wagner [233] have shown that recurrence of (X, P) implies absence of phase transition. These impressive results would merit a careful mathematical exposition. Unfortunately, this goes well beyond the scope of the present book.

CHAPTER II

THE SPECTRAL RADIUS

7. Superharmonic functions and ρ-recurrence

Recall the definition (1.8) of the *spectral radius*

$$\rho = \rho(P) = \limsup_{n\to\infty} p^{(n)}(x,y)^{1/n}$$

of our irreducible Markov chain (X, P). The terminology (although common) may be slightly misleading when we do not specify a suitable space where P acts as a linear operator. When (X, P) is a random walk on a group and μ its law, then we also write $\rho(\mu)$.

A. The spectral radius and superharmonic functions

We start with a characterization of $\rho(P)$ in terms of *t-superharmonic* functions, that is, functions $f : X \to \mathbb{R}$ satisfying $Pf \le t \cdot f$, where $t > 0$. Analogously, *t-harmonic* functions are those satisfying $Pf = t \cdot f$. We write $\mathcal{S}(P,t)$ and $\mathcal{H}(P,t)$ for the collections of all t-superharmonic and t-harmonic functions, respectively, and $\mathcal{S}^+(P,t)$, $\mathcal{H}^+(P,t)$ for the corresponding positive cones. By the minimum principle (1.15), extended to this situation, we have that a function in $\mathcal{S}^+(P,t)$ is either strictly positive at each point or constant equal to 0. Now fix a reference point $o \in X$ and define

$$\mathcal{B}(P,t) = \{f \in \mathcal{S}^+(P,t) : f(o) = 1\}.$$

This is a base of the cone $\mathcal{S}^+(P,t)$.

(7.1) Lemma. *If $\mathcal{B}(P,t)$ is non-empty then it is compact in the topology of pointwise convergence, and $\mathcal{S}^+(P,t)$ is a convex cone with compact base.*

Proof. By Fatou's lemma $\mathcal{B}(P,t)$ is closed. If $x \in X$ then by irreducibility there is n_x such that $p^{(n_x)}(o,x) > 0$. Set $C_x = t^n / p^{(n_x)}(o,x)$. If $f \in \mathcal{B}(P,t)$ then

$$p^{(n)}(o,x) f(x) \le P^n f(o) \le t^n \cdot f(o) = t^n.$$

Hence $f(x) \le C_x$ for all $f \in \mathcal{B}(P,t)$, and compactness follows. □

(7.2) Lemma. $\rho(P) = \min\{t > 0 : \mathcal{S}^+(P,t) \ne \{\mathbf{0}\}\}$.

Proof. If there is $f \ne 0$ in $\mathcal{S}^+(P,t)$, then $p^{(n)}(x,x) f(x) \le P^n f(x) \le t^n \cdot f(x)$, so that $\rho(P) = \limsup_n p^{(n)}(x,x)^{1/n} \le t$.

Conversely, if $t > \rho(P)$, then the function $f(x) = G(x, o|1/t)$ is non-zero and in $\mathcal{S}^+(P,t)$. Hence, $\mathcal{B}^+(P,t) \ne \emptyset$. If $t_1 < t_2$ then $\mathcal{B}^+(P,t_1) \subset \mathcal{B}^+(P,t_2)$. By compactness, it must be that $\mathcal{B}^+(P,\rho) = \bigcap_{t>\rho} \mathcal{B}^+(P,t) \ne \emptyset$. □

(7.3) Exercise. *Let $P = c\,P_1 \otimes I_2 + (1-c)I_1 \otimes P_2$ be a Cartesian product of two irreducible transition operators. Then $\rho(P) = c\,\rho(P_1) + (1-c)\,\rho(P_2)$.*

[Hint: see formula (18.2) in Chapter III.]

B. ρ-Recurrence

From Lemma 1.7 we know that the series $G(x,y|1/\rho)$ either converge or diverge simultaneously for all x, y. In the first case we say that (X, P) is *ρ-transient,* and in the second that it is *ρ-recurrent;* ρ-recurrence is further subdivided into positive and null, according to whether $U'(x,x|1/\rho-)$ is finite or not (this is also independent of x).

If (X, P) is recurrent in the ordinary sense, then $\rho(P) = 1$ and the chain is ρ-recurrent. If (X, P) is transient, then each of the following cases is possible: (1) $\rho(P) = 1$ (for example, the simple random walk on $\mathbb{Z}^3$); (2) $\rho(P) < 1$ and ρ-transience (the simple random walk on $\mathbb{T}_M$ – see Lemma 1.24); (3) $\rho(P) < 1$ and ρ-null-recurrence (the random walk on $\mathbb{Z}$ with law μ given by $\mu(1) = p$, $\mu(-1) = 1-p$, where $p \neq \frac{1}{2}$); (4) $\rho(P) < 1$ and ρ-positive-recurrence. Examples of (3) and (4) for simple random walks on graphs will be given at the end of §9.A.

We have the following analogue of Theorem 1.16.

(7.4) Proposition. *(X, P) is ρ-recurrent if and only if there is a non-zero function h in $\mathcal{H}^+(P,\rho)$ such that every function in $\mathcal{S}^+(P,\rho)$ is a constant multiple of h.*

Proof. If (X, P) is ρ-transient then $f(x) = G(x,o|1/\rho)$ is in $\mathcal{S}^+(P,\rho)$, but not ρ-harmonic.

Conversely, assume ρ-recurrence and choose $h \neq 0$ in $\mathcal{S}^+(P,\rho)$. (Existence follows from Lemma 7.2.) Setting $g = \rho \cdot h - Ph$, a straightforward modification of the argument used for Theorem 1.16 yields $g \equiv 0$, and h is ρ-harmonic. We introduce the transition matrix P_h of the *h-process* on X:

$$p_h(x,y) = \frac{p(x,y)h(y)}{\rho \cdot h(x)}. \tag{7.5}$$

(If $h \in \mathcal{H}^+(P,t)$, where $t \geq \rho$, then the h-process is defined in the same way with t in place of ρ.) We compute

$$p_h^{(n)}(x,y) = \frac{p^{(n)}(x,y)h(y)}{\rho^n \cdot h(x)},$$

so that the Green function of the h-process is

$$G_h(x,y|1) = G(x,y|1/\rho)h(y)/h(x).$$

By our assumption, the h-process is recurrent. Now one checks that $f \in \mathcal{S}(P,\rho)$ if and only if $f/h \in \mathcal{S}(P_h,1)$, so that f/h must be constant. □

Concerning positive eigenfunctions of P, the infinite case is significantly different from finite Markov chains:

(7.6) Lemma. *If X is infinite and P has finite range, then there are non-zero positive t-harmonic functions for every $t \ge \rho(P)$.*

Proof. The ρ-recurrent case, with $t = \rho(P)$, is covered by Proposition 7.4. So assume that $t \ge \rho$ and $G(x,y|t^{-1}) < \infty$. Choose a sequence of distinct points y_n in X, and consider the functions $g_n(x) = G(x,y_n|t^{-1})/G(o,y_n|t^{-1})$. They are in $\mathcal{B}(P,t)$, which is compact, and hence must accumulate. So assume without loss of generality that $g_n(x) \to h(x)$ pointwise. Because of the finite range, we may exchange limit and application of P. Now observe that for every x, $Pg_n(x) = t \cdot g_n(x)$ for all but finitely many n. Therefore

$$Ph(x) = \lim_{n\to\infty} Pg_n(x) = t \cdot h(x)\,. \qquad \square$$

In all that has been said so far in this section, we may replace positive t-superharmonic and t-harmonic functions with *t-excessive* and *t-invariant* measures, that is, measures satisfying $\nu P \le t \cdot \nu$ or $\nu P = t \cdot \nu$, respectively. Checking this is a straightforward exercise. For a t-invariant measure ν, the ν-process is (by abuse of the above notation)

$$p^\nu(x,y) = \frac{\nu(y)p(y,x)}{t \cdot \nu(x)}\,. \tag{7.7}$$

If σ is any t-excessive measure for P, then the *function* (density) σ/ν is in $\mathcal{S}(P^\nu,1)$, and conversely.

In conclusion, we show that only graphs which grow at most quadratically may carry a ρ-recurrent, quasi-transitive random walk.

(7.8) Theorem. *Let X be a graph, and suppose that (X,P) is quasi-transitive. If P is ρ-recurrent, then the simple random walk on X is recurrent, and X is a generalized lattice with dimension 1 or 2.*

Proof. According to Proposition 7.4, $\mathcal{B}(P,\rho)$ consists of a unique ρ-harmonic function h. The associated h-process is recurrent. In view of Theorem 5.13, all we have to show is that (X,P_h) is quasi-transitive. Choose $\gamma \in \mathrm{AUT}(X,P)$. Then $h_\gamma(x) = h(\gamma x)$ is also ρ-harmonic, whence $h_\gamma = \phi(\gamma) \cdot h$ for some constant $\phi(\gamma) > 0$. The function $\phi : \mathrm{AUT}(X,P) \to \mathbb{R}^+$ is a positive exponential (multiplicative homomorphism). But then

$$p_h(\gamma x,\gamma y) = \frac{p(\gamma x,\gamma y)h(\gamma y)}{\rho \cdot h(\gamma x)} = p_h(x,y)\,. \qquad \square$$

8. The spectral radius, the rate of escape, and generalized lattices

In this section, we shall show that $\rho(P) < 1$ implies for a typical random walk that it tends to ∞ at linear speed at least. This is the right occasion for an intermezzo on the behaviour of $d(Z_n, Z_0)/n$, that is, variants of the law of large numbers. We shall then show how the first of these results leads to a method for calculating the spectral radius of random walks on $\mathbb{Z}^d$ and on generalized lattices.

A. The rate of escape

Recall Lemma 1.9, where we showed that $p^{(n)}(x,x) \le \rho(P)^n$. For random walks on graphs, we now extend this in two ways.

(8.1) Lemma. (a) *If (X, P) is uniformly irreducible then there is $A > 0$ such that $p^{(n)}(x,y) \le A^{d(x,y)}\rho(P)^n$.*

(b) *If (X, P) is strongly reversible then $p^{(n)}(x,y) \le C\,\rho(P)^n$ with C as in (3.4).*

Proof. For (a), let ε_0 be as in (1.20). Let $x, y \in X$, $d(x,y) = r > 0$. Applying (1.20) to the successive neighbours on a path of length r from y to x, we find $k_1, \dots, k_r \ge 1$ such that $p^{(k_1+\cdots+k_r)}(y,x) \ge \varepsilon_0^r$. Set $k = k_1 + \cdots + k_r$. Then, using Lemma 1.9 and the fact that $\rho(P) \le 1$,

$$p^{(n)}(x,y)\varepsilon_0^r \le p^{(n)}(x,y)p^{(k)}(y,x) \le p^{(n+k)}(x,x) \le \rho(P)^{n+k} \le \rho(P)^{n+r}\,.$$

We may set $A = \rho(P)/\varepsilon_0$. To see (b), write

$$p^{(n)}(x,y)^2 = \frac{m(y)}{m(x)}p^{(n)}(x,y)p^{(n)}(y,x) \le C^2 p^{(2n)}(x,x) \le C^2\rho(P)^{2n}\,. \qquad \square$$

Note that in (a) we did not use the K from uniform irreducibility (1.20). While (b) will be used in Chapter IV only, we shall give a simple, but useful, application of (a).

(8.2) Proposition. *Suppose that (X, P) is uniformly irreducible and that $\rho(P) < 1$. Then there is a constant $\underline{\mathfrak{m}} > 0$ such that*

$$\mathbb{P}_x\left[\liminf_{n\to\infty} \frac{1}{n} d(Z_n, Z_0) \ge \underline{\mathfrak{m}}\right] = 1 \quad \textit{for every } x, \textit{ and}$$

$$\lim_{n\to\infty} \mathbb{P}_x\left[\inf_{k\ge n} \frac{1}{k} d(Z_k, Z_0) < \underline{\mathfrak{m}}\right] = 0 \quad \textit{uniformly in } x \in X.$$

Proof. Let M be an upper bound on the vertex degrees, and let A be the constant of Lemma 8.1(a). We may suppose $MA > 1$, and choose $\underline{\mathfrak{m}} > 0$

such that $(MA)^{\underline{\mathrm{m}}}\rho(P) < 1$. The events $\mathcal{A}_n = [\inf_{k\geq n} \frac{1}{k} d(Z_k, Z_0) < \underline{\mathrm{m}}\, n]$ in the trajectory space satisfy

$$\begin{aligned}\mathbb{P}_x(\mathcal{A}_n) &\leq \sum_{k=n}^{\infty} \sum_{y:d(y,x)<\underline{\mathrm{m}}\, k} p^{(k)}(x,y) \\ &\leq \sum_{k=n}^{\infty} \sum_{y:d(y,x)\leq\underline{\mathrm{m}}\, k} A^{d(x,y)}\rho(P)^k \leq c\Big((MA)^{\underline{\mathrm{m}}}\rho(P)\Big)^n,\end{aligned}$$

where $c > 0$. Thus $\sum_n \mathbb{P}_x(\mathcal{A}_n)$ converges uniformly in x, and the two statements follow. □

Next, we look for upper bounds on $d(Z_n, Z_0)/n$. It is clear that this will require a moment condition. The step length distributions (1.22) are *tight,* if their tails $\sigma_x\big([n\,,\,\infty)\big)$ are bounded above by the tails of a single distribution σ on $\mathbb{N}_0$. If in addition σ has finite mean, then we say that (X, P) has *uniform first moment.* This means that

$$\text{(8.3)}\qquad \overline{\mathfrak{m}} = \sum_{n=1}^{\infty} \phi(n) < \infty\,, \quad \text{where} \quad \phi(n) = \phi_P(n) = \sup_{x\in X} \sigma_x\big([n\,,\,\infty)\big)\,.$$

(If (8.3) holds, then we can find σ by setting $\sigma(n) = \phi(n) - \phi(n+1)$, $n \geq 0$.) Tightness allows us to compare $d(Z_n, Z_0)$ with $S_n = Y_1 + \cdots + Y_n$, where the Y_i are i.i.d. $\mathbb{N}_0$-valued random variables with distribution σ. This can be achieved by *coupling* the random walk (X, P) with $(Y_n)_{n\in\mathbb{N}}$. That is, instead of the trajectory space, we define a bigger probability space where we can define and compare both sequences of random variables. By abuse of notation, we write $(\Omega, \mathbb{P})$ for this probability space, with $\Omega = [0\,,\,1)^{\mathbb{N}}$ carrying the usual Borel product sigma-algebra, and $\mathbb{P}$ the infinite product of Lebesgue measure on $[0\,,\,1)$. We subdivide $[0\,,\,1)$ into consecutive intervals $I_0 = [0 = a_0\,,\,a_1)\,,\; I_1 = [a_1\,,\,a_2)\,,\ldots,$ with lengths $a_{i+1} - a_i = \sigma(i)\,,\; i \geq 0$. We define $Y_n : \Omega \to \mathbb{N}_0$ as follows. If $\omega = (\omega_n)_{n\in\mathbb{N}} \in \Omega$, then

$$\text{(8.4)}\qquad Y_n(\omega) = i \iff \omega_n \in I_i\,.$$

Then the Y_n are i.i.d. with distribution σ. Next, given $x \in X$, we choose an enumeration $X = \{x_k : k \in \mathbb{N}_0\}$ of X (depending on x), in such a way that $x_0 = x$ and $k' \geq k$ implies $d(x_{k'}, x) \geq d(x_k, x)$. As above, we subdivide $[0\,,\,1)$ into successive intervals J_{x,x_k} of length $p(x, x_k)$, $k \geq 0$. We do this for each $x \in X$.

By construction, the set $\bigcup\{J_{x,y} : y \in B(x,n)\}$ is an interval $[0\,,\,b_{x,n})$ of length $b_{x,n} = \sigma_x([0\,,\,n]) \geq \sigma([0\,,\,n])$, and must contain the interval $\bigcup_{i\leq n} I_i$. In other terms,

$$\text{(8.5)}\qquad d(y,x) \geq n \quad \text{implies} \quad J_{x,y} \subset \bigcup_{i=n}^{\infty} I_i\,.$$

We now define $Z_n^x : \Omega \to X$ by

(8.6) $\qquad Z_0^x(\omega) = x \quad$ and, for $n \geq 1$, $\quad Z_n^x(\omega) = y \iff \omega_n \in J_{x_{n-1},y}$.

It is then straightforward to show that Z_n^x is (a model of) the Markov chain on X with transition matrix P starting at x.

We now arrive at the goal of this construction. Let $\omega \in \Omega$ and $x_n = Z_n^x(\omega)$, $n \geq 0$. Suppose that $d(x_{n-1}, x_n) = \ell$. Then $\omega_n \in J_{x_{n-1},x_n} \subset \bigcup_{i \geq \ell} I_i$ by (8.5), and (8.4) yields $Y_n(\omega) \geq \ell$. We have shown that

(8.7) $\qquad d(Z_{n-1}^x, Z_n^x) \leq Y_n \quad$ for all $x \in X$ and $n \in \mathbb{N}$.

(8.8) Proposition. *Suppose that (X,P) has a uniform first moment. Then we have the following.*

(a) $$\lim_{n\to\infty} \frac{1}{n} d(Z_{n-1}, Z_n) = 0 \quad \textit{and} \quad \limsup_{n\to\infty} \frac{1}{n} \sup_{k \leq n} d(Z_0, Z_k) \leq \overline{\mathfrak{m}}$$

$\mathbb{P}_x$-almost surely for every $x \in X$.

(b) $$\lim_{n\to\infty} \mathbb{P}_x \left[\sup_{k \geq n} \frac{1}{k} d(Z_{k-1}, Z_k) > \varepsilon \right] = 0\,,$$

$$\lim_{n\to\infty} \mathbb{P}_x \left[\frac{1}{n} \sup_{k \leq n} d(Z_{k-1}, Z_k) > \varepsilon \right] = 0\,, \quad \textit{and}$$

$$\lim_{n\to\infty} \mathbb{P}_x \left[\frac{1}{n} \sup_{k \leq n} d(Z_0, Z_k) > \overline{\mathfrak{m}} + \varepsilon \right] = 0\,,$$

uniformly in $x \in X$ for every $\varepsilon > 0$.

Proof. Using the above construction, the law of large numbers yields $\lim_n \frac{1}{n} Y_n = 0$ and $\lim_n \frac{1}{n} S_n = \overline{\mathfrak{m}}$ almost surely. By (8.7), $\sup_{k \leq n} d(Z_0^x, Z_k^x) \leq S_n$ for all n and x. This proves (a) and the first and last statements of (b). For the second statement of (b), with $\phi(n)$ as in (8.3) we have

$$\mathbb{P}\left[\frac{1}{n} \sup_{k \leq n} d(Z_{k-1}^x, Z_k^x) > \varepsilon \right] \leq \sum_{k=1}^{n} \mathbb{P}[Y_k > \varepsilon n] \leq n\, \phi(\lceil \varepsilon n \rceil)\,,$$

which tends to 0, as ϕ is decreasing and $\sum_n \phi(n) < \infty$. $\square$

In particular, we get the following variant of the law of large numbers for the rate of escape (setting $|x| = d(x,o)$ for $x \in X$).

(8.9) Corollary. *If (X,P) is uniformly irreducible and has a uniform first moment, and $\rho(P) < 1$, then*

$$0 < \underline{\mathfrak{m}} \leq \liminf_{n\to\infty} \frac{1}{n} |Z_n| \leq \limsup_{n\to\infty} \frac{1}{n} |Z_n| \leq \overline{\mathfrak{m}} < \infty \quad \mathbb{P}_x\text{-a.s. } \textit{for every } x \in X.$$

The uniform estimates of Propositions 8.2 and 8.8 will be used in Chapter IV. For random walks on groups there is a more precise result than (8.9). The basic tool is *Kingman's subadditive ergodic theorem;* see Kingman [206] and – in the formulation that is most suitable for us – Derriennic [94].

(8.10) Theorem. *Let $(\Omega, \mathbb{P})$ be a probability space and $T : \Omega \to \Omega$ be a measure-preserving transformation. If W_n is a* subadditive *sequence of non-negative real-valued random variables on Ω, that is, $W_{n+k} \le W_n + W_k \circ T^n$ for all $k, n \in \mathbb{N}$, and W_1 is integrable, then there is a T-invariant random variable W_∞ such that*

$$\lim_{n\to\infty} \frac{1}{n} W_n = W_\infty \quad \mathbb{P}\text{-almost surely and in } L^1 .$$

We want to apply this in the vertex-transitive case. Recall the considerations of §1.F regarding the topology of $\mathrm{AUT}(X)$. Let Γ be a closed subgroup of $\mathrm{AUT}(X)$ and μ a regular probability measure on Γ. Consider a sequence $(X_n)_{n\in\mathbb{N}}$ of i.i.d. Γ-valued random variables with common distribution μ. The X_n may be modelled as the projections of the product space $\Gamma^{\mathbb{N}}$ onto Γ, where $\Gamma^{\mathbb{N}}$ is equipped with the product Borel σ-algebra and the probability measure $\mathbb{P} = \mu^{\mathbb{N}}$. (We are using yet another probability space!) The *left* and *right random walks* on Γ with law μ are given respectively by

$$(8.11) \qquad L_n = X_n \cdots X_2 X_1 \quad \text{and} \quad R_n = X_1 X_2 \cdots X_n \quad (L_0 = R_0 = \iota) .$$

Contrary to finitely generated groups acting on their Cayley graphs, in general there is a significant difference between studying R_n and L_n in terms of their action on X. We shall use R_n. The reason is that, when considering $R_n x$, where $x \in X$, we can cancel on the left: $d(R_k x, R_n x) = d(x, X_{k+1} \cdots X_n x)$, when $k \le n$.

Now let Γ be a closed subgroup of $\mathrm{AUT}(X, P)$ which acts transitively on X. We lift the random walk P on X to one on Γ, as follows. Choose a reference point $o \in X$. As the stabilizer Γ_o is open and compact, we may normalize the Haar measure $d\gamma$ on Γ so that $|\Gamma_o| = 1$. It will be convenient to choose and fix, for every $x \in X$, an automorphism $\gamma_x \in \Gamma$ such that $\gamma_x o = x$. We also write $\Gamma^x = \gamma_x \Gamma_o = \{\gamma \in \Gamma : \gamma o = x\}$. Note that X can be identified with Γ/Γ_o via $x \leftrightarrow \Gamma^x$. With the random walk we now associate a measure μ on Γ, absolutely continuous with respect to $d\gamma$, by

$$(8.12) \qquad \mu(d\gamma) = p(o, \gamma o)\, d\gamma .$$

If Γ is finitely generated, X a Cayley graph of Γ, and μ the measure that defines P via $p(x, y) = \mu(x^{-1}y)$, then of course (8.12) reconstructs μ.

(8.13) Proposition. *The measure μ is a probability measure on Γ. For every $\alpha \in \Gamma$, the sequence $(\alpha R_n o)$ is a model of the random walk (Z_n) on X starting at $Z_0 = \alpha o$. In particular, $\bigcup_{n\ge1} (\operatorname{supp} \mu)^n = \Gamma$.*

Proof. First, μ has total mass 1:

$$\mu(\Gamma) = \sum_{x \in X} \int_{\gamma_x \Gamma_o} p(o, \gamma o)\, d\gamma = \sum_{x \in X} p(o,x)\,|\Gamma_o| = 1\,.$$

Next, we prove that

$$\mathbb{P}[\alpha R_1 = x_1, \ldots, \alpha R_n = x_n] = p(\alpha o, x_1) p(x_1, x_2) \cdots p(x_{n-1}, x_n)$$

for every choice of $\alpha \in \Gamma$ and $x_1, \ldots, x_n \in X$. Note that it is enough to prove this for $\alpha = \iota$. For $n = 1$, $\mathbb{P}[R_1 = x_1] = \mu(\Gamma^{x_1}) = p(o, x_1)$. By induction,

$$\begin{aligned}
\mathbb{P}[R_1 = x_1, \ldots, R_n = x_n] &= \int_{\Gamma^{x_1}} \mathbb{P}[R_2 = x_2, \ldots, R_n = x_n \mid X_1 = \gamma]\, \mu(d\gamma) \\
&= \int_{\Gamma^{x_1}} \mathbb{P}[R_1 = \gamma^{-1} x_2, \ldots, R_{n-1} = \gamma^{-1} x_n]\, p(o, \gamma o)\, d\gamma \\
&= \int_{\Gamma^{x_1}} p(o, \gamma^{-1} x_2) \cdots p(\gamma^{-1} x_{n-1}, \gamma^{-1} x_n)\, p(o, x_1)\, d\gamma \\
&= p(o, x_1) p(x_1, x_2) \cdots p(x_{n-1}, x_n)\, |\Gamma^x|\,.
\end{aligned}$$

We have $|\Gamma^x| = 1$. It is now obvious that $\bigcup_{n \ge 1} (\operatorname{supp} \mu)^n = \Gamma$. □

(8.14) Theorem. *If (X, P) is transitive and has finite first moment, then there is a constant $\mathfrak{m}$ such that*

$$\lim_{n \to \infty} \frac{1}{n} d(Z_0, Z_n) = \mathfrak{m} \quad \mathbb{P}_x\text{-a.s. for every } x \in X.$$

When $\rho(P) < 1$ then $\mathfrak{m} > 0$.

Proof. Set $\Gamma = \mathrm{AUT}(X, P)$. Let μ be as in (8.12), and $R_n = X_1 \cdots X_n$ the random walk lifted to Γ according to Proposition 8.13. We use the model $\Omega = \Gamma^{\mathbb{N}}$ with $\mathbb{P} = \mu^{\mathbb{N}}$ and X_n the n-th projection. For T we choose the shift on Ω. Given the starting point $x \in X$, we set $W_n = d(Z_n, x) = d(\gamma_x R_n, x) = d(R_n o, o)$. Now

$$d(R_{n+k} o, o) \le d(R_n o, o) + d(X_{n+1} \cdots X_{n+k} o, o) = W_n + W_k \circ T^n\,,$$

and we may apply Theorem 8.10. As $\mathbb{E}(W_1) < \infty$ by assumption, also $\mathbb{E}(W_\infty) < \infty$, and by the zero–one law, W_∞ must be constant a.s. Proposition 8.2 implies that $\mathfrak{m} > 0$ when $\rho(P) < 1$. □

B. Application to generalized lattices

Let X be a generalized lattice of dimension d, written in "coordinates" $i\mathbf{k}$, where $i \in \mathcal{I}$ (finite) and $\mathbf{k} \in \mathbb{Z}^d$; see §6.A. Suppose that P is adapted to the structure of X by (6.2), and recall the definition of the drift $\mathfrak{m}(P)$ given in Theorem 6.7. From the latter and Proposition 8.2, we get the following.

(8.15) Corollary. *If P has finite first moment and zero drift, then* $\rho(P) = 1$.

We now want to apply this result in order to find a general formula for $\rho(P)$. For the sake of simplicity, we assume that P has finite range. For any non-negative, finite-range irreducible matrix Q over X which satisfies (6.2) (invariance under $\mathbb{Z}^d$), we define matrices $\widetilde{Q}$ over $\mathcal{I}$ and $Q_{\mathbf{c}}$ over X by

$$\tilde{q}(i,j) = \sum_{\mathbf{k}} q(i\mathbf{0}, j\mathbf{k}) \quad \text{and} \quad q_{\mathbf{c}}(i\mathbf{k}, j\mathbf{l}) = q(i\mathbf{k}, j\mathbf{l})\, e^{\mathbf{c}\cdot(\mathbf{l}-\mathbf{k})}, \tag{8.16}$$

where $\mathbf{c} \in \mathbb{R}^d$ and the "$\cdot$" in the exponent is scalar product. By $\widetilde{Q}_{\mathbf{c}}$ we shall mean $(Q_{\mathbf{c}})^{\sim}$. The number $\rho(\widetilde{Q}) = \limsup_n \tilde{q}^{(n)}(i,j)^{1/n}$ is the largest positive eigenvalue of $\widetilde{Q}$, and we can choose the associated positive left eigenvector $\tilde{\nu}_Q$, normalized to become a probability measure on $\mathcal{I}$. Analogously, $\tilde{g}_Q$ stands for the associated positive right eigenvector, normalized to have value 1 in $i_1 \in \mathcal{I}$. (This is the Perron–Frobenius theorem; see Seneta [297]). Verification of the following relations is a straightforward exercise.

$$(PQ)_{\mathbf{c}} = P_{\mathbf{c}} Q_{\mathbf{c}}\,, \quad (P+Q)_{\mathbf{c}} = P_{\mathbf{c}} + Q_{\mathbf{c}}\,, \quad \widetilde{PQ} = \widetilde{P}\widetilde{Q}\,, \tag{8.17}$$

where P, Q are $\mathbb{Z}^d$-invariant as in (6.2). Below, we shall also use that for any irreducible matrix P one has $\rho\big(\lambda I + (1-\lambda)P\big) = \lambda + (1-\lambda)\rho(P)$, where I is the identity matrix. (This follows from Lemma 7.2.)

Now let P be $\mathbb{Z}^d$-invariant as in (6.2), with finite range and stochastic. Let $\tilde{\nu}_{\mathbf{c}} = \tilde{\nu}_{P_{\mathbf{c}}}$ and $\tilde{g}_{\mathbf{c}} = \tilde{g}_{P_{\mathbf{c}}}$. For $\mathbf{c} \in \mathbb{R}^d$ define

$$\varphi_P(\mathbf{c}) = \rho(\widetilde{P}_{\mathbf{c}})\,, \quad \nu_{\mathbf{c}}(i\mathbf{k}) = \tilde{\nu}_{\mathbf{c}}(i) e^{-\mathbf{c}\cdot\mathbf{k}} \quad \text{and} \quad g_{\mathbf{c}}(i\mathbf{k}) = \tilde{g}_{\mathbf{c}}(i) e^{\mathbf{c}\cdot\mathbf{k}}\,. \tag{8.18}$$

Note that $\varphi_P(\mathbf{0}) = 1$ and $g_{\mathbf{0}} \equiv 1$. We call $g_{\mathbf{c}}$ an *extended exponential* on X. When $|\mathcal{I}| = 1$ and $X = \mathbb{Z}^d$, this is a (pure) exponential, that is, a homomorphism from $\mathbb{Z}^d$ to the multiplicative group $\mathbb{R}^+$. Extended exponentials will be important in Section 25.

(8.19) Lemma. $\quad \nu_{\mathbf{c}} P = \varphi_P(\mathbf{c})\, \nu_{\mathbf{c}} \quad$ *and* $\quad P g_{\mathbf{c}} = \varphi_P(\mathbf{c})\, g_{\mathbf{c}}$.

Proof. This is straightforward:

$$\begin{aligned} \nu_{\mathbf{c}} P(j\mathbf{l}) &= \sum_{i,\mathbf{k}} \tilde{\nu}_{\mathbf{c}}(i)\, e^{-\mathbf{c}\cdot\mathbf{k}}\, p(i\mathbf{k}, j\mathbf{l}) = e^{-\mathbf{c}\cdot\mathbf{l}} \sum_i \tilde{\nu}_{\mathbf{c}}(i) \sum_{\mathbf{k}} p_{\mathbf{c}}(i\mathbf{k}, j\mathbf{l}) \\ &= e^{-\mathbf{c}\cdot\mathbf{l}} \sum_i \tilde{\nu}_{\mathbf{c}}(i) \tilde{p}_{\mathbf{c}}(i,j) = \varphi_P(\mathbf{c})\, \tilde{\nu}_{\mathbf{c}}(j)\, e^{-\mathbf{c}\cdot\mathbf{l}}\,. \end{aligned}$$

The second identity is analogous. □

(8.20) Proposition. *If P satisfies (6.2) and has finite range, then the function $\varphi = \varphi_P : \mathbb{R}^d \to (0\,,\infty)$ is infinitely differentiable and strictly convex on $\mathbb{R}^d$,*

$$\lim_{|\mathbf{c}|\to\infty} \varphi(\mathbf{c}) = \infty\,, \quad \operatorname{grad}\varphi(\mathbf{0}) = \mathfrak{m}(P)\,,$$

and the matrix $\Sigma = \Sigma(P)$ of second order derivatives of φ_P at $\mathbf{0}$ is given by

$$\Sigma = \sum_i \widetilde{\nu}(i)\,\mathbb{E}_{i\mathbf{0}}(Y_1 Y_1^t).$$

Here, Y_1 is the $\mathbb{Z}^d$-component of Z_1, as in §6.A. Note that we usually think of elements $\mathbf{y} \in \mathbb{R}^d$ as column vectors, so that $\mathbf{y}\mathbf{y}^t$ is the matrix with elements $y_i y_j$.

Proof of Proposition 8.20. The mapping $\mathbf{c} \mapsto \widetilde{P}_{\mathbf{c}}$ is convex from $\mathbb{R}^d$ to $\mathbb{R}^{|\mathcal{I}|^2}$. By the Perron–Frobenius theorem [297], the largest eigenvalue $\varphi(\mathbf{c})$ is a simple root of the characteristic polynomial $\det(\lambda \widetilde{I} - \widetilde{P}_{\mathbf{c}})$, which is analytic in $(\lambda, \mathbf{c})$. By the implicit function theorem, $\varphi(\mathbf{c})$ is infinitely differentiable. The same is true for the mapping $\mathbf{c} \mapsto \widetilde{\nu}_{\mathbf{c}}$.

Next, let $\mathbf{e}_k$ be the k-th unit vector in $\mathbb{Z}^d$. Then irreducibility yields that there are m and $\varepsilon_0 > 0$ such that $q\big(i\mathbf{0}, j(\pm\mathbf{e}_k)\big) \ge \varepsilon_0$ for all $i, j \in \mathcal{I}$ and $k = 1, \dots, d$, where $Q = \big(\frac{1}{2}(I + P)\big)^m$ (with $I =$ the identity over X in the last expression). Therefore, writing $\mathbf{c} = (c_1, \dots, c_d)$,

$$\widetilde{q}_{\mathbf{c}}(i,j) \ge \varepsilon_0 \sum_{k=1}^{d} (e^{\mathbf{c}\cdot\mathbf{e}_k} + e^{-\mathbf{c}\cdot\mathbf{e}_k}) \ge 2\varepsilon_0 \sum_{k=1}^{d} |c_k|\,,$$

which tends to ∞ as $|\mathbf{c}| \to \infty$. Consequently, $\lim_{|\mathbf{c}|\to\infty} \varphi_Q(\mathbf{c}) \to \infty$. By (8.17) and the remark thereafter, $\varphi_Q = \big(\frac{1}{2}(1 + \varphi_P)\big)^m$. This shows that $\varphi_P(\mathbf{c})$ tends to ∞ as $|\mathbf{c}| \to \infty$.

For the second identity, use the definition of $\widetilde{\nu}_{\mathbf{c}}$ and write

$$\varphi_P(\mathbf{c}) = \sum_{i,j\in\mathcal{I},\mathbf{k}\in\mathbb{Z}^d} \widetilde{\nu}_{\mathbf{c}}(i)\, p(i\mathbf{0}, j\mathbf{k})\, e^{\mathbf{c}\cdot\mathbf{k}}\,.$$

Thus

$$\operatorname{grad}\varphi_P(\mathbf{c}) = \sum_{i,j,\mathbf{k}} \mathbf{k}\, \widetilde{\nu}_{\mathbf{c}}(i)\, p(i\mathbf{0}, j\mathbf{k})\, e^{\mathbf{c}\cdot\mathbf{k}} + \sum_{i,j,\mathbf{k}} \big(\operatorname{grad}_{\mathbf{c}} \widetilde{\nu}_{\mathbf{c}}(i)\big)\, p(i\mathbf{0}, j\mathbf{k})\, e^{\mathbf{c}\cdot\mathbf{k}}\,,$$

and, as $\sum_{\mathcal{I}} \widetilde{\nu}_{\mathbf{c}}(i) = 1$ for all $\mathbf{c}$,

$$\operatorname{grad}\varphi_P(\mathbf{0}) = \mathfrak{m}(P) + \sum_{i\in\mathcal{I}} \big(\operatorname{grad}_{\mathbf{c}} \widetilde{\nu}_{\mathbf{c}}(i)\big) = \mathfrak{m}(P)\,.$$

The computation of the matrix Σ is analogous. We now show that Σ is positive definite. Let $\mathbf{x} \in \mathbb{R}^d \setminus \{\mathbf{0}\}$. Then

$$\mathbf{x}^t \, \Sigma \, \mathbf{x} = \sum_i \widetilde{\nu}(i) \, \mathbb{E}_{i\mathbf{0}}\big((\mathbf{x}^t Y_1)^2\big) \,,$$

which is certainly non-negative. Suppose this is $= 0$. Then Y_1 is orthogonal to $\mathbf{x}$ almost surely for every starting point $i\mathbf{0}$. Inductively, we now get the same for all Y_n. But this contradicts irreducibility.

To complete the proof of strict convexity, we now show that the same holds for the matrix of second order derivatives at any $\mathbf{c}_0 \in \mathbb{R}^d$. Fix $\mathbf{c}_0$ and consider the new transition matrix R defined by

$$r(i\mathbf{k}, j\mathbf{l}) = \frac{p(i\mathbf{k}, j\mathbf{l}) \, e^{\mathbf{c}_0 \cdot (\mathbf{l} - \mathbf{k})} \widetilde{g}_{\mathbf{c}_0}(j)}{\varphi_P(\mathbf{c}_0) \, \widetilde{g}_{\mathbf{c}_0}(i)} \,. \tag{8.21}$$

It is $\mathbb{Z}^d$-invariant, and stochastic by Lemma 8.19. Passing to $R_{\mathbf{c}}$ and $\widetilde{R}_{\mathbf{c}}$ as in (8.16), we find that the latter is given by

$$\widetilde{r}_{\mathbf{c}}(i, j) = \frac{\widetilde{p}_{\mathbf{c}_0 + \mathbf{c}}(i, j) \, \widetilde{g}_{\mathbf{c}_0}(j)}{\varphi_P(\mathbf{c}_0) \, \widetilde{g}_{\mathbf{c}_0}(i)} \,,$$

whose left Perron–Frobenius eigenvector (up to normalization) has i-th component $\widetilde{\nu}_{\mathbf{c}_0 + \mathbf{c}}(i) \, \widetilde{g}_{\mathbf{c}_0}(i)$, with corresponding eigenvalue

$$\varphi_R(\mathbf{c}) = \varphi_P(\mathbf{c}_0 + \mathbf{c}) / \varphi_P(\mathbf{c}_0) \,. \tag{8.22}$$

Therefore the matrix of second order derivatives of φ_P at $\mathbf{c}_0$ coincides with $\varphi_P(\mathbf{c}_0) \, \Sigma(R)$, which is positive definite. □

From the last proposition, we obtain the following two facts: (1) If $C > 0$ and $\{\mathbf{c} : \varphi_P(\mathbf{c}) \le C\}$ is non-empty, then this is a compact, convex set; (2) φ_P assumes its minimal value at its unique stationary point. We can now determine $\rho(P)$.

(8.23) Theorem. *If P satisfies (6.2) ($\mathbb{Z}^d$-invariance) and has finite range, then $\rho(P) = \min\{\varphi_P(\mathbf{c}) : \mathbf{c} \in \mathbb{R}^d\}$.*

Proof. Let $\mathbf{c}_0$ be the minimum and hence stationary point of φ_P. Define R as in (8.21) with respect to this $\mathbf{c}_0$. Then $R = \frac{1}{\varphi_P(\mathbf{c}_0)} D^{-1} P D$, where D is the diagonal matrix over X with entries $g_{\mathbf{c}_0}(i\mathbf{k})$, so that

$$\rho(R) = \rho(P) / \varphi_P(\mathbf{c}_0) \,.$$

On the other hand, (8.22) implies $\operatorname{grad} \varphi_R(\mathbf{0}) = \mathbf{0}$. Consequently $\mathfrak{m}(R) = 0$ by Proposition 8.20, applied to R. Now Corollary 8.15 yields $\rho(R) = 1$, so that $\rho(P) = \varphi_P(\mathbf{c}_0)$. □

For the following exercise, we remark that Proposition 8.20 remains valid without assuming finite range or a moment condition. The only difference is that φ_P will be finite not on the whole of $\mathbb{R}^d$, but on a convex subset containing the origin, and infinitely differentiable in its interior (when the latter is non-empty).

(8.24) Exercise. *In Theorem 8.23, the finite range assumption may be dropped.*

(1) Define $p^\varepsilon(i\mathbf{k}, j\mathbf{l}) = p(i\mathbf{k}, j\mathbf{l})\, e^{-\varepsilon|\mathbf{l}-\mathbf{k}|^2}$. Check that, although P^ε is not stochastic, φ_{P^ε} is in $C^\infty(\mathbb{R}^d)$.

(2) Let $\mathbf{c}_\varepsilon$ be the point where φ_{P^ε} is minimal. Then Theorem 8.23 remains true for P^ε, that is, $\rho(P^\varepsilon) = \varphi_{P^\varepsilon}(\mathbf{c}_\varepsilon)$.

(3) Since $P \geq P^\varepsilon$ elementwise, $\rho(P) \geq \rho(P^\varepsilon)$.

(4) On the other hand, $\rho(P) \leq \varphi_P(\mathbf{c})$ for any $\mathbf{c} \in \mathbb{R}^d$ by Lemma 7.2. Thus, it will be sufficient to show that $\liminf_{\varepsilon\to 0} \varphi_{P^\varepsilon}(\mathbf{c}_\varepsilon) \geq \varphi_P(\mathbf{c}_0)$ for some $\mathbf{c}_0 \in \mathbb{R}^d$; this is then necessarily the minimum of φ_P, and the "$\geq$" must be an "$=$".

To prove the inequality, consider the normalized left Perron–Frobenius eigenvector $\widetilde{\nu}^\varepsilon$ associated as in (8.18) with the matrix $\widetilde{P}^\varepsilon_{\mathbf{c}_\varepsilon}$. Choose a sequence $\varepsilon_n \to 0$ that realizes the lim inf, such that $\mathbf{c}_{\varepsilon_n} \to \mathbf{c}_0$ (note that $\mathbf{c}_\varepsilon$ lies in the compact set $\{\mathbf{c} : \varphi_{P^{1/2}}(\mathbf{c}) \leq 1\}$ when $\varepsilon \leq 1/2$), and also such that $\widetilde{\nu}^{\varepsilon_n}$ tends to a limit probability $\widetilde{\nu}$ on $\mathcal{I}$. Then show that

$$\widetilde{\nu}(j) \lim_{n\to\infty} \varphi_{P^{\varepsilon_n}}(\mathbf{c}_{\varepsilon_n}) \geq \sum_{i\in\mathcal{I}} \widetilde{\nu}(j)\, \widetilde{p}_{\mathbf{c}_0}(i,j) \quad \text{for all } j\,,$$

whence $\lim_{n\to\infty} \varphi_{P^{\varepsilon_n}}(\mathbf{c}_{\varepsilon_n}) \geq \rho(\widetilde{P}_{\mathbf{c}_0})$. This will conclude the proof. Compare with Stone [310], where random walks on $\mathbb{Z}^d$ are considered. □

We now give a simple example in order to illustrate the use of Theorem 8.23.

(8.25) Example. Consider the grid $\mathbb{Z}^2$. Choose $p_0, p_1 \in (0\,,\,1)$ and set $q_i = 1 - p_i$. Consider the following nearest neighbour random walk on the grid: in a point whose coordinates have even (odd) sum, the probabilities to go one step East or North are $p_0/2$ $(p_1/2)$, respectively, and the probabilities to go West or South are $q_0/2$ $(q_1/2)$, respectively; see Figure 8.

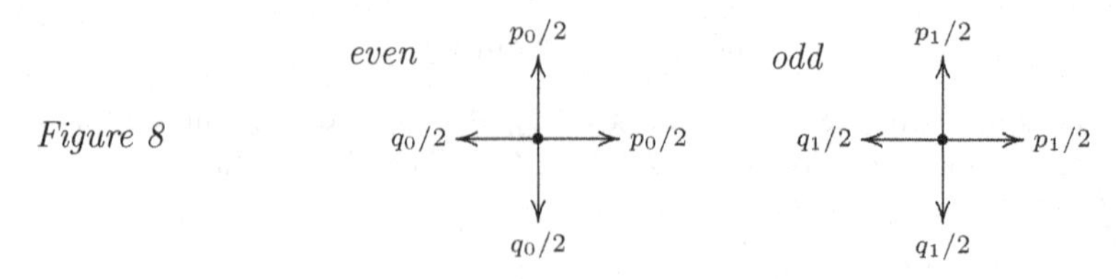

Figure 8

P is $\mathbb{Z}^2$-invariant with two orbits. Before applying Theorem 8.23, we have to introduce the right "coordinates". We set $\mathcal{I} = \{0, 1\}$. With the point $(m_1, m_2) \in \mathbb{Z}^2$, we associate the pair $i\mathbf{k} \in \mathcal{I} \times \mathbb{Z}^2$, where $i = 0$, $\mathbf{k} = \left(\frac{m_1+m_2}{2}, \frac{m_1-m_2}{2}\right)$ if $m_1 + m_2$ is even, and $i = 1$, $\mathbf{k} = \left(\frac{m_1+m_2-1}{2}, \frac{m_1-m_2-1}{2}\right)$ if $m_1 + m_2$ is odd. We then have

$$\begin{aligned} p(0\mathbf{k}, 1\mathbf{k}) &= p\big(0\mathbf{k}, 1(\mathbf{k} - \mathbf{e}_2)\big) = p_0/2\,, \\ p\big(0\mathbf{k}, 1(\mathbf{k} - \mathbf{e}_1)\big) &= p\big(0\mathbf{k}, 1(\mathbf{k} - \mathbf{e}_1 - \mathbf{e}_2)\big) = q_0/2\,, \\ p\big(1\mathbf{k}, 0(\mathbf{k} + \mathbf{e}_1)\big) &= p\big(1\mathbf{k}, 0(\mathbf{k} + \mathbf{e}_1 + \mathbf{e}_2)\big) = p_1/2\,, \\ p(1\mathbf{k}, 0\mathbf{k}) &= p\big(1\mathbf{k}, 1(\mathbf{k} + \mathbf{e}_2)\big) = q_1/2\,. \end{aligned}$$

We now compute $\widetilde{P}_{\mathbf{c}}$, where $\mathbf{c} = (c_1, c_2)$:

$$\widetilde{P}_{\mathbf{c}} = \frac{1}{2}\begin{pmatrix} 0 & (p_0 + q_0 e^{-c_1})(1 + e^{-c_2}) \\ (p_1 e^{c_1} + q_1)(1 + e^{c_2}) & 0 \end{pmatrix}.$$

Thus $\varphi_P(\mathbf{c})^2 = (p_0 + q_0 e^{-c_1})(p_1 e^{c_1} + q_1)(1 + e^{-c_2})(1 + e^{c_2})$, which attains its minimum for $c_1 = \log\sqrt{\frac{q_0 q_1}{p_0 p_1}}$ and $c_2 = 0$. We get $\rho(P) = \sqrt{p_0 q_1} + \sqrt{p_1 q_0}$. □

(8.26) Exercise. *A finite-range quasi-transitive random walk on a generalized lattice is ρ-recurrent if and only if the dimension is 1 or 2.*

(The "only if" is Theorem 7.8.)

9. Computing the Green function

In this section we use computations with generating functions, involving the Green function (1.6), for finding the spectral radius and some of its properties. A prototype of this way of reasoning is Lemma 1.24 in the introductory section. Later on, in Chapter III, we shall extend the methods developed here in order to determine the asymptotic behaviour of transition probabilities.

A. Singularities of the Green function

For our irreducible Markov chain (X, P), the Green function $G(x, y|z)$ is a power series with non-negative coefficients and radius of convergence $\mathtt{r}(P) = 1/\rho(P)$. By Pringsheim's theorem (Hille [173], p. 133), $\mathtt{r}(P)$ is a singularity – the smallest positive singularity – of $G(x, y|z)$. We shall now look for further singularities on the circle of convergence $|z| = \mathtt{r}(P)$, or, conversely, we shall study where G can be analytically extended beyond the circle of convergence.

Recall the definition (§1.A) of the period $\mathtt{d} = \mathtt{d}(P)$ of P. We say that (X, P) is *strongly periodic,* if for some integer n_0

$$\inf\{p^{(n\mathtt{d})}(x, x) : x \in X\} > 0 \quad \text{for all } n \geq n_0\,. \tag{9.1}$$

If this holds with $\mathtt{d} = 1$, then (X, P) is called *strongly aperiodic.*

Typical examples where (9.1) holds are random walks on groups: if μ is the law of the random walk, then $p^{(n)}(x,x) = \mu^{(n)}(o)$ is independent of x.

The simple random walk on a bipartite graph with bounded geometry is strongly periodic with period 2. If the graph is not bipartite, then it has an odd cycle but the simple random walk does not necessarily have to be strongly aperiodic; see Example 9.6 below. The simple random walk will be strongly aperiodic if and only if odd cycles with bounded length are nicely distributed in the graph, or equivalently, if there is an odd ℓ such that every vertex is contained in a closed (not necessarily simple) path with length ℓ.

In the following lemma, we do not suppose that the transition matrices P and Q are irreducible. We set $\rho_{x,y}(P) = \limsup_n p^{(n)}(x,y)^{1/n}$.

(9.2) Lemma. *If $P = cI + (1-c)Q$, where $0 < c < 1$, and $x, y \in X$ are such that $G_Q(x,y) > 0$, then $\rho_{x,y}(P) = c + (1-c)\rho_{x,y}(Q)$, and for $|z| < 1/\rho_{x,y}(P)$,*

$$G_P(x,y|z) = \frac{1}{1-cz}\, G_Q\Big(x,y\Big|\frac{(1-c)z}{1-cz}\Big).$$

Proof. We have

$$p^{(n)}(x,y) = \sum_{k=0}^{n} \binom{n}{k} c^{n-k}(1-c)^k q^{(k)}(x,y).$$

Assuming $q^{(k_0)}(x,y) > 0$, we get $p^{(n)}(x,y) \ge c^{n-k_o}(1-c)^{k_o} q^{(k_0)}(x,y)$ for $n \ge k_0$, so that $\rho_{x,y}(P) \ge c$. For real $t \in [0\,,\,1/c)$,

$$\text{(9.3)} \qquad \sum_{n=0}^{\infty} p^{(n)}(x,y)\, t^n = \sum_{k=0}^{\infty} q^{(k)}(x,y)\Big(\frac{1-c}{c}\Big)^k \sum_{n=k}^{\infty} \binom{n}{k} (ct)^n$$
$$= \frac{1}{1-ct} \sum_{k=0}^{\infty} q^{(k)}(x,y)\Big(\frac{(1-c)t}{1-ct}\Big)^k,$$

and all terms in the series involved are non-negative. The last series converges if $\frac{(1-c)t}{1-ct} < \frac{1}{\rho_{x,y}(Q)}$ and diverges if $\frac{(1-c)t}{1-ct} > \frac{1}{\rho_{x,y}(Q)}$. This implies the first identity of the lemma. If $z \in \mathbb{C}$, $|z| < 1/\rho_{x,y}(P)$, then this yields

$$\Big|\frac{(1-c)z}{1-cz}\Big| \le \frac{(1-c)|z|}{1-c|z|} < \frac{(1-c)\rho_{x,y}(P)^{-1}}{1-c\rho_{x,y}(P)^{-1}} = \frac{1}{\rho_{x,y}(Q)}.$$

We now see that the calculation (9.3) is valid with z in the place of t, and the proof is complete. □

(9.4) Theorem. *If (X,P) is irreducible and strongly periodic with period $\mathtt{d}$, then the singularities of $G_P(x,y|z)$ on the circle of convergence are precisely the numbers $\mathtt{r}(P)e^{2\pi i k/\mathtt{d}}$, $k = 0, \dots, \mathtt{d}-1$.*

Proof. We first show that the numbers $\mathtt{r}(P)e^{2\pi i k/\mathtt{d}}$ are indeed singularities of $G_P(x,y|z)$. Let $j \in \{0,\dots,\mathtt{d}-1\}$ be such that $p^{(k)}(x,y) > 0$ only if $k \equiv j(\mathtt{d})$; see §1.B. Then $G_P(x,y|z) = z^j\, H(z^{\mathtt{d}})$, where $H(w) = \sum_n p^{(n\mathtt{d}+j)}(x,y)\, w^n$. Now, the power series $H(w)$ has non-negative coefficients, and by Pringsheim's theorem its radius of convergence $R = \mathtt{r}(P)^{\mathtt{d}}$ is a singularity of $H(w)$. Therefore the $\mathtt{d}$-th roots of R must be singularities of $G_P(x,y|z)$.

We prove that there are no other singularities. With n_0 as in (9.1), let $m = n_0\mathtt{d}$ and $c = \inf_x p^{(m)}(x,x)$. We can write $P^m = c\,I + (1-c)Q$, where Q is a transition matrix over X. Let $x,y \in X$, and write Y_0 for the periodic class of x with respect to P; see §1.B. Now P^m is a power of an irreducible transition matrix. Therefore $\rho_{x,w}(P^m) = \rho(P)^m = c + (1-c)\rho_{x,w}(Q)$ whenever $w \in Y_0$, and we may write $\mathtt{r}(Q) = 1/\rho_{x,w}(Q)$. If $|z| < \mathtt{r}(P)$ then, using Lemma 9.2,

$$
\begin{aligned}
G_P(x,y|z) &= \sum_{n=0}^{\infty}\sum_{k=0}^{m-1} p^{(nm+k)}(x,y)\, z^{nm+k} \\
&= \sum_{w\in Y_0}\sum_{k=0}^{m-1} G_{P^m}(x,w|z^m)p^{(k)}(w,y)\, z^k \\
&= \sum_{w\in Y_0}\sum_{k=0}^{m-1} \frac{z^k}{1-cz^m} G_Q\left(x,w\Big|\frac{(1-c)z^m}{1-cz^m}\right) p^{(k)}(w,y)\,. \qquad (9.5)
\end{aligned}
$$

Consider the sets

$\mathcal{U} = \left\{z \in \mathbb{C} : \left|\frac{(1-c)z^m}{1-cz^m}\right| < \mathtt{r}(Q)\right\}$ and $\mathcal{U}_t = \left\{z \in \mathbb{C} : \left|\frac{(1-c)z^m}{1-cz^m}\right| < \frac{(1-c)t^m}{1-ct^m}\right\}$,

where $0 < t < \mathtt{r}(P)$. Then $\mathcal{U}$ is the union of the sets $\mathcal{U}_t$, and it is an open set containing $\{z \in \mathbb{C} : |z| \le \mathtt{r}(P)\} \setminus \{z \in \mathbb{C} : z^m = \mathtt{r}(P)^m\}$.

If $z \in \mathcal{U}_t$ then (subdividing into the cases $|z| \le t$ and $|z| > t$) one checks that $|z^k/(1-cz^m)| \le t^k/(1-ct^m)$ for $k = 0,\dots,m-1$. For $z = t$, all terms occurring in (9.5) are non-negative, and their sum $G_P(x,y|t)$ is finite. Thus, the series in (9.5) converges absolutely and uniformly in $\mathcal{U}_t$, and so defines an analytic function there. We conclude that (9.5) defines an analytic function in $\mathcal{U}$, whence $G_P(x,y|z)$ extends to an analytic function on $\mathcal{U}$. This reasoning shows that if $|z| = \mathtt{r}(P)$ and z is a singularity of $G_P(x,y|z)$, then it must be that $z^m = \mathtt{r}(P)^m$. We can replace m by $m+\mathtt{d} = (n_0+1)\mathtt{d}$, and it must be that $z^{m+\mathtt{d}} = \mathtt{r}(P)^{m+\mathtt{d}}$ too, so that $z^{\mathtt{d}} = \mathtt{r}(P)^{\mathtt{d}}$. □

This applies, for example, to random walks on groups and to quasi-transitive random walks on graphs. In particular, for the simple random walk on a quasi-transitive graph X, either $-\mathtt{r}(P)$ is a singularity and X is bipartite or else $-\mathtt{r}(P)$ is not a singularity of the Green function; there are no complex singularities.

For general irreducible P, we can also define the *strong period* by $\mathtt{d}_s = \gcd\{n : \inf_x p^{(n)}(x,x) > 0\}$, if the latter set is non-empty. The proof of Theorem 9.4 shows that the singularities of the Green function on the circle $|z| = \mathtt{r}(P)$ are contained in $\{\mathtt{r}(P)e^{2\pi i k/\mathtt{d}_s} : k = 0, \dots, \mathtt{d}-1\}$. Strongly periodic means that $\mathtt{d} = \mathtt{d}_s$.

(9.6) Example. Take the half-line $\mathbb{N}_0$ with edges $[j, j+1]$ and add a loop at 0, so that each vertex has degree 2. The simple random walk on this graph is aperiodic, but its strong period is 2. Using Lemmas 1.13 and 1.23, we compute

$$G(0,0|z) = \frac{2}{1 - z + \sqrt{1-z^2}}\,.$$

We have $\rho(P) = 1$, and the singularities of $G(0,0|z)$ are $z = \pm 1$. □

In conclusion, we give a class of examples which show how ρ-recurrence may arise for graphs which are not quasi-transitive.

(9.7) Connecting graphs at a common root. Let (X_j, P_j), $j \in \mathcal{I}$, be irreducible Markov chains (finite or infinite). We choose a root in each X_j, and we link the X_j by identifying all these roots while keeping the rest of the X_j disjoint. This gives a set $X = \bigcup_j X_j$ with root o, $\{o\} = \bigcap_j X_j$. We also choose constants $\alpha_j > 0$ such that $\sum_j \alpha_j = 1$ and define a transition matrix P on X by

$$(9.8) \qquad p(x,y) = \begin{cases} p_j(x,y)\,, & \text{if } x, y \in X_j\,,\ x \neq o\,,\\ \alpha_j\, p_j(o,y)\,, & \text{if } x = o\,,\ y \in X_j \setminus \{o\}\,,\\ \sum_j \alpha_j\, p_j(o,o)\,, & \text{if } x = y = o\,,\end{cases}$$

while $p(x,y) = 0$ in all other cases. When each X_j is a graph and P_j is the simple random walk on X_j, then X is the graph obtained by connecting the X_j at the common root o, and the simple random walk on X is obtained by choosing $\alpha_j = \deg_{X_j}(o)/\deg_X(o)$.

By decomposing according to the first step, it is immediate that

$$(9.9) \qquad U(o,o|z) = \sum_j \alpha_j\, U_j(o,o|z) \quad \text{and} \quad \frac{1}{G(o,o|z)} = \sum_j \frac{\alpha_j}{G_j(o,o|z)}\,,$$

where U and G refer to (X,P) and U_j and G_j to (X_j, P_j) (compare with Lemma 1.13).

Suppose that each P_j is reversible with conductances $a_j(x,y)$ $(x, y \in X_j)$ and associated invariant measure $m_j(x) = \sum_{y \in X_j} a_j(x,y)$, so that $p_j(x,y) = a_j(x,y)/m_j(x)$. Then we can consider the network over X obtained by linking the networks over the X_j, leaving the conductances as they are. (We do not admit a loop at o.) We get $m(x) = m_j(x)$, if $x \in X_j \setminus \{o\}$, and $m(o) = \sum_j m_j(o)$. Thus, the associated P on X is obtained as above with $\alpha_j = m_j(o)/m(o)$. For $z = 1$, the second formula in (9.9) now reads $\text{cap}(o) = \sum_j \text{cap}_j(o)$, or in other words, the conductance from o to ∞ in X is the sum of the conductances from o to ∞ in the X_j (the parallel law).

Returning to the general setting of (9.8), let us now determine $\rho(P)$. Let $\mathtt{s}_j$ and $\mathtt{s}$ denote the radii of convergence of the power series $U_j(o,o|z)$ and $U(o,o|z)$, respectively. Then $\mathtt{r}(P_j) \le \mathtt{s}_j$ and $\mathtt{r}(P) \le \mathtt{s}$. From (9.9), $\mathtt{s} = \min_j \mathtt{s}_j$. Recall that $G(o,o|z) = 1/\bigl(1 - U(o,o|z)\bigr)$ and that $\mathtt{r}(P)$ is the smallest positive singularity of this function. Also, for $0 \le z < \mathtt{s}$, the function U is strictly increasing. Thus, we have the following.

Case 1. $U(o,o|\mathtt{s}-) < 1$. Then $G(o,o|z)$ is analytic for $0 \le z < \mathtt{s}$, so that it must be that $\mathtt{r}(P) \ge \mathtt{s}$. Therefore $\rho(P) = \mathtt{s}^{-1}$ and (X,P) is ρ-transient.

Case 2. $U(o,o|\mathtt{s}-) = 1$. Then $\rho(P) = \mathtt{s}^{-1}$, and (X,P) is ρ-positive- or ρ-null-recurrent according to whether $U'(o,o|\mathtt{s}-)$ is finite or infinite.

Case 3. $U(o,o|\mathtt{s}-) > 1$. Then $\rho(P) = t^{-1}$, where t is the unique solution of $U(o,o|t) = 1$ in the interval $(0\,,\,\mathtt{s})$, and (X,P) is ρ-positive-recurrent.

We consider a concrete class of examples of simple random walks on trees. Let X_1 be the tree where each vertex has degree $M \ge 3$, with the exception of the root o, which has degree $M-1$. As $X_2 = X_2^{(k)}$, we choose the finite path $[0,1,\dots,k]$, where $k \ge 1$. The graph X is obtained by identifying o with 0. This is a tree with a "hair" of length k sticking out at o. The simple random walk on X is obtained from the simple random walks on the X_j by setting $\alpha_1 = \frac{M-1}{M}$ and $\alpha_2 = \frac{1}{M}$. We have

$$U_1(o,o|z) = \frac{1}{2(M-1)}\Bigl(M - \sqrt{M^2 - 4(M-1)z^2}\Bigr),$$

$\mathtt{s}_1 = \frac{M}{2\sqrt{M-1}}$ and $U_1(\mathtt{s}_1) = \frac{M}{2(M-1)}$, compare with Lemma 1.24. Using Lemmas 1.13 and 1.23, we can determine $U_2 = U_2^{(k)}$ recursively as follows. For $k \ge 2$, $U_2^{(k)}(0,0|z) = z\,F_2^{(k)}(1,0|z)$ and $F_2^{(k)}(1,0|z) = \frac{z}{2} + \frac{z}{2}F_2^{(k)}(2,1|z)F_2^{(k)}(1,0|z)$. Now $F_2^{(k)}(2,1|z) = F_2^{(k-1)}(1,0|z)$, and

$$U_2^{(k)}(0,0|z) = \frac{z^2}{2 - U_2^{(k-1)}(0,0|z)}.$$

This means that one can obtain $U_2^{(k)}$ as a (finite) continued fraction, but we shall not pursue this here. We shall only consider $k = 1,2,3$:

$$U_2^{(1)}(0,0|z) = z^2\,,\ U_2^{(2)}(0,0|z) = \frac{z^2}{2 - z^2}\,,\quad \text{and}\quad U_2^{(3)}(0,0|z) = \frac{(2-z^2)z^2}{4-3z^2}\,.$$

$\underline{k=1}$: We have $\mathtt{s}_2 = \infty$, so that $\mathtt{s} = \mathtt{s}_1$, and compute $U(\mathtt{s}) = \frac{3}{4} + \frac{1}{4(M-1)} < 1$. Therefore $\rho(P) = \frac{2\sqrt{M-1}}{M}$ and the random walk is ρ-transient.

<u>$k=2$</u>: We have $\mathbf{s}_2=\sqrt{2}$ and $U_2(\mathbf{s}_2)=\infty$. We get $\mathbf{s}=\mathbf{s}_1$ for $3\le M\le 6$ and $\mathbf{s}=\mathbf{s}_2$ for $M\ge 7$. For <u>$M=3$</u>, $U(\mathbf{s})<1$, $\rho(P)=\mathbf{s}_1^{-1}$ and the random walk is ρ-transient. For <u>$M=4$</u>, $U(\mathbf{s})=1$, $\rho(P)=\mathbf{s}_1^{-1}$, $U'(o,o|\mathbf{s}-)=\infty$, and the random walk is ρ-null-recurrent. Finally, for <u>$M>4$</u>, $U(\mathbf{s})>1$, and the random walk is ρ-positive-recurrent. In order to compute $\rho(P)$ in this last case, we have to solve $U(0,0|t)=1$ in $t>0$. Elementary computations yield the result:

$$\rho(P)=\frac{2\sqrt{M-1}}{M}\quad\text{for } M=3,4 \quad\text{and}\quad \rho(P)=\sqrt{\frac{M-1}{2(M-2)}}\quad\text{for } M\ge 5\,.$$

<u>$k=3$</u>: We have $\mathbf{s}_2=2/\sqrt{3}$ and $U_2(\mathbf{s}_2)=\infty$. Therefore $\mathbf{s}=\mathbf{s}_1$ only when $M=3$, while $\mathbf{s}=\mathbf{s}_2$ for $M\ge 4$. For all M, $U(o,o|\mathbf{s}-)>1$, and the random walk is ρ-positive-recurrent. We omit the elementary, but boring, computation of $\rho(P)$.

For <u>$k>3$</u>, one always gets $\mathbf{s}=\mathbf{s}_2$ and $U(\mathbf{s})=U_2(\mathbf{s}_2)=\infty$, and the random walk is ρ-positive with $U(o,o|\rho^{-1})=1$. □

B. A functional equation

Let (X,P,o) be an (irreducible) Markov chain together with a root $o\in X$, fixed throughout this subsection. We shall describe the function $G(z)=G(o,o|z)$ in terms of a functional equation which is going to be very useful. In the sequel, $\mathfrak{U}_a$ $(0<a\le\infty)$ will indicate an open neighbourhood of the real interval (line segment) $[0\,,\,a)$ in the complex plane. We define $\theta=\theta(P)=\mathbf{r}G(\mathbf{r})\in(\mathbf{r}\,,\,\infty]$, where $\mathbf{r}=\mathbf{r}(P)$. Of course, θ also depends on o in general (but not for random walks on groups and vertex-transitive random walks on graphs).

(9.10) Proposition. *There are sets $\mathfrak{U}_\mathbf{r}$ and $\mathfrak{U}_\theta$ and a function $\Phi(\cdot)$, analytic in $\mathfrak{U}_\theta$, such that $zG(z)\in\mathfrak{U}_\theta$ whenever $z\in\mathfrak{U}_\mathbf{r}$ and*

$$G(z)=\Phi\big(zG(z)\big)\,,\quad z\in\mathfrak{U}_\mathbf{r}\,.$$

The function $\Phi(\cdot)$ is unique up to analytic continuation.

For t in the real interval $[0\,,\,\theta)$, $\Phi(t)$ is strictly increasing and strictly convex, $\Phi(t)\le 1+\rho(P)\,t$, $\Phi'(0)=p(o,o)$, and $\Phi'(\theta-)\le\rho(P)$.

Proof. Consider the function

$$W(z)=zG(z)\,. \tag{9.11}$$

For real $z\in[0\,,\,\mathbf{r})$, it is analytic and strictly increasing, and $W'(z)\ge W'(0)=1$. Therefore the inverse function $V(t)$ satisfying $V\big(W(z)\big)=z$

($z \in \mathfrak{U}_{\mathbf{r}}$) exists, is analytic in a $\mathfrak{U}_\theta$, non-zero except for $t = 0$, and unique up to analytic continuation. We define $\Phi(t) = t/V(t)$, $\Phi(0) = 1$. This function is analytic in $\mathfrak{U}_\theta$ and satisfies the proposed equation. Uniqueness of $\Phi(\cdot)$ follows from uniqueness of $V(\cdot)$.

We have $\Phi\big(W(z)\big) = G(z)$, whence $\Phi'\big(W(z)\big) = G'(z)/W'(z)$, in particular $\Phi'(0) = G'(0) = p(o,o)$. If $t = W(z)$, $0 < z < \mathbf{r}$, then $\Phi'(t) = G'(z)/W'(z) = 1/\big(z + \frac{G(z)}{G'(z)}\big) > 0$. Using $U(z) = 1 - 1/G(z)$, where $U(z) = U(o,o|z)$ is the generating function of first return probabilities as in (1.12), one computes $\Phi''(t) = \big(G(z)/W'(z)\big)^3 U''(z) > 0$. Hence, $\Phi(t)$ is strictly increasing and strictly convex for $t \in [0\,,\theta)$. For $z \in [0\,,\mathbf{r})$, $\Phi'\big(W(z)\big) < 1/z$, and $\Phi'(\theta-) \le \rho(P)$. Convexity now yields $\Phi(t) \le \rho(P)\, t$. □

Thus, for real $z \in [0\,,\mathbf{r})$, we can describe $G(z)$ by a figure in the real (t,y)-plane ($0 \le t < \theta$): we obtain $G(z)$ as the ordinate of the point of intersection of the line $y = \frac{1}{z}t$ with the curve $y = \Phi(t)$. See Figure 10 in the proof of Theorem 9.22 below. In particular,

$$\rho(P) = \lim_{t\to\theta-} \frac{\Phi(t)}{t}\,. \tag{9.12}$$

We shall need to know where the tangent to $y = \Phi(t)$ at a point $\big(t_0, \Phi(t_0)\big)$ intersects the y-axis in this figure. This is at $y = \Psi(t_0)$, where

$$\Psi(t) = \Phi(t) - t\Phi'(t)\,. \tag{9.13}$$

If $t = W(z)$, $0 \le z < \mathbf{r}$, then one computes

$$\Psi(t) = \frac{1}{zU'(z) + 1 - U(z)} = \frac{1}{1 + \sum_{n=1}^{\infty}(n-1)\,\mathbb{P}_o[\mathbf{t}^o = n]\, z^n}\,. \tag{9.14}$$

Therefore $\Psi(t)$ is strictly decreasing and positive for $0 \le t < \theta$ (recall that $U(\mathbf{r}) \le 1$). We shall be particularly interested in the limit value $\Psi(\theta-)$. If (X,P) is ρ-recurrent then $\theta = \infty$. In the ρ-null-recurrent case, $\Psi(\theta-) = 0$. In the positive recurrent case ($\rho = 1$, $\theta = \infty$), we have $\Psi(\theta-) = \nu(o)$, where ν is the invariant probability measure for (X,P); compare with Theorem 1.18. We now give some examples, including finite graphs, which will be useful later on.

(9.15) Examples. (1) Let $\mathbb{K}_m$ be the complete graph on m vertices, that is, all pairs of distinct points are neighbours. On $\mathbb{K}_m$, we consider the simple random walk. Choose $o \in \mathbb{K}_m$. The stabilizer of o in $\mathrm{AUT}(\mathbb{K}_m)$ acts transitively on the remaining vertices, and the corresponding factor chain has two states 0 and 1 corresponding to $\{o\}$ and $\mathbb{K}_m \setminus \{o\}$, respectively. Its transition probabilities are $\widetilde{p}(0,1) = 1$, $\widetilde{p}(1,0) = \frac{1}{m-1}$ and $\widetilde{p}(1,1) = \frac{m-2}{m-1}$. We have $G(o,o|z) = \widetilde{G}(0,0|z)$. We compute

$$\Phi(t) = \frac{1}{2}\left(1 + \frac{m-2}{m-1}t + \sqrt{1 - 2\frac{m-2}{m-1}t + \left(\frac{m}{m-1}t\right)^2}\right) \quad \text{and} \quad \Psi(\theta-) = \frac{1}{m}.$$

Note that the Cayley graph of any finite group Γ with respect to $\Gamma \setminus \{o\}$ is $\mathbb{K}_m$, where $m = |\Gamma|$.

(2) The cycle of length ℓ can be seen as the Cayley graph of the cyclic group $\mathbb{Z}_\ell$ of order ℓ with respect to the generator and its inverse. We write $\mathbb{Z}_\ell$ for this graph as well. For the simple random walk on $\mathbb{Z}_\ell$ we cannot calculate $\Phi(t)$ explicitly. With some standard harmonic analysis, one computes

$$G(o,o|z) = \frac{1}{\ell}\sum_{j=0}^{\ell-1} \frac{1}{1 - z\cos(2\pi j/\ell)} \quad \text{and} \quad W\left(\frac{1}{\cosh(s)}\right) = \frac{1+\cosh(\ell s)}{\sinh(\ell s)\sinh(s)},$$

as $\ell\, W(\frac{1}{z}) = T_\ell'(t)/\bigl(T_\ell(t) - 1\bigr)$, where T_ℓ is the ℓ-th Chebyshev polynomial (and using $\cosh s = \cos is$). Once more, $\theta = \infty$ and $\Psi(\theta-) = 1/\ell$. Indeed, for any random walk on a finite group Γ, the invariant probability measure is equidistribution on Γ.

(3) For the simple random walk on $\mathbb{Z}^d$, we also cannot compute $\Phi(t)$ explicitly unless $d = 1$. We have $\rho(P) = 1$ and, via Fourier transformation,

$$W(z) = \frac{1}{(2\pi)^d}\int_{(-\pi\,,\,\pi]^d} \frac{z\,d}{d - z\sum_{k=1}^d \cos s_k}\, d\mathbf{s}\,,$$

where $\mathbf{s} = (s_1, \dots, s_d)$. For $d = 1, 2$, the random walk is recurrent, $\theta = \infty$ and $\Psi(\theta-) = 0$. For $d \ge 3$, $\theta < \infty$. We know that $p^{(2n)}(\mathbf{0},\mathbf{0}) \sim C_d\, n^{-d/2}$ (a complete computation will be done in Chapter III). Therefore, for $d = 3, 4$, $U'(1) = G'(1)/G(1)^2 = \infty$ and $\Psi(\theta-) = 0$. Cartwright [56] has used Fourier analysis and numerical integration to compute some values of θ (for $d \ge 3$) and $\Psi(\theta-)$ (for $d \ge 5$):

d	1	2	3	4	5	6	7	8	9
θ	∞	∞	1.517	1.239	1.156	1.117	1.094	1.079	1.067
$\Psi(\theta-)$	0	0	0	0	0.691	0.824	0.876	0.903	0.920

In [56], he also shows that $\Psi(\theta-) \to 1$ as $d \to \infty$.

(4) Consider the random walk on the additive group $\mathbb{Z}$ whose law is $\mu = p\cdot\delta_1 + q\cdot\delta_{-1}$, where $p + q = 1$. Then $\theta = \infty$,

$$G(z) = \frac{1}{\sqrt{1 - 4pqz^2}}\,, \quad \Phi(t) = \sqrt{1 + 4pqt^2} \quad \text{and} \quad \Psi(\theta-) = 0\,. \qquad \square$$

In the next subsection, we shall see how $\Phi(\cdot)$ may become useful for computing $\rho(P)$, while in Chapter III it will be useful for studying the asymptotic behaviour of transition probabilities.

C. Free products

Let (X_i, o_i), $i \in \mathcal{I}$, be a family of sets with roots $o_i \in X_i$. Here we shall assume that $\mathcal{I}$ is finite, but most of what we are going to do remains valid for countable $\mathcal{I}$. We construct the *free product* $(X, o) = \mathop{*}\limits_{i \in \mathcal{I}} (X_i, o_i)$, as follows. We identify all the o_i with o, as in (9.7), and write $X_i' = X_i \setminus \{o_i\}$; if $x \in X_i'$ then we write $\mathfrak{i}(x) = i$. We think of X as all "words" with "letters" from the X_i', such that no two successive letters come from the same X_i'. Thus,

$$X = \{x_1 x_2 \cdots x_n : n \geq 0\,,\ x_j \in \textstyle\bigcup_i X_i'\,,\ \mathfrak{i}(x_j) \neq \mathfrak{i}(x_{j-1})\}\,; \tag{9.16}$$

for $n = 0$, we intend the empty word o. In particular, $X_i \subset X$. Write $X_i^{\top} = \{x_1 \cdots x_n \text{ as in (9.16)} : \mathfrak{i}(x_n) \neq i\} \cup \{o\}$ and $X_i^{\perp} = \{x_1 \cdots x_n : n \geq 1\,, \mathfrak{i}(x_1) \neq i\}$. If $u \in X_i^{\top}$ and $x \in X_i$, then ux stands for their concatenation as words in X, in particular $uo = u$.

If the X_i carry a *graph* structure (without loops, for simplicity), then neighbourhood in X is given as follows: if $x, y \in X_i$ are neighbours in X_i then $ux \sim uy$ in X for all $u \in X_i^{\top}$. Thus, X looks like an infinite "cactus" whose leaves are copies of the X_i. At the root o, the X_i are joined by their respective roots, as in (9.7). At each other point of X_i, we attach copies of all the X_j, $j \neq i$, by their roots. At each of the new points $x_i x_j$ we then attach copies of the X_k, $k \neq j$, and so on (inductively). Thus, the whole "branch" attached at any $x \in X_i'$ is isomorphic with $X_i^{\perp}$. If the X_i are vertex-transitive then so is X, and one does not have to specify the roots. This does not remain true with "quasi-transitive" in place of "vertex-transitive".

If the $X_i = \Gamma_i$ are *groups*, then we always choose for o_i the respective group identities. The free product $\mathop{*}\limits_{i \in \mathcal{I}} \Gamma_i$ carries a natural group structure: the identity element is o, and the product is concatenation with possible cancellation in the middle to reach the normal form as in (9.16). The Γ_i are subgroups of Γ. If X_i is the Cayley graph of Γ_i with respect to S_i, then X is the Cayley graph of Γ with respect to $\bigcup_{i \in \mathcal{I}} S_i$.

The homogeneous tree $\mathbb{T}_M$ is the free product of M copies of $\mathbb{K}_2$ (the two-element graph); it is the Cayley graph of the group $\mathbb{Z}_2 * \cdots * \mathbb{Z}_2$ (M times). Also, $\mathbb{T}_{2M}$ is the free product of M copies of $\mathbb{Z}$ and the Cayley graph of the free group $\mathbb{F}_M = \mathbb{Z} * \cdots * \mathbb{Z}$ (M times). Figure 9 shows a piece of the graph $\mathbb{K}_2 * \mathbb{K}_2 * \mathbb{Z}_4$, which is the Cayley graph of the group $\Gamma = \langle a_1, a_2, b \mid a_1^2 = a_2^2 = b^4 = o \rangle$ with respect to $S = \{a_1, a_2, b, b^{-1}\}$.

Starting from irreducible transition matrices P_i over X_i, $i \in \mathcal{I}$, we now want to define their free "sum" (more precisely, convex combination) on X. We lift P_i to a (non-irreducible) transition matrix $\overline{P}_i$ over X: if $u \in X_i^{\top}$ and $v, w \in X_i$ then $\bar{p}_i(uv, uw) = p_i(v, w)$, while $\bar{p}_i(x, y) = 0$ in all other

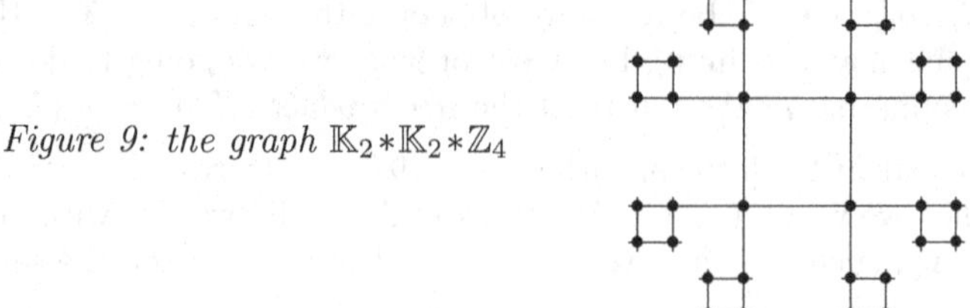

Figure 9: the graph $\mathbb{K}_2 * \mathbb{K}_2 * \mathbb{Z}_4$

cases. Now choose $\alpha_i > 0$ such that $\sum_i \alpha_i = 1$ and define

$$P = \sum_{i \in \mathcal{I}} \alpha_i \overline{P}_i \,. \tag{9.17}$$

This defines an irreducible Markov chain over X which is adapted to the free product structure and arises naturally from the P_i. It is for chains of this type that the function Φ will reveal its usefulness.

If the X_i are *regular* connected graphs with degrees M_i, then X is regular with degree $M = \sum_i M_i$, and the simple random walk on X arises from the simple random walks on the X_i as in (9.17) by choosing $\alpha_i = M_i/M$. Let the $X_i = \Gamma_i$ be groups and each P_i a random walk arising from a probability measure μ_i. Then $\overline{P}_i$ arises from the same μ_i, seen as a probability measure on the free product Γ with $\mu_i(\Gamma \setminus \Gamma_i) = 0$. Thus we obtain a free-sum-type random walk on Γ as in (9.17) with law $\mu = \sum_i \alpha_i \cdot \mu_i$.

In the setting of (9.17), we now want to describe $G(z) = G_P(o,o|z)$ and $\Phi(t) = \Phi_P(t)$ in terms of $G_i(z) = G_{P_i}(o_i,o_i|z)$ and $\Phi_i(t) = \Phi_{P_i}(t)$. We also write $U(z) = U_P(o,o|z)$, $U_i(z) = U_{P_i}(o_i,o_i|z)$, $F_i(x,y|z) = F_{P_i}(x,y|z)$, $\Psi(t) = \Psi_P(t)$ and $\Psi_i(t) = \Psi_{P_i}(t)$. It will be useful to think of the random walk $(Z_n)_{n\ge 0}$ given by P as follows: at each step, we first select a random $i \in \mathcal{I}$ subject to the probability distribution $(\alpha_i)_{i\in\mathcal{I}}$ and then choose the next state according to $\overline{P}_i$. We write $\mathfrak{i}(Z_n) = i$ for the element of $\mathcal{I}$ chosen at step n. Now let $\mathbf{t}^o$ be the "first return" stopping time at o in X, as defined in (1.12), and consider

$$H_i(z) = \sum_{n=1}^{\infty} \mathbb{P}_o[\mathbf{t}^o = n\,,\ \mathfrak{i}(Z_1) = i]\, z^n \quad \text{and} \quad \zeta_i(z) = \frac{\alpha_i z}{1 - U(z) + H_i(z)}\,.$$

(9.18) Proposition. *If $|z| < \mathbf{r}(P)$ then $|\zeta_i(z)| < \mathbf{r}(P_i)$,*

(a) $U(z) = \sum_{i\in\mathcal{I}} H_i(z)$, (b) $\dfrac{H_i(z)}{1 - U(z) + H_i(z)} = U_i\big(\zeta_i(z)\big)$, *and*

(c) *for $u, v \in X_i$,* $F(u,v|z) = F_i\big(u,v|\zeta_i(z)\big)$.

Proof. Relation (a) is obvious. The proof of (b) and (c) needs more work. We remark that we do not exclude $p(x,x) > 0$, which causes slight complications in the formulation of the proof. We start with (c). We fix $i \in \mathcal{I}$ and omit it in the notation for the stopping times

$$\mathbf{s}(0) = 0\,, \quad \mathbf{s}(k) = \min\{n > \mathbf{s}(k-1) : Z_n \in X_i \text{ and } \mathfrak{i}(Z_n) = i\}\,,$$

defined as long as they exist; the starting point is $Z_0 = u \in X_i$. Thus, $(Z_{\mathbf{s}(k)})_{k \ge 0}$ is the "shadow" in X_i of the random walk. Our key argument is the following: if at time n the random walk is at some $x \in X_i$, then it may take a "detour" into the copy of $X_i^{\perp}$ attached at x before performing the next step within X_i subject to $\alpha_i \overline{P}_i$. By the free product structure, the walk has to return to x before performing this next step. That is,

$$Z_{\mathbf{s}(k)-1} = Z_{\mathbf{s}(k-1)} \quad \text{if} \quad \mathbf{s}(k) < \infty\,.$$

As the transition probabilities within the copy of $X_i^{\perp}$ attached at x are the same for each $x \in X_i$, we see that the increments $\mathbf{s}(k) - \mathbf{s}(k-1)$ are i.i.d. Also, $\mathbb{P}_u[Z_n = u\,,\ \mathbf{s}(1) > n]$ is the probability of returning to $u \in X_i$ at time n without allowing steps within X_i that are subject to $\alpha_i \overline{P}_i$. The generating function associated with *first* return probabilities to x of this type, in analogy with (1.12), is $U(z) - H_i(z)$. Therefore, as in Lemma 1.13(a),

$$\sum_{n=0}^{\infty} z^n\, \mathbb{P}_u[Z_n = u\,,\ \mathbf{s}(1) > n] = \frac{1}{1 - U(z) + H_i(z)}\,.$$

Now let $n \ge 1$ and $x_1, \dots, x_n \in X_i$. Write

$$\mathtt{w}(x_1, \dots, x_n) = \mathbb{E}_u\bigl(z^{\mathbf{s}(n)} \mathbf{1}_{[Z_{\mathbf{s}(k)} = x_k\,,\ k=1,\dots,n]}\bigr)\,.$$

Then the above considerations show that $[\mathbf{s}(1) = n+1\,,\ Z_{n+1} = x_1] = [Z_{n+1} = x_1\,,\ \mathfrak{i}(Z_{n+1}) = i\,,\ Z_n = u\,,\ \mathbf{s}(1) > n]$, and

$$\begin{aligned}
\mathtt{w}(x_1) &= \sum_{n=0}^{\infty} z^{n+1}\, \mathbb{P}_u[\mathbf{s}(1) = n+1\,,\ Z_{n+1} = x_1] \\
&= \alpha_i z\, p(u, x_1) \sum_{n=0}^{\infty} z^n\, \mathbb{P}_u[Z_n = u\,,\ \mathbf{s}(1) > n] = \zeta_i(z)\, p_i(u, x_1)\,,
\end{aligned}$$

and inductively

$$\begin{aligned}
\mathtt{w}(x_1, \dots, x_n) &= \mathbb{E}_u\bigl(z^{\mathbf{s}(n) - \mathbf{s}(n-1)} \mathbf{1}_{[Z_{\mathbf{s}(n)} = x_n]} \,|\, Z_{\mathbf{s}(n-1)} = x_{n-1}\bigr)\, \mathtt{w}(x_1, \dots, x_{n-1}) \\
&= \zeta_i(z)\, p_i(x_{n-1}, x_n)\, \mathtt{w}(x_1, \dots, x_{n-1}) = \zeta_i(z)^n \prod_{k=1}^{n} p_i(x_{k-1}, x_k)\,.
\end{aligned}$$

Now let $v \in X_i$, $v \neq u$. The time $\mathbf{s}^v$ of the first arrival of the random walk at v must coincide with some instant $\mathbf{s}(n)$, with random n. (Indeed, the last step must take place within X_i.) Therefore, writing $\mathbf{s}_i^v$ for the first arrival time at v of the Markov chain (X_i, P_i),

$$F(u,v|z) = \sum_{n=1}^{\infty} \mathbb{E}_u\big(z^{\mathbf{s}(n)}\,\mathbf{1}_{[\mathbf{s}^v=\mathbf{s}(n)]}\big) = \sum_{n=1}^{\infty} \sum_{\substack{x_1,\dots,x_n \in X_i \\ x_n=v\,,\ x_k\neq v\ (k<n)}} \mathtt{w}(x_1,\dots,x_n)$$

$$= \sum_{n=1}^{\infty} \zeta_i(z)^n\, \mathbb{P}_u[\mathbf{s}_i^v = n] = F\big(u,v|\zeta_i(z)\big)\,.$$

Statement (b) follows quite easily from (c). This time, $Z_0 = o$, and conditioning on the first step as in Lemma 1.13(c),

$$H_i(z) = \sum_{x\in X_i} \alpha_i z\, p_i(o,x)\, F(x,o|z) = \alpha_i z \sum_{x \in X_i} p_i(o,x)\, F_i\big(x,o_i|\zeta_i(z)\big)$$

$$= \frac{\alpha_i z}{\zeta_i(z)} U_i\big(\zeta_i(z)\big) = \big(1 - U(z) + H_i(z)\big) U_i\big(\zeta_i(z)\big)\,.$$

When z is real, $0 \le z < \mathbf{r}(P)$, all the above are manipulations with series with non-negative terms. Hence (a), (b) and (c) are true for such z and both $\zeta_i(z)$ and $U_i\big(\zeta_i(z)\big)$ must be finite and non-negative. We have $\zeta_i(0) = 0$, and this function is strictly increasing for $z < \mathbf{r}(P)$. As $G(z) = 1/\big(1 - U(z)\big) < \infty$, it must be that $U(z) < 1$, implying that $U_i\big(\zeta_i(z)\big) < 1$ which in turn yields $G_i\big(\zeta_i(z)\big) < \infty$ and $\zeta_i(z) < \mathbf{r}(P_i)$.

If $z \in \mathbb{C}$ is arbitrary, then $|U(z)| \le U(|z|)$, $|F(o,x|z)| \le F\big(o,x\big||z|\big)$ and $|\zeta_i(z)| \le \zeta_i(|z|)$, showing that all the series involved in our computations converge absolutely for $|z| < \mathbf{r}(P)$, making sure that (a), (b) and (c) hold. □

We can now deduce the main theorem of this section.

(9.19) Theorem. *For $P = \sum_{i\in\mathcal{I}} \alpha_i \overline{P}_i$ we have*

$$\Phi(t) = 1 + \sum_{i\in\mathcal{I}} \big(\Phi_i(\alpha_i t) - 1\big)\,.$$

$\Phi(t)$ is analytic for t in a $\mathfrak{U}_{\bar\theta}$, where

$$\bar\theta = \min\{\theta(P_i)/\alpha_i : i \in \mathcal{I}\} \ge \theta(P)\,.$$

Proof. Multiplying numerator and denominator in (9.18.b) by $G(z)$, we obtain $1 - 1/\big(1 + H_i(z)G(z)\big) = U_i\big(\zeta_i(z)\big)$, so that by (1.13.a)

$$1 + H_i(z)G(z) = G_i\big(\zeta_i(z)\big) = \Phi_i\Big(\zeta_i(z) G_i\big(\zeta_i(z)\big)\Big)\,.$$

Therefore

$$(9.20)\qquad \begin{cases} \zeta_i(z)G_i\big(\zeta_i(z)\big) = \zeta_i(z)\big(1 + H_i(z)G(z)\big) = \alpha_i zG(z)\,, \quad \text{and} \\ H_i(z)G(z) = \Phi_i\big(\alpha_i zG(z)\big) - 1\,. \end{cases}$$

Now (9.18.a) gives $\sum_{i\in\mathcal{I}} H_i(z)G(z) = U(z)G(z) = G(z) - 1$, so that

$$\Phi\big(zG(z)\big) - 1 = G(z) - 1 = \sum_{i\in I}\Big(\Phi_i\big(\alpha_i zG(z)\big) - 1\Big)\,.$$

All these relations are true for z in a $\mathfrak{U}_{\mathbf{r}(P)}$. In particular, setting $t = zG(z)$ for $0 \le z < \mathbf{r}(P)$ it must be that $\alpha_i t < \theta(P_i)$, so that $\theta(P) \le \bar\theta$ as asserted, thus completing the proof. □

We now explain the use of this theorem for computing $\rho(P)$. We suppose we have good knowledge of the functions $\Phi_i(\cdot)$ and of $\theta(P_i)$, $i \in \mathcal{I}$. Thus we can find $\bar\theta$ and subsequently $\theta(P)$ in the interval $(0\,,\,\bar\theta]$. The point is that in many typical cases, we will have $\bar\theta > \theta(P)$, so that $\Phi(t)$ extends analytically beyond $\theta(P)$. From Theorem 9.19 we also get

$$(9.21)\qquad \Psi(t) = 1 + \sum_{i\in\mathcal{I}}\big(\Psi_i(\alpha_i t) - 1\big)\,,\quad t \in \mathfrak{U}_{\bar\theta}\,.$$

(9.22) Theorem. *The following two cases can occur.*

(i) *If $\Psi(\bar\theta-) < 0$ then $\theta(P)$ is the unique solution in $(0\,,\,\bar\theta)$ of $\Psi(t) = 0$,*

$$\rho(P) = \min\{\Phi(t)/t : 0 < t < \bar\theta\} = \Phi'(\theta) < 1\,,$$

and (X, P) is ρ-transient.

(ii) *If $\Psi(\bar\theta-) \ge 0$ then $\theta(P) = \bar\theta$, $\rho(P) = \lim_{t\to\bar\theta-}\big(\Phi(t)/t\big)$, and (X, P) is ρ-recurrent if and only if $\bar\theta = \infty$.*

Proof. (i) Once more, we use the fact that $\mathbf{r}(P)$ is the smallest positive singularity of $G(z)$. Let τ be the unique solution in $(0\,,\,\bar\theta)$ of $\Psi(t) = 0$ and let $\mathbf{s} = \tau/\Phi(\tau)$.

Start with small $z \in (0\,,\,\mathbf{s})$. The line $y = \frac{1}{z}t$ intersects $y = \Phi(t)$ at a unique point of the first quadrant; see Figure 10. This is $\big(zG(z), G(z)\big)$. The angle of intersection is non-zero because the tangent to $y = \Phi(t)$ at this point intersects the ordinate axis at $\Psi\big(zG(z)\big) > 0$. By the theorem on implicit functions, $G(\cdot)$ is analytic at z. Increasing z, from a certain point onwards we will find two intersection points, again with non-zero angles, one to the left and the other to the right of τ. By continuity, it is the left one which determines $G(z)$. On reaching $z = \mathbf{s}$, the two points coincide, and the angle of intersection becomes 0, which means that $\mathbf{s}$ is a singularity of $G(\cdot)$. Indeed, consider the complex function

$$\mathcal{F}(z,w) = \Phi(z\,w) - w\,. \tag{9.23}$$

It is analytic in $\{(z,w) \in \mathbb{C}^2 : z\,w \in \mathfrak{U}_{\bar\theta}\}$, and

$$\mathcal{F}\big(z, G(z)\big) = 0 \quad \text{for } z \in \mathfrak{U}_{\mathbf{r}(P)}\,. \tag{9.24}$$

By the theorem on implicit functions, singularities of $G(z)$ are determined by $\mathcal{F}\big(z,G(z)\big) = 0$ and $\mathcal{F}_w\big(z,G(z)\big) = 0$, that is, $\Psi\big(zG(z)\big) = 0$, and (as we are looking for a positive singularity) $z = \mathbf{s}$.

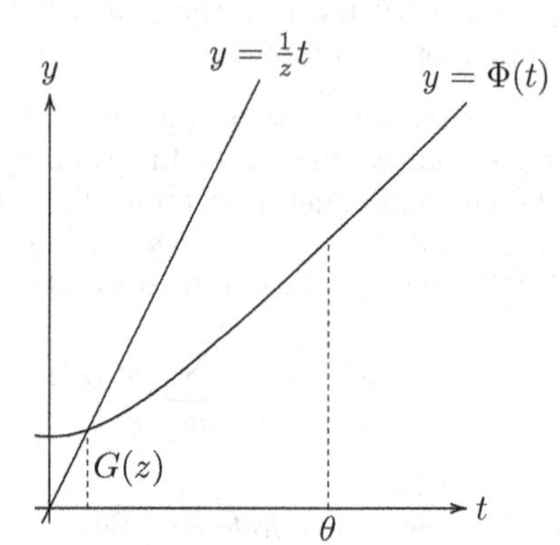

Figure 10

For $z > \mathbf{s}$, the line $y = \frac{1}{z}t$ does not intersect $y = \Phi(t)$, so that we find no real solution for $G(z)$. This shows that $\mathbf{r}(P) = \mathbf{s}$, $\theta(P) = \tau$ and $G\big(\mathbf{r}(P)\big) < \infty$, as asserted. Finally,

$$\rho(P) = \Phi'\big(\theta(P)\big) = \sum_{i\in\mathcal{I}} \alpha_i\, \Phi_i'\big(\alpha_i\theta(P)\big) < \sum_{i\in\mathcal{I}} \alpha_i\, \rho(P_i) \le 1\,.$$

(ii) The function $\Phi(t)/t$ is decreasing and $\Psi(t) > 0$ in $[0\,,\bar\theta)$. Set $\mathbf{s} = \lim_{t\to\bar\theta-}\big(t/\Phi(t)\big)$. Then for $0 < z < \mathbf{s}$, the angle of intersection of $y = \frac{1}{z}t$ with $y = \Phi(t)$ is always non-zero, giving rise to an analytic solution for $G(z)$. Therefore $\mathbf{r}(P) \ge \mathbf{s}$ and $\theta \ge \bar\theta$. Inequalities cannot be strict, and $\mathbf{r}(P) = \mathbf{s}$, as asserted. In particular, $G\big(\mathbf{r}(P)\big) = \infty$ if and only if $\bar\theta = \infty$. □

We now give various applications.

(9.25) The free product of identical pieces. An interesting special case arises when $X_i = X_0$, $o_i = o_0$ and $P_i = P_0$ all coincide and $\alpha_i = 1/M$, where $M = |\mathcal{I}|$. We write $\Phi_i = \Phi_0$, $\Psi_i = \Psi_0$ and $\theta_0 = \theta(P_0)$. Thus,

$$\bar\theta = M\theta_0\,,\quad \Phi(t) = M\Phi_0(t/M) - (M-1) \;\text{ and }\; \Psi(t) = M\Psi_0(t/M) - (M-1)\,.$$

(i) If M is sufficiently large then we will have $\Psi_0(\theta_0-) < (M-1)/M$, that is, $\Psi(\bar\theta-) < 0$. Then $\theta(P)$ is found by solving $\Psi_0(t/M) = (M-1)/M$. One may also set $t/M = W_0(z) = zG_0(z)$, where $0 < t < \bar\theta$ and $0 < z < \mathtt{r}(P_0)$, and substitute this into $\Phi(t)/t$ to obtain

$$\rho(P) = \min\Big\{\frac{1}{z} - \frac{M-1}{MzG_0(z)} : 0 < z < \mathtt{r}(P_0)\Big\}. \tag{9.26}$$

(ii) If $\Psi_0(\theta_0-) \geq (M-1)/M$ then $\theta(P) = M\theta_0$ and

$$\rho(P) = \lim_{t\to\theta_0-} \frac{M\Phi_0(t) - (M-1)}{Mt} = \rho(P_0) - \frac{M-1}{M\theta_0}. \tag{9.27}$$

When in addition (X_0, P_0) is ρ-recurrent we get $\rho(P) = \rho(P_0)$.

(1) For the simple random walk on $\mathbb{D}_{m,M} = \mathbb{K}_m * \cdots * \mathbb{K}_m$ (M times), we have $\Psi_0(\theta_0-) = 1/m$, which is equal to $(M-1)/M$ only when $M = m = 2$. In this case, we obtain $\mathbb{K}_2 * \mathbb{K}_2$, which (as a graph) is isomorphic with the two-way-infinite path $\mathbb{Z}$ (on the group level of course $\mathbb{Z} \not\cong \mathbb{Z}_2 * \mathbb{Z}_2$, the infinite dihedral group). In this case, $\rho(P) = 1$. In all other cases, one may use Example 9.15(1) to compute $\Psi_0(t)$ and solve the equation $\Psi_0(t/M) = (M-1)/M$. However, there is a simpler way of computing $G(z)$ and $\rho(P)$: the stabilizer of o in $\mathrm{AUT}(\mathbb{D}_{M,m})$ acts transitively on each sphere $S(o,k)$, and the factor chain becomes a nearest neighbour random walk on $\mathbb{N}_0$ with transition probabilities $\widetilde{p}(0,1) = 1$ and, for $k \geq 1$, $\widetilde{p}(k,k-1) = \frac{1}{M(m-1)}$, $\widetilde{p}(k,k) = \frac{m-2}{M(m-1)}$ and $\widetilde{p}(k,k+1) = \frac{M-1}{M}$. We leave the easy computations of $\widetilde{U}(0,0|z)$ and $G(z) = \widetilde{G}(0,0|z)$ as an exercise. The result is

$$\rho(P) = \frac{m-2+2\sqrt{(M-1)(m-1)}}{M(m-1)}.$$

For $m = 2$, $\mathbb{D}_{m,M} = \mathbb{T}_M$ and we recover the formula of Lemma 1.24.

(2) For the simple random walk on $\mathbb{Z}_\ell * \cdots * \mathbb{Z}_\ell$ (M times), with $\ell \geq 3$ and $M \geq 2$, we have $\Psi_0(\theta_0-) = 1/\ell < (M-1)/M$. We use (9.26) and substitute $z = 1/\cosh s$, where $0 < z < 1$ and $0 < s < \infty$, to obtain

$$\rho(P) = \min\Big\{\cosh(s) - \frac{(M-1)\sinh(\ell s)\sinh(s)}{M(1+\cosh(\ell s))} : s > 0\Big\}.$$

For $\ell = 3$, this coincides with the preceding example ($m = 3$). For fixed M and $\ell \to \infty$, $\rho(P) \to \frac{\sqrt{2M-1}}{M}$. Here are some numerical values of $\rho(P)$.

	ℓ	3	4	5	6	7	8	9
$M=2:$	$\rho(P)$	0.957	0.919	0.896	0.884	0.877	0.872	0.870
$M=3:$	$\rho(P)$	0.833	0.785	0.764	0.754	0.749	0.747	0.746

(3) An interesting example is the simple random walk on $\mathbb{Z}^d * \mathbb{Z}^d$. From Example 9.15(3) we know that $\Psi_0(\theta_0-) = 0$ for $d \le 4$, and $\rho(P) = \min\{\frac{1}{z} - \frac{1}{2W_0(z)} : 0 < z < 1\}$. For $d = 1$, we have the simple random walk on $\mathbb{T}_4$, and $\rho(P) = \sqrt{3}/2$. For $d = 2, 3, 4$, the minimum can be evaluated numerically. On the other hand, for $d \ge 5$ we have $\Psi_0(\theta_0-) > 1/2$; see the table in Example 9.15(3). In these cases, $\rho(P) = 1 - 1/2\theta_0$:

d	1	2	3	4	5	6	7	8	9
$\rho(P)$	0.866	0.728	0.646	0.596	0.568	0.553	0.543	0.537	0.532

In the same way, one can compute $\rho(P)$ for the simple random walk on $\mathbb{Z}^d * \cdots * \mathbb{Z}^d$ (M times). □

(9.28) Exercise. On the circle $\mathbb{Z}_\ell$, consider the (deterministic) walk P_0 which goes one step clockwise with probability 1. Compute $W_0(z)$. Now take the free product of M identical copies and calculate $\rho(P)$ explicitly.

(9.29) Other examples involving groups. Suppose that μ_i defines a random walk on the group Γ_i, $i \in \mathcal{I}$. Let $\Gamma = \mathop{*}_{i\in\mathcal{I}} \Gamma_i$ and $\mu = \sum_{i\in\mathcal{I}} \alpha_i \cdot \mu_i$, as explained after (9.17). *Suppose that for each $i \in \mathcal{I}$, either $\theta_i = \infty$ (ρ-recurrence) or $\Psi_i(\theta_i-) = 0$. Then $\Psi(\bar\theta-) < 0$, unless $|\mathcal{I}| = 2$ and $\Gamma_1 = \Gamma_2 = \mathbb{Z}_2$.* Indeed, if $\theta_i = \infty$ for all i then $\bar\theta = \infty$ and

$$\Psi(\bar\theta-) = 1 + \sum_i \Big(\frac{1}{|\Gamma_i|} - 1\Big) < 0$$

unless $\Gamma = \mathbb{Z}_2 * \mathbb{Z}_2$. If some θ_i is finite then there is i_0 such that $\bar\theta = \theta_{i_0}/\alpha_{i_0} < \infty$, and

$$\Psi(\bar\theta-) = \sum_{i\ne i_0} \big(\Psi(\alpha_i\bar\theta-) - 1\big) < 0\,.$$

In these cases, $\theta(P) < \bar\theta$, and $\rho(P)$ is obtained according to case (i) of Theorem 9.22.

(1) Let $\Gamma = \langle a_i, i \in \mathcal{I} \mid a_i^2 = o\rangle$, and define the random walk law μ by $\mu(a_i) = \mathsf{p}_i$. Then Γ is the free product of $M = |\mathcal{I}|$ copies of $\mathbb{Z}_2$, where the i-th copy is generated by a_i. The corresponding Cayley graph is $\mathbb{T}_M$, and $\mu = \sum_{i\in\mathcal{I}} \mathsf{p}_i \cdot \delta_{a_i}$. From Example 9.15(1) (with $m = 2$) we get $\bar\theta = \infty$ and

$$\Phi(t) = 1 + \frac{1}{2} \sum_{i\in\mathcal{I}} \Big(\sqrt{1 + 4\mathsf{p}_i^2 t^2} - 1\Big)\,.$$

If $M = 2$ then $\Psi(\bar\theta-) = 0$, $\rho(\mu) = 1$, and the random walk is recurrent. If $M \ge 3$ then $\rho(\mu) = \min\big(\Phi(t)/t\big)$.

(2) A very similar example is that of the free group $\mathbb{F}_M = \langle a_i, i \in \mathcal{I} \mid \ \rangle$, (with $M = |\mathcal{I}|$), where we write a_{-i} for a_i^{-1}. If μ is defined by $\mu(a_{\pm i}) = \mathsf{p}_{\pm i}$, then we have $\mu = \sum_{i\in\mathcal{I}} \alpha_i \cdot \mu_i$, where μ_i is supported on $\{a_{\pm i}\}$ in $\langle a_i \rangle \cong \mathbb{Z}$: $\mu_i(a_{\pm i}) = \mathsf{p}_{\pm i}/(\mathsf{p}_i + \mathsf{p}_{-i})$ and $\alpha_i = \mathsf{p}_i + \mathsf{p}_{-i}$. Using Example 9.15(4), we obtain $\bar{\theta} = \infty$,

$$\Phi(t) = 1 + \sum_{i\in\mathcal{I}} \Bigl(\sqrt{1 + 4\mathsf{p}_i\mathsf{p}_{-i}t^2} - 1\Bigr),$$

and $\rho(\mu) = \min\bigl(\Phi(t)/t\bigr)$ whenever $|\mathcal{I}| \geq 2$.

(3) Let Γ_i, $i \in \mathcal{I}$, be finite groups, and let μ_i be equidistribution on $\Gamma_i \setminus \{o_i\}$. This gives the simple random walk on $\mathbb{K}_{m_i}$, where $m_i = |\Gamma_i|$. Thus $\mu = \sum_i \alpha_i \cdot \mu_i$ describes a nearest neighbour random walk on $\mathop{*}_{i\in\mathcal{I}} \mathbb{K}_{m_i}$, which is the corresponding Cayley graph of $\mathop{*}_{i\in\mathcal{I}} \Gamma_i$. Using Example 9.15(1), we compute

$$\Phi(t) = 1 + \frac{1}{2}\sum_{i\in\mathcal{I}} \left(\sqrt{1 - 2\frac{m_i-2}{m_i-1}\alpha_i t + \Bigl(\frac{m_i}{m_i-1}\alpha_i t\Bigr)^2} + \frac{m_i-2}{m_i-1}\alpha_i t - 1\right).$$

Once more, unless $|\mathcal{I}| = 2$ and both $|\Gamma_i| = 2$, we get $\rho(\mu) = \min\bigl(\Phi(t)/t\bigr)$. In the particular case when $|\mathcal{I}| = 2$, we can solve $\Psi(t) = 0$ and obtain

$$\begin{aligned}\rho(P) =& \frac{1}{2}\Bigl(\frac{m_1-2}{m_1-1}\alpha_1 + \frac{m_2-2}{m_2-1}\alpha_2\Bigr)\\ &+ \frac{1}{2}\sqrt{\Bigl(\frac{m_1-2}{m_1-1}\alpha_1 - \frac{m_2-2}{m_2-1}\alpha_2\Bigr)^2 + 4\Bigl(\frac{\alpha_1}{\sqrt{m_1-1}} + \frac{\alpha_2}{\sqrt{m_2-1}}\Bigr)^2}.\end{aligned}$$

(4) Finally, consider $\Gamma = \mathbb{Z}_2 * \mathbb{Z}^2 = \langle a, b_1, b_2 \mid a^2 = o,\ b_1b_2 = b_2b_1 \rangle$. Let μ_1 be the point mass at a and μ_2 the law of the simple random walk on $\langle b_1, b_2 \rangle \cong \mathbb{Z}^2$. Setting $\mu = \alpha_1 \cdot \mu_1 + \alpha_2 \cdot \mu_2$, we cannot compute $\Phi(t)$ explicitly. Substituting $\alpha_2 t = W_2(z)$, $0 < z < 1$, where $W_2(z)$ is as in 9.15(3), we get

$$\Phi(t) = \frac{1}{2}\Bigl(\sqrt{1 + 4\alpha_1^2 t^2} - 1\Bigr) + \Phi_2(\alpha_2 t) = \frac{1}{2}\left(\sqrt{1 + 4\Bigl(\frac{\alpha_1}{\alpha_2}W_2(z)\Bigr)^2} - 1\right) + G_2(z),$$

so that

$$\rho(\mu) = \min\left\{\frac{\sqrt{\alpha_2^2 + 4\alpha_1^2 W_2(z)^2} - \alpha_2}{2W_2(z)} + \frac{\alpha_2}{z} : 0 < z < 1\right\}.$$

For the simple random walk, $\alpha_1 = 1/5$ and $\alpha_2 = 4/5$, and numerical approximation gives $\rho = 0.89416$. □

The graph of Figure 9 (and more general versions) will be considered in Section 11. The reader is encouraged to "invent" a variety of further examples.

(9.30) Exercise. *Recurrence and ρ-recurrence.*

(1) Show that $\rho(P) \le \sum_i \alpha_i \rho(P_i)$ when $\bar\theta = \infty$. [Hint: use the shapes of Φ and Ψ.]

(2) Show that recurrence of (X, P) implies recurrence of all (X_i, P_i) and that $\sum_{i \in \mathcal{I}'} \nu_i(o_i) \ge |\mathcal{I}| - 1$, where $\mathcal{I}'$ is the set of $i \in \mathcal{I}$ for which (X_i, P_i) is positive recurrent, and ν_i is the corresponding invariant probability measure on X_i.

(3) Show that these conditions are also sufficient for recurrence of (X, P).

(4) Characterize positive recurrence of (X, P) in these terms.

10. The spectral radius and strong isoperimetric inequalities

The main topic of this section will be the question whether $\rho(P) = 1$ or < 1, and how this is related to the underlying structure. In what follows, $\rho(P)$ will be a "true" spectral radius.

A. The spectral radius of reversible Markov chains

Before turning our attention to reversible Markov chains, we start with a lemma on non-negative matrices. Let $Q = \big(q(i,j)\big)_{i,j \in \mathcal{I}}$ be a non-negative real matrix over a finite or countable index set, not necessarily with finite row sums. We assume that Q is irreducible in the sense of (1.5). Even if some power $Q^n = \big(q^{(n)}(i,j)\big)_{i,j \in \mathcal{I}}$ has some infinite entry, the argument of Lemma 1.7 shows that $\rho(Q) = \limsup_n q^{(n)}(i,j)^{1/n}$ is independent of $i, j \in \mathcal{I}$. If $\rho(Q) < \infty$ then all Q^n are finite in each entry. Q acts on functions (vectors) $\mathcal{I} \to \mathbb{R}$ in the usual way by matrix multiplication on the left.

(10.1) Lemma. *If Q is symmetric and $\rho(Q) < \infty$ then Q acts on $\ell^2(\mathcal{I})$ as a bounded linear operator with norm $\|Q\| = \rho(Q)$.*

Proof. Denote by $(\cdot,\cdot)$ the standard inner product in $\ell^2(\mathcal{I})$. Let $f \in \ell_0^+(\mathcal{I})$ be non-zero. Then $(Q^n f, Q^n f) = (f, Q^{2n} f)$ is finite for each n, and

$$(Q^{n+1}f, Q^{n+1}f)^2 = (Q^n f, Q^{n+2} f)^2 \le (Q^n f, Q^n f)\,(Q^{n+2}f, Q^{n+2}f)\,,$$

so that the sequence $\frac{(Q^{n+1}f,Q^{n+1}f)}{(Q^n f,Q^n f)}$ is increasing. Its limit is the same as that of $(Q^n f, Q^n f)^{1/n}$. The latter must be $\rho(Q)^2$. Hence

$$\frac{(Qf,Qf)}{(f,f)} \le \frac{(Q^{n+1}f, Q^{n+1}f)}{(Q^n f, Q^n f)} \le \rho(Q)^2\,.$$

If $f \in \ell_0(\mathcal{I})$ is arbitrary, then $(Qf, Qf) \le (Q|f|, Q|f|) \le \rho(Q)^2 (f,f)$. Consequently, Q is bounded on $\ell^2(\mathcal{I})$ and $\|Q\| \le \rho(Q)$. It is clear that this must be an equality. $\square$

Now let (X, P) be a reversible Markov chain with associated invariant measure $m(\cdot)$ and conductance $a(\cdot,\cdot)$. Recall from §2.A that P acts as a self-adjoint bounded operator on $\ell^2(X,m)$. By using the isomorphism $\ell^2(X,m) \to \ell^2(X)$ which sends $f \in \ell^2(X,m)$ to the function $x \mapsto f(x)\sqrt{m(x)}$, we transform P into the symmetric operator (matrix) with entries $\sqrt{m(x)}p(x,y)/\sqrt{m(y)}$ with the same norm and spectral radius. Thus, Lemma 10.1 yields the following.

(10.2) Corollary. *The operator P on $\ell^2(X,m)$ satisfies*

$$\|P\| = \lim_{n\to\infty} \|P^n\|^{1/n} = \rho(P).$$

Next, recall from §4.A that (X,P) satisfies a strong isoperimetric inequality *IS*, if there is $\kappa > 0$ such that $m(A) \le \kappa\, a(\partial_P A)$ for every finite $A \subset X$, where $\partial_P A$ is the set of edges in $E(P)$ going out of A. Also, recall the definition (4.2) of the Sobolev norm of a function on X.

(10.3) Theorem. *The following statements are equivalent for reversible (X,P).*

(a) *(X,P) satisfies a strong isoperimetric inequality.*

(b) *(Dirichlet inequality) There is $\bar\kappa > 0$ such that*

$$\|f\|_2^2 \le \bar\kappa\, D_P(f) \quad \text{for every } f \in \ell_0(X).$$

(c) *The spectral radius $\rho(P)$ is strictly smaller than 1.*

(d) *The Green kernel defines a bounded linear operator on $\ell^2(X,m)$ by $Gf(x) = \sum_y G(x,y)f(y)$.*

Proof. (a) $\Longrightarrow$ (b). We know from Proposition 4.3 that *IS* is equivalent to the Sobolev inequality $\|f\|_1 \le \kappa\, S_P(f)$ for all $f \in \ell_0(X)$. We show that this implies the proposed inequality for 2-norms. Let $f \in \ell_0(X)$. Then

$$\begin{aligned}
\|f\|_2^4 &= \|f^2\|_1^2 \le \kappa^2\, S_P(f^2)^2 \\
&\le \kappa^2 \Big(\frac{1}{2} \sum_{x,y\in X} a(x,y)|f(x)-f(y)|\big(|f(x)|+|f(y)|\big)\Big)^2 \\
&\le \kappa^2\, D_P(f) \sum_{x,y\in X} a(x,y)\big(|f(x)|+|f(y)|\big)^2/2 \\
&\le \kappa^2\, D_P(f) \sum_{x,y\in X} a(x,y)\big(f(x)^2+f(y)^2\big) = 2\kappa^2\, D_P(f)\|f\|_2^2 .
\end{aligned}$$

(We have used Cauchy–Schwarz in the third inequality.)

(b) $\Longrightarrow$ (a). This is obvious, setting $f = \mathbf{1}_A$ for finite $A \subset X$.

(b) $\Longrightarrow$ (c). From (2.2), we get $(f, Pf) = \|f\|_2^2 - D_P(f) \le (1-\bar{\kappa})\|f\|_2^2$ for $f \in \ell_0(X)$. Hence, P being self-adjoint, its norm on $\ell^2(X,m)$ satisfies $\|P\| \le 1 - \bar{\kappa}$.

(c) $\Longrightarrow$ (b). In the same way, if $\|P\| = \rho(P) < 1$ then $D_P(f) = \|f\|^2 - (f, Pf) \ge (1-\rho(P))\|f\|_2^2$.

(c) $\Longleftrightarrow$ (d). Since $(I-P)Gf = f$ for $f \in \ell_0(X)$, we have $\|G\| < \infty$ on $\ell_2(X,m)$ if and only if $I - P$ is invertible, that is, $\|P\| < 1$. $\square$

For use in the next subsection, we remark that the implications (a) $\Longleftrightarrow$ (b) $\Longrightarrow$ (c) $\Longleftrightarrow$ (d) also remain valid when P is reversible with $a(x,y) = m(x)p(x,y) = m(y)p(y,x)$, but only *substochastic.* In this case, $m(\cdot)$ is excessive, but not necessarily invariant. The main ingredients of the above proof, that is, Proposition 4.3 and Lemma 10.1, remain valid in this setting, but (2.2) is different: one gets $\nabla^*\nabla f(x) = p(x)f(x) - Pf(x)$, where $p(x) = \sum_y p(x,y) \le 1$. Hence $(f, Pf) \le \|f\|^2 - D_P(f)$, but we do not have equality, as needed in (c) $\Longrightarrow$ (b).

B. Application to random walks on graphs

Now let X be a locally finite, connected, infinite graph. Theorem 10.3 applies to the simple random walk. We say that X itself satisfies a strong isoperimetric inequality, if *IS* holds for the simple random walk. Recall that this means existence of $\kappa > 0$ such that $|\partial A| \ge \kappa \cdot m(A)$ for every finite $A \subset X$, where ∂A is the set of edges in $E(X)$ going out of A, and $m(A)$ is the sum of the vertex degrees in A. This implies that X has exponential growth:

$$V_X(n) \ge \left(1 + \tfrac{1}{\kappa}\right)^n, \tag{10.4}$$

where κ is the constant appearing in *IS*. If X has bounded geometry, then *IS* is equivalent to the existence of $\widetilde{\kappa} > 0$ such that

$$|A| \le \widetilde{\kappa}\,|dA| \quad \text{for every finite } A \subset X, \tag{10.5}$$

where dA is the set of vertices in A having a neighbour in $X \setminus A$. A graph which does not satisfy *IS* is often called *amenable.* The converse of the statement about (10.4) is not true: there are amenable (Cayley) graphs which contain an infinite binary tree; see Rosenblatt [280]. This also shows that amenability is not inherited by subgraphs in general. On the other hand, it is follows from Theorem 4.10 that the Cartesian product of two graphs is amenable if and only if this holds for both factors.

Three classes of examples of graphs satisfying *IS* will be provided in the next subsection; groups and vertex transitive graphs will be considered in Section 12.

We now look for adaptedness properties for more general random walks (X, P) such that $\rho(P) < 1$ implies *IS* for X and conversely. In the next theorem, we do not assume that P is reversible.

(10.6) Theorem. *Let P be the transition matrix of a random walk which is uniformly irreducible, and which has an excessive measure ν satisfying $C^{-1} \le \nu(x) \le C$ for all $x \in X$ $(C > 0)$. If $\rho(P) = 1$ then X is amenable.*

Proof. Recall that the assumptions imply that X has bounded geometry. As in the proof of Theorem 3.1, $\hat{P} = \big(\frac{1}{2}(I+P)\big)^K$ has excessive measure ν and satisfies $\hat{p}(x,y) \ge \hat{\varepsilon}_0 > 0$ for every pair of neighbours x, y. If $\rho(P) = 1$ then $\rho(\hat{P}) = 1$ and consequently $\|\hat{P}\| = 1$ on $\ell^2(X,\nu)$. The adjoint $\hat{P}^*$ of $\hat{P}$, given by $\hat{p}^*(x,y) = \nu(y)\hat{p}(y,x)/\nu(x)$, is substochastic by excessiveness of the measure ν. Let $Q = \hat{P}^* P$. Then Q is substochastic and strongly reversible with respect to the excessive measure ν, and $q(x,y) > \hat{\varepsilon}_0/2$ whenever $x \sim y$. This implies $\partial_Q A \supset \partial A$ for every finite $A \subset X$ (the index Q referring to the associated network). Also, using Lemma 10.1, $\rho(Q) = \|Q\| = 1$. Now the remarks after the proof of Theorem 10.3 yield that (X, Q) does not satisfy *IS*. That is, if $\varepsilon > 0$ then there is a finite set $A = A_\varepsilon \subset X$ such that

$$a_Q(\partial_Q A) \le \varepsilon\, \nu(A)\,.$$

Now,

$$a_Q(\partial_Q A) \ge a_Q(\partial A) \ge \frac{\hat{\varepsilon}_0}{2C}|\partial A| \quad \text{and} \quad \nu(A) \le C\,|A|\,.$$

Hence the ratio $|\partial A|/|A|$ becomes arbitrarily small, when A varies over finite subsets of X, and the graph is amenable. □

More generally, the last theorem remains valid when instead of $\rho(P) = 1$, one has that $\rho_2(P) = 1$, where

$$\rho_2(P) = \rho_2(P,\nu) = \lim_n \|P^n\|^{1/n} \tag{10.7}$$

is the "true" spectral radius of P on $\ell^2(X,\nu)$. Indeed, this still implies $\rho(Q) = 1$. Note that in general (unless P is reversible – see Corollary 10.2), each of the inequalities $\rho(P) \le \rho_2(P) \le \|P\|$ may be strict. Later on we shall see an example which shows that the boundedness condition on the excessive measure cannot be dropped.

For the following, recall the definitions (1.22) of the step length distributions σ_x, $x \in X$, and that the latter are tight, if $\phi_P(n) = \sup_{x\in X} \sigma_x\big([n\,,\,\infty)\big) \to 0$ as $n \to \infty$.

(10.8) Theorem. *Suppose that (X,P) is irreducible and has an invariant measure ν satisfying $C^{-1} \le \nu(\cdot) \le C$, and that both P and its ν-adjoint P^* (that is, $p^*(x,y) = \nu(y)p(y,x)/\nu(x)$) have tight step length distributions.*

If the graph X has bounded geometry and is amenable then $\|P\| = 1$ as an operator on $\ell^2(X,\nu)$. If in addition P is strongly reversible with respect to ν, then $\rho(P) = 1$.

Proof. The transition operator $Q = \frac{1}{2}(P + P^*)$ is irreducible and reversible with invariant measure ν. Also, $\phi_Q(n) \le \big(\phi_P(n) + \phi_{P^*}(n)\big)/2$, so that Q has tight step length distributions. As operators on $\ell^2(X,\nu)$, we have (applying Corollary 10.2 to Q)

$$\rho(Q) = \|Q\| \le \tfrac{1}{2}(\|P\| + \|P^*\|) = \|P\| \le 1\,.$$

We now show that (X,Q) does not satisfy *IS*. Theorem 10.3 then yields $\rho(Q) = 1$, so that $\|P\| = 1$.

We use (10.5): given any $\varepsilon > 0$, there is some finite $A = A_\varepsilon \subset X$ such that $|\partial A| < \varepsilon\,|A|$. For $k \ge 0$, define the k-th "interior" of A

$$A_k = \{x \in A : d(x, X \setminus A) > k\}\,.$$

Let M be an upper bound on the vertex degrees in X. If $x \in A \setminus A_k$, $k \ge 1$, then $d(x, A \setminus A_1) \le k-1$. Hence

$$|A \setminus A_k| \le M^k |A \setminus A_1| \le M^k\,|\partial A| < M^k\,\varepsilon\,|A|\,.$$

For $x \in A_k$, we have (setting $\phi = \phi_Q$)

$$q(x, X \setminus A) \le \sum_{d(y,x)\ge k+1} q(x,y) \le \phi(k+1)\,.$$

Let $a(x,y) = \nu(x)q(x,y)$ be the conductance associated with Q. Then for any $k \ge 1$,

$$\begin{aligned} a(\partial_Q A) &= \sum_{x\in A} \nu(x) q(x, X\setminus A) \le \phi(k+1)\,\nu(A_k) + \nu(A\setminus A_k) \\ &\le \phi(k+1)\,\nu(A) + C\,|A\setminus A_k| < \phi(k+1)\,\nu(A) + C\,M^k \varepsilon |A| \\ &\le \big(\phi(k+1) + C^2\,M^k\,\varepsilon\big)\nu(A)\,. \end{aligned}$$

By tightness, k, ε and A (in this order) can be chosen such that the ratio $a(\partial_Q A)/\nu(A)$ becomes arbitrarily small. □

C. Examples: trees, strongly ramified graphs, and tilings

In this subsection we give three classes of examples of graphs satisfying *IS*. We start with trees.

(10.9) Theorem. *A locally finite tree with minimum degree 2 satisfies IS if and only if there is a finite upper bound on the lengths of its unbranched paths.*

By an unbranched path with length n we mean a path $[x_0, x_1, \ldots, x_n]$ such that $\deg(x_i) = 2$ for $1 \le i \le n-1$. Note that we do not assume that the tree has bounded vertex degrees.

Proof of Theorem 10.9. Write T for our tree. If $[x_0, x_1, \ldots, x_n]$ is an unbranched path in T, then set $A = \{x_1, \ldots, x_{n-1}\}$. With respect to the simple random walk, we have $m(A) = 2(n-1)$ and $a(\partial A) = |\partial A| = 2$. Thus, T is amenable when n can be arbitrarily large. (T does not have to be a tree for this argument.)

Now suppose that T has bounded unbranched paths. Apply to T the construction described in Example 3.8: every vertex x with degree $k \ge 4$ is replaced with a path $[x^{(1)}, \cdots, x^{(k-2)}]$, whose first (last) vertex is incident with the first (last) two edges originally incident with x, while each other vertex is incident with one of the remaining edges originally emanating from x. We obtain a new tree T', which has minimum degree 2, maximum degree 3 and bounded unbranched paths. Contracting each maximal unbranched path to a single edge gives rise to a rough isometry $T' \to \mathbb{T}_3$. By Lemma 1.24, $\rho(\mathbb{T}_3) < 1$, and $\mathbb{T}_3$ satisfies *IS* by Theorem 10.3. Theorem 4.7 (with $d = \infty$) now yields that T' satisfies *IS*.

The natural projection $\varphi' : T' \to T$, with $\varphi' x^{(i)} = x$, is not a rough isometry (unless T has bounded geometry). However, it is a *contraction:* each of the paths $[x^{(1)}, \ldots, x^{(k-2)}]$ is contracted to a single point. Thus, φ' is finite-to-one onto vertices and one-to-one onto edges of T (the contracted edges disappear). With this in mind, let $f \in \ell_0(T)$ and $\bar{f} = f \circ \varphi$. Then

$$S_{T'}(\bar{f}) = S_T(f) \quad \text{and} \quad \|\bar{f}\|_1 \ge \|f\|_1$$

(as $\sum_{\varphi' x' = x} \deg'(x') \ge \deg(x)$). Therefore, using Proposition 4.3, we see that together with T', T also satisfies *IS*. □

It should be clear that the above method of comparing Sobolev (and Dirichlet) norms applies more generally to contractions of arbitrary graphs which are finite-to-one onto vertices and bounded-to-one onto edges of the contracted graph. This is of course closely related to shorting of networks; see §2.B. We shall also use this method in the following result, where, however, it works only when vertex degrees are bounded.

(10.10) Theorem. *Let X be a graph with bounded geometry. Suppose that there is $r \ge 0$ such that $X \setminus B(x, r)$ has at least three connected components for each $x \in X$. Then the graph satisfies IS.*

Proof. We start by supposing $r = 0$, that is, deletion of any vertex leaves at least three components. We choose an origin o and define a map $\tau : X \to X$ as follows. We set $\tau(o) = o$, and if $d(x, o) = n + 1$, then we choose $y \sim x$ such that $d(y, o) = n$ and define $\tau(x) = y$.

Let $y \in X$, $n = d(x,y) > 0$. If we delete y, then one of the components of $X \setminus \{y\}$ contains o, and there must be at least two other components. Each of them contains a neighbour of y at distance $n+1$ from o, which must be mapped onto y under τ. Thus, $|\tau^{-1}\{y\}| \ge 2$ for every $y \in X$.

Now let $A \subset X$, and consider $A^0 = A \setminus dA$. Then $|\tau^{-1}(A^0)| \ge 2|A^0|$, and $\tau^{-1}(A^0) \subset A$ by construction. Therefore $2|A^0| \le |A|$, and $|dA| \ge |A|/2$.

Next, suppose that $r \ge 1$. We can find a subset V of X which is maximal with respect to the property that $d(v,w) \ge 2r+1$ for all $v, w \in V$, $v \ne w$. This means that adding any $x \in X \setminus V$, this property fails, and $d(x,w) \le 2r$ for some $w \in V$. Now, the balls $B(w,r)$, $w \in V$, are all pairwise disjoint. We can expand each of them to obtain a system of connected subgraphs A_w ($w \in V$) of X such that (1) $B(w,r) \subset A_w \subset B(w,2r)$, (2) the A_w are pairwise disjoint, and (3) $\bigcup_{w \in V} A_w = X$. (For each x not contained in some $B(w,r)$, one has to decide to which A_w it will go. One may proceed by induction on $d(x,V) = r+1, \dots, 2r$.) Now construct a new graph $\bar{V}$ by contracting each A_w to a single point $\bar{w}$; two vertices $\bar{v}, \bar{w}$ are neighbours if there is an edge between A_v and A_w. The contraction mapping $\varphi : X \to \bar{V}$ is obviously a rough isometry (if $d_{\bar{V}}(\varphi x, \varphi y) = 1$ then $d_X(x,y) \le 8r+1$).

By construction, $\bar{V}$ splits into at least three components when deleting any vertex. Therefore $\bar{V}$ and consequently (Theorem 4.7) also X satisfy *IS*. $\square$

The last theorem applies, in particular, to finitely generated groups and quasi-transitive graphs with infinitely many ($\Longleftrightarrow$ more than two) ends; see Section 21.

The third class of examples regards tilings in the plane. Recall what we have said in §6.C about tilings, and let $\mathcal{T}$ be a locally finite tiling of an open, simply connected set $\mathcal{O} \subset \mathbb{R}^2$, with at least three edges on each tile. Consider the associated edge graph $X = X(\mathcal{T})$. We shall also think of a tile T as the subgraph induced by its vertices, so that $x \in T$ will mean that x is a vertex, $|T|$ is the number of vertices (= number of edges) and $E(T)$ the set of edges of T.

We define the *characteristic numbers* of edges e and vertices x of X by

$$\phi(e) = 1 - \sum_{x \in e} \frac{1}{\deg(x)} - \sum_{T : E(T) \ni e} \frac{1}{|T|} \quad \text{and} \quad \psi(x) = \frac{\deg(x)}{2} - 1 - \sum_{T \ni x} \frac{1}{|T|}.$$

(10.11) Theorem. *Each of the conditions*

$$\text{(a) } \inf\{\phi(e) : e \in E(X)\} > 0, \qquad \text{(b) } \inf\{\psi(x) : x \in X\} > 0$$

implies that the edge graph of $\mathcal{T}$ satisfies IS.

Proof. As usual, we identify subsets of X with induced subgraphs. We first reduce the class of finite subgraphs A for which *IS* needs to be verified.

(1) If A is disconnected, with components $A_1, \dots, A_k$, then $\frac{a(\partial A)}{m(A)} \geq \frac{a(\partial A_1)}{m(A_1)} + \dots + \frac{a(\partial A_k)}{m(A_k)}$. Hence, it is sufficient to work with connected graphs.

(2) Now assume that A is connected. We set $\mathcal{P} = \mathcal{P}(A) = \{T \in \mathcal{T} : E(T) \subset E(A)\}$. The infinite face F_∞ of A is the closure of the component of ∞ in $\widehat{\mathbb{C}} \setminus E(A)$. We say that A is *simply connected,* if every tile $T \in \mathcal{T}$ that is not contained in F_∞ belongs to $\mathcal{P}$. Now, if A is connected, but not simply connected, then we can fill in the "holes", that is, the tiles not contained in F_∞, to obtain a new subgraph $\bar{A} \supset A$. Then $\partial\bar{A} \subset \partial A$, so that $\frac{a(\partial A)}{m(A)} \geq \frac{a(\partial \bar{A})}{m(\bar{A})}$. Therefore it is enough to consider only simply connected, finite subgraphs.

Let A be such a subgraph. We shall use Euler's formula, which says that $|A| - |E(A)| + |\mathcal{P}(A)| = 1$.

Suppose that condition (a) holds: $\phi(e) \geq \varepsilon > 0$ for all $e \in E(X)$. Recall that dA is the set of vertices in A having a neighbour outside A. For $x \in A$, we write $\deg_A(x)$ for the number of neighbours of x in A. Then $x \in dA$ if and only if $\deg_A(x) < \deg(x)$. We calculate

$$\begin{aligned}\sum_{e \in E(A)} \phi(e) &= |E(A)| - \sum_{x \in A} \frac{\deg_A(x)}{\deg(x)} - \sum_{T \in \mathcal{T}} \frac{|E(A) \cap E(T)|}{|T|} \\ &= |E(A)| - |A| - |\mathcal{P}| + \sum_{x \in dA} \left(1 - \frac{\deg_A(x)}{\deg(x)}\right) - \sum_{T \in \mathcal{T} \setminus \mathcal{P}} \frac{|E(A) \cap E(T)|}{|T|} \\ &\leq -1 + |dA| \leq a(\partial A).\end{aligned}$$

Now note that $m(A) = 2|E(A)| + a(\partial A)$. Hence,

$$a(\partial A) \geq \varepsilon \cdot |E(A)| = \frac{\varepsilon}{2}\big(m(A) - a(\partial A)\big),$$

and *IS* holds with $\kappa = \varepsilon/(\varepsilon + 2)$.

Next, assume that (b) holds: $\psi(x) \geq \varepsilon > 0$ for all $x \in X$. In the same way as above,

$$\begin{aligned}\sum_{x \in A} \psi(x) &= \sum_{x \in A} \frac{\deg(x) - \deg_A(x)}{2} + |E(A)| - |A| - |\mathcal{P}| - \sum_{T \in \mathcal{T} \setminus \mathcal{P}} \frac{|T \cap A|}{|T|} \\ &\leq \frac{1}{2} a(\partial A).\end{aligned}$$

Since $|E(T)| \geq 3$ for every $T \in \mathcal{P}$ and each $e \in E(A)$ lies on no more than two $T \in \mathcal{P}$, we have $3|\mathcal{P}| \leq 2|E(A)|$, so that Euler's formula yields $|A| \geq |E(A)|/3$,

$$a(\partial A) \geq 2\varepsilon \cdot |A| \geq \frac{2\varepsilon}{3} \cdot |E(A)| = \frac{\varepsilon}{3}\big(m(A) - a(\partial A)\big),$$

and *IS* holds with $\kappa = \varepsilon/(\varepsilon+3)$. □

From Theorem 10.11 we obtain the following simple condition. Let $k = \min\{\deg(x) : x \in X\}$ and $m = \min\{|T| : T \in \mathcal{T}\}$. We have $k, m \geq 3$. If $km > 2(k+m)$ then $X(\mathcal{T})$ satisfies *IS*. Indeed, we then have $\phi(e) \geq 1 - \frac{2}{k} - \frac{2}{m} > 0$ for every edge. Thus, if $k \geq 7$ or $m \geq 7$, then *IS* must hold. For tilings quadrilaterals, this condition becomes $k \geq 5$, and so on.

We illustrate the use of Theorem 10.11 with the tiling of the upper half plane shown in Figure 11. We have $|T| = 5$ for each tile. We get $k = 3$ and $m = 5$, so that $km < 2(k+m)$. However, for any edge e, either both endpoints have degree 4, in which case $\phi(e) = \frac{1}{10}$, or one of them has degree 4 and the other degree 3, so that $\phi(e) = \frac{1}{60}$. The lower bound in condition (a) is $\frac{1}{60}$. Also, for a vertex x, we have $\psi(x) = \frac{4}{2} - 1 - \frac{2}{5} = \frac{3}{5}$ when $\deg(x) = 4$ and $\psi(x) = \frac{1}{10}$ when $\deg(x) = 3$, so that condition (b) holds with lower bound $\frac{1}{10}$.

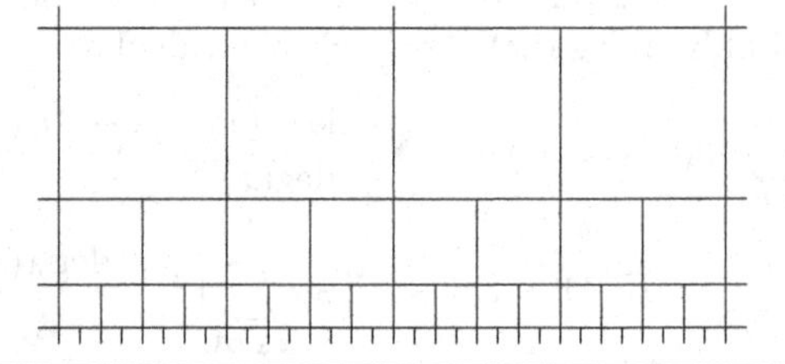

Figure 11: a tiling of the upper half plane

11. A lower bound for simple random walks

Applying Corollary 10.2, in this section we derive a lower bound for $\rho(P)$, where P is the simple random walk on a graph with bounded geometry. For vertex-transitive graphs we then use coverings and Theorem 9.22 to show that the bound is attained precisely when the graph is the homogeneous tree.

(11.1) Theorem. *Let X be a graph with degrees bounded by M. Then the simple random walk on X satisfies*

$$\rho(P) \geq \frac{2\sqrt{M-1}}{M}.$$

Note that the lower bound is just the spectral radius of the simple random walk on the homogeneous tree $\mathbb{T}_M$. Preparing for the proof of the theorem, we introduce the function $g : \mathbb{N}_0 \to \mathbb{R}$,

$$g(n) = \left(1 + \tfrac{M-2}{M}n\right)/\sqrt{M-1}^{\,n}.$$

Now let o be a reference vertex in our graph X, and define $f = f_X : X \to \mathbb{R}$ by $f(x) = g\big(d(x,o)\big)$.

We remark that f arises naturally from the tree: if $X = \mathbb{T}$ (with $\mathbb{T} = \mathbb{T}_M$) then $f_{\mathbb{T}}$ is the unique function on $\mathbb{T}$ which (i) is radial ($f_{\mathbb{T}}(x)$ depends only on $d(x,o)$), and satisfies (ii) $f_{\mathbb{T}}(o) = 1$ and (iii) $Pf_{\mathbb{T}} = \frac{2\sqrt{M-1}}{M} f_{\mathbb{T}}$, where P is the simple random walk on $\mathbb{T}$. This is a straightforward combinatorial exercise, but these facts will not be needed here. We shall meet $f_{\mathbb{T}}$ again in §19.C and §28.B as a spherical function.

(11.2) Lemma. *With respect to the simple random walk on X, the function $f = f_X$ satisfies $Pf \ge \frac{2\sqrt{M-1}}{M} f$.*

Proof. We have $Pf(o) = g(1) = \frac{2\sqrt{M-1}}{M} f(o)$. Let $x \in X$ with $d(x,o) = n \ge 1$. Let $\deg^-(x)$, $\deg^0(x)$ and $\deg^+(x)$ denote the numbers of neighbours of x at distances $n-1$, n and $n+1$ from o, respectively. Then

$$\begin{aligned}
Pf(x) &= \tfrac{1}{\deg(x)}\Big(\deg^-(x)g(n-1) + \deg^0(x)g(n) + \deg^+(x)g(n+1)\Big)\\
&\ge \tfrac{1}{\deg(x)}\Big(g(n-1) + \big(\deg(x)-1\big)g(n+1)\Big)\\
&\ge \tfrac{1}{M}\Big(g(n-1) + (M-1)g(n+1)\Big) = \tfrac{2\sqrt{M-1}}{M} g(n) = \tfrac{2\sqrt{M-1}}{M} f(x),
\end{aligned}$$

as g is decreasing in n. □

Proof of Theorem 11.1. In the graph X, let $B_n = B(o,n)$, $S_n = B_n \setminus B_{n-1}$, and write $f_n = f \cdot \mathbf{1}_{B_n}$. If $x \in B_{n-1}$ then $Pf_n(x) = Pf(x)$. If $x \in S_n$ then $Pf_n(x) = Pf(x) - \frac{\deg^+(x)}{\deg(x)} g(n+1)$. Recall that the invariant measure is given by the vertex degrees. Therefore, using Lemma 11.2,

$$\begin{aligned}
(Pf_n, f_n) &= \sum_{x\in B_n} Pf(x)\, f(x)\deg(x) - \sum_{x \in S_n} \tfrac{\deg^+(x)}{\deg(x)} g(n+1)\, f(x)\deg(x)\\
&\ge \tfrac{2\sqrt{M-1}}{M}(f_n, f_n) - (M-1)|S_n|\, g(n)\, g(n+1)\,.
\end{aligned}$$

Now observe that $(f_n, f_n) \ge \sum_{k=0}^n |S_k|\, g(k)^2$. Dividing, and using $g(n) \ge g(n+1)$, we now obtain

$$\rho(P) \ge \frac{(Pf_n, f_n)}{(f_n, f_n)} \ge \frac{2\sqrt{M-1}}{M} - (M-1)\frac{|S_n| g(n)^2}{\sum_{k=0}^n |S_k| g(k)^2}\,.$$

The proof will be completed when we show that the last ratio tends to 0 as $n \to \infty$. Now for $k \le n$,

$$|S_n|\, g(n)^2 \le |S_k|(M-1)^{n-k} g(n)^2 = \left(\frac{M+(M-2)n}{M+(M-2)k}\right)^2 |S_k|\, g(k)^2\,.$$

Therefore

$$\frac{|S_n|\,g(n)^2}{\sum_{k=0}^n |S_k|\,g(k)^2} \le \frac{\big(M+(M-2)n\big)^2}{\sum_{k=0}^n \big(M+(M-2)k\big)^2},$$

which tends to 0 as required. □

(11.3) Exercise. *Let T be a tree with minimum degree m. Then the simple random walk on X satisfies $\rho(P) \le 2\sqrt{m-1}/m$.*

[Hint: To prove this, first read about horocycles in (12.13) below. Consider the functions $g(x) = \lambda^{-\mathfrak{h}(x)}$, where $\lambda \ge 1$, and apply Lemma 7.2.]

Before turning to vertex-transitive graphs, we briefly introduce coverings. For the rest of this subsection, we only consider graphs without loops. For a graph X and a vertex $x \in X$, we write $N(x) = N_X(x) = \{v \in X : v \sim x\}$. Given two connected graphs X, Y, we say that Y *covers* X if there is a mapping $\phi : Y \to X$ such that

(i) $y \sim w$ in Y implies $\phi y \sim \phi w$ in X, and
(ii) for any $y \in Y$, the restriction of ϕ to $N_Y(y)$ is bijective onto $N_X(\phi y)$.

The *covering map* ϕ is necessarily surjective ($x \in \phi Y \implies N(x) \subset \phi Y$).

The *universal cover* of X is a tree T (it covers every cover of X) which can be constructed as follows. A *non-reversing path* in X is a path $\pi = [x_0, x_1, \dots, x_k]$ such that $x_{j+1} \ne x_{j-1}$ for all j. Choose a root $o \in X$ and define (the vertex set of) T as the set of all non-reversing paths in X starting at o, including the trivial path $[o]$. In T, two paths π, π' are neighbours if one of them extends the other by one vertex (edge). The covering map onto X is then given by $\phi[o = x_0, x_1, \dots, x_k] = x_k$.

(11.4) Lemma. *Let P_X and P_Y denote the transition matrices of the simple random walk on the graphs X and Y, respectively. If Y covers X then $\rho(P_Y) \le \rho(P_X)$.*

Proof. Let $y, w \in Y$. If $p_Y(y,w) > 0$ then the two are neighbours in Y, whence $p_X(\phi y, \phi w) = \frac{1}{\deg(\phi y)} = p_Y(y,w)$, where ϕ is the covering map. If $p_Y(y,w) = 0$ then it may still be that $\phi y \sim \phi w$ in X. In any case, $p_Y(y,w) \le p_X(\phi y, \phi w)$. By induction, and using (ii), one now proves immediately that

$$p_Y^{(n)}(y,w) \le p_X^{(n)}(\phi y, \phi w) \quad \text{for all } y, w \in Y,\ n \in \mathbb{N}.$$

Taking n-th roots and lim sup, we obtain the proposed inequality. □

We now want to show that the simple random walk on a vertex-transitive graph X with degree M which is not a tree must have spectral radius strictly larger than $\frac{2\sqrt{M-1}}{M}$. Such a graph must have a cycle of length $\ell \geq 3$. We first construct a "universal" graph with degree M and a cycle of length ℓ, which covers every vertex-transitive X with these properties. Then we shall show that the simple random walk on this universal graph satisfies $\rho > \frac{2\sqrt{M-1}}{M}$.

Given $M, \ell \geq 3$, consider the group with presentation

$$\Gamma_{M,\ell} = \langle a_1, \ldots, a_{M-2}, b \mid a_i^2 = o, b^\ell = o \rangle \tag{11.5}$$

and its "natural" set of generators $S = \{a_1, \ldots, a_{M-2}, b, b^{-1}\}$. We write $\mathbb{X}_{M,\ell}$ for the Cayley graph of $\Gamma_{M,\ell}$ with respect to S. $\Gamma_{M,\ell}$ is the free product of the $M-2$ two-element groups generated by the a_i and the cyclic group of order ℓ generated by b. The graph $\mathbb{X}_{M,\ell}$ is the Cayley graph of the group with respect to S, that is, $\mathbb{X}_{M,\ell} = \underbrace{\mathbb{K}_2 * \cdots * \mathbb{K}_2}_{M-2} * \mathbb{Z}_\ell$. It has infinitely many cycles of length ℓ whose edges are of the form $[x, xb]$ ($x \in \Gamma_{M,\ell}$), which we call the b-edges. In addition, there are the a-edges, i.e., those of the form $[x, xa_i]$. Each vertex lies on a cycle and $M-2$ a-edges. If we contract each cycle to a single point, then we obtain a homogeneous tree with degree $\ell(M-2)$. Figure 9 (in §9.C) shows a piece of $\mathbb{X}_{4,4}$.

(11.6) Theorem. *If X is a vertex-transitive graph with degree M and without loops, and the minimal length of a cycle in X is ℓ ($3 \leq \ell < \infty$), then X is covered by $\mathbb{X}_{M,\ell}$.*

Proof. Throughout this proof, "cycle" will always indicate a cycle of length ℓ. For $y, w \in \mathbb{X}_{M,\ell}$, we write $d_a(y, w)$ for the minimum number of a-edges on a path from y to w. Thus, $d_a(y, w) = 0$ if the two lie on the same cycle, $d_a(y, w) = 1$, if they lie on different cycles which are linked by an a-edge, and so on. By C_y we denote the (unique) cycle containing y. Consider the following finite subgraphs of $\mathbb{X}_{M,\ell}$ and their boundaries:

$$B_n = \{y \in \mathbb{X}_{M,\ell} : d_a(y, o) \leq n\} \quad \text{and} \quad dB_n = \{y \in B_n : y \sim X \setminus B_n\},$$

where o is the identity of $\Gamma_{M,\ell}$. We shall inductively construct a sequence of mappings $\phi_n : B_n \to X$ such that for each n,

(i′) ϕ_n preserves neighbourhood,
(ii′) for any $y \in B_n$, the restriction of ϕ_n to $N_{B_n}(y)$ is injective into $N_X(\phi y)$,
(ii″) each cycle in B_n is mapped bijectively onto a cycle in X, and
(iii′) ϕ_n coincides with ϕ_{n-1} on B_{n-1}.

Note that in (ii′), injectivity automatically becomes bijectivity, if $y \in B_n \setminus dB_n$. The direct limit ϕ of the ϕ_n (i.e., $\phi y = \phi_n y$, if $y \in B_n$) must satisfy (i) and (ii).

$\underline{n=0}$: We have that B_0 is the cycle in $\mathbb{X}_{M,\ell}$ containing o. We can choose a cycle in X and map B_0 bijectively onto that cycle, so that (i′) holds. This gives ϕ_0.

$\underline{n-1 \to n}$: On B_{n-1}, ϕ_n is defined by (iii′). $B_n \setminus B_{n-1}$ is a collection of disjoint cycles $\mathcal{C}$. Each of them is connected by precisely one edge (an a-edge) to B_{n-1}, and there are no further edges between them.

Pick $y \in dB_{n-1}$. We have the edges $[y, ya_i] = [y, w_i]$ $(i = 1, \ldots, M-2)$ with $w_i \in B_n \setminus B_{n-1}$. We have to say where ϕ_n maps the cycles $\mathcal{C}_{w_i}$. Consider $x = \phi_{n-1}y$ $(= \phi_n y)$ and the cycle $\phi_{n-1}\mathcal{C}_y$ in X. Two of the M neighbours of x lie on $\phi_{n-1}\mathcal{C}_y$. Let $v_1, \ldots, v_{M-2}$ be the remaining neighbours of x in X. None of the v_i can lie on $\phi_{n-1}\mathcal{C}_y$: otherwise, we would get an edge cutting the cycle $\phi_{n-1}\mathcal{C}_y$ into two pieces, giving rise to cycles of length $< \ell$ in X.

We claim that for each v_i, there is some cycle $\bar{\mathcal{C}}_i$ in X which contains v_i, but not x. Indeed, let $cy(x)$ denote the number of cycles in X going through x and $cy(x, v_i)$ the number of cycles going through both x and v_i. By vertex-transitivity, $cy(x) = cy(v_i)$, and $cy(x, v_i) < cy(x)$, as $\phi_{n-1}\mathcal{C}_y$ does not meet v_i. Thus, $cy(x, v_i) < cy(v_i)$, as proposed.

We now define $\phi_n w_i = v_i$, thereby mapping $\mathcal{C}_{w_i}$ onto $\bar{\mathcal{C}}_i$ in such a way that neighbourhood is preserved $(i = 1, \ldots, M-2)$.

Because of the "tree-like" structure of $\mathbb{X}_{M,\ell}$ and B_n, we can do this for each $y \in dB_{n-1}$ independently without any violation. This completes the construction of ϕ_n. □

One now has to study the spectral radius of the simple random walk on $\mathbb{X}_{M,\ell}$. By combining Lemma 11.4 with Theorem 11.6 we obtain the main result of this subsection:

(11.7) Corollary. *If X is a vertex-transitive graph with degree M and a cycle of length ℓ ($\ell \geq 3$, minimal) then, setting $g(t) = \left(\sqrt{1+t^2} - 1\right)/t$,*

$$\rho(P_X) \geq \min\left\{\frac{M-2}{M}\, g\left(\frac{1+\cosh(\ell s)}{\sinh(\ell s)\sinh(s)}\right) + \frac{2}{M}\cosh(s) : s > 0\right\}$$
$$> \frac{2\sqrt{M-1}}{M}.$$

Proof. We first show that the middle term is $\rho(P_{\mathbb{X}_{M,\ell}})$. Using the method of Section 9, we can write the function $\Phi(t)$ corresponding to the simple random walk on $\mathbb{X}_{M,\ell}$ in terms of the one corresponding to $\mathbb{Z}_2$ (Example 9.15(1) with $m = 2$) and the one corresponding to $\mathbb{Z}_\ell$. We write $\Phi_\ell(t)$ and $W_\ell(z)$ for the functions given for $\mathbb{Z}_\ell$ in Example 9.15(2). Then $\bar{\theta} = \infty$ and $\Phi(t) = \frac{M-2}{2}\left(\sqrt{1+(2t/M)^2} - 1\right) + \Phi_\ell(2t/M)$.

In $\Phi(t)/t$, we substitute $2t/M = W_\ell(1/\cosh(s))$. Passing to the minimum in $t > 0$ ($\iff s > 0$), we obtain the proposed expression for $\rho(P_{\mathbb{X}_{M,\ell}})$.

Lemma 11.4 now yields the first inequality. To see that the second inequality is strict, observe that $g(\cdot)$ is strictly increasing, while the argument is strictly decreasing with increasing ℓ for fixed s. Therefore $\rho(P_{\mathbb{X}_{M,\ell}}) > \rho(P_{\mathbb{X}_{M,\ell+1}})$, which is $\geq \rho(\mathbb{T}_M)$, because $\mathbb{T}_M$ covers each of our graphs. As a matter of fact, $\rho(\mathbb{T}_M) = \lim_{\ell\to\infty} \rho(P_{\mathbb{X}_{M,\ell}})$. □

12. The spectral radius and amenability

We shall now see how geometric (*IS*) and analytic/probabilistic (norm and spectral radius) features are related to the algebraic structure of groups.

A. Amenable groups

(12.1) Definition. A discrete group Γ is called *amenable* if there is a finitely additive probability measure defined on the family of all subsets of Γ which is invariant under left multiplication by elements of Γ.

This is the original definition, going back to von Neumann. We next list classes of countable groups which are known to be (non-)amenable.

(12.2) Theorem. (a) *Finite groups and abelian groups are amenable.*

(b) *If Γ is the union of an increasing sequence of amenable subgroups then Γ is amenable.*

(c) *Every subgroup of an amenable group is amenable.*

(d) *If Γ has an amenable subgroup with finite index then Γ is amenable.*

(e) *If $\Gamma_1 \trianglelefteq \Gamma$ then Γ is amenable if and only if both Γ_1 and Γ/Γ_1 are amenable.*

(f) *Solvable groups are amenable.*

(g) *Finitely generated groups with* subexponential growth *are amenable, that is, groups with $\lim_n V(n)^{1/n} = 1$ for the growth function with respect to some ($\iff$ every) finite symmetric set of generators.*

(h) *For every $s \geq 2$, the* free group $\mathbb{F}_s = \langle a_1, \dots, a_s \mid \cdot \rangle$ *is non-amenable.*

Within the class of discrete groups, (b) and (c) reduce the problem of amenability to the finitely generated ones, and (f) follows from (a) and (e). Finitely generated solvable groups either are nilpotent-by-finite or have exponential growth (Rosenblatt [280]). An example of a solvable group with exponential growth is $\langle a, b \mid ab^2 = ba\rangle$. Regarding (h), we remark that the Cayley graph of $\mathbb{F}_s$ with respect to $S = \{a_1, a_1^{-1}, \dots, a_s, a_s^{-1}\}$ is the tree $\mathbb{T}_{2s-1}$. From (c), it follows that every group having $\mathbb{F}_2$ as a subgroup is non-amenable. This applies to groups with *infinitely many ends* and to groups which are *hyperbolic* in the sense of Gromov and are "non-elementary" (with infinite boundary). We shall study this in detail in Chapter IV. In particular, *Fuchsian groups* (discrete groups of Möbius transformations of the unit disk) with infinite limit set contain $\mathbb{F}_2$ and are non-amenable. On the other hand, there are non-amenable groups which do not contain $\mathbb{F}_2$ – the Burnside groups $\mathbb{B}(m,n) = \langle a_1, \dots, a_m \mid x^n = o \;\forall x \rangle$ for certain m, n (Olshansky [250] and Adyan [1]).

We shall not prove Theorem 12.2, but with the exception of (b) and (e), the reader should be able to do this as an exercise at the end of this subsection. Good sources for amenability are, for example, the books by Greenleaf [145], Wagon [334] and Pier [264].

Usually, the "construction" of an invariant measure as in (12.1) requires use of the axiom of choice (the Hahn–Banach theorem), while there are various amenability criteria which do not need this. In our context, the most useful one will be the following; see [118].

(12.3) Følner's criterion. *A discrete group Γ is amenable if and only if for every $\varepsilon > 0$ and every finite $K \subset \Gamma$ there is some finite $U = U_{K,\varepsilon} \subset \Gamma$ such that $|Uy \,\triangle\, U| < \varepsilon\,|U|$ for every $y \in K$.*

Here (as usual), $UK = \{xy : x \in U\,,\ y \in K\}$, and $\triangle$ is symmetric difference. The next proposition explains *a posteriori* why a graph which does not satisfy *IS* is called amenable.

(12.4) Proposition. *A finitely generated group is amenable if and only if its Cayley graph with respect to some ($\Longleftrightarrow$ every) finite symmetric set of generators is amenable (does not satisfy IS).*

Proof. Let S be a finite, symmetric set of generators of the group Γ. In the Cayley graph $X(\Gamma, S)$, one has $\sum_{s \in S} |Us \,\triangle\, U| = 2|\partial U|$ for any finite $U \subset \Gamma$. Thus, if Følner's criterion is satisfied, then no Cayley graph of Γ satisfies *IS*.

Conversely, suppose that some Cayley graph of Γ is amenable. As all Cayley graphs of Γ with respect to finite, symmetric generating sets are metrically equivalent, Theorem 4.7 tells us that they are all amenable. Hence Følner's criterion holds for all these generating sets. As every finite $K \subset \Gamma$ is contained in some generating set, the criterion is satisfied for all K. $\square$

From Theorems 10.6 and 10.8 we now deduce the amenability criteria of Kesten and Day for finitely generated groups. If μ is a probability measure on a (so far, discrete) group Γ, then we write L_μ and R_μ for the left and right convolution operators on $\ell^2(\Gamma)$ induced by μ, that is, $L_\mu f = \mu * f$ and $R_\mu f = f * \mu$. For the transition operator $P = P_\mu$ of the right random walk induced by μ, we get $P = R_{\check{\mu}}$, where $\check{\mu}(x) = \mu(x^{-1})$. The counting measure is always P-invariant. In the irreducible case, we write $\rho(P) = \rho(\mu)$. The latter coincides with $\rho(\check{\mu})$. The respective adjoint operators are those induced by $\check{\mu}$ in place of μ.

(12.5) Corollary. *Let Γ be a finitely generated group and μ a probability measure on Γ.*

(a) *If μ defines an irreducible random walk with $\rho(\mu) = 1$ then Γ is amenable.*

(b) *If Γ is amenable and μ is irreducible and symmetric, then $\rho(\mu) = 1$.*

Proof. Part (a) follows from Theorem 10.6 and Proposition 12.4, and (b) is immediate from Theorem 10.8. □

In fact, the last corollary holds under less restrictive assumptions. The most general statement is the one which is formulated without proof in Theorem 12.7 below in the context of locally compact groups.

B. Automorphism groups and the spectral radius

We now want to study what can be said about *IS* and $\rho(P)$ in terms of the group $\mathrm{AUT}(X, P)$ or suitable subgroups, in particular when the latter act transitively or quasi-transitively. To this end, we need some facts concerning amenability of locally compact groups, which we shall state without proof.

In analogy with Definition 12.1, a locally compact group Γ is called amenable if there is a finitely additive probability measure on the family of Borel subsets of Γ which is invariant under left multiplication by group elements. Properties (a)–(g) of Theorem 12.2 remain valid on replacing "finite" with "compact" and considering only closed subgroups. In the topological version of Følner's criterion, K is compact, counting measure is replaced with right Haar measure and U has to have positive finite measure. When working with left Haar measure, as we usually do, one has to replace UK with KU.

Any finite Borel measure on Γ induces left and right convolution operators L_μ and R_μ on $L^2(\Gamma)$. The last denotes the Hilbert space of (equivalence classes of) measurable functions on Γ with integrable squares with respect to left Haar measure $|\cdot| = |\cdot|_\Gamma$. One has always

$$(12.6)\quad \|L_\mu\| \le \mu(\Gamma) \text{ and } \|R_\mu\| = \|L_{\tilde{\mu}}\|\,, \text{ where } d\tilde{\mu}(\gamma) = \sqrt{\Delta(\gamma^{-1})}\, d\check{\mu}(\gamma)\,.$$

Here, as in the discrete case, $\check{\mu}$ is the measure arising from μ by inversion: $\check{\mu}(B) = \mu(B^{-1})$. If μ has density Φ with respect to left Haar measure, then we shall also write L_Φ for L_μ and R_Φ for R_μ. Recall the formula $F * \Phi(\gamma) = \int_\Gamma F(\gamma\alpha)\Phi(\alpha^{-1})\, d\gamma$. In this setting, one has the analogues of the criteria of Kesten and Day, see Berg and Christensen [41], [42] or Derriennic and Guivarc'h [95].

(12.7) Theorem. (1) $\|L_\mu\| = \mu(\Gamma)$ *if and only if the closed subgroup of Γ generated by $(\mathrm{supp}\,\mu)(\mathrm{supp}\,\mu)^{-1}$ is amenable.* (2) *The "true" spectral radius $\rho_2(L_\mu) = \lim_n \|L_\mu^n\|^{1/n}$ satisfies $\rho_2(L_\mu) = \mu(\Gamma)$ if and only if the subgroup $\langle \mathrm{supp}\,\mu\rangle^-$ of Γ is amenable.*

We shall now apply this to reversible transitive random walks. Let (X, P) be reversible with respect to the measure m, and let Γ be a closed, transitive subgroup of $\mathrm{AUT}(X, P)$. Recall that $g(\gamma) = m(\gamma x)/m(x)$ does not depend on x, and that either $m(\cdot)$ is constant and P is symmetric, or m is unbounded. We define $\widetilde{P}$ on X with entries

$$\widetilde{p}(x,y) = \sqrt{m(x)}p(x,y)/\sqrt{m(y)} = \sqrt{p(x,y)p(y,x)}\,. \tag{12.8}$$

$\widetilde{P}$ is symmetric and Γ-invariant, $\rho(\widetilde{P}) = \rho(P)$, and $(P^n)^\sim = \widetilde{P}^n$. However, it is not necessarily a transition operator, as the row sum

$$\tau = \sum_{y\in X} \widetilde{p}(x,y) \tag{12.9}$$

(independent of x) is not always equal to 1; it may even be infinite. Of course, if $m(\cdot)$ is constant then $\widetilde{P} = P$ and $\tau = 1$.

(12.10) Theorem. *If (X, P) is reversible and $\Gamma \le \mathrm{AUT}(X, P)$ acts transitively then*

$$\rho(P) \le \sum_{y\in X} p(x,y)\sqrt{\frac{|\Gamma_y|/m(y)}{|\Gamma_x|/m(x)}} \le \tau\,.$$

The middle term is ≤ 1. The first inequality is an equality if and only if Γ is amenable, and the second inequality is an equality if and only if Γ is unimodular.

Before the proof, we need some preparation. Let $Q = \big(q(x,y)\big)$ be an arbitrary non-negative matrix over X which is Γ-invariant. The row sums of Q are constant, but not assumed to be equal to 1, and perhaps even infinite. We want to relate the action of Q on $\ell^2(X)$ (with weights equal to 1) to the action of the convolution operator $R_{\Phi[Q]}$ on $L^2(\Gamma)$, where

$$\Phi[Q](\gamma) = \frac{1}{|\Gamma_o|}q(o,\gamma^{-1}o)\,, \quad \gamma\in\Gamma\,.$$

(12.11) Lemma. (a) *If Q_1, Q_2 are two Γ-invariant operators, then $\Phi[Q_1Q_2] = \Phi[Q_2] * \Phi[Q_1]$.*

(b) *One has $\|Q\| = \|R_{\Phi[Q]}\|$ and $\rho_2(Q) = \rho_2(R_{\Phi[Q]})$, where norms and spectral radii of the operators are taken on the respective ℓ^2- and L^2-spaces.*

Proof. We may assume that $|\cdot|$ is normalized so that $|\Gamma_o| = 1$. Statement (a) is straightforward from (1.30):

$$\begin{aligned}\Phi[Q_1Q_2](\gamma) &= \sum_{x\in X} q_1(o,x)q_2(x,\gamma^{-1}o)\\ &= \int_\Gamma q_2(\alpha o,\gamma^{-1}o)q_1(o,\alpha o)\,d\alpha = \Phi[Q_2]*\Phi[Q_1](\gamma)\,.\end{aligned}$$

To prove (b), we first define two operators $S : L^2(G) \to \ell^2(X)$ and $T : \ell^2(X) \to L^2(\Gamma)$ by

$$SF(x) = \int_{\{\gamma\in\Gamma:\gamma o = x\}} F(\gamma)\,d\gamma \qquad \text{and} \qquad Tf(\gamma) = f(\gamma o)\,.$$

One immediately verifies that $T = S^*$ (the adjoint), and that $ST = I$, the identity on $\ell^2(X)$. Therefore $\|S\| = \|T\| = 1$. Using (1.30), we compute for $f \in \ell^2(X)$

$$\begin{aligned} S\,R_{\Phi[Q]}\,T\,f(x) &= \int_{\{\gamma\in\Gamma:\gamma o = x\}} Tf * \Phi[Q](\gamma)d\gamma \\ &= \int_{\{\gamma\in\Gamma:\gamma o = x\}} \int_\Gamma f(\gamma\alpha o)q(\gamma o, \gamma\alpha o)\,d\alpha\,d\gamma \\ &= \int_{\{\gamma\in\Gamma:\gamma o = x\}} \int_\Gamma f(\beta o)q(x, \beta o)\,d\beta\,d\gamma \\ &= \sum_{y\in X} \int_{\{\beta\in\Gamma:\beta o = y\}} q(x,y)f(y) = Qf(x) \end{aligned}$$

and for $F \in L^2(\Gamma)$

$$\begin{aligned} T\,Q\,S\,F(\gamma) &= \sum_{x\in X} q(\gamma o, x) \int_{\{\alpha\in\Gamma:\alpha o = x\}} F(\alpha)\,d\alpha = \int_\Gamma q(\gamma o, \alpha o)F(\alpha)\,d\alpha \\ &= \int_\Gamma F(\alpha)\Phi(\alpha^{-1}\gamma)\,d\alpha = R_{\Phi[Q]}F(\gamma)\,. \end{aligned}$$

Thus, $S\,R_{\Phi[Q]}\,T = Q$ and $T\,Q\,S = R_{\Phi[Q]}$, so that $\|R_{\Phi[Q]}\| = \|Q\|$.

Combining this with (a), we see that $\|R^n_{\Phi[Q]}\|^{1/n} = \|Q^n\|^{1/n}$, and passing to the limit, we obtain $\rho_2(R_{\Phi[Q]}) = \rho_2(Q)$. □

Proof of Theorem 12.10. The spectral radius of P acting on $\ell^2(X,m)$ is the same as the spectral radius of $\widetilde{P}$ acting on $\ell^2(X)$. By Lemmas 10.1 and 12.11, $\|P\| = \rho(P) = \rho(\widetilde{P}) = \rho_2(R_\Phi)$, where $\Phi = \Phi[\widetilde{P}]$. Now a standard computation for convolution operators (see e.g. Hewitt and Ross [172], §20) shows that norm and spectral radius of R_Φ coincide with norm and spectral radius of the left convolution operator L_Ψ, where $\Psi(\gamma) = \Phi(\gamma^{-1})\Delta(\gamma^{-1})^{1/2}$. We use (1.30) once more to compute

$$\int_\Gamma \Psi(\gamma)\,d\gamma = \sum_{x\in X} p(o,x)\sqrt{\frac{|\Gamma_x|/m(x)}{|\Gamma_o|/m(o)}}\,.$$

Now, Proposition 8.13 yields that $\operatorname{supp}\Psi = \{\gamma\in\Gamma : p(o,\gamma o) > 0\}$ generates Γ as a semigroup. Therefore, we may apply Theorem 12.7 to see that the first of the two proposed inequalities holds and is an identity if and only if Γ is amenable.

Next, use symmetry of $\widetilde{P}$, Cauchy–Schwarz and – twice – (1.30) to get

$$\left(\sum_{x\in X} p(o,x)\sqrt{\frac{|\Gamma_x|/m(x)}{|\Gamma_o|/m(o)}}\right)^2 \le \sum_{x\in X}\widetilde{p}(o,x)\sum_{x\in X}\widetilde{p}(x,o)\frac{|\Gamma_x|}{|\Gamma_o|}$$
$$= \tau\frac{1}{|\Gamma_o|}\int_\Gamma \widetilde{p}(\gamma o,o)\Delta(\gamma^{-1})\,d\gamma = \tau^2\,.$$

Equality holds if and only if there is a constant c such that $|\Gamma_x|/|\Gamma_o| = c$ for all x with $p(o,x) > 0$, that is, $\Delta(\cdot) \equiv c$ on $\operatorname{supp}\Phi = (\operatorname{supp}\Phi)^{-1}$. Thus $c = 1$, and as $\operatorname{supp}\Phi$ generates Γ, we see that equality holds if and only if Γ is unimodular. Finally, Cauchy–Schwarz and Lemma 3.25 imply that

$$\sum_{x\in X} p(o,x)\sqrt{\frac{|\Gamma_x|/m(x)}{|\Gamma_o|/m(o)}} \le \sum_{x\in X} p(o,x)\sum_{x\in X}\frac{|\Gamma_x|}{|\Gamma_o|}p(x,o) = 1\,. \qquad \square$$

We can apply this to quasi-transitive random walks.

(12.12) Corollary. *Suppose that X is a quasi-transitive graph. Then the following statements are equivalent.*

(a) *X is an amenable graph.*

(b) *Some ($\Longleftrightarrow$ every) quasi-transitive subgroup of* AUT(X) *is amenable and unimodular.*

(c) *$\rho(P) = 1$ for some ($\Longleftrightarrow$ every) strongly reversible, quasi-transitive random walk on X.*

Proof. That (a) $\Longleftrightarrow$ (c) follows from Theorems 10.6 and 10.8 To see (a) $\Longleftrightarrow$ (b), let $\Gamma \le \mathrm{AUT}(X)$ be quasi-transitive with N orbits X_i, $i \in \mathcal{I}$. Consider the graph $X_1^{(2N-1)}$ constructed in the proof of Proposition 3.9. It is roughly isometric with X, so that X is amenable if and only if $X_1^{(2N-1)}$ is amenable (Theorem 4.7). Now Γ acts on $X_1^{(2N-1)}$, and $\Gamma_1 = \bigcap_{x\in X_1}\Gamma_x$ is a compact normal subgroup of Γ. We have that $\Gamma/\Gamma_1 \le \mathrm{AUT}(X_1^{(2N-1)})$, so that $X_1^{(2N-1)}$ is an amenable graph if and only if Γ/Γ_1 is amenable and unimodular (Theorem 12.10 applied to the simple random walk). Γ_1 being a compact normal subgroup, the structure theory of locally compact groups (see Pier [264] or Paterson [255]– compare with Theorem 12.2) tells us that Γ/Γ_1 is amenable and unimodular if and only if Γ itself has these properties. $\square$

C. Some explicit computations

Besides the qualitative result of Corollary 12.12, Theorem 12.10 also contains a formula which may be used to calculate $\rho(P)$ in the case when $\mathrm{AUT}(X, P)$ contains a transitive subgroup which is amenable. Here we shall give three examples of computations by use of this method.

(12.13) Homogeneous trees. Let T be a locally finite tree, and let ϑT be its boundary, as defined in §6.B. We now select an end $\omega \in \vartheta T$. In analogy with §6.B, we define the *confluent* $x \curlywedge y$ of two vertices x, y with respect to ω: this is the first common vertex on the rays $\pi(x, \omega)$ and $\pi(y, \omega)$. We now choose and fix, as usual, a root $o \in T$ and define for $x \in T$ its *height* with respect to ω by

$$\mathfrak{h}(x) = \mathfrak{h}(x, \omega) = d(x, x \curlywedge o) - d(o, x \curlywedge o) \in \mathbb{Z}\,.$$

The k-th *horocycle* of T with respect to ω and o is the set $H_k = \{x \in T : \mathfrak{h}(x) = k\}$. We now look at T as an infinite "genealogical" tree, where ω is the "mythical ancestor", the horocycles H_k ($k \in \mathbb{Z}$) are successive "generations", and each $x \in H_k$ has precisely one neighbour x^- (its "father") in H_{k-1} and $\deg(x) - 1$ neighbours (the "sons") in H_{k+1}. When thinking of the tree in this way, it will be best to view the horocycles in horizontal lines. Figure 12 is a "horocyclic" drawing of the homogeneous tree $\mathbb{T}_3$. Each H_k is infinite.

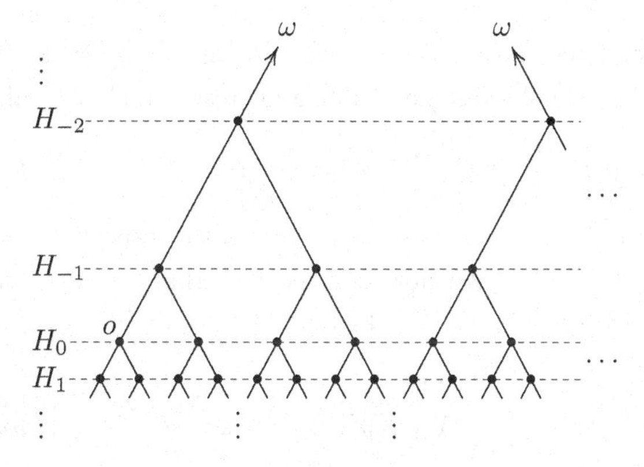

Figure 12: $\mathbb{T}_3$ in horocyclic layers

It is clear that the action of $\mathrm{AUT}(T)$ extends to the boundary: if $\xi \in \vartheta T$ is represented by a ray $\pi = [x_0, x_1, x_2, \dots]$ and $\gamma \in \mathrm{AUT}(T)$, then $\gamma\xi$ is the end of T represented by $\gamma\pi = [\gamma x_0, \gamma x_1, \gamma x_2, \dots]$. (Equivalence of rays is preserved under γ.) Now consider the group $\Gamma = \mathrm{AUT}(T)_\omega$ of all $\gamma \in \mathrm{AUT}(T)$ which fix ω. This is precisely the group of automorphisms γ which preserve the "genealogical" order, that is, $\gamma x^- = (\gamma x)^-$ for all x.

(12.14) Lemma. *The group $\Gamma = \mathrm{AUT}(T)_\omega$ is amenable.*

Proof. First check that $\mathfrak{h}(x) - \mathfrak{h}(y) = d(x, x \curlywedge y) - d(y, x \curlywedge y)$. If $\gamma \in \Gamma$ then $\gamma x \curlywedge \gamma y = \gamma(x \curlywedge y)$. These two facts yield that $\Phi(\gamma) = \mathfrak{h}(\gamma x) - \mathfrak{h}(x)$ is independent of x and defines a homomorphism $\Gamma \to \mathbb{Z}$. Being a subgroup of $\mathbb{Z}$, $\Phi(\Gamma)$ is amenable. Using Theorem 12.2(e) in its version for locally compact groups (see e.g. Pier [264]), we see that we only have to prove amenability of the group $\Gamma_{\mathfrak{h}} = \ker \Phi$. This is the subgroup of Γ which leaves H_0, and hence all horocycles, fixed as sets. Now write $\pi(o,\omega) = [o = x_0, x_1, x_2, \dots]$. If $\gamma \in \Gamma$ then $\pi(\gamma o, \omega) = [\gamma x_0, \gamma x_1, \gamma x_2, \dots]$. In particular, if $\gamma \in \Gamma_{\mathfrak{h}}$ and $n_0 = d(o, \gamma o \curlywedge o)$ then $\gamma x_n = x_n$ for all $n \geq n_0$. This means that $\Gamma_{\mathfrak{h}} = \bigcup_n \Gamma_{x_n}$, where (recall) Γ_{x_n} is the stabilizer of x_n in Γ. Obviously, $\Gamma_{x_n} \subset \Gamma_{x_{n+1}}$. That is, $\Gamma_{\mathfrak{h}}$ is the union of an increasing sequence of compact, whence amenable, subgroups. Therefore (see again [264]) $\Gamma_{\mathfrak{h}}$ is amenable. $\square$

For the following, we shall consider only $T = \mathbb{T}_{q+1}$, the homogeneous tree of degree $M = q+1 \geq 3$. The group Γ acts transitively. We know that $\mathbb{T}_{q+1}$ satisfies *IS*, so that Corollary 12.12 tells us that Γ must be non-unimodular. It is very easy to compute the modular function. If $x, y \in \mathbb{T}_{q+1}$ and $w = x \curlywedge y$, then set $k = d(x,w)$ and $\ell = d(y,w)$. We have $|\Gamma_x w| = |\Gamma_y w| = 1$, while $|\Gamma_w x| = q^k$ and $|\Gamma_w y| = q^\ell$. Therefore, using Lemma 1.29,

$$\frac{|\Gamma_x|}{|\Gamma_y|} = \frac{|\Gamma_x|}{|\Gamma_w|}\frac{|\Gamma_w|}{|\Gamma_y|} = q^{\ell - k} = q^{\mathfrak{h}(y) - \mathfrak{h}(x)}. \tag{12.15}$$

From this and (1.28), we find that $\Delta(\gamma) = q^{\mathfrak{h}(\gamma o)}$ for $\gamma \in \Gamma$.

As a first example, define a nearest neighbour walk $P = P_a$ on $\mathbb{T}_{q+1}$ by

$$p(x^-, x) = a/q, \quad p(x, x^-) = 1 - a, \quad \text{where } 0 < a < 1. \tag{12.16}$$

It is Γ-invariant and reversible with respect to the measure $m(x) = \left(\frac{a}{q(1-a)}\right)^{\mathfrak{h}(x)}$. We now use the fact that the first inequality in Theorem 12.10 is an equality, and compute via (12.15)

$$\rho(P) = p(o, o^-)\sqrt{\frac{|\Gamma_{o^-}|/m(o^-)}{|\Gamma_o|/m(o)}} + \sum_{x : x^- = o} p(o,x)\sqrt{\frac{|\Gamma_x|/m(x)}{|\Gamma_o|/m(o)}} = 2\sqrt{a(1-a)}\,.$$

When $a = \frac{1}{2}$, $\rho(P) = 1$. For $a = \frac{q}{q+1}$, we have the simple random walk and recover the formula for $\rho(P)$ found in Lemma 1.24. Of course, these results can also be obtained by the more elementary technique used there.

As a second example, consider an arbitrary *radial* random walk on $\mathbb{T} = \mathbb{T}_{q+1}$, that is, $p(x,y)$ depends only on the distance between x and y. Write

$S_n = \{x : d(x,o) = n\}$, and let $a_n = p(o, S_n)$, so that $p(x,y) = a_n/|S_n|$ when $d(x,y) = n$. As P is Γ-invariant, Theorem 12.10 and formula (12.15) give

$$\rho(P) = \sum_{x \in \mathbb{T}} p(o,x)\, q^{-\mathfrak{h}(x)/2} = \sum_{n=0}^{\infty} \frac{a_n}{|S_n|} \sum_{k=-n}^{n} |S_n \cap H_k|\, q^{-k/2}\,.$$

One computes $|S_0| = 1$ and $|S_n| = (q+1)q^{n-1}$. Also, $S_n \cap H_k \neq \emptyset$ if and only if $|k| \le n$ and $n-k$ is even,

$$|S_n \cap H_{-n}| = 1\,, \quad |S_n \cap H_n| = q^n \quad \text{and}$$
$$|S_n \cap H_{-n+2j}| = (q-1)q^{j-1} \quad \text{for } j = 1, \dots, n-1\,.$$

An elementary calculation now yields

$$\rho(P) = \sum_{n=0}^{\infty} a_n\, q^{-n/2} \left(1 + n\frac{q-1}{q+1}\right). \tag{12.17}$$

(12.18) Coupling two trees. Consider $\mathbb{T}_{q+1}$ and $\mathbb{T}_{r+1}$ with $q, r \ge 2$. As in (12.13), we select ends ω_1 of $\mathbb{T}_{q+1}$ and ω_2 of $\mathbb{T}_{r+1}$, respectively, and construct the following subgraph of the direct product of $\mathbb{T}_{q+1}$ and $\mathbb{T}_{r+1}$.

$$\mathrm{DL}_{q,r} = \{x_1x_2 \in \mathbb{T}_{q+1} \times \mathbb{T}_{r+1} : \mathfrak{h}(x_1) + \mathfrak{h}(x_2) = 0\}\,.$$

Here, $\mathfrak{h}(\cdot)$ stands of course for the height with respect to the fixed end in the corresponding tree. Recall that $x_1x_2 \sim y_1y_2$ when $x_1 \sim y_1$ and $x_2 \sim y_2$. To visualize $\mathrm{DL}_{q,r}$, draw $\mathbb{T}_{r+1}$ as in Figure 12, and (on the left) $\mathbb{T}_{q+1}$ in the same way, but upside down, with the respective horocycles $H_k(\mathbb{T}_{q+1})$ and $H_{-k}(\mathbb{T}_{r+1})$ on the same level. Connect the two origins by an elastic spring. It can move along each of the two trees, may expand infinitely, but always has to remain horizontal. The vertex set of $\mathrm{DL}_{q,r}$ is the collection of all admissible positions of the spring. From a given position x_1x_2 with $\mathfrak{h}(x_1) + \mathfrak{h}(x_2) = 0$, it may move upwards to one of the "sons" of x_1 and simultaneously to the "father" of x_2, or downwards in the analogous way. Such a move corresponds to going to a neighbour of x_1x_2, and this may also serve to visualize the simple random walk on $\mathrm{DL}_{q,r}$. We invite the reader to draw a figure according to these "instructions".

Now let $\Gamma_1 = \mathrm{AUT}(\mathbb{T}_{q+1})_{\omega_1}$ and $\Gamma_2 = \mathrm{AUT}(\mathbb{T}_{r+1})_{\omega_2}$. The group

$$\Gamma = \{\gamma_1\gamma_2 : \gamma_i \in \Gamma_i\,,\ \mathfrak{h}(\gamma_1 o_1) + \mathfrak{h}(\gamma_2 o_2) = 0\}$$

acts canonically on $\mathrm{DL}_{q,r}$: if $\gamma_i x_i = y_i$, $i = 1, 2$, where $x_1x_2 \in \mathrm{DL}_{q,r}$, then also $y_1y_2 \in \mathrm{DL}_{q,r}$, and we define $\gamma_1\gamma_2(x_1x_2) = y_1y_2$. The action is transitive. Being a closed subgroup of the amenable group $\Gamma_1 \times \Gamma_2$, Γ is also amenable. Now let $o = o_1 o_2$ and $x = x_1 x_2$, where x_1 is a "son" of o_1 and $x_2 = o_2^-$. Then $|\Gamma_o x| = q$ and $|\Gamma_x o| = r$. Using Theorem 12.10, we find the spectral radius of the simple random walk:

$$\rho(P) = \frac{2\sqrt{qr}}{q+r}. \tag{12.19}$$

In particular, the graph $\mathrm{DL}_{q,r}$ is amenable if and only if $q = r$.

The next example is more complicated. We will not lay out each detail of the background, and the reader might have to consult the relevant literature.

(12.20) The buildings of $PGL(n, \mathfrak{F})$, $\mathfrak{F}$ a local field. For the necessary background on local fields, the reader may consult Cassels [67] or Serre [299]. Let $\mathfrak{F}$ be a (commutative) non-archimedean local field with valuation $v : \mathfrak{F}^* \to \mathbb{Z}$ and $v(0) = \infty$ ($\mathfrak{F}^*$ denotes the multiplicative group, and v is a homomorphism). Denote by $\mathfrak{O} = \{\mathfrak{a} \in \mathfrak{F} : v(\mathfrak{a}) \geq 0\}$ the ring of *integers* and $\mathfrak{P} = \{\mathfrak{a} \in \mathfrak{F} : v(\mathfrak{a}) \geq 1\}$ the maximal ideal in $\mathfrak{O}$. Let q (a prime power) be the – finite – order of the *residual field* $\mathfrak{K} = \mathfrak{O}/\mathfrak{P}$. The *absolute value* of $\mathfrak{a} \in \mathfrak{F}$ is $|\mathfrak{a}| = q^{-v(\mathfrak{a})}$. It induces an ultrametric. Choose a *uniformizer* $\mathfrak{p} \in \mathfrak{P}$, that is, $|\mathfrak{p}| = q^{-1}$.

The best-known example is the field $\mathbb{Q}_p$ of p-adic numbers, where p is a prime. Each element $\mathfrak{a} \in \mathbb{Q}_p \setminus \{0\}$ can be written as $\mathfrak{a} = \sum_{n \geq k} s_n\, q^n$, where $s_n \in \{0, \dots, p-1\}$ and $s_k \neq 0$; we have $v(\mathfrak{a}) = k \in \mathbb{Z}$. The operations in $\mathbb{Q}_p$ are the natural extensions of addition and multiplication in $\mathbb{Q}^+$ (for which the series are finite). $\mathbb{Q}$ is dense, $\mathbb{Q}_p$ is complete, $\mathfrak{O} = \mathbb{Z}_p$ is the closure of $\mathbb{Z}$, and one typically takes $\mathfrak{p} = p$.

$GL(n, \mathfrak{F})$ is the group of invertible $n \times n$ matrices over $\mathfrak{F}$, and $PGL(n, \mathfrak{F}) = GL(n, \mathfrak{F})/\mathfrak{F}^*$, where, more precisely, $\mathfrak{F}^*$ stands for all non-zero multiples of the identity matrix. Associated with $PGL(n, \mathfrak{F})$, there is a homogeneous space X which is called a *(linear) building of type* $\widetilde{A}_{n-1}$. This is an $(n-1)$-dimensional simplicial complex with several particular features; see Ronan [279] or Brown [48]. In order to understand its structure, for us it will be enough to describe its 1-skeleton, which is a countable graph. X will stand for this graph, and $PGL(n, \mathfrak{F}) \subset \mathrm{AUT}(X)$ is closed and acts transitively; the topology of $PGL(n, \mathfrak{F})$ inherited from $\mathfrak{F}$ coincides with the topology of pointwise convergence on X. We now describe X.

A *lattice* is an $\mathfrak{O}$-submodule of $\mathfrak{F}^n$ of the form $L = \mathfrak{O}\mathbf{v}_1 + \cdots + \mathfrak{O}\mathbf{v}_n$, where $\{\mathbf{v}_1, \dots, \mathbf{v}_n\}$ is a basis of $\mathfrak{F}^n$. Two lattices L, L' are equivalent if $L' = \mathfrak{a}\, L$ for some $\mathfrak{a} \in \mathfrak{F}^*$. The equivalence class of L is denoted by $[L]$. Then $X = \{[L] : L \text{ a lattice}\}$. Two points $x, y \in X$ with $x = [L]$ are neighbours in the graph X if we can write $y = [L']$ such that $\mathfrak{p}L \subset L' \subset L$ strictly. For such L', the mapping $[L'] \mapsto L'/\mathfrak{p}L$ is an isomorphism from the neighbourhood graph $N(x) = \{y \in X : y \sim x\}$ of $x \in X$ onto the family of non-trivial subspaces of $\mathfrak{K}^n$, where two elements are neighbours if one strictly contains the other. Thus,

$$\deg(x) = \sum_{k=1}^{n-1} \binom{n}{k}_q, \quad \text{where} \quad \binom{n}{k}_q = \frac{(q^n-1)\cdots(q^{n-k+1}-1)}{(q^k-1)\cdots(q-1)},$$

as the number of subspaces of $\mathfrak{K}^n$ is $W(n,q) = \sum_{k=0}^{n} \binom{n}{k}_q$ (see e.g. Goldman and Rota [143]). The $W(n,q)$ satisfy the recurrence relation $W(n+1,q) = 2W(n,q-1) + (q^n-1)W(n-1,q)$, with $W(0,q) = 1$ and $W(1,q) = 2$; see Andrews [7]. For $n = 2$ we get $\deg(x) = q+1$, and for $n = 3$ one finds $\deg(x) = 2(q^2+q+1)$.

The building associated with $PGL(2,\mathfrak{F})$ is the tree $\mathbb{T}_{q+1}$. For any $n \geq 2$, the *apartment* of X associated with a given basis $\{\mathbf{v}_1, \ldots, \mathbf{v}_n\}$ of $\mathfrak{F}^n$ is

$$\mathfrak{A} = \mathfrak{A}(\mathbf{v}_1, \ldots, \mathbf{v}_n) = \{[\mathfrak{p}^{\ell_1}\mathfrak{O}\mathbf{v}_1 + \cdots + \mathfrak{p}^{\ell_n}\mathfrak{O}\mathbf{v}_n] : \ell_i \in \mathbb{Z}\}.$$

For $n = 2$, this is a two-sided infinite geodesic path in the tree.

For $n = 3$, the subgraph of X induced by $\mathfrak{A}$ is the tiling of the plane by equilateral triangles. Furthermore, every edge in X is common to $q+1$ triangles; see Figure 13.

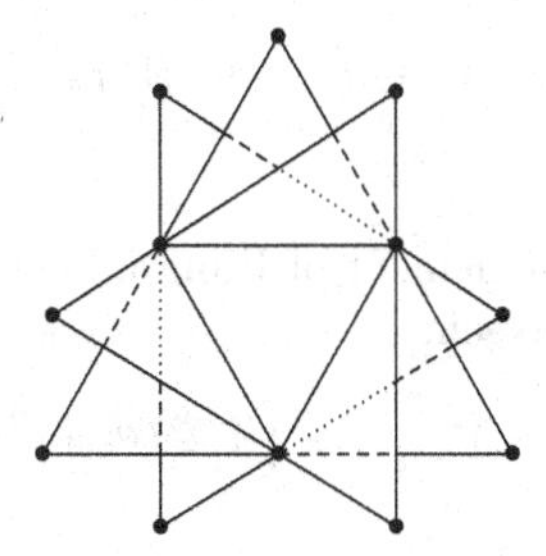

Figure 13: a triangle and its neighbours ($q = 3$)

For any n, the group $PGL(n,\mathfrak{F})$ acts on X by matrix multiplication on the left, i.e., $\gamma[L] = [\gamma L] = [\mathfrak{O}\gamma\mathbf{v}_1 + \cdots + \mathfrak{O}\gamma\mathbf{v}_n]$. Here, we typically write g as a matrix in $GL(n,\mathfrak{F})$, while thinking of it as an element of $PGL(n,\mathfrak{F})$ consisting of all its non-zero multiples.

We now want to compute the spectral radius of the simple random walk P on X by use of Theorem 12.10. Let Γ be the image in $PGL(n,\mathfrak{F})$ of the upper-triangular subgroup of $GL(n,\mathfrak{F})$. It acts transitively on X.

To see this, we choose $o = [\mathfrak{O}^n] = [\mathfrak{O}\mathbf{e}_1 + \cdots + \mathfrak{O}\mathbf{e}_n]$ as our "root" in X, where $\{\mathbf{e}_1, \ldots, \mathbf{e}_n\}$ is the standard basis of $\mathfrak{F}^n$. The stabilizer of o in $PGL(n,\mathfrak{F})$ is (isomorphic to) the group $PGL(n,\mathfrak{O}) = GL(n,\mathfrak{O})/\mathfrak{O}^*$, where $\mathfrak{O}^* = \mathfrak{O} \setminus \mathfrak{P}$ is the set of invertibles in $\mathfrak{O}$, and $GL(n,\mathfrak{O})$ consists of all matrices $\gamma \in GL(n, \mathfrak{F})$ such that γ and γ^{-1} have all their entries in $\mathfrak{O}$. Now consider $\beta \in GL(n, \mathfrak{F})$. Locate an entry $\beta_{n,j}$ in row n whose absolute value is largest in that row. By right multiplication by a permutation matrix (an element of $GL(n, \mathfrak{O})$), we can move this entry to position (n, n). Let $E_{i,j}$ be the matrix with entry 1 in position (i, j) and 0 elsewhere. If $\mathfrak{a} \in \mathfrak{O}$, then $I + \mathfrak{a}E_{i,j} \in GL(n, \mathfrak{O})$. By right multiplication by $n - 1$ matrices $I + \mathfrak{a}_j E_{n,j}$, $j = 1, \ldots, n-1$, we can replace all entries in row n, except that in position (n, n), by 0.

Now repeat with the $(n-1) \times (n-1)$ matrix at the upper left of the new β obtained in this way, and continue. At the end, we get $\alpha \in GL(n, \mathfrak{O})$ such that $\beta\alpha = \gamma$ is upper triangular, i.e., in Γ. So $\alpha\, o = o$ and $\beta\, o = \gamma\, o$. As every element of X is of the form $\beta\, o$ for some β as above, we get transitivity of Γ.

Also, it is well known that Γ is amenable, being solvable. (In particular when $n = 2$, this Γ is a subgroup of $\mathrm{AUT}(\mathbb{T})_\omega$ for some end ω of $\mathbb{T} = \mathbb{T}_{q+1}$.) In order to apply Theorem 12.10, we have to understand the action of Γ_o on $N(o)$. Write $\mathcal{E} = \{0,1\}^n \setminus \{\underline{0}, \underline{1}\}$, where $\underline{0} = (0, \ldots, 0)$ and $\underline{1} = (1, \ldots, 1)$. Let $\mathfrak{A} = \mathfrak{A}(\mathbf{e}_1, \ldots, \mathbf{e}_n)$. The neighbours of o in $\mathfrak{A}$ are of the form

$$x_{\underline{\varepsilon}} = [\mathfrak{p}^{\varepsilon_1}\mathfrak{O}\mathbf{e}_1 + \cdots + \mathfrak{p}^{\varepsilon_n}\mathfrak{O}\mathbf{e}_n], \quad \text{where } \underline{\varepsilon} \in \mathcal{E}.$$

We can represent $\Gamma_o = \Gamma \cap PGL(n, \mathfrak{O})$ by the group

$$\Gamma_o = \{\gamma = (\gamma_{ij})_{i,j=1,\ldots,n} : \gamma_{ij} \in \mathfrak{F},\ \gamma_{ij} = 0\ (i > j),\\ |\gamma_{ii}| = 1,\ |\gamma_{ij}| \le 1\ (i < j)\}.$$

Also, an element of Γ mapping o to $x_{\underline{\varepsilon}}$ is $\gamma_{\underline{\varepsilon}} = \mathrm{diag}(\mathfrak{p}^{\varepsilon_i})_{i=1,\ldots,n}$. Hence, we can represent

$$\begin{aligned} \Gamma_{x_{\underline{\varepsilon}}} = \gamma_{\underline{\varepsilon}}\Gamma_o\gamma_{\underline{\varepsilon}}^{-1} &= \{\beta = (\beta_{ij}) : \beta_{ij} \in \mathfrak{F},\ \beta_{ij} = 0\ (i > j),\\ &\qquad |\gamma_{ii}| = 1,\ |\beta_{ij}| \le q^{\varepsilon_j - \varepsilon_i}\ (i < j)\} \quad \text{and} \\ \Gamma_o \cap \Gamma_{x_{\underline{\varepsilon}}} &= \{\alpha = (\alpha_{ij}) : \alpha_{ij} \in \mathfrak{F},\ \alpha_{ij} = 0\ (i > j),\\ &\qquad |\alpha_{ii}| = 1,\ |\alpha_{ij}| \le q^{-\max\{0, \varepsilon_i - \varepsilon_j\}}\ (i < j)\}. \end{aligned}$$

The left Haar measure on the group of upper-triangular invertible $n \times n$ matrices over $\mathfrak{F}$ is given by

$$d\gamma = |\gamma_{11}^n \gamma_{22}^{n-1} \cdots \gamma_{nn}|^{-1} \prod_{i \le j} d\gamma_{ij},$$

where $d\gamma_{ij}$ stands for the Lebesgue (Haar) measure λ on the additive group $\mathfrak{F}$; compare with Hewitt and Ross [172], p. 209, where this is stated for matrices over $\mathbb{R}$. The above two stabilizers and their intersections are compact open subgroups, so that their Haar measure is the restriction of the Haar measure on the whole group. If we normalize λ so that $\lambda(\mathfrak{O}) = 1$ and hence $\lambda(\mathfrak{P}) = 1/q$, $\lambda(\mathfrak{O}^*) = 1 - 1/q$, the measures of our stabilizers are $|\Gamma_o| = (1-1/q)^n$ and $|\Gamma_o \cap \Gamma_{x_{\underline{\varepsilon}}}| = (1-1/q)^n q^{-M(\underline{\varepsilon})}$, where

$$M(\underline{\varepsilon}) = \sum_{1 \le i < j \le n} \max\{0, \varepsilon_i - \varepsilon_j\}$$

is the number of times that a 1 comes before a 0 in the vector $\underline{\varepsilon}$. With this notation, we get

$$|\Gamma_o x_{\underline{\varepsilon}}| = |\Gamma_o| \big/ |\Gamma_o \cap \Gamma_{x_{\underline{\varepsilon}}}| = q^{M(\underline{\varepsilon})}\,.$$

Write $|\underline{\varepsilon}| = \sum_i \varepsilon_i$ and let $W(n,k;q) = \sum_{\underline{\varepsilon} \in \{0,1\}^n : |\underline{\varepsilon}| = k} M(\underline{\varepsilon})$. Then $W(n,0;q) = W(n,n;q) = 1$, and expanding with respect to the value of ε_n, one obtains $W(n,k;q) = W(n-1,k-1;q) + q^k W(n-1,k;q)$. The q-binomial coefficients satisfy the same recursion, whence they coincide with the $W(n,k;q)$. Therefore $\sum_{\underline{\varepsilon} \in \mathcal{E}} q^{M(\underline{\varepsilon})} = \deg(o)$. Now one checks easily that no two different $x_{\underline{\varepsilon}}, x_{\underline{\varepsilon}'}$ can stay in the same Γ_o-orbit. Thus, each $x_{\underline{\varepsilon}}$ is the representative of some Γ_o-orbit, and the last identity tells us that we get all orbits in this way, that is, $N(x) = \bigcup_{\underline{\varepsilon} \in \mathcal{E}} \Gamma_o x_{\underline{\varepsilon}}$ (disjoint union). Next, $|\Gamma_{x_{\underline{\varepsilon}}} o| = |\Gamma_o \gamma_{\underline{\varepsilon}}^{-1} x_{\underline{\varepsilon}}| = |\Gamma_o x_{\underline{1}-\underline{\varepsilon}}| = q^{M(\underline{1}-\underline{\varepsilon})}$. From this and Theorem 12.10, via Lemma 1.29, we get $\rho(P) = \frac{1}{\deg(o)} \sum_{\underline{\varepsilon} \in \mathcal{E}} |\Gamma_o x_{\underline{\varepsilon}}| \sqrt{q^{M(\underline{1}-\underline{\varepsilon})}/q^{M(\underline{\varepsilon})}}$, and finally

$$\rho(P) = \frac{\sum_{\underline{\varepsilon} \in \mathcal{E}} q^{(M(\underline{\varepsilon}) + M(\underline{1}-\underline{\varepsilon}))/2}}{\sum_{\underline{\varepsilon} \in \mathcal{E}} q^{M(\underline{\varepsilon})}} = \frac{\sum_{k=1}^{n-1} \binom{n}{k} q^{k(n-k)/2}}{\sum_{k=1}^{n-1} \binom{n}{k}_q} \tag{12.21}$$

(with ordinary binomial coefficients in the numerator of the second term). For $n = 2$, we obtain the spectral radius of the simple random walk operator on $\mathbb{T}_{q+1}$. For $n = 3$ and $n = 4$, we find

$$\rho(P) = \frac{3q}{1 + q + q^2} \quad \text{and} \quad \rho(P) = \frac{6q^2 + 8q^{3/2}}{q^4 + 3q^3 + 4q^2 + 3q + 3}\,, \tag{12.22}$$

respectively. □

Notes and remarks

7. Superharmonic functions and ρ-recurrence

§**A.** These results are due to Pruitt [271].

§**B.** The study of ρ-recurrence goes back to Pruitt [271] and Vere-Jones [327], [328], [329]. Theorem 7.8 has its root in a result for locally compact groups indicated by Guivarc'h [156], p. 85.

8. The spectral radius, the rate of escape, and generalized lattices

§**A.** To my knowledge, Proposition 8.2 first appears explicitly in Kaimanovich and Woess [191]. For groups, compare with Guivarc'h' [156]. The coupling is due to Kaimanovich [191], and the distance estimates of Proposition 8.8 and Corollary 8.9 are also taken from [191]. The lifting of the random walk to Γ in the transitive case was first introduced by Soardi and Woess [305] (for the simple random walk) and then systematically exploited by Woess [345]. For Theorem 8.14 in the context of random walks on locally compact groups, see Derriennic [94] and Guivarc'h [156]. With the exception of abelian groups (compare with Theorem 6.7), it is not easy to compute the limit $\mathfrak{m}$ explicitly. The existence of $\mathfrak{m}$ was first proved for random matrix products by Furstenberg and Kesten [128], and the first formula for $\mathfrak{m}$ – also in the context of random walks on matrix groups – is due to Furstenberg [124]. For free groups and homogeneous trees, computing $\mathfrak{m}$ involves the use of boundary theory; see Sawyer [290], Derriennic [94], Sawyer and Steger [292], Lalley [212] and Cartwright, Kaimanovich and Woess [59]. The last three references also contain central limit theorems for $d(Z_n, Z_0)$ and – in [292] – other functions of Z_n.

§**B.** For random walks on $\mathbb{Z}^d$, this method for determining $\rho(P)$ has its roots in Kesten [200]; see Stone [310], Lemma 5.

9. Computing the Green function

§**A.** Theorem 9.4 was proved by Cartwright [58].

§§**B,C.** The main result here is Theorem 9.19, which has a "story". After several special cases considered previously by Gerl [133], Woess [338], Picardello and Woess [257] and Soardi [301], this was first proved for free products of groups simultaneously and independently by Cartwright and Soardi [302], [63] and by Woess [341]. (It is amusing to remember the reciprocal embarrassment when we met a couple of weeks later.) The methods are in principle the same, [63] being more harmonic-analysis-flavoured and [341] more oriented towards local limit theorems.

A couple of months later, again independently, this appears as the main basic result in Voiculescu's fundamental paper on freely independent non-commuting random variables [330] in a different and in many respects more general setting; see also [331] and the recent book [332].

At about the same time, the same result (for simple random walks on free products of groups) was also proved in the PhD-thesis of McLaughlin [231]. The latter has been taken up very recently by Quenell [272], and subsequently by Gutkin [161], who considered free products of arbitrary rooted sets and extended the results on Green functions and spectra to this setting.

The shortest proof of Theorem 9.19 was given by Cartwright [56], but it needs some ingredients which we shall touch only marginally, in Chapter IV. The proof given here via Proposition 9.18 is based on [161], with some input from [341], and gives the most elementary access. Note that the stopping time arguments used here become completely elementary when rewritten in terms of the combinatorics of paths. Donald Cartwright pointed out to me that in Proposition 9.10 it is better to work with $S(z) = W(1/z)$, which is true in many respects [56]. I have chosen to stick to the formulation as given here in order to maintain continuity in the use of generating functions which converge in a disk around the origin.

A few words on the various examples: the closed formula for $W(\cdot)$ in Example 9.15(2) is due to Paschke [254]. The formula for $\rho(P)$ in 9.25(1) was first computed by Gerl [133], the ones in 9.29(3) by Woess ([336] and [338]), see also Cartwright and Soardi [62]. I am grateful to Klaus Schiefermayr for the computation (using *Mathematica*) of $\rho(P)$ in 9.25(3), when $d = 2, 3, 4$. For 9.29(2), see Gerl [131] (where $\mathbb{F}_2$ is considered – in this case, the equation for $\theta(P)$ can be solved), Woess [337] and Gerl and Woess [140]; the first implicit appearance of the formula for $\Phi(t)$ can be traced back to Dynkin and Malyutov [111]; see also Levit and Molchanov [215]. Example 9.29(1) is very similar; compare with Figà-Talamanca and Steger [116]. Example 9.29(4) is from Soardi [301]. In Chapter III we shall see that one of the most interesting of these examples is 9.25(3), based on Cartwright [56].

10. Spectral radius and strong isoperimetric inequality

§**A.** Lemma 10.1 and Corollary 10.2 are extracted from Gerl [139]. (For groups, compare also with Avez [10].) Previous versions (Kesten [198], Papangelou [252]) used the "heavy machinery" of spectral resolution.

The main equivalence (a) $\Longleftrightarrow$ (c) in Theorem 10.3 is due to Dodziuk [98] and Dodziuk and Kendall [99] for simple random walks on graphs with bounded geometry and for arbitrary locally finite graphs, respectively. This is the discrete analogue of Cheeger's inequality [70] regarding the bottom of the spectrum of the Laplacian on a Riemannian manifold. For regular graphs, see also Biggs, Mohar and Shawe-Taylor [43]. Theorem 10.3 as stated here for general reversible Markov chains is taken from Gerl [139]; see also Ancona [5], [6] and, for graphs, Mohar [237].

§**B.** Theorems 10.6 and 10.8 were first proved for random walks on groups, see below (§12.A). The extension to strongly reversible random walks is due to Kaimanovich [186]. Only slight adaptations were necessary for the "non-reversible" versions given here.

§**C.** Theorem 10.9 is adapted from Gerl [137]. Theorem 10.10 is due to Pittet [265]. Compare also with Mohar [238]. Theorem 10.11 was proved by Woess [352]. We have omitted the (slightly more complicated) proof of a third condition which implies IS: (c) $\inf\{\chi(T) : T \in \mathcal{T}\} > 0$, where

$$\chi(T) = \frac{|T|}{2} - 1 - \sum_{x \in T} \frac{1}{\deg(x)}$$

is the characteristic number of tile $T \in \mathcal{T}$. This was also proved by Zuk [357]. Forerunners are due to Dodziuk [98], Soardi [303], Calogero [49] and He and Schramm [169], see also Mohar [239]. For the graph in Figure 11, there are other ways to prove IS, for example by showing that it is roughly isometric with the hyperbolic upper half plane.

11. A lower bound for simple random walks

Theorem 11.1 is due to Zuk [357]. Theorem 11.6 and Corollary 11.7 were proved by Paschke [254]. The proof also works when ℓ is not assumed to be minimal, but more effort is needed then. As a matter of fact, the lower bound provided by Corollary 11.7 is only very slightly above $2\sqrt{M-1}/M$. However, the significance of this result is more qualitative than quantitative.

12. The spectral radius and amenability

§**A.** Corollary 12.5 was originally proved in the seminal papers of Kesten [198], [199] and Day [88].

§**B.** Theorem 12.10 and Corollary 12.12 are due to Soardi and Woess [305]. These results also extend to quasi-transitive random walks; see Salvatori [289] and Saloff-Coste and Woess [286], [287]. An example where $\tau = \infty$ is given in [289].

§**C.** The example (12.13) is from [305] and is interesting in many respects; see Cartwright, Kaimanovich and Woess [59]. The graph of (12.18) was "invented" by R. Diestel and I. Leader in an attempt to solve the following problem of [305]: is there a vertex-transitive graph which is not roughly isometric with a Cayley graph? We believe that $\mathrm{DL}_{q,r}$ is a counterexample. The example (12.20) is from [287], with a little help from Donald Cartwright and Tim Steger.

Further results

There are many further results concerning the spectrum of P on suitable spaces, in particular $\ell^p(X)$. For references up to 1988, see Mohar and Woess [240]. Concerning the spectrum of the simple random walk on Cayley graphs, more recent, significant references are the two papers by de la Harpe, Robertson and Valette [166], [167]; see also the references given there.

There is a recent series of papers concerning approximations of the spectral radius, in particular for surface groups: Cherix and Valette [72], Bartholdi, Cantat, Ceccherini-Silberstein and de la Harpe [26], Zuk [356], [357], Nagnibeda [243], [244], and Bartholdi and Ceccherini-Silberstein [25]. For example, consider the genus-two-surface group $\Gamma = \langle a_1, a_2, b_1, b_2 \mid [a_1, b_1][a_2, b_2] = o \rangle$ and its Cayley graph with respect to these generators and their inverses: the best-known estimate for the simple random walk is $0.6624 \le \rho(P) \le 0.6628$ [243], [25].

Regarding quasi-transitive planar graphs, there is an important theorem obtained by combining recent results by Medolla and Soardi [232] and Benjamini and Schramm [37] with Theorem 5.13. Call a graph *quasi-planar* if it can be embedded in the plane so that every edge is crossed by at most K other edges, where K is some fixed constant.

Theorem. *A quasi-planar, quasi-transitive graph either is a generalized lattice of dimension ≤ 2 or satisfies IS.*

The proof is as follows: [37] shows that a quasi-planar graph which is transient must have non-constant harmonic functions with finite Dirichlet sum; [232] has proved that on a quasi-transitive, amenable graph, all Dirichlet-finite harmonic functions are constant. For (purely) planar graphs, a direct proof of the above theorem is due to Babai [12]; see also [13].

Compare this with the following result of Woess [347] regarding Gromov-hyperbolic graphs (the latter will be considered in Chapter IV); see Exercise 22.17.

Theorem. *A quasi-transitive hyperbolic graph either has linear growth or satisfies IS.*

For (Cayley graphs of) finitely generated groups this is well known (Gromov [151]) but it does not extend immediately to the quasi-transitive case; the proof uses Corollary 12.12.

Still more recently, Benjamini and Schramm [39] have proved the following significant result, whose flavour parallels in some sense Theorem 5.2 of Thomassen (saying that under a mild additional hypothesis, a transient graph contains a transient subtree).

Theorem. *If a locally finite graph satisfies IS then it has a subtree satisfying IS.*

CHAPTER III

THE ASYMPTOTIC BEHAVIOUR OF TRANSITION PROBABILITIES

The theme of this chapter is the asymptotic evaluation of $p^{(n)}(x,y)$ as $n \to \infty$, and the question how this is related to the structure of the underlying graph or group. We have already encountered such a result at the very beginning, when considering Pólya's walk: for the simple random walk in the d-dimensional grid, $p^{(2n)}(0,0)$ is of the order of $n^{-d/2}$. There is a large body of work concerning this type of problem, and we will not be able to cover all aspects.

A result which gives a numerical estimate of $p^{(n)}(x,y)$ for fixed x,y as $n \to \infty$ is usually called a *local limit theorem.* In this chapter we shall often assume that (X,P), besides being irreducible, is also aperiodic, so that $p^{(n)}(x,y) > 0$ for all $n \geq n_{x,y}$. This is no severe restriction and serves only to avoid discussions involving the parity of n.

Throughout this chapter, for real sequences (a_n) and (b_n), we indicate by $a_n \sim b_n$ that the quotient tends to 1.

13. The local central limit theorem on the grid

The aim of this section is to generalize Pólya's example. Let P (irreducible) be defined by a probability measure μ on $\mathbb{Z}^d$ with finite second moment, mean $\mathfrak{m} = \mathfrak{m}(\mu)$ and covariance matrix Σ. The corresponding quadratic form is

$$\Sigma[\mathbf{x}] = \mathbf{x} \cdot \Sigma\mathbf{x} = \sum_{\mathbf{k}} |\mathbf{x} \cdot (\mathbf{k} - \mathfrak{m})|^2 \mu(\mathbf{k}).$$

The following is well known; compare also with Proposition 8.20.

(13.1) Lemma. *Σ is positive definite.*

Proof. Suppose that $\Sigma[\mathbf{x}] = 0$ for $\mathbf{x} \in \mathbb{R}^d$, $\mathbf{x} \neq \mathbf{0}$. Then $\mathbf{k} - \mathfrak{m}$ is orthogonal to $\mathbf{x}$ for every $\mathbf{k} \in \operatorname{supp}\mu$. Irreducibility yields $-\mathbf{k} \in \operatorname{supp}\mu^{(n)}$ for some n, so that $-\mathbf{k} - n\mathfrak{m}$ is orthogonal to $\mathbf{x}$. Summing, we see that $(1+n)\mathfrak{m}$ and therefore $\mathbf{k}$ itself must be orthogonal to $\mathbf{x}$. Thus $\operatorname{supp}\mu$ is orthogonal to $\mathbf{x}$, in contradiction with irreducibility. □

(Indeed, the conclusion holds under weaker conditions.) In particular, Σ is invertible, and we shall write $\Sigma^{-1}[\mathbf{x}]$ for the corresponding quadratic form. We use the centred characteristic function

$$\psi(\mathbf{x}) = \psi_\mu(\mathbf{x}) = \sum_{\mathbf{k} \in \mathbb{Z}^d} \mu(\mathbf{k}) \exp\big(i(\mathbf{k} - \mathfrak{m}) \cdot \mathbf{x}\big). \tag{13.2}$$

It satisfies $\operatorname{grad}\psi(\mathbf{0}) = \mathbf{0}$, and its second order derivatives at 0 are given by $-\Sigma$. Fourier inversion gives

$$\mu^{(n)}(\mathbf{k}) = \frac{1}{(2\pi)^d} \int_{\mathbb{W}_d} \psi(\mathbf{x})^n \exp\bigl(-i\mathbf{x} \cdot (\mathbf{k} - n\mathbb{m})\bigr)\, d\mathbf{x}\,,$$

where $\mathbb{W}_d = [-\pi\,,\,\pi]^d$.

(13.3) Lemma. *If μ is irreducible and aperiodic, then $|\psi(\mathbf{x})| < 1$ for all $\mathbf{x} \in \mathbb{W}_d \setminus \{\mathbf{0}\}$.*

Proof. Suppose $|\psi(\mathbf{x})| = 1$ for some $\mathbf{x} \in \mathbb{W}_d$, that is, $\psi(\mathbf{x}) = e^{it_0}$, where $t_0 \in \mathbb{R}$. Set $t = t_0 + \mathbf{x} \cdot \mathbb{m}$. Then, for each n,

$$e^{int} = \sum_{\mathbf{k} \in \mathbb{Z}^d} \mu^{(n)}(\mathbf{k})\, e^{i\mathbf{x}\cdot\mathbf{k}}\,,$$

a convex combination of points on the unit circle. Therefore $e^{i\mathbf{x}\cdot\mathbf{k}} = e^{int}$ and $\mathbf{x} \cdot \mathbf{k} \equiv nt$ modulo 2π for all $\mathbf{k} \in \operatorname{supp}\mu^{(n)}$. Aperiodicity yields that $\mathbf{0} \in \operatorname{supp}\mu^{(n)}$ for all but finitely many n. Consequently, $t \equiv 0$, and by irreducibility, $\mathbf{x} \cdot \mathbf{k} \equiv 0$ modulo 2π for all $\mathbf{k} \in \mathbb{Z}^d$. Substituting the integer unit vectors for $\mathbf{k}$, we get that all coordinates of $\mathbf{x}$ are multiples of 2π. As $\mathbf{x} \in \mathbb{W}_d$, we must have $\mathbf{x} = \mathbf{0}$. □

Now suppose that μ has finite moment of order $2r$, where $r \geq 0$ is integer. Then ψ has continuous derivatives of order $2r$, and one can differentiate inside the sum of (13.2). Let $\Delta = \sum_{j=1}^d (\partial^2/\partial x_j^2)$ denote the Laplace operator with respect to the variable $\mathbf{x} \in \mathbb{R}^d$, and Δ^r its r-th iterate.

(13.4) Lemma. *Setting $\psi_n(\mathbf{x}) = \psi(\mathbf{x}/\sqrt{n})$,*

$$\left(\frac{|\mathbf{k} - n\mathbb{m}|}{\sqrt{n}}\right)^{2r} \bigl(2\pi\sqrt{n}\bigr)^d \mu^{(n)}(\mathbf{k})$$
$$= (-1)^r \int_{\sqrt{n}\mathbb{W}_d} \exp\left(-i\mathbf{x} \cdot \frac{\mathbf{k} - n\mathbb{m}}{\sqrt{n}}\right) \Delta^r \psi_n(\mathbf{x})^n\, d\mathbf{x}\,,$$

where $\sqrt{n}\,\mathbb{W}_d$ is the cube $\mathbb{W}_d$ rescaled by $\sqrt{n}$.

Proof[2]**.** We expand and apply Green's second identity to compute for fixed $\mathbf{y} \in \mathbb{R}^d$:

$$\int_{\mathbb{W}_d} \Bigl(\psi(\mathbf{x})^n\, \Delta^r e^{-i\mathbf{x}\cdot\mathbf{y}} - e^{-i\mathbf{x}\cdot\mathbf{y}}\, \Delta^r \psi(\mathbf{x})^n\Bigr)\, d\mathbf{x}$$
$$= \sum_{k=0}^{r-1} \int_{\mathbb{W}_d} \Bigl(\Delta^k \psi(\mathbf{x})^n\, \Delta^{r-k} e^{-i\mathbf{x}\cdot\mathbf{y}} - \Delta^{k+1}\psi(\mathbf{x})^n\, \Delta^{r-k-1} e^{-i\mathbf{x}\cdot\mathbf{y}}\Bigr)\, d\mathbf{x}$$
$$= \sum_{k=0}^{r-1} I(k, r-k-1)\,,$$

[2] See comments on page 317

where for $k, l \in \mathbb{N}_0$

$$I(k,l) = \int_{\partial \mathbb{W}_d} \Big(\Delta^k \psi(\mathbf{x})^n \frac{\partial}{\partial \mathbf{n}} \Delta^l e^{-i\mathbf{x}\cdot\mathbf{y}} - \Delta^l e^{-i\mathbf{x}\cdot\mathbf{y}} \frac{\partial}{\partial \mathbf{n}} \Delta^k \psi(\mathbf{x})^n \Big)\, d_\partial \mathbf{x}.$$

Here, $\frac{\partial}{\partial \mathbf{n}}$ is the outer normal derivative and d_∂ is Lebesgue measure on the boundary $\partial \mathbb{W}_d$. By induction on k,

$$\Delta^k \psi(\mathbf{x})^n = \varphi_{n,k}(\mathbf{x}) e^{-i\,n\,\mathbb{m}\cdot\mathbf{x}}, \quad \text{and} \quad \Delta^l e^{-i\mathbf{x}\cdot\mathbf{y}} = (-1)^l |\mathbf{y}|^{2l} e^{-i\mathbf{x}\cdot\mathbf{y}},$$

where $\varphi_{n,k}(\mathbf{x})$ is 2π-periodic in each variable. Now we decompose $I(k,l)$ as the sum of d pairs of integrals over opposite faces of $\mathbb{W}_d$. Let us consider the two faces corresponding to $x_d = \pm\pi$. Their respective contribution is

$$(-1)^l |\mathbf{y}|^{2l} \times$$
$$\pm \int_{\mathbb{W}_{d-1}} e^{-i\,(\mathbf{x}',\pm\pi)\cdot(n\mathbb{m}+\mathbf{y})} \Big(i(n\,\mathbb{m}_d - y_d)\varphi_{n,k}(\mathbf{x}',\pm\pi) - \frac{\partial}{\partial x_d}\varphi_{n,k}(\mathbf{x}',\pm\pi) \Big)\, d\mathbf{x}'.$$

If $n\mathbb{m} + \mathbf{y} = \mathbf{k} \in \mathbb{Z}^d$ then those two contributions sum to 0 by periodicity of $\varphi_{n,k}$, and the same is true for all other pairs of opposite faces, whence for such $\mathbf{y}$,

$$|\mathbf{y}|^{2r} \int_{\mathbb{W}_d} \psi(\mathbf{x})^n\, e^{-i\mathbf{x}\cdot\mathbf{y}}\, d\mathbf{x} = (-1)^r \int_{\mathbb{W}_d} e^{-i\mathbf{x}\cdot\mathbf{y}}\, \Delta^r \psi(\mathbf{x})^n\, d\mathbf{x}.$$

Replacing $\mathbf{x}$ with $\frac{1}{\sqrt{n}}\mathbf{x}$, where the "new" $\mathbf{x}$ is in $\sqrt{n}\,\mathbb{W}_d$, we get the proposed statement. □

(13.5) Lemma. *For any $a > 0$,*

$$\lim_{n\to\infty} \Delta^r \psi_n(\mathbf{x})^n = \Delta^r \exp\big(-\tfrac{1}{2}\Sigma[\mathbf{x}]\big)$$

uniformly for $\mathbf{x} \in a\mathbb{W}_d$.

Proof. We shall show by induction on $s \le 2r$ that for each fixed $\ell \ge 0$

$$\text{(13.6)} \qquad \lim_{n\to\infty} \frac{\partial^s}{\partial x_{j_1} \cdots \partial x_{j_s}} \big[\psi_n(\mathbf{x})^{n-\ell} - \exp\big(-\tfrac{1}{2}\Sigma[\mathbf{x}]\big)\big] = 0$$

uniformly in $\mathbf{x} \in a\mathbb{W}_d$ for all mixed derivatives of order s. For $s = 0$ this is true by the classical central limit theorem. To proceed, first observe that we can write

$$\text{(13.7)} \qquad \psi(\mathbf{x}) = 1 - \frac{1}{2}\Sigma[x] + \sum_{k=3}^{2r} H_k(\mathbf{x}) + \mathfrak{o}(|x|^{2r})_{\mathbf{x}\to 0},$$

where H_k is a homogeneous polynomial of degree k in $\mathbf{x} = (x_1, \ldots, x_d)$. Finiteness of $M_{2r}(\mu)$ is used here, and yields more generally that

$$\frac{\partial^s \psi(\mathbf{x})}{\partial x_{j_1} \cdots \partial x_{j_s}} = \sum_{k=0}^{2r-s} H_{k;j_1,\ldots,j_s}(\mathbf{x}) + \mathfrak{o}(|x|^{2r-s})_{\mathbf{x}\to 0}\,,$$

with $H_{k;j_1,\ldots,j_s}$ homogeneous of degree k. In particular, we compute

$$H_{0;j_1}(\mathbf{x}) = 0\,, \quad H_{1;j_1}(\mathbf{x}) = -\Sigma_{j_1} \cdot \mathbf{x} \quad \text{and} \quad H_{0;j_1,j_2} = -\Sigma_{j_1,j_2}\,,$$

where Σ_j denotes the j-th row of Σ. We obtain

$$\frac{\partial \psi_n(\mathbf{x})}{\partial x_{j_1}} = -\tfrac{1}{n}\Sigma_{j_1} \cdot \mathbf{x} + \mathfrak{o}\big(\tfrac{1}{n}\big)\,, \qquad \frac{\partial^2 \psi_n(\mathbf{x})}{\partial x_{j_1} x_{j_2}} = -\tfrac{1}{n}\Sigma_{j_1,j_2} + \mathfrak{o}\big(\tfrac{1}{n}\big)\,, \quad \text{and}$$

$$\frac{\partial^s \psi_n(\mathbf{x})}{\partial x_{j_1} \cdots \partial x_{j_s}} = \mathcal{O}\big(\tfrac{1}{n^{s/2}}\big) \quad \text{for } 3 \le s \le 2r\,,$$

where the $\mathfrak{o}(\cdot)$ and $\mathcal{O}(\cdot)$ are uniform in $\mathbf{x} \in a\mathbb{W}_d$. Now suppose that (13.6) is true for derivatives of order $< s$, and write

$$\text{(13.8)} \quad \frac{\partial^s \psi_n(\mathbf{x})^{n-\ell}}{\partial x_{j_1} \cdots \partial x_{j_s}} = \frac{\partial^{s-1}}{\partial x_{j_1} \cdots \partial x_{j_{s-1}}} \left[\psi_n(\mathbf{x})^{n-\ell-1} (n-\ell) \frac{\partial \psi_n(\mathbf{x})}{\partial x_{j_s}} \right]$$

$$= \sum_{\{1,\ldots,s-1\} = A \cup B} \frac{\partial^{|A|} \psi_n(\mathbf{x})^{n-\ell-1}}{\partial \mathbf{x}_A} (n-\ell) \frac{\partial^{|B|+1} \psi_n(\mathbf{x})}{\partial \mathbf{x}_{B \cup \{s\}}}\,,$$

where for a subset A of $\{1, \ldots, s\}$, the symbol $\partial^{|A|}/\partial \mathbf{x}_A$ stands for the mixed partial derivatives of order $|A|$ with respect to variables x_{j_k}, $k \in A$. By the above, if $|B| \ge 2$, then

$$\lim_{n\to\infty} (n-\ell) \frac{\partial^{|B|+1} \psi_n(\mathbf{x})}{\partial \mathbf{x}_{B \cup \{s\}}} = 0$$

uniformly in $\mathbf{x} \in a\mathbb{W}_d$. Writing $A_k = \{1, \ldots, s-1\} \setminus \{k\}$, we therefore get for $s \ge 2$

$$\lim_{n\to\infty} \frac{\partial^s \psi_n(\mathbf{x})^{n-\ell}}{\partial x_{j_1} \cdots \partial x_{j_s}}$$

$$= -\frac{\partial^{s-1} \exp\big(-\frac{1}{2}\Sigma[\mathbf{x}]\big)}{\partial x_{j_1} \cdots \partial x_{j_{s-1}}} \Sigma_{j_s} \cdot \mathbf{x} - \sum_{k=1}^{s-1} \frac{\partial^{s-2} \exp\big(-\frac{1}{2}\Sigma[\mathbf{x}]\big)}{\partial \mathbf{x}_{A_k}} \Sigma_{j_k,j_s}\,.$$

An immediate computation shows that this is the required mixed partial derivative of $\exp(-\frac{1}{2}\Sigma[\mathbf{x}])$. The – simpler – case $s = 1$ has to be treated separately, and is left to the reader. □

Replacing $\psi_n(\mathbf{x})^n$ by its limit in (13.4), and integrating over $\mathbb{R}^d$, one computes

$$
\begin{aligned}
(13.9)\quad (-1)^r \int_{\mathbb{R}^d} &\exp\left(-i\mathbf{x}\cdot\frac{\mathbf{k}-n\mathfrak{m}}{\sqrt{n}}\right) \Delta^r \exp(-\tfrac{1}{2}\Sigma[\mathbf{x}])\,d\mathbf{x} \\
&= \left(\frac{|\mathbf{k}-n\mathfrak{m}|}{\sqrt{n}}\right)^{2r} \int_{\mathbb{R}^d} \exp\left(-i\mathbf{x}\cdot\frac{\mathbf{k}-n\mathfrak{m}}{\sqrt{n}}\right) \exp(-\tfrac{1}{2}\Sigma[\mathbf{x}])\,d\mathbf{x} \\
&= (2\pi)^{d/2}\left(\frac{|\mathbf{k}-n\mathfrak{m}|}{\sqrt{n}}\right)^{2r} \frac{1}{\sqrt{\det\Sigma}} \exp\left(-\frac{\Sigma^{-1}[\mathbf{k}-n\mathfrak{m}]}{2n}\right).
\end{aligned}
$$

Combining (13.4) and (13.9), we can now deduce the following form of the local central limit theorem.

(13.10) Theorem. *If μ is irreducible and aperiodic with mean $\mathfrak{m}$ and finite covariance matrix Σ, and if $M_{2r}(\mu) < \infty$, then*

$$
\left(\frac{|\mathbf{k}-n\mathfrak{m}|}{\sqrt{n}}\right)^{2r} \left[\sqrt{2n\pi}^{\,d}\,\mu^{(n)}(\mathbf{k}) - \frac{1}{\sqrt{\det\Sigma}} \exp\left(-\frac{\Sigma^{-1}[\mathbf{k}-n\mathfrak{m}]}{2n}\right)\right]
$$

tends to 0 as $n \to \infty$, uniformly for $\mathbf{k} \in \mathbb{Z}^d$.

Proof. In view of (13.4) and (13.9), the expression in the theorem is $(-1)^r/\sqrt{2\pi}^{\,d}$ times the difference

$$
\begin{aligned}
\int_{\sqrt{n}\mathbb{W}_d} &\exp\left(-i\mathbf{x}\cdot\frac{\mathbf{k}-n\mathfrak{m}}{\sqrt{n}}\right) \Delta^r \psi_n(\mathbf{x})^n\,d\mathbf{x} \\
&- \int_{\mathbb{R}^d} \exp\left(-i\mathbf{x}\cdot\frac{\mathbf{k}-n\mathfrak{m}}{\sqrt{n}}\right) \Delta^r \exp(-\tfrac{1}{2}\Sigma[\mathbf{x}])\,d\mathbf{x}.
\end{aligned}
$$

We choose $a > 0$ and $0 < b < 1$ and decompose this difference as $I_1(\mathbf{k},n,a) + I_2(\mathbf{k},n,a) + I_3(\mathbf{k},n,a,b) + I_4(\mathbf{k},n,b)$, where

$$
\begin{aligned}
I_1(\mathbf{k},n,a) &= \int_{a\mathbb{W}_d} \exp\left(-i\mathbf{x}\cdot\frac{\mathbf{k}-n\mathfrak{m}}{\sqrt{n}}\right) \Delta^r \left[\psi_n(\mathbf{x})^n - \exp(-\tfrac{1}{2}\Sigma[\mathbf{x}])\right] d\mathbf{x}, \\
I_2(\mathbf{k},n,a) &= -\int_{\mathbb{R}^d\setminus a\mathbb{W}_d} \exp\left(-i\mathbf{x}\cdot\frac{\mathbf{k}-n\mathfrak{m}}{\sqrt{n}}\right) \Delta^r \exp(-\tfrac{1}{2}\Sigma[\mathbf{x}])\,d\mathbf{x}, \\
I_3(\mathbf{k},n,a,b) &= \int_{b\sqrt{n}\mathbb{W}_d\setminus a\mathbb{W}_d} \exp\left(-i\mathbf{x}\cdot\frac{\mathbf{k}-n\mathfrak{m}}{\sqrt{n}}\right) \Delta^r \psi_n(\mathbf{x})^n\,d\mathbf{x}, \quad \text{and} \\
I_4(\mathbf{k},n,b) &= \int_{\sqrt{n}\mathbb{W}_d\setminus b\sqrt{n}\mathbb{W}_d} \exp\left(-i\mathbf{x}\cdot\frac{\mathbf{k}-n\mathfrak{m}}{\sqrt{n}}\right) \Delta^r \psi_n(\mathbf{x})^n\,d\mathbf{x}.
\end{aligned}
$$

We replace each integrand by its absolute value; the resulting integrals do not depend on $\mathbf{k}$.

For any choice of a (which will depend on I_2 and I_3), $I_1(\mathbf{k}, n, a) \to 0$ uniformly in $\mathbf{k}$ by Lemma 13.5. Next,

$$|I_2(\mathbf{k}, n, a)| \le \int_{\mathbb{R}^d \setminus a\mathbb{W}_d} \Delta^r \exp(-\tfrac{1}{2}\Sigma[\mathbf{x}])\, d\mathbf{x}\,,$$

which becomes small for a sufficiently large.

To deal with I_3, we show that for each $s \le 2r$, $0 < \varepsilon \le 1/4$ and $\ell \ge 0$ we can find positive constants $b = b_{s,\varepsilon,\ell}$ and $C = C_{s,\varepsilon,\ell}$ such that

$$\left|\frac{\partial^s \psi_n(\mathbf{x})^{n-\ell}}{\partial x_{j_1} \cdots \partial x_{j_s}}\right| \le C \exp\Big(-(\tfrac{1}{2} - \varepsilon)\Sigma[\mathbf{x}]\Big) \quad \text{for all } \mathbf{x} \in b\sqrt{n}\mathbb{W}_d\,.$$

If $s = 0$, then we can use Lemma 13.1 and (13.7) to guarantee the existence of $b > 0$ such that $\frac{1}{2} \le |\psi(\mathbf{x})| \le 1 - (\frac{1}{2} - \varepsilon)\Sigma[\mathbf{x}]$ for all $\mathbf{x} \in b\mathbb{W}_d$. Therefore

$$|\psi_n(\mathbf{x})|^{n-\ell} \le |\psi_n(\mathbf{x})|^{-\ell} \exp\Big(-(\tfrac{1}{2} - \varepsilon)\Sigma[\mathbf{x}]\Big) \le 2^\ell \exp\Big(-(\tfrac{1}{2} - \varepsilon)\Sigma[\mathbf{x}]\Big)$$

for $\mathbf{x} \in b\sqrt{n}\mathbb{W}_d$. Now suppose that the statement is true for all partial derivatives up to order $s-1$. The computations of Lemma 13.5 show that we can find a constant C_0 such that for all $j_1, \ldots, j_s$,

$$\left|\frac{\partial \psi(\mathbf{x})}{\partial x_{j_1}}\right| \le C_0\, |\mathbf{x}| \quad \text{and} \quad \left|\frac{\partial^s \psi(\mathbf{x})}{\partial x_{j_1} \cdots \partial x_{j_s}}\right| \le C_0 \quad \text{for } 2 \le s \le 2r\,.$$

Therefore

$$n\left|\frac{\partial \psi_n(\mathbf{x})}{\partial x_{j_1}}\right| = \sqrt{n}\left|\frac{\partial \psi(\mathbf{x}/\sqrt{n})}{\partial x_{j_1}}\right| \le C_0\, |\mathbf{x}| \quad \text{and}$$

$$n\left|\frac{\partial^s \psi_n(\mathbf{x})}{\partial x_{j_1} \cdots \partial x_{j_s}}\right| = \frac{n}{\sqrt{n}^s}\left|\frac{\partial^s \psi(\mathbf{x}/\sqrt{n})}{\partial x_{j_1} \cdots \partial x_{j_s}}\right| \le C_0 \quad \text{for } 2 \le s \le 2r\,.$$

We now apply the induction hypothesis, with $\varepsilon/2$ in the place of ε, to formula (13.8) and obtain

$$\left|\frac{\partial^s \psi_n(\mathbf{x})^{n-\ell}}{\partial x_{j_1} \cdots \partial x_{j_s}}\right| \le (C_1|\mathbf{x}| + C_2) \exp\Big(-\tfrac{1-\varepsilon}{2}\Sigma[\mathbf{x}]\Big)$$

with C_1 and C_2 depending on s, ε and ℓ. The last expression is bounded above (on the whole of $\mathbb{R}^d$) by $C_{s,\varepsilon,\ell} \exp\Big(-(\frac{1}{2} - \varepsilon)\Sigma[\mathbf{x}]\Big)$ for a suitable constant $C_{s,\varepsilon,\ell}$.

From this we obtain, in particular, that for suitable constants b and C_3,

$$|\Delta^{2r}\psi_n(\mathbf{x})^n| \le C_3 \exp\Big(-\tfrac{1}{4}\Sigma[\mathbf{x}]\Big), \quad \mathbf{x} \in b\sqrt{n}\mathbb{W}_d .$$

We now can bound

$$|I_3(\mathbf{k},n,a,b)| \le C_3 \int_{\mathbb{R}^d \setminus a\mathbb{W}_d} \exp\Big(-\tfrac{1}{4}\Sigma[\mathbf{x}]\Big),$$

and choose a sufficiently large to make this small.

Finally, to deal with I_4, we use aperiodicity: by Lemma 13.3, there is $\delta > 0$ such that $|\psi(\mathbf{x})| \le 1 - 2\delta$ for $\mathbf{x} \in \mathbb{W}_d \setminus b\mathbb{W}_d$. A slight modification of the arguments used for controlling I_3 (using $n|\partial\psi_n(\mathbf{x})/\partial x_j| \le C_0\sqrt{n}$ and induction on s) shows that

$$|\Delta^r\psi_n(\mathbf{x})^n| \le C_4\,(1-\delta)^n \qquad \text{for all} \quad \mathbf{x} \in \sqrt{n}\mathbb{W}_d \setminus b\sqrt{n}\mathbb{W}_d\,, \quad \text{and}$$
$$|I_4(\mathbf{k},n,b)| \le C_4\,\big(2\pi\sqrt{n}\big)^d\,(1-\delta)^n\,,$$

which also tends to 0 as $n \to \infty$. □

Note that the theorem holds for every r such that $M_{2r}(\mu) < \infty$, in particular also for $r = 0$.

(13.11) Corollary. *When $\mathfrak{m}(\mu) = \mathbf{0}$,*

$$\mu^{(n)}(\mathbf{k}) \sim C\,n^{-d/2}\,\exp\left(-\tfrac{1}{2n}\Sigma^{-1}[\mathbf{k}]\right)$$

uniformly for $\mathbf{k}/\sqrt{n}$ bounded, where $C = (2\pi)^{-d/2}(\det\Sigma)^{-1/2}$.

An analogous local limit theorem when $\mathfrak{m}(\mu) \ne \mathbf{0}$ does not follow directly from Theorem 13.10. For this purpose, we require a finite exponential moment, see (1.22).

(13.12) Theorem. *Assume that μ is irreducible and aperiodic and has some finite exponential moment. Then*

$$\mu^{(n)}(\mathbf{k}) \sim C_{\mathbf{k}}\,\rho(\mu)^n\,n^{-d/2} \qquad \text{as } n \to \infty\,.$$

Proof. We write φ_μ for the function studied in §8.B: here we are in the situation of a lattice with one orbit only, and $\varphi_\mu(\mathbf{c}) = \sum_{\mathbf{k}} \mu(\mathbf{k})\,e^{\mathbf{c}\cdot\mathbf{k}}$. Let $\mathbf{c}_0$ be the unique point where this function is minimal (Proposition 8.20). Then $\rho(\mu) = \varphi_\mu(\mathbf{c}_0)$ by Theorem 8.23. We consider the probability measure

$$\bar{\mu}(\mathbf{k}) = \mu(\mathbf{k})\,e^{\mathbf{c}_0\cdot\mathbf{k}}/\rho(\mu)\,.$$

It is irreducible and aperiodic, satisfies an exponential moment condition, and $\mathfrak{m}(\bar{\mu}) = \frac{1}{\rho(\mu)}\operatorname{grad}\varphi_\mu(\mathbf{c}_0) = \mathbf{0}$. Write $\bar{\Sigma}$ for its covariance matrix. By Corollary 13.11,

$$(13.13) \qquad \frac{\bar{\mu}^{(n)}(\mathbf{k})}{C\,n^{-d/2}\exp\left(-\frac{1}{2n}\bar{\Sigma}^{-1}[\mathbf{k}]\right)} = \frac{\mu^{(n)}(\mathbf{k})}{C\,n^{-d/2}\,\rho(\mu)^n\exp\left(-\mathbf{c}_0\cdot\mathbf{k}-\frac{1}{2n}\bar{\Sigma}^{-1}[\mathbf{k}]\right)} \to 1$$

as $n\to\infty$, uniformly for $\mathbf{k}/\sqrt{n}$ bounded. □

14. Growth, isoperimetric inequalities, and the asymptotic type of random walk

The results we shall deal with in this section are not as sharp as asymptotic equivalence $\sim$. If (a_n) and (b_n) are two non-negative sequences then we write $a_n \preccurlyeq b_n$ if there are $C \ge c > 0$ such that for all sufficiently large n, $a_n \le C\sup\{b_k : cn \le k \le Cn\}$. If also $b_n \preccurlyeq a_n$ then we write $a_n \approx b_n$. (In the same way, we shall compare non-negative functions on $\mathbb{R}^+$.) An equivalence class of sequences under this relation is called an *asymptotic type.* By the asymptotic type of a random walk (X,P) we mean the asymptotic type of the sequence $\left(p^{(n)}(x,x)\right)$. By irreducibility, it is independent of x. We want to determine this type by finding a concrete numerical sequence in the same equivalence class. For example, we know from Corollary 13.11 that for an aperiodic random walk on $\mathbb{Z}^d$ with mean 0 and finite variance, the asymptotic type is $(n^{-d/2})$.

Clearly, asymptotic type is less precise than asymptotic equivalence. For example, sequences of the form $\left(e^{-\lambda n}Q(n)\right)$ (where $\lambda > 0$ and Q is a polynomial) are all of asymptotic type (e^{-n}). Throughout this section we shall deal with reversible (X,P) which do not satisfy *IS*.

A. Upper bounds and Nash inequalities

We shall take up the material from Section 4. Our first aim is to link $\mathfrak{F}$-isoperimetric inequalities with the asymptotic type of (X,P), which is assumed to be reversible with invariant measure $m(\cdot)$ and conductance $a(\cdot,\cdot)$. The analytic link is the following.

(14.1) Proposition. *Let $\mathfrak{F} : \mathbb{R}^+ \to \mathbb{R}^+$ be such that $\mathfrak{f}(t) = t/\mathfrak{F}(t)$ is increasing. If (X,P) satisfies $IS_{\mathfrak{F}}$ (with constant κ) then the* Nash inequality

$$NA_{\mathfrak{g}}: \qquad \|f\|_2^2 \le \mathfrak{g}\left(\|f\|_1^2/\|f\|_2^2\right) D_P(f),$$

with $\mathfrak{g}(t) = 4\kappa^2\mathfrak{f}(4t)^2$, holds for all $f \in \ell_0(X)$.

Proof. In terms of $\mathfrak{f}$, the isoperimetric inequality reads $m(A) \le \kappa\,\mathfrak{f}\big(m(A)\big)\,a(\partial A)$.

We start with non-negative $f \in \ell_0(X)$. We can write $f(x) = \int_0^\infty \mathbf{1}_{[f>t]}(x)\,dt$. We have $[f > t] \subset \operatorname{supp} f$, and monotonicity of $\mathfrak{f}$ (this property is crucial !) implies $m[f > t] \le \kappa\,\mathfrak{f}\big(m(\operatorname{supp} f)\big)\,a(\partial[f > t])$. We compute

$$\begin{aligned}\|f\|_1 = \sum_x f(x)m(x) &= \int_0^\infty m[f > t]\,dt \\ &\le \kappa\,\mathfrak{f}\big(m(\operatorname{supp} f)\big)\int_0^\infty a(\partial[f > t]) = \kappa\,\mathfrak{f}\big(m(\operatorname{supp} f)\big)\,S_P(f)\,.\end{aligned}$$

(For the last identity, see the proof of Proposition 4.3.) We now apply this to f^2. In the proof of Theorem 10.3, we have seen that $S_P(f^2)^2 \le 2D_P(f)\|f\|_2^2$. Therefore

$$\|f\|_2^2 = \|f^2\|_1^2/\|f\|_2^2 \le 2\kappa^2\,\mathfrak{f}\big(m(\operatorname{supp} f)\big)^2\,D_P(f)\,.$$

Now let $t > 0$ and set $f_t(x) = \max\{f(x) - t, 0\}$. One checks that $f^2 \le f_t^2 + 2tf$. Also, $D_P(f_t) \le D_P(f)$ and $m(\operatorname{supp} f_t) = m[f > t] \le \|f\|_1/t$ (Markov's inequality). Therefore – using monotonicity of $\mathfrak{f}$ once more –

$$\begin{aligned}\|f\|_2^2 \le \|f_t\|^2 + 2t\|f\|_1 &\le \tfrac{1}{2}\mathfrak{g}\big(m(\operatorname{supp} f_t)/4\big)\,D_P(f_t) + 2t\|f\|_1 \\ &\le \tfrac{1}{2}\mathfrak{g}\big(\|f\|_1/4t\big)\,D_P(f) + 2t\|f\|_1\,.\end{aligned}$$

If we now choose $4t = \|f\|_2^2/\|f\|_1$ then we get the proposed inequality for non-negative f. The general case is immediate from $D_P(|f|) \le D_P(f)$. □

We shall use this to obtain upper bounds on $p^{(n)}(x,y)/m(y)$. We need some preparatory facts.

(14.2) Lemma. (a) *For $f \in \ell^2(X,m)$, the sequences*

$$\|P^n f\|_2^2 \qquad \text{and} \qquad \|P^n f\|_2^2 - \|P^{n+1} f\|_2^2$$

are decreasing.

$$\text{(b)} \qquad \sup_{x,y}\frac{p^{(2n)}(x,y)}{m(y)} = \sup_x \frac{p^{(2n)}(x,x)}{m(x)} = \|P^{2n}\|_{1\to\infty} = \|P^n\|_{1\to 2}^2\,,$$

where $\|\cdot\|_{p\to q}$ denotes norms of operators from $\ell^p(X,m)$ to $\ell^q(X,m)$.

(c) *Setting $\bar{P} = \frac{1}{2}(I + P)$, we have*

$$\begin{aligned}p^{(2n)}(x,x) &\le 2\bar{p}^{(2n)}(x,x) \quad \text{for all } x\,, \text{ and} \\ D_P(f) &\le 2(\|f\|_2^2 - \|\bar{P}f\|_2^2) \quad \text{for all } f \in \ell_0(X)\,.\end{aligned}$$

Proof. (a) We have seen in the proof of Lemma 10.1 that the sequence $(P^{n+1}f, P^{n+1}f)/(P^n f, P^n f)$ is increasing and bounded by $\rho(P)^2 \le 1$. For the second part, note that the operator $I - P^2$ is non-negative definite and has a square root on $\ell^2(X, m)$, which commutes with P. Hence, as $\|P\|_{2\to 2} \le 1$,

$$\begin{aligned}\|P^n f\|_2^2 - \|P^{n+1} f\|_2^2 &= ((I - P^2)P^n f, P^n f) = \|(I - P^2)^{1/2} P^n f\|_2^2 \\ &\le \|(I - P^2)^{1/2} P^{n-1} f\|_2^2 = \|P^{n-1} f\|_2^2 - \|P^n f\|_2^2 .\end{aligned}$$

(b) For the first identity, observe that $m(x)p^{(2n)}(x, y) = (P^n \delta_x, P^n \delta_y)$ and apply the Cauchy–Schwarz inequality. The other identities are straightforward exercises.

(c) Use the fact that $p^{(2n)}(x, x)$ is decreasing, and write

$$\begin{aligned}\bar{p}^{(2n)}(x, x) &= \frac{1}{2^{2n}} \sum_{i=0}^{2n} \binom{2n}{i} p^{(i)}(x, x) \\ &\ge \frac{1}{2^{2n}} \sum_{k=0}^{n} \binom{2n}{2k} p^{(2n)}(x, x) = \frac{1}{2} p^{(2n)}(x, x) .\end{aligned}$$

For the above inequality, note that $\bar{P}^2 = \frac{1}{4}(I + 2P + P^2) \ge \frac{1}{2}P$ elementwise, and that $\bar{P}^2$ is also reversible with respect to m, whence

$$\|f\|_2^2 - \|\bar{P} f\|_2^2 = ((I - \bar{P}^2)f, f) = D_{\bar{P}^2}(f) \ge \tfrac{1}{2} D_P(f, f) . \qquad \square$$

We can now show how the Nash inequality determines an upper bound on the transition probabilities.

(14.3) Theorem. *Suppose that (X, P) is reversible and that $m_0 = \inf_x m(x) > 0$. If $NA_{\mathfrak{g}}$ holds for a increasing, continuous function $\mathfrak{g} : \mathbb{R}^+ \to \mathbb{R}^+$ then*

$$\sup_{x,y} \frac{p^{(2n)}(x, y)}{m(y)} \le 2\mathfrak{a}(n) ,$$

where $\mathfrak{a} : [0, \infty) \to (0, \frac{1}{m_0}]$ is the solution of the differential equation

$$\mathfrak{a}'(t) = -\frac{\mathfrak{a}(t)}{2\mathfrak{g}(1/\mathfrak{a}(t))}, \quad \mathfrak{a}(0) = 1/m_0 . \tag{14.4}$$

Proof. In view of Lemma 14.2, we work with $\bar{P}$: the Nash inequality becomes

$$\|f\|_2^2 \le 2\,\mathfrak{g}(\|f\|_1^2/\|f\|_2^2)\,(\|f\|_2^2 - \|\bar{P} f\|_2^2) .$$

Let $f \in \ell_0(X)$ with $\|f\|_1 = 1$. We apply the inequality with $\bar{P}^n f$ in place of f. Setting $u(n) = \|\bar{P}^n f\|_2^2$ and $u(t)$ for the piecewise linear extension to $\mathbb{R}^+$, we have that $u(t)$ is decreasing and convex by (14.2.a), and

$$u(n) \le 2\mathfrak{g}\big(1/u(n)\big)\big(u(n) - u(n+1)\big)\,.$$

By monotonicity of $\mathfrak{g}$,

$$u(t) \le -2\mathfrak{g}\big(1/u(t)\big)\,u'(t+)\,.$$

Also, $u(0) \le 1/m_0$. Therefore, elementary calculus shows that $u(t) \le \mathfrak{a}(t)$ for all $t \ge 0$, where $\mathfrak{a}(t)$ is as stated. The result follows; see Lemma 14.2(b). □

This yields concrete upper bounds in the presence of nice isoperimetric inequalities, and, if (X, P) is quasi-homogeneous as in (4.14), good upper bounds can also be extracted from the growth function. In the sequel, when involving the growth functions $V_P(x, n)$ and $V_P(n)$ corresponding to P, we shall omit the subscript P and assume implicitly that the graph associated with (X, P) (see §2.A) is locally finite. Also, we always suppose that $m_0 = \inf_X m(x) > 0$.

(14.5) Corollary. (a) *If (X, P) satisfies IS_d – and in particular, if (X, P) is quasi-homogeneous and $V(n) \succcurlyeq n^d$ – then*

$$\sup_{x,y} \frac{p^{(n)}(x,y)}{m(y)} \preccurlyeq n^{-d/2}\,.$$

(b) *If (X, P) satisfies $IS_{\mathfrak{F}}$ with $\mathfrak{F}(t) = t/[\log(2+t)]^{1/\alpha}$, where $0 < \alpha \le \infty$ – and in particular, if (X, P) is quasi-homogeneous and $V(n) \succcurlyeq \exp(n^\alpha)$, where $0 < \alpha \le 1$ – then*

$$\sup_{x,y} \frac{p^{(n)}(x,y)}{m(y)} \preccurlyeq \exp(-n^{\frac{\alpha}{\alpha+2}})\,.$$

Proof. First of all, note that under the assumptions of Theorem 14.3, we also have

$$\frac{p^{(2n+1)}(x,y)}{m(y)} = \sum_{w \in X} p(x,w) \frac{p^{(2n)}(w,y)}{m(y)} \le 2\mathfrak{a}(n)\,.$$

(a) IS_d implies $NA_{\mathfrak{g}}$ with $\mathfrak{g}(t) \approx (1+t)^{2/d}$, and the solution of (14.4) satisfies $\mathfrak{a}(t) \approx (1+t)^{-d/2}$. The "and in particular" follows from Proposition 4.15.

(b) This time, $IS_{\mathfrak{F}}$ implies $NA_{\mathfrak{g}}$ with $\mathfrak{g}(t) = [\log(2+t)]^{2/\alpha}$, and the solution of (4.14) satisfies $\mathfrak{a}(t) \approx \exp(-t^{\frac{\alpha}{\alpha+2}})$. The special case where $\mathfrak{F}(t) = t$ ($\alpha = \infty$) is covered by Theorem 10.3. Once more, the "and in particular" follows from Proposition 4.15. □

As typical examples, (a) applies to symmetric random walks with finite first moment on groups with polynomial growth of degree d, and also to quasi-transitive random walks with analogous properties on graphs; see Corollary 4.6. This yields yet another way of classifying the recurrent quasi-transitive graphs. Also, (b) applies to quasi-transitive, symmetric random walks with bounded range on amenable graphs with exponential growth: in this case, we get $\alpha = 1$. Grigorchuk [147] has given examples of finitely generated groups satisfying $\exp(n^\alpha) \preccurlyeq V(n) \preccurlyeq \exp(n^\beta)$, where $0 < \alpha \le \beta < 1$. (In those examples, it is not known whether $\alpha = \beta$.) This gives an upper bound on transition probabilities with $\alpha \ne 1$. In Section 15, we shall consider several examples in more detail. There is one general consequence which is worth while noting.

(14.6) Corollary. *If* (X,P) *satisfies* $\inf_X m(x) > 0$ *and* $\sup_E r(e) < \infty$ *then*

$$\sup_{x,y} \frac{p^{(n)}(x,y)}{m(y)} \preccurlyeq n^{-1/2}.$$

This holds, in particular, for the simple random walk on a locally finite, infinite graph.

Indeed, under these assumptions, IS_1 is satisfied. Theorem 14.3 also has a converse:

(14.7) Proposition. *Suppose* (X,P) *is reversible and* $\lim_{n\to\infty} \|P^n\|_{1\to 2} = 0$. *Define*

$$\mathfrak{g}(t) = \inf\left\{1 + \frac{n}{-\log(t\,\|P^n\|_{1\to 2})} : n \ge 0\,,\ \|P^n\|_{1\to 2} < 1/t\right\}.$$

Then (X,P) *satisfies* $NA_{4\mathfrak{g}}$.

Proof. Let $f \in \ell_0(X)$ with $\|f\|_1 = 1$. Set $v(n) = \|P^n f\|_2^2$. Recall that $\|P^{n+1}f\|_2^2/\|P^n f\|_2^2$ is increasing (Lemma 10.1), whence $\|f\|_2^2/\|P^n f\|_2^2 = \prod_{k=1}^n (\|P^{k-1}f\|_2^2/\|P^k f\|_2^2) \le (\|f\|_2^2/\|Pf\|_2^2)^n$. Using $\log x \le x-1$, we obtain

$$\log\left(\frac{\|f\|_2^2}{\|P^n\|_{1\to 2}}\right) \le \log\left(\frac{\|f\|_2^2}{\|P^n f\|_2^2}\right) \le n\log\left(\frac{\|f\|_2^2}{\|Pf\|_2^2}\right) \le n\,\frac{\|f\|_2^2 - \|Pf\|_2^2}{\|Pf\|_2^2},$$

and, if n is such that $\|f\|_2^2 > \|P^n\|_{1\to 2}$ (here we use that the latter tends to 0), then

$$\begin{aligned}\|f\|_2^2 &= \left(\|f\|_2^2 - \|Pf\|_2^2\right) + \|Pf\|_2^2 \\ &\le \left(1 + \frac{n}{\log\left(\|f\|_2^2/\|P^n\|_{1\to 2}\right)}\right)\left(\|f\|_2^2 - \|Pf\|_2^2\right).\end{aligned}$$

Now recall Lemma 2.5:

$$\left(\|f\|_2^2 - \|Pf\|_2^2\right) = D_{P^2}(f) \le 4\, D_P(f).$$

If $\|f\|_1 \neq 1$ then we normalize it. Thus, when $\|f\|_2^2/\|f\|_1^2 > \|P^n\|_{1\to 2}$ we have

$$\|f\|_2^2 \le 4\mathfrak{g}_n\left(\|f\|_1^2/\|f\|_2^2\right) D_P(f),$$

where $\mathfrak{g}_n(t) = 1 + \dfrac{n}{-\log(t\|P^n\|_{1\to 2})}$. □

We remark that in reality, we have proved the following slightly stronger version of the Nash inequality:

$$\|f\|_2^2 \le \mathfrak{g}\left(\|f\|_1^2/\|f\|_2^2\right)\left(\|f\|_2^2 - \|Pf\|_2^2\right) \quad \text{for all } f \in \ell_0(X), \tag{14.8}$$

with $\mathfrak{g}$ as in the proposition. In view of Lemma 14.2, it makes no essential difference if we work with (14.8) or with the Nash inequality, as the latter implies (14.8) for $\bar{P}^2$. In the following we suppose again that the graph associated with (X, P) is locally finite.

(14.9) Theorem. (a) *If $m_0 > 0$ and $NA_{\mathfrak{g}}$ holds with $\mathfrak{g}(t) \preccurlyeq t^{2/d}$, where $d > 0$, then $V(n) \succcurlyeq n^d$.*

(b) *If $m_0 > 0$ and $NA_{\mathfrak{g}}$ holds with $\mathfrak{g}(t) \preccurlyeq [\log(2+t)]^{2/\alpha}$, where $\alpha > 0$, then $V(n) \succcurlyeq \exp\left(n^{\alpha/(\alpha+1)}\right)$.*

Proof. Suppose more generally that $NA_{\mathfrak{g}}$ holds with $\mathfrak{g}$ continuous and strictly increasing. Fix $x \in X$, $r, s \in \mathbb{N}$ and consider the function $f(y) = \max\{r + s - d(y,x), 0\}$. If $d(v,w) = 1$ then $|f(v) - f(w)| \le 1$, so that $D_P(f) \le V(x, r+s)/2$. Also, $f \ge s$ on $B(x,r)$. Note that every f satisfies $\|f\|_1^2/\|f\|_2^2 \le m(\operatorname{supp} f)$. Therefore

$$\begin{aligned} s^2\, V(x,r) &\le \mathfrak{g}\big(V(x,r+s)\big)\, V(x,r+s)/2, \quad \text{or equivalently} \\ V(x,r+s) &\ge \mathfrak{g}^{-1}\big(2s^2 V(x,r)/V(x,r+s)\big), \end{aligned}$$

where $\mathfrak{g}^{-1}$ denotes the inverse function of $\mathfrak{g}$. Now let $n \ge s$, set $q = \lfloor n/s \rfloor$ (the next lower or equal integer) and $r_j = js$, $j = 0, \dots, q$.

If there is $j \in \{1,\dots,q\}$ such that $V(x,r_{j-1})/V(x,r_j) \ge 1/2$ then

$$V(x,n) \ge V(x,r_j) \ge \mathfrak{g}^{-1}(s^2)\,.$$

Otherwise, $V(x,r_{j-1})/V(x,r_j) < 1/2$ for all j, and

$$V(x,n) \ge V(x,r_q) \ge 2^q V(x,0)\,.$$

We have proved that

$$\text{(14.10)}\quad V(x,n) \ge \min\{\mathfrak{g}^{-1}(s^2)\,,\, 2^{\lfloor n/s \rfloor} m_0\} \quad \text{for all } n,s \in \mathbb{N} \text{ with } n \ge s\,.$$

From this general result we can deduce (b): if $\mathfrak{g}(t) \preccurlyeq [\log(2+t)]^{2/\alpha}$ then $\mathfrak{g}^{-1}(t) \succcurlyeq \exp(t^{\alpha/2})$. Therefore, if we set $s = \lfloor n^{1/(1+\alpha)} \rfloor$ then

$$V(x,n) \ge \min\{c_1 \exp(c_2\, s^\alpha)\,,\, 2^{\lfloor n/s \rfloor} m_0\} \succcurlyeq \exp\bigl(n^{\alpha/(\alpha+1)}\bigr)\,.$$

For (a), we proceed differently, starting with $r = s = 2^k$. In this case we get $4^k V(x,2^k) \le C\,V(x,2^{k+1})^{1+2/d}$, where $C \ge 1$. Similarly as for Lemma 4.13, we show inductively that one can find a constant $c_0 > 0$ such that $V(x,2^k) \ge c_0\, 2^{kd}$. To have this for $k = 0$, we choose $c_0 \le 2m_0$. Now suppose that we have the lower bound for k, and write

$$\begin{aligned} V(x,2^{k+1}) &\ge \bigl(4^k V(x,2^k)/C\bigr)^{d/(d+2)} \ge \bigl(c_0\, 2^{(d+2)k}/C\bigr)^{d/(d+2)} \\ &= \frac{1}{2^d}(c_0/C)^{d/(d+2)} 2^{(k+1)d}\,. \end{aligned}$$

Thus, induction works if $c_0 \le 2^{-d}(c_0/C)^{d/(d+2)}$, that is, we have to choose $c_0 \le \min\{2m_0\,,\, 2^{-d(d+2)/2} C^{-d/2}\}$.

Finally, if $2^{k-1} \le n < 2^k$ then $V(x,n) \ge c_0\, 2^{(k-1)d} \ge c_0 2^{-d}\, n^d$. □

(14.11) Corollary. (a) *If* $m_0 > 0$ *and* $d > 0$ *then*

$$\sup_{x,y} \frac{p^{(n)}(x,y)}{m(y)} \preccurlyeq n^{-d/2} \implies V(n) \succcurlyeq n^d\,.$$

(b) *If* $m_0 > 0$ *and* $0 < \beta < 1$ *then*

$$\sup_{x,y} \frac{p^{(n)}(x,y)}{m(y)} \preccurlyeq \exp(-n^\beta) \implies V(n) \succcurlyeq \exp\bigl(n^{2\beta/(\beta+1)}\bigr)\,.$$

Proof. In both cases, we bound the function $\mathfrak{g}(t)$ of Proposition 14.7 from above and then apply Theorem 14.9.

(a) We have $\|P^n\|_{1\to 2} \le C\,n^{-d/2}$. Therefore

$$\mathfrak{g}(t) \le \inf\left\{1 + \frac{2n}{d\log(n) - 2\log(Ct)} : n > (Ct)^{2/d}\right\}.$$

Replacing n by a real variable x, we find that the infimum is attained at $x_0 = e\,(Ct)^{2/d}$. Setting $n = \lfloor x_0 \rfloor + 1$, we obtain $\mathfrak{g}(t) \le 1 + \frac{2}{d}\big(e\,(Ct)^{2/d} + 1\big) \preccurlyeq t^{2/d}$.

(b) This time $\|P^n\|_{1\to 2} \le c_1 \exp(-c_2 n^\beta)$, and

$$\mathfrak{g}(t) \le \inf\left\{1 + \frac{n}{c_2 n^\beta - \log(c_1 t)} : n > \left(\frac{\log(c_1 t)}{c_2}\right)^{1/\beta}\right\}.$$

Proceeding as above, we find $x_0 = \Big(\big(\log(c_1 t)\big)/\big(c_2(1-\beta)\big)\Big)^{1/\beta}$, and consequently $\mathfrak{g}(t) \preccurlyeq [\log(2+t)]^{(1-\beta)/\beta}$. □

In particular, we see from Corollary 4.16 and the above results that for quasi-homogeneous (X,P) with $m_0 = \inf_X m(x) > 0$, one has

$$IS_d \iff V_P(n) \succcurlyeq n^d \iff \sup_{x,y} \frac{p^{(n)}(x,y)}{m(y)} \preccurlyeq n^{-d/2}$$
$$\iff NA_{\mathfrak{g}} \quad \text{with} \quad \mathfrak{g}(t) = t^{2/d} \qquad (d > 0).$$

In general, it is very easy to see that $NA_{\mathfrak{g}}$ implies $IS_{\mathfrak{F}}$ with $\mathfrak{F}(t) = t/\mathfrak{g}(t)$, but in view of Proposition 14.1, the two conditions are not equivalent; compare with the example of Coulhon and Ledoux [83].

B. Gaussian upper bounds

Let (X,P) be reversible, and denote by $d(\cdot,\cdot)$ the distance in the graph associated with the chain (§2.B). The aim of this subsection is to prove the following off-diagonal estimate.

(14.12) Theorem. *If* $\sup_x \dfrac{p^{(n)}(x,x)}{m(x)} \le C_0\,n^{-d/2}$ *for all* $n \ge 2$ *then*

$$\frac{p^{(n)}(x,y)}{m(y)} \le C_1\,n^{-d/2} \exp\left(-\frac{d(x,y)^2}{C_2 n}\right).$$

This needs some preparation. Fix $o \in X$, define $w_s(x) = e^{s\,d(x,o)}$, where $s \in \mathbb{R}$, and a new (non-stochastic) operator P_s by $p_s(x,y) = p(x,y)w_s(y)/w_s(x)$; compare with (8.16). We have $(P_s)^n = (P^n)_s$. Also, $w_s(y)/w_s(x) \le e^{|s|d(y,x)}$, so that the row sums of P_s^n are bounded by $e^{n|s|}$. Consequently,

(14.13) $$\|P_s^n\|_{p\to p} \le e^{n|s|} \quad \text{for every } p \ge 1.$$

We shall need a better estimate for small $|s|$.

(14.14) Lemma. $\langle \nabla w_s f, \nabla w_{-s} f \rangle \geq -s^2(1+e^{2|s|})\|f\|_2^2$ *for all $f \in \ell_0(X)$ and $s \in \mathbb{R}$.*

Proof. We fix s and set $w(x) = w_s(x)$. Replacing f with wf, we have to show $\langle \nabla w^2 f, \nabla f \rangle \geq -s^2(1+e^{2|s|})\|wf\|_2^2$. We write

$$\begin{aligned}
4\langle \nabla w^2 f, \nabla f \rangle &= 2\sum_{x,y} \big(f(x)-f(y)\big)\big(w(x)^2 f(x) - w(y)^2 f(y)\big) a(x,y) \\
&= A_1 + A_2\,, \qquad \text{where} \\
A_1 &= \sum_{x,y} \big(f(x)-f(y)\big)^2 \big(w(x)^2 + w(y)^2\big) a(x,y) \quad \text{and} \\
A_2 &= \sum_{x,y} \big(f(x)^2 - f(y)^2\big)\big(w(x)^2 - w(y)^2\big) a(x,y)\,.
\end{aligned}$$

Now $|1-e^s|/(1+e^s) \leq |s|$, so that $|w(x)-w(y)| \leq |s|\big(w(x)+w(y)\big)$ when $d(x,y) \leq 1$. We use this, $(a \pm b)^2 \leq 2(a^2+b^2)$ and Cauchy–Schwarz to bound

$$\begin{aligned}
A_2^2 &\leq \sum_{x,y} \big(f(x)-f(y)\big)^2 \big(w(x)+w(y)\big)^2 a(x,y) \\
&\qquad \times \sum_{x,y} \big(f(x)+f(y)\big)^2 \big(w(x)-w(y)\big)^2 a(x,y) \\
&\leq 8s^2 A_1 \sum_{x,y} \big(f(x)^2 + f(y)^2\big)\big(w(x)^2 + w(y)^2\big) a(x,y) \\
&= 16 s^2 A_1 \sum_x f(x)^2 w(x)^2 \sum_y \big(1 + w(y)^2/w(x)^2\big) a(x,y) \\
&\leq 16 s^2 (1+e^{2|s|}) A_1 \|wf\|_2^2\,.
\end{aligned}$$

Therefore $|A_2| \leq A_1 + 4s^2(1+e^{2|s|})\|wf\|_2^2$. □

The following is the key estimate.

(14.15) Proposition. *Under the above assumptions there is $C > 0$ such that for all n and all s*

$$\|P_s^n\|_{2\to 2} \leq \exp\big(C(s^2 n + 1)\big)\,.$$

Proof. When $|s| > 1$ this is immediate from (14.13). So suppose $|s| \leq 1$. We use the continuous-time semigroups

$$P_{s,t} = e^{-t} \sum_{n=0}^{\infty} \frac{t^n}{n!} P_s^n \quad (s \in \mathbb{R}\,,\ t > 0)\,,$$

$P_{0,t}$ is stochastic and reversible with respect to $m(\cdot)$, and $P_{s,t} = (P_{0,t})_s$ (conjugation by w_s). We differentiate and apply Lemma 14.14:

$$\begin{aligned}\frac{\partial}{\partial t}\|P_{s,t}f\|_2^2 &= 2\big(w_{-s}P_{s,t}f, (P-I)w_sP_{s,t}f\big)\\ &= -2\big\langle \nabla w_{-s}P_{s,t}f, \nabla w_sP_{s,t}f\big\rangle \le C_3s^2\|P_{s,t}f\|_2^2\end{aligned}$$

with $C_3 = 2+2e^2$. Elementary calculus yields $\|P_{s,t}f\|_2^2 \le e^{C_3s^2}\|f\|_2^2$, since $P_{s,0}f = f$. We now have to transform this into the desired inequality for P_s^n. Let $f \in \ell_0(X)$ be non-negative. Note that the adjoint of P_s on $\ell^2(X,m)$ is P_{-s}, and also that $P_{-s}^j \le e^{2|s|j}P_s^j$ elementwise. In particular,

$$\big(P_s^jf, P_s^{j+2r}f\big) \ge e^{-2|s|j}\big(P_s^jf, P_{-s}^rP_s^{j+r}f\big) = e^{-2|s|r}\|P_s^{j+r}f\|_2^2\,.$$

Write $E(n) = \{j \in \mathbb{N}\cap[n-\sqrt{n}\,, n] : j \text{ even}\}$. We obtain

$$\begin{aligned}\exp^{C_3s^2}\|f\|_2^2 &\ge \left\|e^{-t}\sum_{j=0}^{\infty}\frac{t^j}{j!}P_s^jf\right\|_2^2 \ge e^{-2t}\sum_{j,k\in E(n)}\frac{t^{j+k}}{j!k!}(P_s^jf, P_s^kf)\\ &\ge e^{-2t}\sum_{j,k\in E(n)}\frac{t^{j+k}}{j!k!}e^{-|s|\,|k-j|}\|P_s^{(j+k)/2}f\|_2^2\\ &\ge e^{-3|s|\sqrt{n}}\|P_s^nf\|_2^2\Big(e^{-t}\sum_{j\in E(n)}\frac{t^j}{j!}\Big)^2.\end{aligned}$$

(We have used (14.13) in the last inequality.) We now set $t=n$. The term in the large parenthesis is $\ge \frac{1}{2}\mathbb{P}[n-\sqrt{n} \le S_n < n]$, where S_n is a Poisson random variable with mean n. By the central limit theorem (or directly), this has a positive limit. Therefore we find C_4 not depending on f such that

$$\|P_s^nf\|_2^2 \le C_4\exp(C_3s^2 + 3|s|\sqrt{n})\|f\|_2^2\,.$$

As $\|P_s^nf\| \le \big\|P_s^n|f|\big\|$ for arbitrary f, the proposed estimate follows. □

Proof of Theorem 14.12. The assumption implies $\|P^{2n}\|_{1\to\infty} \le C_0(2n)^{-d/2}$ for all $n \ge 1$ (Lemma 14.2). If the proposed statement holds for even n, then it is easily seen to hold with suitably modified constants for all odd $n > 2$ as well (exercise!). Therefore, we may replace P with P^2, or – which amounts to the same – we assume that $\|P^n\|_{1\to\infty} \le C_0n^{-d/2}$ for all $n \ge 1$.

First of all, this implies $\|P_s\|_{1\to\infty} \le C_0e^{|s|}$. Also, $\|P_s\|_{\infty\to\infty} \le e^{|s|}$, and Riesz–Thorin interpolation (see e.g. Bennett and Sharpley [40]) implies

$$\|P_s\|_{p\to\infty} \le C_0^{1/p}e^{|s|}\,,\quad 1 \le p \le \infty\,. \tag{14.16}$$

Analogously, we combine $\|P^n\|_{1\to\infty} \le C_0 n^{-d/2}$ and $\|P^n\|_{\infty\to\infty} \le 1$ to get

$$(14.17)\qquad \|P^n\|_{p\to\infty} \le C_0^{1/p} n^{-d/(2p)}, \quad 1 \le p \le \infty.$$

Next, we use Stein's interpolation theorem (see [40], Theorem 3.3). If $z = \pm s + it$, where $s \in [0,1]$ and $t \in \mathbb{R}$, then P_z (defined like P_s with z in the place of s) gives rise to an analytic family of operators with admissible growth. Therefore we can interpolate between $\|P_0^n\|_{\infty\to\infty} \le 1$ and the estimate of Proposition 14.15 to obtain $\|P_{\theta s}^n\|_{p\to p} \le \exp\bigl(C\theta(s^2n+1)\bigr)$, where $\frac{1}{p} = \frac{\theta}{2} + \frac{1-\theta}{\infty}$, $p \ge 2$. Replacing θs with s, this becomes

$$(14.18)\qquad \|P_s^n\|_{p\to p} \le \exp\Bigl(C\bigl(\tfrac{p}{2}s^2n + \tfrac{2}{p}\bigr)\Bigr), \quad 2 \le p < \infty.$$

We conclude this *tour de force* through interpolation by applying the Riesz–Thorin theorem to (14.17) and (14.18), and get

$$\|P_s^n\|_{p\to q} \le (C_0 n^{-d/2})^{\frac{1}{p}-\frac{1}{q}} \exp\bigl(C(\tfrac{q}{2}s^2n + \tfrac{2}{q})\bigr) \quad \text{for } 2 \le p \le q < \infty.$$

The last inequality will now be combined with (14.16): decompose $n = 1 + \sum_{j=1}^N n_j$ and choose $2 = p_1 < p_2 < \cdots < p_N < \infty$. Then, assuming $C \ge 1/2$ and using $|s| \le C(s^2+1)$,

$$\begin{aligned}\|P_s^n\|_{2\to\infty} &\le \|P_s\|_{p_N\to\infty} \prod_{j=1}^{N-1} \|P_s^{n_j}\|_{p_j\to p_{j+1}} \\ &\le C_0^{\frac{1}{2}} n^{-\frac{d}{2}\left(\frac{1}{2}-\frac{1}{p_N}\right)} \left(\prod_{j=1}^{N-1}\left(\frac{n_j}{n}\right)^{\frac{1}{p_j}-\frac{1}{p_{j+1}}}\right)^{-\frac{d}{2}} \\ &\quad \times \exp\left(Cs^2n\left(\frac{1}{n} + \sum_{j=1}^{N-1}\frac{p_{j+1}}{2}\frac{n_j}{n}\right) + C\left(1 + \sum_{j=1}^{N-1}\frac{2}{p_{j+1}}\right)\right).\end{aligned}$$

Now we have to make the right choices. We may suppose $n \ge 2$. Set $p_j = 2j^2$, $\eta_j = cj^{-5}$ with $c = \left(\sum_1^\infty j^{-5}\right)^{-1}$, and $N = N(n) = \max\{j : n\eta_j \ge 1\} = \lfloor (cn)^{1/5} \rfloor$. Also, we choose $n_j = \lfloor n\eta_j \rfloor$ for $j \ge 2$, and $n_1 = n - 1 - \sum_2^N n_j$. Then

$$\frac{1}{n} + \sum_{j=1}^{N-1}\frac{p_{j+1}}{2}\frac{n_j}{n} \le 1 + \sum_{j=1}^{\infty}(j+1)^2\eta_j = A_1 < \infty, \qquad 1 + \sum_{j=1}^{N-1}\frac{2}{p_{j+1}} \le A_2 < \infty,$$

and by construction, $\frac{n_1}{n} \geq c - \frac{1}{n} \geq \frac{4}{5} - \frac{1}{2}$ (as $n \geq 2$), while for $j \geq 2$, $n\eta_j \leq j^5 \lfloor n\eta_j \rfloor$, so that $\frac{n_j}{n} \geq \eta_j^2$. Therefore

$$\left(\prod_{j=1}^{N-1} \left(\frac{n_j}{n}\right)^{\frac{1}{p_j} - \frac{1}{p_{j+1}}} \right)^{-\frac{d}{2}} \leq \left(\tfrac{10}{3}\right)^{3d/16} \prod_{j=2}^{\infty} \eta_j^{-d/p_j} = B_1 < \infty\,.$$

Finally, the choice of $N(n)$ implies that $n^{d/(2p_N)} \leq B_2 < \infty$ for all n. Altogether, we get for $|s| \leq 1$

$$\|P_s^n\|_{2\to\infty} \leq C_0^{1/2} B_1 B_2 n^{-d/4} \exp\bigl(CA_1 s^2 n + CA_2\bigr)\,.$$

The adjoint of P_s is P_{-s}. Therefore we get the same upper bound for $\|P_s^n\|_{1\to 2}$, and combining the two,

$$\|P_s^n\|_{1\to\infty} \leq K_1 n^{-d/2} \exp\bigl(K_2(s^2 n + 1)\bigr)\,, \quad |s| \leq 1\,,$$

where $K_1, K_2 > 0$. (By (14.16) this is also true for $n = 1$.) For $|s| > 1$, it is immediate that $\|P_s^n\|_{1\to\infty} \leq e^{|s|n} \|P^n\|_{1\to\infty}$, and with suitable choices of K_1 and K_2, the above inequality will hold as well. It is equivalent to

$$\frac{p^{(n)}(x,y)}{m(y)} \leq K_1 n^{-d/2} \exp\Bigl(K_2(s^2 n + 1) + s\bigl(d(x,o) - d(y,o)\bigr)\Bigr)\,.$$

The choice of o and s was arbitrary. Now let $o = x$ and $s = d(x,y)/(2K_2 n)$. Then we obtain

$$\frac{p^{(n)}(x,y)}{m(y)} \leq K_1 n^{-d/2} \exp\left(K_2 - \frac{d(x,y)^2}{4K_2 n}\right)\,. \qquad \square$$

If X is any connected graph and $d(\cdot,\cdot)$ its graph metric, then Theorem 14.12 remains valid if one considers random walks with arbitrary bounded range instead of nearest neighbour type. The constants will of course change. In particular, the result applies to almost transitive, strongly reversible random walks with bounded range on graphs with polynomial growth.

C. Lower bounds

In principle, §A shows us the following (under suitable conditions): if we have a lower bound on $V_P(n)$ then we get an upper bound on $\sup_{x,y}\bigl(p^{(n)}(x,y)/m(y)\bigr)$, and vice versa. In this subsection, we shall use upper bounds on the growth function to get lower bounds on the transition probabilities. As in the preceding subsection, we shall restrict ourselves to the case where the graph structure on X is the one induced by (X,P), which is assumed to be reversible. Again, generalization to reversible random walks with bounded range will be immediate via equivalence of the metrics of X and its k-fuzz. The sharpest results concern the case of polynomial growth. We start with the following simple result.

(14.19) Theorem. *If $p^{(n)}(x,y) \le C_1\, m(y)\, n^{-d/2} \exp\left(-\frac{d(x,y)^2}{C_2 n}\right)$ and $V_P(x,n) \le C_0\, n^d$ for all n and y, then there is a constant C_3 depending on C_0, C_1, C_2 and d such that*

$$p^{(2n)}(x,x) \ge C_3\, m(x)\, n^{-d/2} \quad \text{for all } n\,.$$

Proof. As one would expect, the Gaussian upper bound implies existence of $\delta > 0$, depending only on C_0, C_1, C_2 and d, such that

$$\mathbb{P}_x[Z_n \in B(x, \delta\sqrt{n})] \ge 1/2 \quad \text{for all } n\,.$$

Indeed,

$$\begin{aligned}
\mathbb{P}_x[Z_n \notin B(x,\delta\sqrt{n})] &\le C_1 \sum_{r>\delta\sqrt{n}} n^{-d/2}\, e^{-r^2/C_2 n}\, m\big(B(x,r)\setminus B(x,r-1)\big)\\
&\le C_1 \sum_{r>\delta\sqrt{n}} n^{-d/2}\Big(e^{-r^2/C_2 n} - e^{-(r+1)^2/C_2 n}\Big) m\big(B(x,r)\big)\\
&\le (3C_0C_1/C_2) \sum_{r>\delta\sqrt{n}} n^{-(d+2)/2}\, e^{-r^2/C_2 n}\, r^{d+1}\\
&\le \int_{\delta\sqrt{n}-1}^{\infty} \Big(\tfrac{s}{\sqrt{n}}\Big)^{d+1} e^{-s^2/C_2 n}\, \tfrac{ds}{\sqrt{n}} \le \int_{\delta/2}^{\infty} t^{d+1} e^{-t^2/C_2}\, dt\,.
\end{aligned}$$

In the second inequality convergence of $\sum_{r\ge 0} e^{-r^2/C_2 n} m\big(B(x,r)\big)$ has been used to separate the differences, shift one index and reassemble. In the third, we used $e^x - e^y \le (x-y)e^x$ for $x > y$. Now we only have to apply Cauchy–Schwarz:

$$\frac{p^{(2n)}(x,x)}{m(x)} \ge \sum_{y\in B(x,\delta\sqrt{n})} \frac{p^{(n)}(x,y)^2}{m(y)} \ge \Bigg(\sum_{y\in B(x,\delta\sqrt{n})} p^{(n)}(x,y)\Bigg)^2 \frac{1}{V(x,\delta\sqrt{n})}\,,$$

which is larger than $(4C_0)^{-1}(\delta\sqrt{n})^{-d}$. □

(14.20) Exercise. *If $p^{(n)}(x,y) \le C_1\, m(y)\, n^{-d/2} \exp\left(-\frac{d(x,y)^2}{C_2 n}\right)$ for all n and y then $V_P(x,n) \ge C_4\, n^d$*

[Hint: bound $\mathbb{P}_o[Z_n \in B(o,\delta\sqrt{n})]$ from above.]

In particular, Theorem 14.12 yields yet another – more complicated – way to prove Corollary 14.11(a). Also, combining Theorems 14.12 and 14.19 we see that

$$\bar{V}_P(n) \preccurlyeq n^d \quad\text{and}\quad \sup_x \frac{p^{(n)}(x,x)}{m(x)} \preccurlyeq n^{-d/2} \implies \inf_x \frac{p^{(2n)}(x,x)}{m(x)} \succcurlyeq n^{-d/2}\,.$$

While in the upper bound it is easy to get rid of the restriction to even n, this is not immediate for the lower bound. First of all, one has to have $p^{(n)}(x,x) > 0$ for all $n \geq n_x$, that is, aperiodicity. This will yield lower bounds of the form $C_x\, n^{-d/2}$. A global lower bound of the same order will follow from strong aperiodicity; recall (9.1).

We now want to generalize the method used in the proof of Theorem 14.19, which consisted in translating an upper bound on $\mathbb{P}_o[Z_n \notin B(o,k)]$, where $k = k(n)$ is chosen suitably, into a lower bound on $p^{(2n)}(o,o)$. We start with the following nice identity plus by-products.

(14.21) Lemma. $$P^n = \sum_{k=0}^{n} \mathbb{P}_0[S_n = k]\, T_k(P)\,,$$
where S_n is the simple random walk on $\mathbb{Z}$ (with law $\mu = \frac{1}{2}(\delta_{-1}+\delta_1)$) and T_k is the k-th Chebyshev polynomial. Also, $\|T_k(P)\|_{2\to2}^2 \leq 1$, and $\mathbb{P}_0[S_n \geq k] \leq e^{-k^2/2n}$.

Proof. The Chebyshev polynomials satisfy $T_n(\cos x) = \cos nx$. Replacing $\cos x$ by $z \in \mathbb{C}$, we write $z = \frac{1}{2}(w + w^{-1})$ with $w = e^{ix}$. Now, the binomial theorem and symmetry of S_n give

$$z^n = \sum_{k=-n}^{n} \mathbb{P}_0[S_n = k]\, w^k = \sum_{k=0}^{n} \mathbb{P}_0[S_n = k] \frac{w^k + w^{-k}}{2} = \sum_{k=0}^{n} \mathbb{P}_0[S_n = k]\, T_k(z)\,.$$

Replacing z with P, we get the identity. The explicit formula for T_n is

$$T_n(z) = \frac{1}{2}\Big(\big(z + i(1-z^2)^{1/2}\big)^k + \big(z - i(1-z^2)^{1/2}\big)^k\Big)\,.$$

The operator $I - P^2$ on $\ell^2(X,m)$ is non-negative definite, so that it has a square root. For a moment, think of $\ell^2(X,m)$ as a complex Hilbert space of complex-valued functions with inner product $(f,g) = \sum_X f(x)\overline{g(x)}m(x)$. We can consider the operator $R = P + i(I - P^2)^{1/2}$. As P is self-adjoint, $R^* = \overline{R}$, and $R^*R = RR^* = I$, that is, R is an isometry. Therefore $T_n(P) = \frac{1}{2}(R^n + \overline{R}^n)$ is a contraction of $\ell^2(X,m)$.

For the statement regarding S_n, let $t > 0$. Then, by Markov's inequality,

$$\mathbb{P}_0[S_n \geq k] = \mathbb{P}_0[e^{tS_n} \geq e^{tk}] \leq e^{-tk}\, \mathbb{E}_0[e^{tS_n}] = e^{-tk}(\cosh t)^n \leq e^{-tk+nt^2/2}\,.$$

Setting $t = k/n$, we obtain the proposed bound. □

(14.22) Theorem. *Suppose that $V_P(x,n) \leq m(x)\mathfrak{v}(n)$, where $\mathfrak{v} : \mathbb{N} \to [2\,,\infty)$ is increasing, and such that the function $n \mapsto n^2/\log \mathfrak{v}(n)$ is increasing and unbounded. Then*

$$\frac{p^{(2n)}(x,x)}{m(x)} \geq \frac{1}{3V\big(x, \mathfrak{w}(6n)\big)}\,,$$

where $\mathfrak{w}(n) = \min\{k : n \le k^2/\log \mathfrak{v}(k)\}$. In particular,

$$\frac{V_P(x,n)}{m(x)} \preccurlyeq n^d \quad \textit{implies} \quad \frac{p^{(2n)}(x,x)}{m(x)} \succcurlyeq (n\log n)^{-d/2} \quad (d \ge 0), \text{ and}$$

$$\frac{V_P(x,n)}{m(x)} \preccurlyeq \exp(n^\beta) \quad \textit{implies} \quad \frac{p^{(2n)}(x,x)}{m(x)} \succcurlyeq \exp\big(-n^{\beta/(2-\beta)}\big) \quad (0 < \beta < 2).$$

Proof. Fix $k \in \mathbb{N}$, and define the annuli $A_j = B(x, r_{j+1}) \setminus B(x, r_j)$, where $r_j = 2^j k$, $j \ge 0$. Using Lemma 14.21, we have the following for our Markov chain Z_n:

$$\mathbb{P}_x[Z_n \in A_j] = \frac{(P^n \mathbf{1}_{A_j}, \delta_x)}{m(x)} = \sum_{k=0}^{n} \mathbb{P}_0[S_n = k] \frac{(T_k(P)\mathbf{1}_{A_j}, \delta_x)}{m(x)} .$$

As $T_k(P)(x,y) = 0$ when $d(x,y) > k$, and applying Cauchy–Schwarz, this is

$$\le \sum_{k=r_j}^{n} \mathbb{P}_0[S_n = k] \frac{\|T_k(P)\mathbf{1}_{A_j}\|_2 \, \|\delta_x\|_2}{m(x)} \le \sqrt{\frac{m(A_j)}{m(x)}}\, \mathbb{P}_0[S_n \ge r_j]$$
$$\le \exp\left(\frac{1}{2}\left(\log \mathfrak{v}(r_{j+1}) - \frac{r_j^2}{n}\right)\right),$$

and thus

$$\mathbb{P}_x[Z_n \notin B(x,k)] \le \sum_{j=0}^{\infty} a_j \qquad \text{with} \qquad a_j = \exp\left(\frac{1}{2}\left(\log \mathfrak{v}(r_{j+1}) - \frac{r_j^2}{n}\right)\right).$$

We want to bound this sum above by a number < 1 via an appropriate choice of k, depending on n. Choose $k = \mathfrak{w}(Cn)$ with $C > 4$. By our monotonicity assumption,

$$\frac{a_j + 1}{a_j} \le \exp\left(\frac{1}{2}\left((r_{j+2}^2 - r_{j+1}^2)\frac{\log \mathfrak{v}(r_{j+1})}{r_{j+1}^2} - \frac{r_{j+1}^2 - r_j^2}{n}\right)\right)$$
$$\le \exp\left(\frac{1}{2}4^j\left(12\log \mathfrak{v}(k) - \frac{3k^2}{n}\right)\right) \le \exp\left(-\frac{3C-12}{2}\log \mathfrak{v}(k)\right).$$

Analogously, $a_0 \le \exp\left(-\frac{C-4}{2}\log \mathfrak{v}(k)\right)$, and we get

$$\mathbb{P}_x[Z_n \notin B(x,k)] \le \frac{\mathfrak{v}(k)^{C-4}}{\mathfrak{v}(k)^{(3C-12)/2} - 1} .$$

Using our assumption that $\mathfrak{v}(k) \geq 2$, this is $\leq \frac{4}{7}$ if we choose $C = 6$. The proof now concludes exactly as for Theorem 14.19. □

We note that the assumption $\mathfrak{v}(0) \geq 2$ (which is sufficient for all our purposes) can be replaced by $\mathfrak{v}(0) > 1$. The choice of a suitable C in the proof will then depend only on the value of $\mathfrak{v}(0)$.

Theorem 14.22 is nice because of its generality. The requirements on $\mathfrak{v}(n)$ will hold in all typical cases, and nothing else is needed to obtain the lower bounds. The last statement of the theorem is of course only interesting when $0 < \beta < 1$. In typical examples, the bounds will of course not be sharp.

15. The asymptotic type of random walks on amenable groups

For a symmetric random walk with finite range on a group with polynomial growth of degree d, Corollary 14.5(a) and Theorem 14.19 give the precise asymptotic type $p^{(2n)}(x,x) \approx n^{-d/2}$. If the group is non-amenable, then $p^{(2n)}(x,x) \approx e^{-n}$ by Corollary 12.5. For groups in between these two extremes, the upper and lower bounds given in Section 14 are sharp only in a few cases. In this section, we continue to pursue the asymptotic type for amenable groups. After a general result (§15.A), we present groups with exponential growth where the upper bound of Corollary 14.5(b) is sharp, that is, $p^{(2n)}(x,x) \approx \exp(-n^{1/3})$. Then we proceed to exhibit a class of groups, also having exponential growth, where $p^{(2n)}(x,x) \approx \exp(-n^{k/(k+2)})$ for $k \in \mathbb{N}$.

A. Comparison and stability of asymptotic type on groups

Let Γ_1 and Γ_2 be finitely generated groups. We fix finite, symmetric generating sets S_i, $i = 1,2$, of the two, and consider the corresponding Cayley graphs with their metrics $d_i(\cdot,\cdot)$. Suppose that $\varphi : \Gamma_1 \to \Gamma_2$ is a rough isometry. Then we have from Theorem 3.10 that the associated Dirichlet norms satisfy $D_2(f) \geq \varepsilon_2 D_1(f \circ \varphi)$ for all $f \in \ell_0(\Gamma_2)$, where $\varepsilon_2 > 0$. Also, rough isometry does not depend on the particular choice of the generating sets, since all Cayley graphs of a given group with respect to finite generating sets are metrically equivalent.

(15.1) Theorem. *Suppose that Γ_1 is amenable and that $\varphi : \Gamma_1 \to \Gamma_2$ is a rough isometry. For $i = 1,2$, let μ_i be symmetric probability measures on Γ_i such that $M_2(\mu_1) < \infty$ and μ_2 is irreducible on Γ_2. Then there is a constant $\varepsilon > 0$ such that*

$$\mu_2^{(2n+2)}(o) \leq 2(1-\varepsilon)^{2n} + 2\mu_1^{(2\lfloor \varepsilon n \rfloor)}(o)$$

for all n. In particular, $\mu_2^{(2n)}(o) \preccurlyeq \mu_1^{(2n)}(o)$.

Proof. Note that we do not require μ_1 to be irreducible, that is, it may be supported by a subgroup of Γ_1. Nevertheless, we may consider the associated Dirichlet norm over Γ_1 in the usual way, and the proof in Proposition 3.20 plus Theorem 3.2, showing that $D_1(\cdot) \geq \varepsilon_1 D_{\mu_1}(\cdot)$ on $\ell_0(\Gamma_1)$, remains valid without any change. Combining this with the above considerations and Theorem 3.1, we find a constant $\varepsilon > 0$ such that

$$D_{\mu_2}(f) \geq 2\varepsilon\, D_{\mu_1}(f \circ \varphi) \quad \text{for all } f \in \ell_0(\Gamma_2)\,. \tag{15.2}$$

First, we suppose that φ is surjective. Then there is a constant $N < \infty$ such that $1 \leq |\varphi^{-1}\{y\}| \leq N$ for every $y \in \Gamma_2$. Now let $A_2 \subset \Gamma_2$ be finite, and set $A_1 = \varphi^{-1}(A_2)$. Then $|A_2| \leq |A_1| \leq N|A_2|$. Consider the truncated transition matrices $Q_i = Q_{A_i}$ defined by

$$q_i(x,y) = \begin{cases} \mu_i(x^{-1}y)\,, & \text{if } x,y \in A_i\,, \\ 0\,, & \text{otherwise.} \end{cases}$$

They are symmetric and substochastic. Recall that in the symmetric case, we choose $m(\cdot)$ to be the counting measure. We think of the finite-dimensional space $\ell^2(A_i)$ as a subspace of $\ell_0(\Gamma_i)$, consisting of all functions with support in A_i. The Dirichlet norm associated with Q_i is

$$D_{Q_i}(f) = \big((I_i - Q_i)f, f\big) = \sum_{x \in A_i} \big(f(x) - Q_i f(x)\big)\, f(x)\,,$$

where $f \in \ell^2(A_i)$, and I_i is the identity matrix over A_i (alternatively seen as a matrix over Γ_i equal to 0 outside of A_i). As $\operatorname{supp} f \subset A_i$, we have $D_{Q_i}(f) = D_{\mu_i}(f)$. If $f \in \ell^2(A_2)$ then $f \circ \varphi$ has its support in A_1. Therefore we have

$$D_{Q_2}(f) \geq 2\varepsilon\, D_{Q_1}(f \circ \varphi) \quad \text{and} \quad \|f\|_2 \leq \|f \circ \varphi\|_2 \tag{15.3}$$

for all $f \in \ell^2(A_2)$. (From this point onwards it may be better to think of Q_i and I_i as finite-dimensional matrices). Let $\lambda_i(j)$, $j = 1, \ldots, |A_i|$, be the eigenvalues of Q_i in descending order (including multiplicities). They lie in the interval $[-1\,,\,1]$, and $1 - \lambda_i(j)$ are the eigenvalues of $I_i - Q_i$ in increasing order. The min-max theorem (see e.g. Horn and Johnson [176]) says that

$$1 - \lambda_i(j) = \min_{\dim \mathcal{W} = j} \max_{f \in \mathcal{W}} \frac{D_{Q_i}(f)}{\|f\|_2^2}\,,$$

where the minimum is over subspaces $\mathcal{W}$ of $\ell^2(A_i)$. Now, if $\mathcal{W}$ is a subspace of $\ell_2(A_2)$ then the subspace $\mathcal{W} \circ \varphi$ of $\ell_2(A_1)$ has the same dimension. (It is here that we use surjectivity of φ.) We infer from (15.3) that

$$1 - \lambda_2(j) \geq 2\varepsilon\big(1 - \lambda_1(j)\big) \quad \text{for } j = 1, \ldots, |A_2|\,.$$

(We may assume that $\varepsilon \le 1/2$.) Now suppose that $\lambda_1(j) \ge 1/2$. Then, using the inequalities $1 - 2\varepsilon(1-t) \le e^{-2\varepsilon(1-t)}$ and $e^{t-1} \le \sqrt{t}$ for $1/2 \le t \le 1$, we get

$$\lambda_2(j) \le 1 - 2\varepsilon\bigl(1 - \lambda_1(j)\bigr) \le \lambda_1(j)^{\varepsilon}.$$

On the other hand, if $\lambda_1(j) < 1/2$ then $\lambda_2(j) < 1-\varepsilon$. We now want to compare the traces of Q_i^n, $i = 1, 2$: $\mathrm{Tr}[Q_i^n] = \sum_{x \in A_i} q_i^{(n)}(x,x) = \sum_{j=1}^{|A_i|} \lambda_i(j)^n$ is non-negative. As $|\lambda_i(j)| \le 1$, we have

$$\sum_{\lambda_2(j)<0} \lambda_2(j)^{2n+2} \le - \sum_{\lambda_2(j)<0} \lambda_2(j)^{2n+1} \le \sum_{\lambda_2(j)>0} \lambda_2(j)^{2n},$$

so that

$$\begin{aligned} \mathrm{Tr}[Q_2^{2n+2}] &\le 2 \sum_{\substack{\lambda_2(j)>0 \\ \lambda_1(j)<1/2}} \lambda_2(j)^{2n} + 2 \sum_{\substack{\lambda_2(j)>0 \\ \lambda_1(j)\ge 1/2}} \lambda_2(j)^{2n} \\ &\le 2|A_2|(1-\varepsilon)^{2n} + 2\,\mathrm{Tr}[Q_1^{2\lfloor \varepsilon n \rfloor}]. \end{aligned}$$

Now observe that $q_1^{(n)}(y,y)$ is the probability that the random walk on Γ_1 with law μ_1 is at the starting point $y \in A_1$ after n steps without ever leaving A_1. Therefore $q_1^{(n)}(y,y) \le \mu_1^{(n)}(o)$ and $\mathrm{Tr}[Q_1^n] \le |A_1|\,\mu_1^{(n)}(o) \le N|A_2|\,\mu_1^{(n)}(o)$. We have proved that

$$\mathrm{Tr}[Q_2^{2n+2}] \le 2|A_2|\Bigl((1-\varepsilon)^{2n} + N\,\mu_1^{(2\lfloor \varepsilon n \rfloor)}(o)\Bigr)$$

for any choice of finite $A_2 \subset \Gamma_2$. For surjective φ, the theorem will now follow immediately from the technical result that follows in Proposition 15.4 below. (For the second statement of the theorem, recall that amenability yields $\mu_1^{(2n)}(o) \ge C(1-\varepsilon)^n$ for some $C > 0$; see Corollary 12.5.)

We now explain what to do when φ is not surjective. Recall from Definition 3.7 that it is "almost" surjective. Therefore we can partition Γ_2 into finite sets A_y, $y \in \varphi\Gamma_1$, such that $y \in A_y$ and A_y is contained in the ball of radius B centred at y. ("Ball" refers to the metric of the Cayley graph in consideration.) Let $\ell = \max\{|A_y| : y \in \varphi\Gamma_1\}$. Now consider the group $\widetilde{\Gamma}_1 = \Gamma_1 \times \mathbb{Z}_\ell$. We can extend φ to a mapping $\widetilde{\varphi} : \widetilde{\Gamma}_1 \to \Gamma_2$ such that $\widetilde{\varphi}(\{x\} \times \mathbb{Z}_\ell) = A_{\varphi x}$ for every $x \in \Gamma_1$. It is straightforward that this is a rough isometry of $\widetilde{\Gamma}_1$ *onto* Γ_2, and μ_1 can be considered as a probability measure on $\widetilde{\Gamma}_1$ with finite second moment, 0 outside of $\Gamma_1 \equiv \Gamma_1 \times \{0\}$. □

(15.4) Proposition. *Let Γ be an amenable group, and let μ be a symmetric probability measure on Γ. For any finite $A \subset \Gamma$, let Q_A be the substochastic matrix over A defined by $q_A(x,y) = \mu(x^{-1}y)$, where $x, y \in A$. Then, for each n,*

$$\mu^{(n)}(o) = \sup_A \frac{1}{|A|} \mathrm{Tr}[Q_A^n].$$

Proof. We have seen in the proof of Theorem 15.1 that "$\geq$" holds in every case (without assuming amenability). For the reverse inequality, we fix n, pick $\varepsilon \in (0\,,\,1)$ and choose a finite set $K_\varepsilon \subset \Gamma$ such that $o \in K_\varepsilon$ and $\mu(K_\varepsilon) \geq 1 - \varepsilon$. By Følner's criterion (12.3), for each n there is a sequence of finite sets $U_k \subset \Gamma$ such that

$$\lim_{k\to\infty} \frac{|U_k K_\varepsilon^n|}{|U_k|} = 1.$$

We set $A_k = A_{k,\varepsilon,n} = U_k K_\varepsilon^n$. If $Z_j = xX_1 \cdots X_j$ is the random walk with law μ starting at $x \in U_k \subset A_k$, then

$$\begin{aligned} q_{A_k}^{(n)}(x,x) &= \mathbb{P}_x[Z_n = x\,,\; Z_j \in A_k \text{ for all } j \leq n] \\ &= \mathbb{P}_x[Z_n = x] - \mathbb{P}_x[Z_n = x\,,\; Z_j \notin A_k \text{ for some } j < n] \\ &\geq \mu^{(n)}(o) - \mathbb{P}_x[X_j \notin K_\varepsilon \text{ for some } j < n] \\ &\geq \mu^{(n)}(o) - (n-1)\varepsilon\,. \end{aligned}$$

Indeed, if $X_j \in K_\varepsilon$ for all $j \leq n$ then also $Z_j = xX_1 \cdots X_j \in U_k K_\varepsilon^j \subset A_k$. Therefore

$$\mu^{(n)}(o) - (n-1)\varepsilon \leq \frac{1}{|U_k|}\mathrm{Tr}[Q_{A_k}^n] \leq \frac{|A_k|}{|U_k|} \sup_A \frac{1}{|A|}\mathrm{Tr}[Q_A^n].$$

Letting first k tend to ∞ and then ε to 0, we obtain the proposed result. □

The relevance of group invariance lies in the fact that $p^{(n)}(x,x)$ is independent of x, so that the argument using traces can be used. Theorem 15.1 yields the main result of this subsection.

(15.5) Corollary. *Let Γ_1 and Γ_2 be finitely generated groups that are roughly isometric. For $i = 1, 2$, let μ_i be an irreducible, symmetric probability measure with finite second moment on Γ_i, $i = 1, 2$. Then*

$$\mu_1^{(2n)}(o) \approx \mu_2^{(2n)}(o)\,.$$

Indeed, for non-amenable groups, this follows from Corollary 12.5.

In particular, we see that for the simple random walk on the Cayley graph of a finitely generated group, the asymptotic type of $\mu^{(2n)}(o)$ is independent of the chosen set of generators. However, later on we shall see that the precise asymptotic behaviour is not as stable with respect to perturbations of μ.

(15.6) Exercise. Reformulate (15.1)–(15.5) for transitive graphs and check that the proofs remain the same.

B. Polycyclic groups

There are many equivalent definitions of polycyclic groups. The one which justifies the name is the following: a group Γ is called *polycyclic* if it has a normal series

$$\Gamma = \Gamma_0 \geq \Gamma_1 \geq \cdots \geq \Gamma_r = \{o\}$$

such that every quotient Γ_i/Γ_{i+1} is (finite or infinite) cyclic. For a long list of equivalent properties, see Wolf [353], Raghunathan [273] or Segal [295]. In particular, Γ has a torsion-free subgroup Γ^* with finite index, and the latter has a nilpotent normal subgroup $\mathfrak{N}$ such that $\Gamma^*/\mathfrak{N} \cong \mathbb{Z}^d$, where $d \geq 0$. Furthermore, Wolf has proved that a polycyclic group has either polynomial or exponential growth. This has been generalized by Rosenblatt [280]:

(15.7) Theorem. *A finitely generated solvable group has either polynomial or exponential growth.*

(This has already been mentioned in §3.B.) A simple example of a polycyclic group with exponential growth is the following. Take the matrix $A = \left(\begin{smallmatrix} 2 & 1 \\ 1 & 1 \end{smallmatrix}\right)$. All its positive and negative powers have integer entries, so that the infinite cyclic group generated by A acts on $\mathbb{Z}^2$. Our example is the corresponding semidirect product $\mathbb{Z} \leftthreetimes \mathbb{Z}^2$.

If Γ is polycyclic and not nilpotent-by-finite, then Theorem 15.7 and Corollary 14.5 tell us that any symmetric random walk with finite range on Γ satisfies $\mu^{(2n)}(o) \preccurlyeq \exp(-n^{1/3})$. The aim of this section is to prove that one has the same type of lower bound:

(15.8) Theorem. *If Γ is polycyclic and μ a finitely supported, symmetric (irreducible) probability measure on Γ, then one has the following alternative (with suitable positive constants $C_1 \leq C_2$).*

(a) Γ *has polynomial growth with degree d, and*

$$C_1\, n^{-d/2} \leq \mu^{(2n)}(o) \leq C_2\, n^{-d/2}\,.$$

(b) Γ *has exponential growth, and*

$$C_1 \exp\bigl(-C_2 n^{1/3}\bigr) \leq \mu^{(2n)}(o) \leq C_2 \exp\bigl(-C_1 n^{1/3}\bigr)\,.$$

In view of the results of Section 14 (Corollary 14.5, Theorems 14.12 and 14.19), only the lower bound in (b) has yet to be proved. By Corollary

15.5, we may assume without loss of generality that Γ is torsion-free, and $\Gamma/\mathfrak{N} \cong \mathbb{Z}^d$, where $\mathfrak{N}$ is a (torsion-free) nilpotent normal subgroup of Γ.

We choose elements x_i in Γ which map onto the natural generators $\mathbf{e}_i$ of $\mathbb{Z}^d$, $i = 1, \dots, d$. Also, we choose a symmetric set of generators $S_{\mathfrak{N}} = \{y_1^{\pm 1}, \dots, y_r^{\pm 1}\}$ of $\mathfrak{N}$. Then

$$S_\Gamma = \{x_1^{\pm 1}, \dots, x_d^{\pm 1}, y_1^{\pm 1}, \dots, y_r^{\pm 1}\}$$

generates Γ. We shall refer by $d_{\mathfrak{N}}(\cdot,\cdot)$ and $d_\Gamma(\cdot,\cdot)$ to the corresponding Cayley graph metrics and write $|y|_{\mathfrak{N}} = d_{\mathfrak{N}}(y, o)$ and $|x|_\Gamma = d_\Gamma(x, o)$, where $x \in \Gamma$, $y \in \mathfrak{N}$. Also, if $\mathbf{k} \in \mathbb{Z}^d$ then we denote by $\mathbf{x}^{\mathbf{k}}$ the element $x_1^{k_1} \cdots x_d^{k_d} \in \Gamma$, and $|\mathbf{k}| = \sum_{i=1}^d |k_i|$. We shall need some algebraic observations. By $[x, y]$ (where $x, y \in \Gamma$) we denote the commutator $xyx^{-1}y^{-1}$ (always $\in \mathfrak{N}$). Let

$$M_1 = \max\{|x_i^{\varepsilon_i} y_j^{\pm 1} x_i^{-\varepsilon_i}|_{\mathfrak{N}} : i = 1, \dots, d\,, j = 1, \dots, r\,, \varepsilon_i = \pm 1\}\,,$$
$$M_2 = \max\{|[x_i^{\varepsilon_i}, x_j^{\varepsilon_j}]|_{\mathfrak{N}} : i, j = 1, \dots, d\,, \varepsilon_i, \varepsilon_j = \pm 1\}\,.$$

We may suppose that $M_1 \geq 2$ and set $M = d^2 M_1^2 M_2$.

(15.9) Proposition. *If $\mathbf{k}, \mathbf{l} \in \mathbb{Z}^d$ then $\mathbf{x}^{\mathbf{k}}\,\mathbf{x}^{\mathbf{l}} = b\,\mathbf{x}^{\mathbf{k}+\ell}$ with $b = b(\mathbf{k}, \mathbf{l}) \in \mathfrak{N}$ and $|b|_{\mathfrak{N}} \leq M^{|\mathbf{k}|+|\mathbf{l}|}$.*

Proof. We may suppose that $\mathbf{k} \neq \mathbf{0}$. First of all, observe that for $y \in \mathfrak{N}$ and $\mathbf{k} \in \mathbb{Z}^d$,

$$|\mathbf{x}^{\mathbf{k}} y \mathbf{x}^{-\mathbf{k}}|_{\mathfrak{N}} \leq |y|_{\mathfrak{N}}\, M_1^{|\mathbf{k}|}\,. \tag{15.10}$$

This is straightforward by induction on $|\mathbf{k}|$. Next, verify that for $x, y \in \Gamma$ and $k \geq 0$, we can write

$$[x^k, y] = \prod_{j=1}^{k} a_{k-j}(x, y)\,, \quad \text{where} \quad a_j(x, y) = x^j [x, y] x^{-j}\,.$$

(Use the identity $[x_j, y] = a_{j-1}(x, y)[x^{j-1}, y]$.) In order to prove the statement of the proposition, we first assume that $\mathbf{l} = \mathbf{e}_i$. When $i = d$ then $b = o$ and there is nothing to prove, so assume $i < d$. Let $\mathbf{k}(j) = (k_1, \dots, k_j, 0, \dots, 0)$. Also, if $k_j = 0$ set $u_j = o$, and otherwise

$$u_j = \prod_{\ell=1}^{|k_j|} a_{|k_j|-\ell}(x_j^{\varepsilon_j}, x_i)\,, \quad \text{where} \quad \varepsilon_j = \operatorname{sign} k_j\,.$$

Then $x_j^{k_j} x_i = u_j x_i x_j^{k_j}$, and if $j > i$ then

$$\mathbf{x}^{\mathbf{k}(j)}\, x_i = \left(\mathbf{x}^{\mathbf{k}(j-1)}\, u_j\, \mathbf{x}^{-\mathbf{k}(j-1)}\right) \mathbf{x}^{\mathbf{k}(j-1)}\, x_i\, x_j^{k_j}\,.$$

Therefore we get $\mathbf{x}^{\mathbf{k}}\,x_i = b\,\mathbf{x}^{\mathbf{k}+\mathbf{e}_i}$, where

$$b = b(\mathbf{k},\mathbf{e}_i) = \prod_{\ell=1}^{d-i} \mathbf{x}^{\mathbf{k}(d-\ell)} u_{d-\ell+1} \mathbf{x}^{-\mathbf{k}(d-\ell)}\,.$$

To bound the length of b, we get from (15.10)

$$|u_j|_{\mathfrak{N}} \le M_2 \sum_{\ell=1}^{|k_j|} M_1^{|k_j|-\ell} \le M_2\, M_1^{|k_j|}\,, \quad \text{and}$$

$$|b|_{\mathfrak{N}} \le M_1^{|\mathbf{k}|} \sum_{j=i+1}^{d} |u_j|_{\mathfrak{N}} \le d\, M_2\, M_1^{2|\mathbf{k}|}\,.$$

The proof with $\mathbf{l} = -\mathbf{e}_i$ in place of $\mathbf{e}_i$ is exactly the same and leads to the same bound for $b(\mathbf{k},-\mathbf{e}_i)$ Next, if $\mathbf{l} = \ell_i\mathbf{e}_i$ with $\ell_i \in \mathbb{Z}^d \setminus \{\mathbf{0}\}$ then by induction on $|\ell_i|$,

$$|b(\mathbf{k},\ell_i\mathbf{e}_i)|_{\mathfrak{N}} \le \sum_{j=0}^{|\ell_i|-1} d\, M_2\, M_1^{2(|\mathbf{k}|+j)} \le d\, M_2\, M_1^{2(|\mathbf{k}|+|\ell_i|)}\,.$$

Finally,

$$\begin{aligned}|b(\mathbf{k},\mathbf{l})|_{\mathfrak{N}} = |b(\mathbf{k},\ell_1\mathbf{e}_1 + \cdots + \ell_{d-1}\mathbf{e}_{d-1})|_{\mathfrak{N}} &\le \sum_{i=1}^{d-1} d\, M_2\, M_1^{2(|\mathbf{k}|+|\ell_1|+\cdots+|\ell_{d-1}|)}\\ &\le d^2\, M_2\, M_1^{2(|\mathbf{k}|+|\mathbf{l}|)}\,.\end{aligned}$$

□

As a consequence, if $u,v \in \mathfrak{N}$ and $\mathbf{k},\mathbf{l} \in \mathbb{Z}^d$ then there is $w \in \mathfrak{N}$ such that

$$(15.11)\quad u\,\mathbf{x}^{\mathbf{k}}\,v\,\mathbf{x}^{\mathbf{l}} = w\,\mathbf{x}^{\mathbf{k}+\ell}\,, \quad \text{and} \quad |w|_{\mathfrak{N}} \le |u|_{\mathfrak{N}} + |v|_{\mathfrak{N}}\, M^{|\mathbf{k}|} + M^{|\mathbf{k}|+|\mathbf{l}|}\,.$$

(Apply (15.10) to $\mathbf{x}^{\mathbf{k}}\,v\,\mathbf{x}^{-\mathbf{k}}$ and use Proposition 15.9.)

Proof of Theorem 15.8. In view of Corollary 15.5, it is enough to prove the lower bound in (b) for the random walk $Z_n = X_1 \cdots X_n$ whose law μ is equidistributed on S_Γ. Let K_j be the projection of X_j onto $\mathbb{Z}^d = \Gamma/\mathfrak{N}$, and $S_n = K_1 + \cdots + K_n$ the projection of Z_n. This is a modified simple random walk on the grid, with "holding probability" $r/(r+d)$ at each point. We can write

$$(15.12)\qquad X_j = Y_j\,\mathbf{x}^{K_j}\,, \quad \text{where} \quad Y_j \in \mathfrak{N} \quad \text{and} \quad |Y_j|_{\mathfrak{N}} + |K_j|_{\mathbb{Z}^d} = 1\,.$$

Analogously, we write $Z_n = U_n\,\mathbf{x}^{S_n}$ with $U_n \in \mathfrak{N}$ (in particular, $U_1 = X_1$). Using (15.11) and (15.12), we get (omitting the subscripts in $|\cdot|_{\mathfrak{N}}$ and $|\cdot|_{\mathbb{Z}^d}$)

$$\begin{aligned}|U_n| &\le |U_{n-1}| + |Y_n|\,M^{|S_{n-1}|} + M^{|S_{n-1}|+|K_n|}\\ &\le |U_{n-1}| + M^{|S_{n-1}|+1} \le \text{ (inductively)}\\ &\le 1 + M\sum_{j=1}^{n-1} M^{|S_j|} \le M\sum_{j=1}^{n} M^{|S_j|}\,.\end{aligned}$$

Therefore, if $\max\{|S_j| : j \le n\} \le m$ then $|U_n| \le nM^{1+m}$. We now use the method of Theorem 14.19, but instead of balls, we consider the finite sets

$$A_{n,r} = \{w = u\mathbf{x}^{\mathbf{k}} \in \Gamma : |\mathbf{k}| \le r\,,\ |u|_{\mathfrak{N}} \le nM^{1+r}\}\,.$$

As $\mathfrak{N}$ has polynomial growth, there is a constant $c_1 > 0$ such that $|A_{n,r}| \le e^{c_1(1+r+\log n)}$. Using Cauchy–Schwarz as in the proof of Theorem 14.19, we obtain

$$\mu^{(2n)}(o) \ge \frac{\mathbb{P}_o[Z_n \in A_{n,r}]^2}{|A_{n,r}|} \ge \frac{\mathbb{P}_o[\max\{|S_j| : j \le n\} \le r]^2}{e^{c_1(1+r+\log n)}}\,.$$

We now use the following inequality.

(15.13) Exercise. A variation on Kolmogorov's inequality (Alexopoulos [2]): show that there is a constant $c_2 > 0$ such that

$$\mathbb{P}_o[\max\{|S_j| : j \le n\} \le r] \ge e^{-c_2(1+n/r^2)} \quad \text{for all } r, n \in \mathbb{N}\,.$$

Thus, we have $\mu^{(2n)}(o) \ge \exp\bigl(-c_3(2 + \frac{n}{r^2} + r + \log n)\bigr)$ with $c_3 = \max\{c_1, c_2\}$. Setting $r = n^{1/3}$, we obtain the lower bound proposed in Theorem 15.8. □

C. The solvable Baumslag–Solitar groups

Let q be an integer ≥ 2. The corresponding Baumslag–Solitar group is

$$\mathrm{BS}_q = \left\{\begin{pmatrix} q^k & \frac{m}{q^\ell} \\ 0 & 1\end{pmatrix} : k, \ell, m \in \mathbb{Z}\,,\ \ell \ge 0\right\}.$$

This is a solvable, non-polycyclic group with exponential growth. It is the semidirect product $\mathbb{Z} \ltimes \mathbb{Z}[\frac{1}{q}]$, where the action of $\mathbb{Z}$ on $\mathbb{Z}[\frac{1}{q}]$ is given by $(k, \frac{m}{q^\ell}) \mapsto q^k \frac{m}{q^\ell}$. The elements $a = \left(\begin{smallmatrix} q & 0 \\ 0 & 1\end{smallmatrix}\right)$ and $b = \left(\begin{smallmatrix} 1 & 1 \\ 0 & 1\end{smallmatrix}\right)$ satisfy the relation $ab = b^q a$ and generate BS_q. Indeed, as an abstract group, it has the presentation

$$\mathrm{BS}_q = \langle a, b \mid ab = b^q a \rangle .$$

Every element $x = \begin{pmatrix} q^k & \frac{m}{q^\ell} \\ 0 & 1 \end{pmatrix}$ can be written uniquely as

$$x = \begin{pmatrix} 1 & \frac{m}{q^\ell} \\ 0 & 1 \end{pmatrix} \begin{pmatrix} q^k & 0 \\ 0 & 1 \end{pmatrix} = (a^{-\ell} b^m a^\ell) a^k ,$$

where q does not divide m when $\ell \geq 1$. We write $k(x), \ell(x)$ and $m(x)$ for these numbers; $k(\cdot)$ is the natural projection of BS_q onto $\mathbb{Z}$.

(15.14) Theorem. *Every symmetric random walk with finite second moment on* BS_q *satisfies*

$$\mu^{(2n)}(o) \approx \exp(-n^{1/3}) .$$

Proof. Again, we only have to prove the lower bound and are free to choose which random walk to study. We apply the same method as for polycyclic groups to the simple random walk $Z_n = X_1 \cdots X_n$ on the Cayley graph of BS_q with respect to $\{a^{\pm 1}, b^{\pm 1}\}$. (The starting point is $Z_0 = o$.) We write $L_n = \ell(Z_n)$, $M_n = m(Z_n)$ and $S_n = k(Z_n)$. The latter is the random walk on $\mathbb{Z}$ which in a step moves left or right with probability $1/4$ each, or remains where it is with probability $1/2$. Also, we write $B_n = M_n / q^{L_n}$. We prove by induction that

$$|B_n| \leq \sum_{j=0}^{n-1} q^{S_j} \quad \text{and} \quad L_n \leq \max\{-S_j : 0 \leq j < n\} .$$

This is true for $n = 0$. Suppose that it holds for $n - 1$.

If $X_n = a^{\pm 1}$ then $B_n = B_{n-1}$ and $S_n = S_{n-1} \pm 1$, so that the statement also holds for n.

If $X_n = b^{\pm 1}$ then $S_n = S_{n-1}$ and $B_n = B_{n-1} \pm q^{S_{n-1}}$. The bound on $|B_n|$ is obvious, while the one on L_n follows from $L_n \leq \max\{-S_{n-1}, L_{n-1}\}$.

We now set

$$A_{n,r} = \{x \in \mathrm{BS}_q : |k(x)| \leq r, \ \ell(x) \leq r, \ |m(x)| \leq n\, q^{2r}\}$$

and proceed as before: if $\max\{|S_j| : 0 \leq j < n\} \leq r$ then $Z_n \in A_{n,r}$, whence $\mu^{(2n)}(o) \geq \mathbb{P}_o[\max\{|S_j| : 0 \leq j < n\} \leq r]^2 / |A_{n,r}|$. The proof concludes exactly as for Theorem 15.8 (after (15.11)). □

D. Random walks on lamplighter groups

Suppose that at each site (vertex) of the grid $\mathbb{Z}^d$ there is a lamp which may be switched on or off. Initially, no lamp is lit. A lamplighter starts from the origin and performs random steps. At each step, he has two choices: he may either decide to change the state of the lamp where he stands, or he may move across some edge of the grid to a neighbouring site. In the resulting random process, we have to observe two things: the position of the lamplighter, performing a random walk in the grid, and the sites where the lamps turned on – a finite subset of $\mathbb{Z}^d$. This can be described as a random walk on a group, the *lamplighter group* over $\mathbb{Z}^d$.

The lamplighter groups are examples of *wreath products*. Let Γ and $\mathfrak{A}$ be arbitrary groups. A *configuration* is a function $\eta : \Gamma \to \mathfrak{A}$ with finite support, where $\operatorname{supp} \eta = \{x \in \Gamma : \eta(x) \neq o_{\mathfrak{A}}\}$. Equipped with pointwise multiplication in $\mathfrak{A}$, the set $\mathfrak{B}$ of all configurations is a group – the *direct sum* of copies of $\mathfrak{A}$, indexed by Γ. Now Γ acts on $\mathfrak{B}$ by group automorphisms via $(y, \eta) \mapsto T_y \eta$, where $T_y \eta(x) = \eta(y^{-1}x)$. The resulting semidirect product $\Gamma \leftthreetimes \mathfrak{B}$ is the *wreath product* of Γ with $\mathfrak{A}$, denoted by $\Gamma \wr \mathfrak{A}$. Both $\mathfrak{A}$ and Γ embed into the wreath product: $a \in \mathfrak{A} \mapsto (o_\Gamma, \eta_a)$, where $\eta_a \in \mathfrak{B}$ is defined by $\eta_a(o_\Gamma) = a$ and $\eta_a(x) = o_{\mathfrak{A}}$ if $x \in \Gamma \setminus \{o_\Gamma\}$. The embedding of Γ is given by $y \mapsto (y, \eta_{o_{\mathfrak{A}}})$. Here, $\eta_{o_{\mathfrak{A}}}$ is the empty configuration (constant with value $o_{\mathfrak{A}}$). The unit element of the wreath product is $o = (o_\Gamma, \eta_{o_{\mathfrak{A}}})$. If Γ and $\mathfrak{A}$ are finitely generated then so is their wreath product, which is generated by the union of the embeddings of the respective generating sets.

Here we shall only consider the case when Γ and $\mathfrak{A}$ are abelian, with respective group operations written additively, and the product in $\Gamma \wr \mathfrak{A}$ is given by

$$(y_1, \eta_1)(y_2, \eta_2) = (y_1 + y_2, \eta_1 + T_{y_1} \eta_2).$$

Thus, $\Gamma \wr \mathfrak{A}$ is solvable and has exponential growth. The lamplighter group is $\mathbb{Z}^d \wr \mathbb{Z}_2$. If $\mathfrak{A}$ has more than two elements, then we may imagine each lamp having more than just the two states "on" and "off". In this subsection we want to study the asymptotic type of random walk on these generalized lamplighter groups.

(15.15) Theorem. *Let $\mathfrak{A}$ be a finite abelian group. Every symmetric random walk with finite second moment on $\mathbb{Z}^d \wr \mathfrak{A}$ satisfies*

$$\mu^{(2n)}(o) \approx \exp\big(-n^{d/(d+2)}\big).$$

Proof[3]. Let μ_0 be the equidistribution on $\{\mathbf{0}, \pm\mathbf{e}_i : i = 1, \dots, d\} \subset \mathbb{Z}^d$ and ν_0 the one on the whole of $\mathfrak{A}$. Via the embedding of $\mathbb{Z}^d$ and $\mathfrak{A}$ into

---

[3] See comments on page 317

$\mathbb{Z}^d \wr \mathfrak{A}$, both are also considered as measures on the wreath product. For the proof, in view of Corollary 15.5, it is sufficient to consider the symmetric random walk on $\mathbb{Z}^d \wr \mathfrak{A}$ whose law is $\mu = \nu_0 * \mu_0 * \nu_0$ that is,

$$\mu(y,\eta) = \begin{cases} \mu_0(y)/|\mathfrak{A}|^2\,, & \text{if } \eta \in \{\eta_a + T_y\eta_b : a,b \in \mathfrak{A}\}\,,\\ 0\,, & \text{otherwise.}\end{cases}$$

Since $\nu_0 * \nu_0 = \nu_0$, we have $\mu^{(n)} = (\nu_0 * \mu_0)^{(n)} * \nu_0$. Consider i.i.d. random variables (K_n, V_n), where $K_n \in \mathbb{Z}^d$ has distribution μ_0 and the $\mathfrak{B}$-valued random variables V_n are all equidistributed on the set of configurations $\eta \in \mathfrak{B}$ with $\operatorname{supp}\eta \subset \{\mathbf{0}\}$, and K_n and V_n are independent. Then $\mu^{(n)}$ is the distribution of

$$\left(S_n, \sum_{j=1}^{n+1} T_{S_{j-1}} V_j\right) \in \mathbb{Z}^d \wr \mathfrak{A}\,,$$

where $S_n = K_1 + \cdots + K_n$ is the random walk on $\mathbb{Z}^d$ with law μ_0, with $S_0 = \mathbf{0}$. We start with the (easy) lower bound, proceeding as in the previous two theorems. Set

$$A_r = \{(\mathbf{l},\eta) \in \mathbb{Z}^d \wr \mathfrak{A} : |\mathbf{l}| \le r\,,\ \eta(\mathbf{k}) = o_{\mathfrak{A}} \text{ if } |\mathbf{k}| > r\}\,.$$

Then $|A_r| \le C\,r^d\,|\mathfrak{A}|^{Cr^d}$, and if $\max\{|S_j| : j \le n\} \le r$ then $Z_n \in A_r$. Indeed, the lamplighter cannot turn on any lamp before reaching it. Once more, we obtain

$$\mu^{(2n)}(o) \ge \mathbb{P}_\mathbf{0}[\max\{|S_j| : j \le n\} \le r]^2/|A_r| \ge \exp\Big(-\bar{C}\big(1+\tfrac{n}{r^2}+r^d+\log r\big)\Big)\,.$$

Choosing $r = n^{1/(d+2)}$ leads to the desired result.

The upper bound involves a subtle tool. We first need the following fact. If $\mathbf{k}_0,\dots,\mathbf{k}_n \in \mathbb{Z}^d$ (not necessarily distinct) then $\sum_{j=1}^{n+1} T_{\mathbf{k}_{j-1}} V_j$ is equidistributed on the set of all configurations $\eta \in \mathfrak{B}$ with $\operatorname{supp}\eta \subset \{\mathbf{k}_0,\dots,\mathbf{k}_n\}$. This is straightforward when all $\mathbf{k}_j$ are distinct. If they are not, then observe that for fixed $\mathbf{k} \in \mathbb{Z}^d$ and $i < j$, the distribution of $T_\mathbf{k} V_i + T_\mathbf{k} V_j$ is the same as that of $T_\mathbf{k} V_i$. Indeed, the sum of two or more independent, uniform $\mathfrak{A}$-valued random variables is again uniformly distributed on $\mathfrak{A}$.

We now compute

$$\begin{aligned}
\mathbb{P}_o[Z_n = o] &= \mathbb{P}_o[S_n = \mathbf{0}\,, \sum_{j=1}^{n+1} T_{S_{j-1}} V_j = \eta_{o_\mathfrak{A}}]\\
&= \sum_{\substack{\mathbf{k}_0,\dots,\mathbf{k}_n \in \mathbb{Z}^d\\ \mathbf{k}_0 = \mathbf{k}_n = \mathbf{0}}} \mathbb{P}_o\left[\sum_{j=1}^{n+1} T_{\mathbf{k}_{j-1}} V_j = \eta_{o_\mathfrak{A}}\right] \mathbb{P}_\mathbf{0}[S_0 = \mathbf{k}_0, S_1 = \mathbf{k}_1, \dots, S_n = \mathbf{k}_n]\\
&= \mathbb{E}_\mathbf{0}\big(|\mathfrak{A}|^{-|D_n|} \mid S_n = \mathbf{0}\big)\,\mathbb{P}_\mathbf{0}[S_n = \mathbf{0}] = \mathbb{E}_\mathbf{0}\big(|\mathfrak{A}|^{-|D_n|}\,\mathbf{1}_{[S_n = \mathbf{0}]}\big)\,,
\end{aligned}$$

where $D_n = \{S_0, \dots, S_n\}$ and $|D_n|$ is the number of distinct sites in $\mathbb{Z}^d$ visited by the random walk up to time n. The upper bound now follows directly from the theorem on random walks in the grid that is stated next. □

(15.16) Theorem. *Let S_n be a symmetric random walk with finite second moment in $\mathbb{Z}^d$, and $|D_n|$ the number of distinct sites visited up to time n. Then, for every $z > 0$ there is a constant $\alpha(z) > 0$ such that*

$$\lim_{n\to\infty} n^{-d/(d+2)} \log \mathbb{E}_{\mathbf{0}}\big(\exp(-z|D_n|)\big) = -\alpha(z)\,.$$

This profound result, which rather belongs to the realm of Section 13, is due to Donsker and Varadhan [100]. It uses refined estimates for large deviations of $|D_n|$ which indoubtedly go beyond the scope of the present book. Preliminary results that may lead to a better understanding of this important theorem can be found in the book by Hughes [177].

16. Simple random walks on the Sierpiński graphs

In Section 14 (Corollary 14.5, Theorem 14.19), we have seen that under certain conditions (quasi-homogeneity), polynomial growth with degree r and decay of order $n^{-r'/2}$ for transition probabilities occur with the same exponents $r' = r$. In this section we shall study a class of graphs with polynomial growth, where r' is strictly smaller than r. These are the simplest "fractal" graphs, strongly related to the Sierpiński fractals in $d \geq 2$ dimensions.

We explain the recursive construction of the d-dimensional Sierpiński graph $\mathbf{S}_d$. Let $\mathbf{0} = \mathbf{x}_0, \mathbf{x}_1, \dots, \mathbf{x}_d$ be the vertices of a standard equilateral simplex in $\mathbb{R}^d$ with all vertices in the non-negative cone. We write $S^{(0)} = S_d^{(0)}$ for its 1-skeleton; this is the complete graph with vertices $\mathbf{0}, \mathbf{x}_1, \dots, \mathbf{x}_d$. If we have already constructed $S^{(k)} = S_d^{(k)}$, then we define

$$S^{(k+1)} = S_d^{(k+1)} = \bigcup_{j=0}^{d} \big(2^k\,\mathbf{x}_j + S^{(k)}\big)\,,$$

where $\mathbf{x} + S^{(k)}$ denotes the translate of the graph $S^{(k)}$ by the vector $\mathbf{x} \in \mathbb{R}^d$. Also, we write $-S^{(k)}$ for the reflection of $S^{(k)}$ through the origin, and $\mathbf{S}^{(k)} = \mathbf{S}_d^{(k)} = -S^{(k)} \cup S^{(k)}$. As $\mathbf{x}_0 = \mathbf{0}$, the $\mathbf{S}^{(k)}$, $k \geq 0$, are an increasing family of finite graphs. The *Sierpiński graph* is their union: $\mathbf{S}_d = \bigcup_{k\geq 0} \mathbf{S}_d^{(k)}$. It is regular with vertex degrees equal to $2d$. Figure 14 shows (a finite piece of) $\mathbf{S}_2$; the part with the bigger •'s is $\mathbf{S}_2^{(1)}$.

The subgraph $\mathbf{S}^{(k)}$ is the ball in $\mathbf{S}$ of radius 2^k centred at vertex $\mathbf{0}$. We have $|S^{(k)}| = (d+1)\,|S^{(k-1)}| - \frac{d(d+1)}{2} = \frac{d+1}{2}\big((d+1)^k + 1\big)$. Thus, with

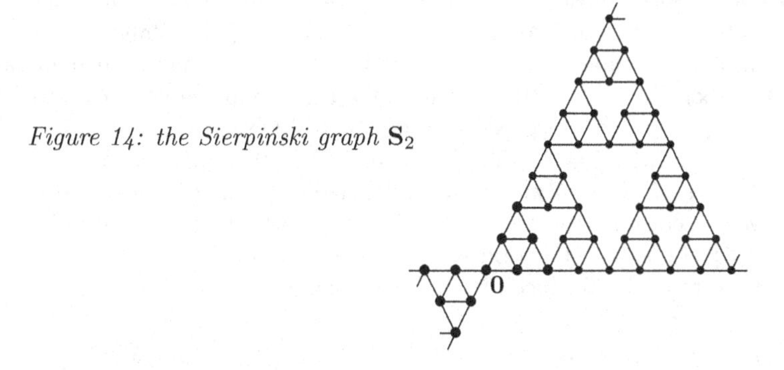

Figure 14: the Sierpiński graph $\mathbf{S}_2$

respect to the counting measure, $V(\mathbf{0}, 2^k) = (d+1)^{k+1} + d$. From this we obtain

$$n^r + d \le V(\mathbf{0}, n) \le (d+1)^2\, n^r + d \qquad \text{with} \qquad r = \frac{\log(d+1)}{\log 2}\,.$$

Let us now consider simple random walk on $\mathbf{S}_d$. From Theorem 14.22 we get $p^{(2n)}(\mathbf{0},\mathbf{0}) \ge C\,(n \log n)^{-r/2}$. However, this lower bound is poor. Without any computation, we can see immediately that the random walk is recurrent for all d. Indeed, the number of edges going out of $\mathbf{S}^{(k)}$ is constant (equal to $2d^2$), so that Nash-Williams' criterion (Corollary 2.20) applies.

A. Stopping times and an equation for the Green function

Consider the graph $2\mathbf{S}_d$. One of the important properties of the Sierpiński graph is that the vertex set of $2\mathbf{S}_d$ is contained in $\mathbf{S}_d$. Two points of $2\mathbf{S}_d$ are neighbours in this graph precisely when they are at distance 2 in $\mathbf{S}_d$.

The simple random walk Z_n on $\mathbf{S}_d$ visits $2\mathbf{S}_d$ infinitely often with probability 1. Suppose that $Z_0 = 2\mathbf{x}$, where $\mathbf{x} \in \mathbf{S}_d$. We can "factor" the random walk with respect to its successive visits in $2\mathbf{S}_d$, that is, we consider the stopping times

$$\mathbf{t}_0 = 0\,, \quad \mathbf{t}_j = \min\{n > \mathbf{t}_{j-1} : Z_n \in 2\mathbf{S}_d\,,\ Z_n \ne Z_{\mathbf{t}_{j-1}}\}\,.$$

The ball $B(2\mathbf{x}, 2)$ in $\mathbf{S}_d$ is connected to the rest of $\mathbf{S}_d$ only through its vertices $2\mathbf{y} \in 2\mathbf{S}_d$, where $\mathbf{y} \sim \mathbf{x}$ in $\mathbf{S}_d$. Furthermore, each $B(2\mathbf{x}, 2)$ is isomorphic with $\mathbf{S}^{(1)}$. This is the basis of the following fundamental lemma.

(16.1) Lemma. *The increments* $\mathbf{t}_j - \mathbf{t}_{j-1}$ *are i.i.d. with probability generating function*

$$\phi(z) = \mathbb{E}_{2\mathbf{x}}(z^{\mathbf{t}_1}) = \frac{z^2}{1 + (z-1)\big((d-2)z - (2d-1)\big)}\,.$$

Proof. Isomorphism of all $B(2\mathbf{x},2)$ implies immediately that the increments are i.i.d. We now compute $\phi(z) = \mathbb{E}_{\mathbf{0}}(z^{\mathbf{t}_1}) = \sum_{n=1}^{\infty} \mathbb{P}_{\mathbf{0}}[\mathbf{t}_1 = n]\, z^n$.

Consider the random walk restricted to $\mathbf{S}^{(1)}$, and make the boundary points $\pm 2\mathbf{x}_j$ $(j = 1,\dots,d)$ absorbing. We can group the vertices into four classes: $C_0 = \{\mathbf{0}\}$, $C_1 = \{\pm\mathbf{x}_j : j \neq 0\}$, $C_2 = \{\pm(\mathbf{x}_j + \mathbf{x}_k), j \neq k, j,k \neq 0\}$ and $C_3 = \{\pm 2\mathbf{x}_j : j \neq 0\}$. Then, for all $i, i' \in \{0,1,2,3\}$ and $\mathbf{x} \in C_i$, the probabilities $\widetilde{p}(i,i') = p(\mathbf{x}, C_{i'})$ are independent of the choice of $\mathbf{x} \in C_i$, so that we can consider the *factor chain* with transition probabilities $\widetilde{p}(\cdot,\cdot)$ on $\{0,1,2,3\}$. (This construction is more general than the factor chains with respect to groups introduced in §1.F.) We compute

$$\widetilde{p}(0,1) = 1\,,\ \widetilde{p}(1,0) = \tfrac{1}{2d}\,,\ \widetilde{p}(1,1) = \widetilde{p}(1,2) = \tfrac{d-1}{2d}\,,\ \widetilde{p}(1,3) = \tfrac{1}{2d}\,,$$
$$\widetilde{p}(2,1) = \widetilde{p}(2,3) = \tfrac{1}{d}\,,\ \widetilde{p}(2,2) = \tfrac{d-2}{d}\,,\quad \text{and} \quad \widetilde{p}(3,3) = 1\,;$$

the other transition probabilities are 0. We have $\phi(z) = \widetilde{F}(0,3|z)$. Using Lemma 1.13(d), we obtain the following system of equations.

$$\begin{aligned}
&\widetilde{F}(0,3|z) = z\,\widetilde{F}(1,3|z)\,,\\
&\widetilde{F}(1,3|z) = \tfrac{1}{2d}z + \tfrac{1}{2d}z\,\widetilde{F}(0,3|z) + \tfrac{d-1}{2d}z\,\widetilde{F}(1,3|z) + \tfrac{d-1}{2d}z\,\widetilde{F}(2,3|z)\,,\\
&\widetilde{F}(2,3|z) = \tfrac{1}{d}z + \tfrac{1}{d}z\,\widetilde{F}(1,3|z) + \tfrac{d-2}{d}z\,\widetilde{F}(2,3|z)\,.
\end{aligned}$$

The rest is straightforward computation. □

Also, we note that for $\mathbf{y} \sim \mathbf{x}$ in $\mathbf{S}_d$,

$$\mathbb{E}_{2\mathbf{x}}\big(z^{\mathbf{t}_1}\,\mathbf{1}_{2\mathbf{y}}(Z_{\mathbf{t}_1})\big) = \tfrac{1}{2d}\phi(z)\,. \tag{16.2}$$

Indeed, each point $2\mathbf{y}$ with $\mathbf{y} \sim \mathbf{x}$ is equally likely to be $Z_{\mathbf{t}_1}$, when $Z_0 = 2\mathbf{x}$.
Furthermore, we shall need $\psi(z) = \sum_{n=1}^{\infty} \mathbb{P}_{2\mathbf{x}}[Z_n = 2\mathbf{x}\,,\ \mathbf{t}_1 > n]\, z^n$.

(16.3) Lemma. $$\psi(z) = \frac{d-(d-2)z}{d+z}\,\phi(z)\,.$$

Proof. For the factor chain introduced in the last proof, $1 + \psi(z) = \widetilde{G}(0,0|z)$, so that $\psi(z) = \widetilde{U}(0,0|z)/\big(1 - \widetilde{U}(0,0|z)\big)$. The computations are now similar as above, using Lemma 1.13(c),(d) and the fact that $\widetilde{F}(2,0|z) = \widetilde{F}(2,1|z)\,\widetilde{F}(1,0|z)$. We leave the details as an exercise. □

We now consider the random variables $J(n) = \max\{j : \mathbf{t}_j \le n\}$ and write, for $\mathbf{x}, \mathbf{y} \in \mathbf{S}_d$,

$$G(2\mathbf{x}, 2\mathbf{y}|z) = \sum_{k=0}^{\infty} G_k(2\mathbf{x}, 2\mathbf{y}|z) = \sum_{k=0}^{\infty}\sum_{n=0}^{\infty} \mathbb{P}_{2\mathbf{x}}[Z_n = 2\mathbf{y}\,,\ J(n) = k]\, z^n\,.$$

Note that $Z_n = 2\mathbf{y} \in 2\mathbf{S}_d$ implies $Z_{J(n)} = 2\mathbf{y}$ and $Z_i \notin 2\mathbf{S}_d$ for $J(n) < i < n$. Therefore

$$
\begin{aligned}
G_k(2\mathbf{x}, 2\mathbf{y}|z) &= \sum_{\ell=0}^{\infty}\sum_{n=\ell}^{\infty} \mathbb{P}_{2\mathbf{x}}[\mathbf{t}_k = \ell\,,\ Z_\ell = Z_n = 2\mathbf{y}\,,\ Z_i \notin 2\mathbf{S}_d\ (\ell < i < n)]\, z^n \\
&= \sum_{\ell=0}^{\infty} \mathbb{P}_{2\mathbf{x}}[\mathbf{t}_k = \ell\,,\ Z_\ell = 2\mathbf{y}]\, z^\ell \\
&\quad \times \sum_{n=\ell}^{\infty} \mathbb{P}_{2\mathbf{x}}[Z_n = 2\mathbf{y}\,,\ Z_i \notin 2\mathbf{S}_d\ (\ell < i < n) \mid Z_\ell = 2\mathbf{y}]\, z^{n-\ell} \\
&= \big(1 + \psi(z)\big)\, \mathbb{E}_{2\mathbf{x}}\big(\mathbf{1}_{\{2\mathbf{y}\}}(Z_{\mathbf{t}_k})\, z^{\mathbf{t}_k}\big)\,.
\end{aligned}
$$

By induction on k, it is now straightforward to show from Lemma 16.1 that

$$\mathbb{E}_{2\mathbf{x}}\big(\mathbf{1}_{\{2\mathbf{y}\}}(Z_{\mathbf{t}_k})\, z^{\mathbf{t}_k}\big) = p^{(k)}(\mathbf{x}, \mathbf{y})\, \phi(z)^k\,;$$

for $k = 1$, this is (16.2). We have proved the following.

(16.4) Proposition. *For all* $\mathbf{x}, \mathbf{y} \in \mathbf{S}_d$ *and* $|z| < 1$,

$$G(2\mathbf{x}, 2\mathbf{y}|z) = \big(1 + \psi(z)\big)\, G\big(\mathbf{x}, \mathbf{y}|\phi(z)\big)\,.$$

We set $G(z) = G(\mathbf{0}, \mathbf{0}|z)$. It satisfies the functional equation

$$G(z) = \big(1 + \psi(z)\big)\, G\big(\phi(z)\big)\,. \tag{16.5}$$

Now, if $|z| < 1$ then $|\phi(z)| \le \phi(|z|) < |z^2| < |z|$. If $\phi^{(n)}$ denotes the n-th iterate of ϕ, that is, $\phi^{(0)}(z) = z$ and $\phi^{(n)}(z) = \phi\big(\phi^{(n-1)}(z)\big)$, then we see that $\phi^{(n)}(z) \to 0$ uniformly in compact subsets of the open unit disk. Recall that $G(0) = 1$, and observe that the Taylor expansion of $\psi(z)$ at 0 starts with $z^2/(2d^2)$. Therefore, iterating (16.5), we obtain

$$
\begin{aligned}
G(z) &= \lim_{N\to\infty} \prod_{n=0}^{N-1} \Big(1 + \psi\big(\phi^{(n)}(z)\big)\Big)\, G\big(\phi^{(N)}(z)\big) \\
&= \prod_{n=0}^{\infty} \Big(1 + \psi\big(\phi^{(n)}(z)\big)\Big)
\end{aligned}
\tag{16.6}
$$

for $|z| < 1$. We now want to show that convergence in (16.6) holds in a much larger subset of the complex plane. Indeed, $G(1/z)/z$ is a matrix element of the resolvent of the symmetric operator P, so that it must extend analytically everywhere outside the ℓ^2-spectrum of P, which is contained in the real interval $[-1\,,\,1]$.

(16.7) Proposition. *We have $\phi^{(n)}(z) \to 0$ except for $1/z$ in a closed subset of the real interval $[\frac{d-3}{2d}\,,\,1]$ containing its two endpoints, and convergence is uniform for $1/z$ in closed subsets of $\mathbb{C} \setminus [\frac{d-3}{2d}\,,\,1]$.*

In particular, in this region the infinite product (16.6) converges to an analytic function.

Proof. We work with the conjugate

$$\check{\phi}(z) = 1/\phi(1/z) = 1 + 2d(z-1)(z - \tfrac{d-3}{2d})$$

and show that its iterates tend to ∞ outside of $[\frac{d-3}{2d}\,,\,1]$.

We already know that $|\check{\phi}(z)| \geq \check{\phi}(|z|) > |z|^2$ for $|z| > 1$. Therefore $\check{\phi}^{(n)}(z) \to \infty$ in this region, uniformly in $\{|z| \geq c\}$ for every $c > 1$.

The function $\check{\phi}(z)$ is symmetric with respect to the point $\frac{3d-3}{4d}$. Consequently we also have $|\check{\phi}(z)| > |z - \frac{3d-3}{2d}|^2$ for $|z - \frac{3d-3}{2d}| > 1$, and the iterates of $\check{\phi}$ tend to ∞ in this region, too.

We are left with studying the intersection D of the two closed disks $\{|z| \leq 1\}$ and $\{|z - \frac{3d-3}{2d}| \leq 1\}$.

Consider the (real) solutions of $\check{\phi}(x) = \frac{d-3}{2d}$: they are $x_{1,2} = \big((3d-3) \pm \sqrt{(d+1)^2 - 4}\big)/(4d)$. If $z = x + iy$ then $\mathrm{Re}\big(\check{\phi}(z)\big) = \check{\phi}(x) - 2dy^2$. Thus, if $\check{\phi}(x) < \frac{d-3}{2d}$ then $|\check{\phi}(z) - \frac{3d-3}{2d}| > 1$ and the iterates tend to ∞. This settles the strip $\{x_1 < \mathrm{Re}(z) < x_2\}$.

Finally, we are left with the intersection of D with the two strips $\{\frac{d-3}{2d} \leq \mathrm{Re}(z) \leq x_1\}$ and $\{x_2 \leq \mathrm{Re}(z) \leq 1\}$. Now $\mathrm{Im}\big(\check{\phi}(z)\big) = y\big(4dx - (3d-3)\big)$. If x is outside the interval $[x_1\,,\,x_2]$ then $|4dx - (3d-3)| > \sqrt{(d+1)^2 - 4} > 1$ [we use $d \geq 2$ here!], and if in addition $\mathrm{Im}(z) \neq 0$ then $\big|\mathrm{Im}\big(\check{\phi}(z)\big)\big| > \sqrt{(d+1)^2 - 4}\,|\mathrm{Im}(z)|$. Thus, after a finite number of iterations, $\check{\phi}^{(n)}(z)$ must leave D.

Altogether, we see that only for real $z \in [\frac{d-3}{2d}\,,\,x_1] \cup [x_2\,,\,1]$ may the iterates of $\check{\phi}(z)$ not tend to ∞.

We have $\check{\phi}'(1) = d + 3$, and $z = 1$ itself is a repelling fixed point of $\check{\phi}(\cdot)$, so that $\frac{d-3}{2d}$, x_1, x_2 and 1 must belong to the Julia set (= set of points which are not attracted by ∞) of $\check{\phi}(\cdot)$. As a matter of fact, it is not hard to see that this is a Cantor set. □

B. Singularity analysis

We now want to use Proposition 16.7 and the representation (16.6) for finding the asymptotic behaviour of the power series coefficients $p^{(n)}(\mathbf{0},\mathbf{0})$ of $G(z)$. To this end, we need good knowledge of the behaviour of $G(z)$ near its principal singularity $z = 1$. In the rest of this chapter, we shall use methods from complex analysis – as typically applied in combinatorial enumeration

– on several occasions. We now present the analytic tool that we shall use for the Sierpiński graphs. This is *singularity analysis,* developed by Flajolet and Odlyzko [117]; see also Odlyzko [249]. It has several variants; the "prototype" considers the coefficients of the power series expansion at 0 of $(1-z)^\zeta$. We cannot reproduce the whole theory with its proofs here, but only present the main tool elaborated by Flajolet and Odlyzko; see in particular Proposition 1 and Corollaries 2 and 3 in [117].

We introduce the sets $D_{\alpha,\delta} = \{z \in \mathbb{C} : |z| \le 1+\delta\,,\ |\arg(z-1)| \ge \alpha\}$, where $0 < \alpha < \frac{\pi}{2}$ and $\delta > 0$.

(16.8) Theorem. *Let $F(z) = \sum_n f_n\, z^n$ be a power series with real coefficients and a singularity at $z = 1$. Assume that it extends analytically to a suitable set $D_{\alpha,\delta} \setminus \{1\}$.*

If $F(z) = C\,(1-z)^\zeta + \mathcal{O}(|1-z|^\beta)$ as $z \to 1$ in $D_{\alpha,\delta}$, where $\zeta \in \mathbb{C} \setminus \mathbb{N}_0$ and $\beta \in \mathbb{R}$, $\beta > \mathrm{Re}(\zeta)$, then

$$f_n = C\binom{n-\zeta-1}{n} + \mathcal{O}(n^{-\beta-1})\,.$$

Note the particular case when $C = 0$. We want to apply (an extension of) this theorem to $G(z) = \sum_n p^{(n)}(\mathbf{0},\mathbf{0})\, z^n$. The first problem is to find the leading *real* exponent η of $G(z)$ at $z = 1$. If we write $G(z) = (1-z)^\eta\, H_+(z)$ and substitute this into (16.5) then we see that H_+ satisfies the functional equation $H_+(z) = \big(1 + f(z)\big)\, H_+\big(\phi(z)\big)$, where

$$1 + f(z) = \big(1 + \psi(z)\big)\left(\frac{1-\phi(z)}{1-z}\right)^\eta.$$

Precisely the same proof as that of Proposition 16.7 (note that $f(0) = 0$) shows that H_+ has the product expansion

$$H_+(z) = \prod_{n=0}^{\infty}\Big(1 + f\big(\phi^{(n)}(z)\big)\Big)\,,$$

which converges to an analytic extension of $H_+(z)$ in the complement of the Julia set of ϕ. Now ϕ has a local inverse $\phi^{(-1)}$ near $z = 1$, analytic and such that in an open disk $U_c = \{z \in \mathbb{C} : |z-1| < c\}$, its derivative is bounded in absolute value by a constant smaller than 1. In particular, $z = 1$ is an attractive fixed point of $\phi^{(-1)}$, and its iterates $\phi^{(-n)}(z)$ tend to 1 exponentially fast in U_c. We now choose η such that $f(1) = 0$, that is, $\eta = \frac{\log(d+1)}{\log(d+3)} - 1$. We shall fix this value of η for the rest of the present section. Then the infinite product

$$H_-(z) = \prod_{n=1}^{\infty}\Big(1 + f\big(\phi^{(-n)}(z)\big)\Big)$$

converges and is analytic in U_c. The "backward completion" $H(z) = H_-(z)\,H_+(z)$ of $H_+(z)$ is analytic in $U_c \setminus [1\,,\infty)$ and satisfies $H(z) = H\big(\phi^{(-1)}(z)\big)$. We now modify $H(z)$ to obtain a function which is periodic. We learn from Beardon's beautiful book on the theory of functional iterations [29], Thm. 6.3.2, that one can extract the linear part of the expansion of $\phi^{(-1)}$ at $z=1$ by conjugating by a function $g(z)$ which is analytic in the disk U_c (with smaller c if necessary), and real-valued if $z \in U_c$ is real:

$$g^{(-1)} \circ \phi^{(-1)} \circ g(z) = 1 + \tfrac{1}{d+3}(z-1)\,, \quad \text{and} \quad g(1) = g'(1) = 1\,.$$

If we now define $K(z) = H\big(g(z)\big)$, this function is analytic in $U_c \setminus [1\,,\infty)$ and satisfies

$$K(z) = K\big(1 + \tfrac{1}{d+3}(z-1)\big)\,.$$

Thus, the function $T_0(w) = K\big(1-(d+3)^w\big)$ is periodic with period 1 and analytic in the semi-infinite strip $\{\mathrm{Re}(w) < \log c/\log(d+3)\,,\ |\mathrm{Im}(w)| < \pi/\log(d+3)\}$. This allows analytic continuation to the whole strip $\{|\mathrm{Im}(w)| < \pi/\log(d+3)\}$, and a rapidly converging Fourier expansion

$$T_0(w) = \sum_{k=-\infty}^{\infty} a_k \exp(2k\,\pi\,i\,w) \quad \text{with} \quad a_k = \mathcal{O}\Big(\exp\Big(-\big(\tfrac{2\pi^2}{\log(d+3)} - \varepsilon\big)|k|\Big)\Big)$$

for all $\varepsilon > 0$. We now consider the approximation error in replacing $H_+(z)$ with $K(z)$. As $z \to 1$, we have $H_-(z) = \mathcal{O}(|z-1|)$, whence $H_+(z) = \big(1+\mathcal{O}(|z-1|)\big)H(z)$, and (as $K(z)$ is bounded)

$$H_+(z) - K(z) = \big(1+\mathcal{O}(|z-1|)\big)\big(H(z)-K(z)\big) + \mathcal{O}(|z-1|)\,.$$

In order to control $H(z) - K(z) = H(z) - H\big(z+\mathcal{O}(|1-z|^2\big)$, we study $H'(z)$.

(16.9) Lemma. *For each $\delta > 0$ and $0 < \alpha < \pi/2$,*

$$H'(z) = \mathcal{O}_{\alpha,\delta}\big(|z-1|^{-\log(d+3+\delta)/\log(d+3-\delta)}\big)\,,$$

as $z \to 1$ and $|\arg(z-1)| \ge \alpha$.

Proof. We can find a disk U_c around 1 where $|\phi'(z) - (d+3)| < \delta$. We write $U_{c,\alpha} = \{z \in U_c : |\arg(z-1)| > \alpha\}$.

If c is suitably small then $\big|\arg\big(\phi(z)-1\big)\big| > \alpha$ for all $z \in U_{c,\alpha}$. This can be seen by elementary computations, for example by observing that near $x=1$, the line $y = a(x-1)$ is mapped to a curve $y = w(x)$ where $w(1) = 1$, $w'(1) = a$ and $w''(1) = a\Big(1 + (1+a^2)\phi''(1)/\phi'(1)^2\Big)$, with the value $\phi''(1) = 2d^2 + 6d + 12$. Thus the image lies above or below the line according to the sign of a.

This implies that $V_{c,\alpha} = U_{c,\alpha} \setminus \phi^{(-1)}(U_{c,\alpha})$ lies in an annulus around 1, and by continuity, $|(z-1)^{\log(d+3+\delta)/\log(d+3-\delta)} H'(z)| \le C < \infty$ in $V_{c,\alpha}^-$. Now for every $z \in U_{c,\alpha} \setminus \{1\}$ there is an $n \ge 0$ such that $\phi^{(n)}(z) \in V_{c,\alpha}$. We show by induction on n that

$$|H'(z)| \le C\,|z-1|^{-\log(d+3+\delta)/\log(d+3-\delta)} \quad \text{for all } z \in U_{c,\alpha} \setminus \{1\}.$$

This is true for $n = 0$. For the induction step, suppose that the statement holds for $\phi(z) \in U_{c,\alpha}$. Then

$$\begin{aligned}|H'(z)| = \bigl|\phi'(z)\,H'\bigl(\phi(z)\bigr)\bigr| &\le C\,(d+3+\delta)\,|\phi(z)-1|^{-\log(d+3+\delta)/\log(d+3-\delta)}\\ &\le C\,(d+3+\delta)\,\bigl(|z-1|(d+3-\delta)\bigr)^{-\log(d+3+\delta)/\log(d+3-\delta)}.\end{aligned}$$

The estimate follows. □

Combining all these computations, we obtain the following expansion.

(16.10) Theorem. *Near $z = 1$ and for $|\arg(z-1)| \ge \alpha$, the function $G(z) = G(\mathbf{0},\mathbf{0}|z)$ has a singular expansion*

$$G(z) = (1-z)^{\eta}\left(T_0\Bigl(\frac{\log(1-z)}{\log(d+3)}\Bigr) + \mathcal{O}_{\alpha,\varepsilon}(|z-1|^{1-\varepsilon})\right)$$

for all $\alpha < \frac{\pi}{2}$ and $\varepsilon > 0$, where $\eta = \frac{\log(d+1)}{\log(d+3)} - 1$ and $T_0(w)$ is a non-constant periodic function with period 1 which is analytic in the strip $\{|\mathrm{Im}(w)| < \frac{\pi}{\log(d+3)}\}$.

Proof. We only have to show that $T_0(w)$ cannot be constant. There are two ways to do this; one of them fails for $d = 2$ and the other for $d = 3$.

First assume that $d \ge 3$. The rational function $\phi(z)$ has two poles $p_2 > p_1 > 1$ and the three fixed points $0, 1, \frac{2d}{d-2}$. In the real interval $[1\,,\,p_1)$, $\phi(z)$ has derivative $> d+3$ and tends to $+\infty$ when $z \to p_1$. The local inverse of ϕ around 1 extends to this interval and has the attracting fixed point $z = 1$. The point $\frac{2d}{d-2}$ lies in $(p_2\,,\,\infty)$. We can find a (unique) $z_1 \in (1\,,\,p_1)$ satisfying $\phi(z_1) = \frac{2d}{d-2}$, and the sequence of (real) points $z_n = \phi^{(-n+1)}(z_1)$ is decreasing with limit 1.

Now suppose that $T_0(w)$ is constant. Then $K(z)$ and hence also $H(z) = K\bigl(g^{(-1)}(z)\bigr)$ are constant. This constant must be real and positive, as H_+ is positive in $(0,1)$ and H_- is positive for real z near 1. Therefore $H_+(z) = H(z)/H_-(z)$ has an analytic extension to a full neighbourhood U_c of $z = 1$ and $H_+(z) > 0$ in $U_c \cap \mathbb{R}$. In $U_c \setminus [1\,,\,\infty)$, we can write

$G(z) = (1-z)^\eta\, H_+(z)$. There is some z_n in U_c, and $\phi^{(n)}$ maps a suitable neighbourhood $V \subset U_c$ to a neighbourhood W of $\frac{2d}{d-2}$, such that in each iterate $\phi^{(j)}(V)$, $j = 0, \dots, n$, both $\phi(z)$ and $\psi(z)$ are well defined (i.e., have no poles), and such that $\phi^{(n)} : V \to W$ is invertible. Also, the iterates of non-real points in V are non-real.

Now (16.5) holds in a set containing $\mathbb{C} \setminus \mathbb{R}$. Therefore we have in $V \setminus \mathbb{R}$

$$(1-z)^\eta\, H_+(z) = \big(1 + \psi_n(z)\big) G\big(\phi^{(n)}(z)\big)$$

with $\psi_n(z) = -1 + \prod_{j=0}^{n-1}\Big(1 + \psi\big(\phi^{(j)}(z)\big)\Big)$. Let $z \to z_n$ in the upper half plane. Then $(1-z)^\eta \to (z_n - 1)^\eta\big(\cos(\eta\pi) - i\sin(\eta\pi)\big)$. If $z \to z_n$ in the lower half plane, then the last "$-$" becomes a "$+$". (Recall that $(1-z)^\eta = \exp\big(\eta \log(1-z)\big)$ and that $\mathrm{Im}(\log w) = \arg(w) \in (-\pi\,,\,\pi)$.) One checks that $1 + \psi_n(z_n) > 0$: indeed, $\psi(z) > 0$ in $(1\,, p_1)$. Therefore we find a constant $a > 0$ such that

$$\lim_{\mathbb{C}\setminus\mathbb{R} \ni z \to \frac{2d}{d-2}} \mathrm{Re}\big(G(z)\big) = a\, \cos(\eta\pi)\,.$$

Now $\phi(W)$ is a neighbourhood of W, and ϕ maps non-real points in W to non-real points. We compute $\psi(\frac{2d}{d-2}) = -2$ and apply (16.5) once more to obtain

$$\lim_{\mathbb{C}\setminus\mathbb{R} \ni z \to \frac{2d}{d-2}} \mathrm{Re}\big(G(z)\big) = - \lim_{\mathbb{C}\setminus\mathbb{R} \ni z \to \frac{2d}{d-2}} \mathrm{Re}\Big(G\big(\phi(z)\big)\Big)\,.$$

That is, $a = -a$, a contradiction.

For $d = 2$, $\phi(z) = z^2/(4-3z)$ has no fixed points besides 0 and 1. There is only one pole $p_1 = 4/3$. As above, ϕ is strictly increasing in $[1\,, p_1)$, and the inverse of ϕ in this interval has the attracting fixed point $z = 1$. We use the point $z_0 = -4$ which is mapped to 1 by ϕ. Observe that $\psi(-4) = -1$, so that (16.5) yields a neighbourhood W of $z = -4$ such that in $W \setminus \mathbb{R}$

$$G(z) = (4+z)^{1+\eta}\, h(z) H_+\big(\phi(z)\big) \quad \text{with} \quad h(z) = \frac{(2-z)(1-z)^\eta}{(2+z)(4-3z)^{1+\eta}}\,.$$

Now $\phi(z_1) = -4$, where $z_1 = 6 - 2\sqrt{5} > p_1$. We can find $z_2 \in (1\,, p_1)$ such that $\phi(z_2) = 6 - 2\sqrt{5}$, and its successive preimages z_n tend to 1 from the right and satisfy $\phi^{(n)}(z_n) = -4$. Assuming that $T_0(w)$ is constant, we proceed as above and find a neighbourhood V of some z_n such that in $V \setminus \mathbb{R}$

$$(1-z)^\eta\, H_+(z) = \big(1 + \psi_n(z)\big)\big(4 + \phi^{(n)}(z)\big)^{1+\eta}\, h\big(\phi^{(n)}(z)\big)\, H_+\big(\phi^{(n+1)}(z)\big)\,.$$

After checking that $\psi_n(z_n) \neq -1$, we take absolute values on both sides and let $z \to z_n$ in $V \setminus \mathbb{R}$. On the left, the limit is non-zero, while on the right it is 0, a contradiction. The same method works for all $d \neq 3$. □

Combining Theorems 16.8 and 16.10, we can now deduce the main result of this subsection.

(16.11) Theorem. *For the simple random walk on* $\mathbf{S}_d$,

$$p^{(n)}(\mathbf{0},\mathbf{0}) = n^{-\log(d+1)/\log(d+3)}\left(T\Big(\frac{\log n}{\log(d+3)}\Big) + \mathcal{O}_\varepsilon(n^{\varepsilon-1})\right)$$

for all $\varepsilon > 0$, *where* $T(w)$ *is a non-constant periodic* C^∞*-function with period* 1. *Its Fourier series is*

$$T(w) = \sum_{k=-\infty}^{\infty} \frac{a_k}{\Gamma\big(1 - \frac{\log(d+1)+2k\pi i}{\log(d+3)}\big)} \exp(-2k\,\pi\, i\, w)\,,$$

where the a_k *are the Fourier coefficients of* $T_0(w)$.

Proof. By Theorem 16.10 we have in a suitable $D_{\alpha,\delta}$

$$G(z) = \sum_{k=-\infty}^{\infty} a_k\,(1-z)^{\eta+2k\pi i/\log(d+3)} + \mathcal{O}(|z-1|^{\eta+1-\varepsilon})\,.$$

The series converges uniformly, so that we may apply Theorem 16.8 to each term and take the sum over k to obtain

$$p^{(n)}(\mathbf{0},\mathbf{0}) = \sum_{k=-\infty}^{\infty} a_k \binom{n - \frac{\log(d+1)+2k\pi i}{\log(d+3)}}{n} + \mathcal{O}\big(n^{-(\eta+2-\varepsilon)}\big)\,.$$

We now claim that, with T as defined above, the sum over k can be written as $n^{-\log(d+1)/\log(d+3)}\Big(T\big(\log n/\log(d+3)\big) + \mathcal{O}\big(n^{-(1-\varepsilon)}\big)\Big)$ for all $\varepsilon > 0$. We write the k-th term in the sum as

$$\frac{a_k}{\Gamma\big(1 - \frac{\log(d+1)+2k\pi\, i}{\log(d+3)}\big)}\;\frac{\Gamma\big(n+1-\frac{\log(d+1)+2k\pi i}{\log(d+3)}\big)}{n!}$$

and use various formulae for the Gamma function. First, $|\Gamma(x+iy)| = \sqrt{2\pi}\,|y|^{x-1/2}\,e^{-|y|\pi/2}\,\big(1+\mathfrak{o}(1)_{|y|\to\infty}\big)$ – see Lebedev [213], p.15 – so that

$$a_k/\Gamma\big(1 - \tfrac{\log(d+1)+2k\pi\, i}{\log(d+3)}\big) = \mathcal{O}\Big(\exp\Big(-\big(\tfrac{\pi^2}{\log(d+3)} - \varepsilon\big)|k|\Big)\Big)$$

for all $\varepsilon > 0$. Also, $|\Gamma(x+iy)| \le \Gamma(x)$ for $x \ge 0$, and with a use of Stirling's formula,

$$\frac{1}{n!}\left|\Gamma\Big(n+1-\frac{\log(d+1)+2k\pi i}{\log(d+3)}\Big)\right| \le C\,n^{-\log(d+1)/\log(d+3)}\,.$$

Therefore, if $b > 0$ is sufficiently large,

$$\sum_{|k| \geq b \log n} a_k \binom{n - \frac{\log(d+1)+2k\pi i}{\log(d+3)}}{n} = \mathcal{O}\left(n^{-\frac{\log(d+1)}{\log(d+3)}-1}\right).$$

The same holds for the sum over $|k| \geq \log n$ in the series defining $T(\log n / \log(d+3))$. Now we only have to compare the two sums in the range $-b \log n < k < b \log n$. We use the following form of Stirling's formula – see [213], (1.4.23):

$$\Gamma(z) = \sqrt{2\pi}\, z^{z-1/2}\, e^{-z} \left(1 + \mathcal{O}_\alpha(|z|^{-1})\right) \quad \text{as } z \to \infty\,,\ |\arg(z)| \leq \alpha < \pi\,.$$

This implies

$$\frac{1}{n!}\Gamma\left(n+1-\frac{\log(d+1)+2k\pi i}{\log(d+3)}\right) = n^{-(\log(d+1)+2k\pi i)/\log(d+3)}\left(1+\mathcal{O}\left(\tfrac{(\log n)^2}{n}\right)\right)$$

with the $\mathcal{O}(\cdot)$ independent of $k \in (-b \log n\,,\ b \log n)$. The result follows. □

17. Local limit theorems on free products

Here we take up the machinery developed in Section 9; the reader is invited to recall §§9.B–C. We shall derive the asymptotic behaviour of transition probabilities on a variety of free products, in particular of groups. Here it will become clear why we insisted on distinguishing the two different cases $\theta < \bar{\theta}$ and $\theta = \bar{\theta}$ in Theorem 9.22. In the sequel, we suppose that the setting is precisely that of (9.16) and (9.17). Also, we shall suppose that the random walk is strongly periodic as in (9.1), which holds automatically for random walks on groups.

We start with some extensions of the composition formulae derived in §9.C. Recall that $F_i(\cdot,\cdot|z)$ denotes the probability generating functions for the first hitting times of the Markov chain (X_i, P_i), and $X_i' = \setminus\{o\}$.

(17.1) Lemma. *Let $P = \sum_{i \in \mathcal{I}} \alpha_i \overline{P}_i$ be defined on $X = \ast_{i \in I} X_i$, and $\mathtt{r} = \mathtt{r}(P) = 1/\rho(P)$.*

(a) *For $i \in \mathcal{I}$, one has $\zeta_i(\mathtt{r}) < \mathtt{r}(P_i)$, unless $\theta = \theta(P_i)/\alpha_i$, in which case $\zeta_i(\mathtt{r}) = \mathtt{r}(P_i)$.*

(b) *If $x = x_1 \cdots x_k \in X$ with $x_j \in X'_{i_j}$, $i_j \neq i_{j-1}$, then*

$$F(x, o|z) = F_{i_1}\left(x_1, o|\zeta_{i_1}(z)\right) \cdots F_{i_k}\left(x_k, o|\zeta_{i_k}(z)\right).$$

Proof. By (9.20), $\zeta_i(\mathtt{r})\, G_i\big(\zeta_i(\mathtt{r})\big) = \alpha_i\, \theta \leq \theta(P_i) = \mathtt{r}(P_i)\, G_i\big(\mathtt{r}(P_i)\big)$. The strict monotonicity of the functions involved for positive real arguments implies (a).

(b) For x as above, set $y_0 = o$ and $y_j = x_1 \cdots x_j$ $(j = 1, \dots, k)$. By the tree-like (or, rather, cactus-like) structure of the free product, the random walk starting at y_j has to pass through y_{j-1} before reaching o. Therefore, the same stopping time argument as used in the proof of Lemma 1.23 yields

$$F(x, o|z) = F(y_k, y_{k-1}|z) \cdots F(y_1, y_0|z).$$

When starting at y_j and before hitting y_{j-1}, our random walk has to perform its steps within the copy of $X_{i_{j-1}}^{\perp}$ attached at y_{j-1} according to the construction of the free product. This is the same as starting at x_j and walking until the first visit to o. Therefore $F(y_j, y_{j-1}|z) = F(x_j, o|z)$, i.e., we can cancel on the left when going towards o, and Proposition 9.18(c) yields the proposed formula. □

Note that the last argument does not in general work in the other direction, going from o to x. However, it does work in the case of random walks on free products of groups.

Using (9.20) and the definition of $\zeta_i(z)$, we see that $\zeta_i(z) = \alpha_i z G(z) / \Phi_i\big(\alpha_i z G(z)\big)$. For $x_i \in X_i$, let

$$\Phi_i(x_i|t) = F_i\big(x_i, o|t/\Phi_i(t)\big).$$

This function is analytic for t in a $\mathfrak{U}_{\theta(P_i)}$, as defined in §9.B. Recalling that $G(x, o|z) = F(x, o|z)G(z)$, we see that for $x = x_1 \cdots x_k$ as in Lemma 17.1 above we have

$$(17.2)\quad G(x, o|z) = \Phi\big(x|zG(z)\big), \quad \text{where} \quad \Phi(x|t) = \Phi(t) \prod_{j=1}^{k} \Phi_{i_j}(x_j|\alpha_{i_j} t).$$

The function $\Phi(x|\cdot)$ is analytic in a $\mathfrak{U}_{\bar{\theta}}$.

To conclude the preparations, let $\mathtt{m} = \mathtt{m}(x)$ be the (unique) number in $\{0, \dots, \mathtt{d} - 1\}$ such that $p^{(n)}(x, o) > 0$ only when $\mathtt{d}$ divides $n - \mathtt{m}$. Finally, recall the definition (9.13) and the composition formula (9.21) for Ψ.

A. The typical case: $n^{-3/2}$

By the "typical" case we intend the one of Theorem 9.22(i), when θ is strictly smaller than $\bar{\theta}$. The basic result of this subsection is the following.

(17.3) Theorem. *Let $P = \sum_{i \in \mathcal{I}} \alpha_i \overline{P}_i$ be strongly periodic with period $\mathtt{d}$ on $X = \mathop{*}_{i \in I} X_i$. Suppose $\Psi(\bar{\theta}-) < 0$. If $x = x_1 \cdots x_k \in X$ then*

$$p^{(n)}(x, o) = \frac{\mathtt{d}}{2\sqrt{\rho\pi}}\, h(x)\, \rho^n\, n^{-3/2} + \mathcal{O}(\rho^n\, n^{-5/2}),$$

as $n \to \infty$, $n \equiv \mathtt{m}(x)$ modulo $\mathtt{d}$, where $\rho = \rho(P)$ is given by Theorem 9.22(i), and setting $i_j = \mathtt{i}(x_j)$,

$$h(x) = g(x)\,\Phi'(\theta)\sqrt{\frac{2\theta}{\Phi''(\theta)}}\left(\frac{\Phi'(\theta)}{\Phi(\theta)} + \sum_{j=1}^{k} \frac{\alpha_{i_j}\Phi'_{i_j}(x_j|\alpha_{i_j}\theta)}{\Phi_{i_j}(x_j|\alpha_{i_j}\theta)}\right), \quad \textit{with}$$

$$g(x) = \Phi(\theta)\prod_{j=1}^{k}\Phi_{i_j}(x_j|\alpha_{i_j}\theta).$$

Before the proof, we proceed as in the previous section and study the behaviour of $G(x,o|z)$ near the principal singularity $z = \mathtt{r}$.

(17.4) Proposition. *Under the assumptions of Theorem 17.3, for each $x \in X$ there exist a neighbourhood $\mathfrak{V} = \mathfrak{V}_x \subset \mathbb{C}$ of $z = \mathtt{r}$ and functions $A(x|z)$ and $B(x|z)$ analytic in $\mathfrak{V}$, such that $A(x|\mathtt{r}) = g(x)$, $B(x|\mathtt{r}) = h(x)$, and for $z \in \mathfrak{V} \setminus (\mathtt{r}, \infty)$,*

$$G(x,o|z) = A(x|z) - B(x|z)\sqrt{\mathtt{r} - z}\,.$$

Proof. We know from Theorem 9.22 that $G(\mathtt{r})$ is finite. In (9.23) we introduced the function $\mathcal{F}(z,w) = \Phi(z\,w) - w$, which is analytic in a neighbourhood of $(\mathtt{r}, G(\mathtt{r}))$ and satisfies $\mathcal{F}(z, G(z)) \equiv 0$. Also, we have the partial derivatives $\mathcal{F}_w(\mathtt{r}, G(\mathtt{r})) = 0$ and $\mathcal{F}_{ww}(\mathtt{r}, G(\mathtt{r})) = \mathtt{r}^2\Phi''(\theta) > 0$. Therefore, the root $G(\mathtt{r})$ of the function $w \mapsto \mathcal{F}(\mathtt{r}, w)$ has multiplicity 2. The Weierstrass preparation theorem (see e.g. Hörmander [175], Thm. 7.5.1) implies that in a neighbourhood $\mathfrak{U}$ of $(\mathtt{r}, G(\mathtt{r}))$ we can decompose

$$\mathcal{F}(z,w) = \mathcal{H}(z,w)\left(a(z) + b(z)\big(w - G(\mathtt{r})\big) + \big(w - G(\mathtt{r})\big)^2\right), \tag{17.5}$$

where $\mathcal{H}$ is analytic and non-zero and $a(z)$ and $b(z)$ are analytic in the z-projection $\mathfrak{V} = \mathfrak{V}_o$ of $\mathfrak{U}$. By evaluating $\mathcal{F}$, $\mathcal{F}_w$, $\mathcal{F}_{ww}$ and $\mathcal{F}_z$ at the point $(\mathtt{r}, G(\mathtt{r}))$, we find $a(\mathtt{r}) = b(\mathtt{r}) = 0$ and $a'(\mathtt{r}) = 2\theta\,\Phi'(\theta)^4/\Phi''(\theta) > 0$.

The decomposition (17.5) yields a quadratic equation for $G(z)$. Its discriminant $d(z) = b(z)^2 - 4a(z)$ is analytic in $\mathfrak{V}$ and has a Taylor expansion at $z = \mathtt{r}$ starting with $4a'(\mathtt{r})(\mathtt{r} - z)$. Solving for $G(z)$, we now get that in $\mathfrak{V} \setminus (\mathtt{r}, \infty)$, $G(z) = G(\mathtt{r}) - \frac{1}{2}b(z) \pm \frac{1}{2}\sqrt{d(z)}$. There are various ways to see that the correct sign is the "$-$". For example, otherwise we would obtain negative transition probabilities for large n at the end of our computations. We can write

$$\sqrt{d(z)} = 2B(z)\sqrt{\mathtt{r} - z}\,, \quad \text{with} \quad B(\mathtt{r}) = \sqrt{a'(\mathtt{r})} = g(o)\,.$$

$B(z) = B(o|z)$ and $A(z) = A(o|z) = G(\mathbf{r}) - \frac{1}{2}b(z)$ are analytic in $\mathfrak{V}$, and we obtain

$$G(z) = A(z) - B(z)\sqrt{\mathbf{r} - z} \tag{17.6}$$

as proposed.

Next, let $x \in X_i' = X_i \setminus \{o\}$. As $\theta < \bar{\theta}$ by assumption, Lemma 17.1(a) implies that the function $t \mapsto \Phi_i(x|\alpha_i t)$ is analytic near $t = \theta$, and

$$\alpha_i \Phi_i'(x|\alpha_i\theta) = F_i'\big(x, o|\alpha_i\theta/\Phi_i(\alpha_i\theta)\big)\,\Psi_i(\alpha_i\theta)/\Phi_i(\alpha_i\theta)^2 > 0\,.$$

In particular, $t \mapsto \Phi_i(x|\alpha_i t)$ is real-valued and strictly increasing for real t near θ. Therefore the simple branching point $z = \mathbf{r}$ of $G(z)$ is also a simple branching point of $F(x,o|z) = \Phi_i\big(x|\alpha_i z G(z)\big)$, and we can find a neighbourhood $\mathfrak{V}_x \subset \mathfrak{V}_o$ of $\mathbf{r}$ and functions $A_i(x|z)$ and $B_i(x|z)$ which are analytic there, such that in $\mathfrak{V}_x \setminus (\mathbf{r}\,,\infty)$

$$F(x,o|z) = A_i(x|z) - B_i(x|z)\sqrt{\mathbf{r} - z}\,. \tag{17.7}$$

In particular, $A_i(x|\mathbf{r}) = \Phi_i(x|\alpha_i\theta)$ and

$$B_i(x|\mathbf{r}) = \lim_{z\to\mathbf{r}-} \frac{\Phi_i(x|\alpha_i\theta) - \Phi_i\big(x|\alpha_i z G(z)\big)}{\sqrt{\mathbf{r} - z}} = \alpha_i \Phi_i'(x|\alpha_i\theta)\,\Phi'(\theta)\sqrt{\frac{2\theta}{\Phi''(\theta)}}\,.$$

Now let finally $x = x_1 \cdots x_k$ with $x_j \in X_{i_j}'$. Combining (17.2), (17.5) and (17.7), we find

$$G(x,o|z) = \big(A(z) - B(z)\sqrt{\mathbf{r} - z}\,\big)\prod_{j=1}^{k}\Big(A_{i_j}(x_j|z) - B_{i_j}(x_j|z)\sqrt{\mathbf{r} - z}\Big)\,.$$

Carrying out the multiplication and collecting all terms which come along with an even or odd power of $\sqrt{\mathbf{r} - z}$, we obtain $A(x|z)$ or $B(x|z)\sqrt{\mathbf{r} - z}$, respectively. In particular,

$$A(x|\mathbf{r}) = A(\mathbf{r})\prod_{j=1}^{k} A_{i_j}(x_j|\mathbf{r}) \quad \text{and}$$

$$B(x|\mathbf{r}) = B(\mathbf{r})\prod_{j=1}^{k} A_{i_j}(x_j|\mathbf{r}) + \sum_{\ell=1}^{k} A(\mathbf{r})B_{i_\ell}(x_\ell|\mathbf{r})\prod_{j\neq\ell} A_{i_j}(x_j|\mathbf{r})$$

as asserted. □

Proof of Theorem 17.3. We know from Theorem 9.4 that the only singularities of $G(x,o|z)$ on the circle $|z| = \mathtt{r}$ are the points $\mathtt{r}\,\xi_\ell$ with $\xi_\ell = e^{2\pi i\ell/\mathtt{d}}$, $\ell = 0,\ldots,\mathtt{d}-1$. We have $G(x,o|z) = \xi_\ell^{\mathtt{m}}\,G(x,o|z/\xi_\ell)$. We can expand $B(x|z)$ at $z = \mathtt{r}$:

$$B(x|z) = h(x) + k(x)(\mathtt{r}-z) + C(x|z)(\mathtt{r}-z)^2\,,$$

with $C(x|z)$ analytic in $\mathfrak{V}_x$. (We do not need $k(x)$ explicitly.) Thus, for $z/\xi_\ell \in \mathfrak{V}_x \setminus (\mathtt{r}\,,\infty)$,

$$G(x,o|z) = \xi_\ell^{\mathtt{m}}\Big(A(x|z/\xi_\ell) - h(x)\sqrt{\mathtt{r}-z/\xi_\ell} \\ - k(x)\sqrt{\mathtt{r}-z/\xi_\ell}^{\,3} - C(x|z/\xi_\ell)\sqrt{\mathtt{r}-z/\xi_\ell}^{\,5}\Big).$$

At this point we could appeal once more to singularity analysis (Theorem 16.8), but we continue in a more "old-fashioned" way. The auxiliary function

$$H(z) = G(x,o|z) \\ +\sum_{\ell=0}^{\mathtt{d}-1}\xi_\ell^{\mathtt{m}}\Big(h(x)\sqrt{\mathtt{r}-z/\xi_\ell} + k(x)\sqrt{\mathtt{r}-z/\xi_\ell}^{\,3} + C(x|\mathtt{r})\sqrt{\mathtt{r}-z/\xi_\ell}^{\,5}\Big)$$

is analytic for $|z| < \mathtt{r}$, and its expansion at the origin has non-zero coefficients

$$h_{n\mathtt{d}} = p^{(n\mathtt{d})}(x,o) - \sum_{\ell=1}^{\mathtt{d}-1}\xi_\ell^{\mathtt{m}-n}\frac{1}{2\sqrt{\rho\,\pi}}\,h(x)\,\rho^{n\mathtt{d}}\,n^{-3/2} + O(\rho^{n\mathtt{d}}\,n^{-5/2})\,.$$

On the other hand, $H(z)$ is three times continuously differentiable on the circle $|z| = r$, and the Riemann–Lebesgue lemma (see e.g. Olver [251], p. 310) tells us that $h_n = \mathfrak{o}(\rho^n\,n^{-3})$. The result now follows by observing that $\sum_{\ell=1}^{\mathtt{d}-1}\xi_\ell^{\mathtt{m}-n} = \mathtt{d}$ if $\mathtt{d}$ divides $n-\mathtt{m}$, and $= 0$ otherwise. □

The technique that we have used in the last proof is called the *method of Darboux:* one identifies all singularities on the circle of convergence and then subtracts part of the expansion near them, so that the remaining part is sufficiently often differentiable on $\{|z| = \mathtt{r}\}$. We refer to Olver [251], Chap. 8, §9.2 for more details. See also Pólya [270].

We now present several examples of classes of random walks on groups where Theorem 17.3. applies.

(17.8) Corollary. *Let μ_i be irreducible probability measures on the groups Γ_i, $i \in \mathcal{I}$, and μ be a convex combination of the μ_i on the free product $\Gamma = \ast_{i \in I} \Gamma_i$. Then*

$$\mu^{(n)}(x) = \frac{\mathtt{d}}{2\sqrt{\rho\,\pi}}\, h(x^{-1})\, \rho^n\, n^{-3/2} + \mathcal{O}(\rho^n\, n^{-5/2})$$

as $n \to \infty$ with $n \equiv \mathtt{m}(x)$ modulo $\mathtt{d}$, in each of the following cases.

(a) *Each Γ_i is finite and the μ_i are arbitrary, with the exception of the case $|\mathcal{I}| = |\Gamma_1| = |\Gamma_2| = 2$.*

(b) *$\Gamma_i = \mathbb{Z}^{d_i}$ with $d_i \le 4$, and the μ_i have finite support, or finite mean and finite moments of order $\min\{d_i, 2\}$.*

(c) *Each Γ_i has polynomial growth with degree $d_i \le 4$, and the μ_i are symmetric with finite moments of order $\min\{d_i, 2\}$.*

(d) *Γ is the free product of identical pieces $\Gamma_i = \Gamma_0$ and $\mu_i = \mu_0$, with $\alpha_i = 1/|\mathcal{I}|$, and $|\mathcal{I}| > 1/\big(1 - \Psi_0(\theta_0-)\big)$.*

Proof. (a) We have $\bar\theta = \infty$ and $\Psi_i(\theta_i-) = 1/|\Gamma_i|$, see (9.14). Therefore $\Psi(\bar\theta-) = 1 + \sum_i \big(\frac{1}{|\Gamma_i|} - 1\big) < 0$, unless $|\mathcal{I}| = |\Gamma_1| = |\Gamma_2| = 2$.

(b) For $d = 3, 4$, Corollary 13.11 and Theorem 13.12 yield that in this case $U_i'\big(\mathtt{r}(\mu_i)\big) = \infty$; see (9.15.3). Therefore $\Psi_i(\theta_i-) = 0$ by (9.14). For $d = 1, 2$, this follows from ρ-recurrence (Theorem 6.1, Exercise 8.26). If $i_0 \in \mathcal{I}$ is such that $\bar\theta = \theta_i/\alpha_i$ then

$$\Psi(\bar\theta) = \sum_{i \ne i_0} \big(\Psi_i(\alpha_i\bar\theta) - 1\big) < 0\,,$$

as $\Psi_i(t) < 1$ for $t > 0$.

(c) The argument is the same as for (b), using Theorem 15.8(a) (for $d_i \ge 2$) and Corollary 6.8(b) (for $d_i = 1$) instead of (13.11) and (13.12).

(d) This follows from (9.25). □

Further examples can be obtained by combining the above four cases in various ways. Finally, we prove another result that explains why the asymptotic behaviour of Theorem 17.3 is "typical" for free products of groups. The following is a preliminary step.

(17.9) Lemma. *Let Γ be an arbitrary discrete group, and let S be a finite, symmetric set of generators. If S contains an element of order ≥ 3 (possibly infinite), then there is a symmetric probability measure μ supported by S such that $\Psi_\mu(\theta-) < 1/2$.*

Proof. By assumption, there is $x_0 \in S$ such that the subgroup Γ_0 generated by x_0 is cyclic of order m with $3 \le m \le \infty$. Let μ_0 be the law of the simple random walk on Γ_0, that is, $\mu_0(x_0^{\pm 1}) = 1/2$. Then $\mathbf{r}(\mu_0) = 1$, $\theta(\mu_0) = \infty$ and $\Psi_{\mu_0}(\theta(\mu_0)-) = 1/m$, see (9.14) and the subsequent lines. Again by virtue of (9.14), we can find $\delta > 0$ such that the denominator in the formula for Ψ_{μ_0} satisfies

$$1 + \sum_{n=1}^{\infty} (n-1)\, \mathbb{P}_o^{\mu_0}[\mathbf{t}^o = n]\, (1-\delta)^n > 2\,.$$

Now, if S is symmetric, generates Γ and contains x_0, then we define μ on Γ with support S by

$$\mu(x) = \begin{cases} (1-\delta)\mu_0(x)\,, & \text{if } x = x_0^{\pm 1}\,, \\ \delta/(|S|-2)\,, & \text{if } x \in S \setminus \{x_0^{\pm 1}\}\,. \end{cases}$$

For the associated function Ψ and numbers $\mathbf{r} = \mathbf{r}(\mu) \geq 1$ and $\theta = \theta(\mu)$, we have $\mathbb{P}_o^{\mu}[\mathbf{t}^o = n] \geq \mathbb{P}_o^{\mu_0}[\mathbf{t}^o = n](1-\delta)^n$, whence

$$1/\Psi(\theta) = 1 + \sum_{n=1}^{\infty} (n-1)\, \mathbb{P}_o^{\mu}[\mathbf{t}^o = n]\, \mathbf{r}^n > 2\,. \qquad \square$$

As a consequence, except for the case when $|\Gamma| = 2$, we can always find a symmetric, irreducible probability measure on Γ with $\Psi(\theta-) < 1/2$: there must be an element of order > 2, and S can be chosen to contain this element. Also, finiteness of S is not really needed in this argument.

(17.10) Corollary. (a) *If $\Gamma = \Gamma_1 * \Gamma_2$, with exception of the case $|\Gamma_1| = |\Gamma_2| = 2$, then there is a symmetric, irreducible probability measure μ on Γ for which the conclusion of Corollary 17.8 holds (with $\mathtt{d} = 1$ or 2).*

(b) *Furthermore, if S_i are symmetric generating sets of Γ_i ($i = 1, 2$) containing elements of degree ≥ 3, then one can choose μ of the form $\alpha_1 \cdot \mu_1 + \alpha_2 \cdot \mu_2$ with the μ_i supported by S_i and suitable coefficients α_i.*

Proof. We start with (b). From Lemma 17.9 we get symmetric probabilities μ_i supported by S_i such that $\Psi_i(\theta_i) < 1/2$, where $\theta_i = \theta(\mu_i)$.

(1) If $\theta_1 = \theta_2 = \infty$ then we may choose $\alpha_1 = \alpha_2 = 1/2$. For $\mu = \alpha_1 \cdot \mu_1 + \alpha_2 \cdot \mu_2$ we get $\bar{\theta} = \infty$, and (9.21) yields $\Psi(\bar{\theta}-) < 0$, as required in Theorem 17.3.

(2) If both θ_1 and θ_2 are finite then we choose $\alpha_i = \theta_i/(\theta_1 + \theta_2)$. Then $\bar{\theta} = \theta_1/\alpha_1 = \theta_2/\alpha_2$ and by (9.21), $\Psi(\bar{\theta}) = \Psi_1(\theta_1) + \Psi_2(\theta_2) - 1 < 0$.

(3) Finally, if $\theta_1 < \infty$ and $\theta_2 = \infty$, say, then we choose α_1 sufficiently small, $\alpha_2 = 1 - \alpha_1$, such that $\Psi_2(\alpha_2\theta_1/\alpha_1) < 1/2$. Then $\bar{\theta} = \theta_1/\alpha_1$ and $\Psi(\bar{\theta}) = \Psi_1(\theta_1) + \Psi_2(\alpha_2\theta_1/\alpha_1) - 1 < 0$.

Part (a) follows from (b), unless precisely one of the two groups, say Γ_2, has order 2. In this case, we can choose μ_2 arbitrarily on Γ_2 (different from δ_o): we get $\theta_2 = \infty$ and $\Psi_2(\theta_2-) = 1/2$. The proof now continues as in (b), case (1) or (3), accordingly. □

Analogously, let $\Gamma = \underset{i \in \mathcal{I}}{*} \Gamma_i$. Then Γ carries a free-sum probability measure $\mu = \sum_i \alpha_i \cdot \mu_i$ for which the conclusion of Corollary 17.8 holds, unless $|\mathcal{I}| = |\Gamma_1| = |\Gamma_2| = 2$.

In the latter, exceptional, case, Γ is (isomorphic with) the *infinite dihedral group* $\langle a, b \mid a^2 = b^2 = o \rangle$, and μ must be supported in $\{o, a, b\}$. In particular, Γ has an infinite cyclic subgroup of index 2, and – as μ is symmetric – the random walk must be recurrent by Theorem 3.24. In the aperiodic case

$$\mu^{(n)}(x) = C\, n^{-1/2} + \mathcal{O}(n^{-3/2})\,. \tag{17.11}$$

(17.12) Exercise. Let $X_1 = \{o, a\}$ and $X_2 = \{o, b\}$ and let P_i be arbitrary irreducible Markov chains on the X_i. Consider the free product $X_1 * X_2$ and $P = \alpha_1 \overline{P_1} + \alpha_2 \overline{P_2}$. Compute $G(x, o|z)$ and the asymptotic behaviour of $p^{(n)}(x, o)$. In particular, prove (17.11).

The reader is invited to consider further examples, also including free products which do not arise from groups. As one such example, take two Sierpiński graphs rooted at the respective origins. The simple random walk on each of the two is null recurrent. Therefore, if X is their free product, then the simple random walk on X satisfies the prerequisites of Theorem 17.3, and we get

$$p^{(n)}(o, o) = C\, \rho^n\, n^{-3/2} + \mathcal{O}(\rho^n\, n^{-5/2})\,,$$

where $\rho < 1$. In particular, the periodic oscillations have disappeared. Intuitively, this can be explained as follows. For the simple random walk on $\mathbf{S}_d$, the oscillations are induced by the larger and larger "holes" around which the random walk will run very often. In the free product, the random walk spends only a finite amount of time in each of the infinitely many copies of the $\mathbf{S}_d$ used to build up X, and in the asymptotics, the information regarding the holes is lost.

B. Instability of the exponent

If we consider the Euclidean case, i.e., random walks on $\mathbb{Z}^d$, then we know that every irreducible probability measure with finite support satisfies a local limit theorem of the form $\mu^{(n)}(o) \sim C\, \rho(\mu)^n\, n^{-d/2}$, if n is divisible by the period of μ. While ρ depends on μ, the exponent $-d/2$ relies only on the underlying group. An analogous phenomenon will be found in Section 18 for finitely supported random walks on free groups. Of course one cannot expect stability of the exponent when considering random walks that do not have finite range or a finite exponential moment. Indeed, on $\mathbb{Z}^d$ the local limit theorem is closely connected with the domain of attraction to which the probability measure μ belongs; see Rvačeva [283]. However, one might be tempted to believe that the following holds for any finitely generated group Γ: if $\mu^{(n)}(o) \sim C\,\rho(\mu)^n\,n^{-\lambda}$ for some irreducible, aperiodic, finitely supported probability measure on Γ, then it holds – with the same λ – for every other μ of that type. In this subsection we shall exhibit an example, due to Cartwright, which shows that this conjecture is *wrong.* To wit, we show the following.

(17.13) Theorem. *For $d \geq 5$, the simple random walk on $\mathbb{Z}^d * \mathbb{Z}^d$ satisfies*

$$\mu^{(2n)}(o) \sim C\,\rho(\mu)^{2n}\,n^{-d/2} \quad \text{as } n \to \infty\,.$$

By the simple random walk we mean of course the random walk on the Cayley graph of $\mathbb{Z}^d * \mathbb{Z}^d$ with respect to the natural generators and their inverses, as considered in (9.25.3). Now we know from Corollary 17.10(b) that there is another symmetric probability measure $\bar{\mu}$ with the same support as μ, but with

$$\bar{\mu}^{(2n)}(o) \sim \bar{C}\,\rho(\bar{\mu})^{2n}\,n^{-3/2} \quad \text{as } n \to \infty\,.$$

In particular, μ and $\bar{\mu}$ have equivalent Dirichlet forms, and we see that comparison of the forms cannot be used to detect a term such as $n^{-\lambda}$, once the spectral radius is strictly less than 1. Note that this is not in contrast with the results on the asymptotic type of Section 14.

We start with some preparatory observations regarding the free product of identical pieces, as studied in (9.25). Here we suppose that the number of factors is 2 (but everything extends immediately to $M \geq 3$), and that the identical factors are given by a group Γ_0 and a probability measure μ_0. As before, we write μ for the resulting probability measure on the free product $\Gamma = \Gamma_0 * \Gamma_0$. Also, $\mathtt{r}_0 = 1/\rho(\mu_0)$, $\mathtt{r} = 1/\rho(\mu)$, and $\theta = \theta(\mu) = \mathtt{r}G(\mathtt{r})$. For the rest, we use the same notation as in (9.25). We are interested in the "degenerate" case $\Psi(\theta) > 0$ of Theorem 9.22(ii).

(17.14) Lemma. *Suppose that μ_0 is symmetric and ρ-transient, and that $\Psi(\theta_0) > 1/2$ (so that $\theta = 2\theta_0$). Then*

(a) $$\frac{\mathtt{r}G(\mathtt{r})}{1 + G(\mathtt{r})} = \mathtt{r}_0\,;$$

(b) *for all* z *in* $\mathbb{C}$ *with* $z \notin (-\infty, -\mathbf{r}) \cup (-\infty, -\mathbf{r})$,

$$G(z) + 1 = 2\,G_0\left(\frac{zG(z)}{1+G(z)}\right) ;$$

(c) $G'(\mathbf{r}-) = \sum_{n=1}^{\infty} n\,\mu^{(n)}(o)\,\mathbf{r}^{n-1} < \infty$, *and for* $|z| \le \mathbf{r}$,

$$G(z) = G(\mathbf{r}) + G'(\mathbf{r}-)(\mathbf{r}-z) + \mathfrak{o}(\mathbf{r}-z) \quad \text{as } z \to \mathbf{r}.$$

Proof. (a) Recall that $\Phi(t) = 2\Phi_0(t/2) - 1$. Thus by (9.12)

$$\frac{\mathbf{r}G(\mathbf{r})}{1+G(\mathbf{r})} = \frac{\theta}{1+\Phi(\theta)} = \frac{2\theta_0}{2\Phi_0(\theta_0)} = \mathbf{r}_0 .$$

(b) Next recall also that $\Phi_0(t) = t/V_0(t)$, where $V_0(t)$ is the inverse function of $W_0(z) = zG_0(z)$ for z in a neighbourhood $\mathfrak{U}_{\mathbf{r}_0}$ of the interval $[0, \mathbf{r}_0)$. We have from Theorem 9.19

$$G(z) + 1 = 2\Phi_0\bigl(zG(z)/2\bigr) = zG(z)/V_0\bigl(zG(z)/2\bigr) .$$

With a little algebra, we obtain the required identity, valid in $\mathfrak{U}_{\mathbf{r}_0}$.

Now observe that $W_0(1/z)$ is a diagonal matrix element of the resolvent of P_{μ_0} (the transition operator given by μ_0) on $\ell^2(\mathbb{Z}^d)$. It must be analytic outside the spectrum (see Dunford and Schwarz [107]), which by symmetry is contained in $[-\rho(\mu_0), \rho(\mu_0)]$. Thus, $G_0(z)$ is analytic in $\mathbb{C} \setminus \{z \in \mathbb{R} : |z| \ge \mathbf{r}_0\}$.

In the same way, $G(z)$ is analytic in $\mathbb{C} \setminus \{z \in \mathbb{R} : |z| \ge \mathbf{r}\}$. Also, if $\mathfrak{m}$ is the *spectral* or *Plancherel measure* associated with P_μ (i.e., the diagonal element at o of the resolution of the identity – see [107] once more), then

$$G(z) = \int_{-\rho(\mu)}^{\rho(\mu)} \frac{1}{1-zt}\, d\mathfrak{m}(t) . \tag{17.15}$$

From this formula we see that $\mathrm{Im}(1/z)$ and $\mathrm{Im}\bigl(1/G(z)\bigr)$ have the same sign. In particular, when z is strictly complex then $zG(z)/\bigl(1+G(z)\bigr)$ is also strictly complex. On the other hand, we know from (a) that $zG(z)/\bigl(1+G(z)\bigr) \in [-\mathbf{r}_0, \mathbf{r}_0]$ when $z \in [-\mathbf{r}, \mathbf{r}]$. Therefore (b) holds by analytic continuation in the whole complex plane with the exception of $(-\infty, -\mathbf{r})$ and $(\mathbf{r}, \infty)$.

(c) For $0 < z < \mathbf{r}$, we differentiate the equation $G(z) = \Phi\bigl(zG(z)\bigr)$ and find $G'(z) = G(z)\,\Phi'\bigl(zG(z)\bigr)\Big/\Bigl(1 - z\Phi'\bigl(zG(z)\bigr)\Bigr)$. Now $\Psi(\theta) > 0$ by assumption, so that $\Phi'\bigl(\mathbf{r}G(\mathbf{r})\bigr) < 1/\mathbf{r}$. Consequently,

$$|G'(z)| \le \sum_{n=0}^{\infty} n\,\mu^{(n)}(o)\,\mathbf{r}^{n-1} = G'(\mathbf{r}-) < \infty \quad \text{for } |z| \le \mathbf{r}. \qquad \square$$

For proving Theorem 17.13, we also need the following result, which, like Theorem 15.16, belongs to the realm of Section 13.

(17.16) Proposition. *The Green function $G_d(z) = G_d(\mathbf{0},\mathbf{0}|z)$ of the simple random walk on $\mathbb{Z}^d$ $(d \ge 1)$ has a singular expansion near $z = 1$ of the form*

$$G_d(z) = \begin{cases} f(z) + g(z)\,(1-z)^{(d-2)/2}\,, & \text{if } d \text{ is odd,} \\ f(z) + g(z)\,(1-z)^{(d-2)/2}\,\log(1-z)\,, & \text{if } d \text{ is even,} \end{cases}$$

where f and g, depending on d, are analytic in a neighbourhood of 1 and $g(1) \ne 0$.

Proof. We proceed as in (17.15) and write

$$G_d(z) = \int_{-1}^{1} \frac{1}{1-zt}\, d\mathfrak{m}_d(t)\,. \tag{17.17}$$

(17.18) Exercise. Show that the exponential generating function

$$E_d(z) = \sum_{n=0}^{\infty} \mu_d^{(n)}(\mathbf{0}) \frac{z^n}{n!} = \int_{-1}^{1} e^{zt}\, d\mathfrak{m}_d(t)$$

satisfies $E_d(z) = E_1(z/d)^d$. (Consider the e.g.f. of a Cartesian product of two Markov chains.)

Continuing the proof of (17.16), we now see that $\mathfrak{m}_d$ is the n-th convolution power of $\mathfrak{m}_1$, rescaled to the interval $[-1\,,\,1]$ by a factor of d. Now, $S_d(z) = G_d(1/z)/z$ is the *Stieltjes transform* of the measure $\mathfrak{m}_d$, and the distribution function of the measure is found via the inversion formula

$$\frac{1}{2}\Big(\mathfrak{m}_d\big((-\infty\,,\,t)\big) + \mathfrak{m}_d\big((-\infty\,,\,t]\big)\Big) = \frac{1}{\pi} \lim_{y\to 0+} \int_{-\infty}^{t} \operatorname{Im}\big(S_d(x-iy)\big)\,dx\,.$$

See Dunford and Schwarz [107], Th. X.6.1, or (in the context of continued fractions) Wall [335]. From Lemma 1.24, we know that $G_1(z) = 1/\sqrt{1-z^2}$. The standard version of the square root is analytic in $\mathbb{C}\setminus(-\infty\,,0]$, and takes all values in the right half plane. Thus, $S_1(z) = \operatorname{sign}\operatorname{Re}(z)/\sqrt{z^2-1}$ for $z \in \mathbb{C}\setminus[-1\,,\,1]$. The inversion formula now yields after a short computation that $\mathfrak{m}_1$ has density

$$\mathfrak{f}_1(t) = \frac{1}{\pi\sqrt{1-t^2}}\,\mathbf{1}_{[-1\,,\,1]}(t)$$

with respect to Lebesgue measure. Thus $\mathfrak{m}_d$ has density $\mathfrak{f}_d(t) = d\,\mathfrak{f}_1^{(d)}(dt)$. We now prove the following. For $1-\varepsilon < t \le d$,

$$\mathfrak{f}_d(t) = (1-t)^{(d-2)/2}\,\mathfrak{g}_d(t)\,, \tag{17.19}$$

where $\mathfrak{g}_d$ is a real-analytic function in a neighbourhood of 1, and $\mathfrak{g}_d(1) \neq 0$.

We set $\bar{\mathfrak{f}}(t) = \mathfrak{f}_1(1-t)$ and show by induction on d that for $0 \le t < \bar{\varepsilon}$ one has $\bar{\mathfrak{f}}^{(d)}(t) = t^{(d-2)/2}\bar{\mathfrak{g}}_d(t)$ with $\bar{\mathfrak{g}}_d$ real-analytic near 0 and $\bar{\mathfrak{g}}_d(0) \neq 0$. This is obvious for $d = 1$. Suppose it is true for d. We write $\bar{\mathfrak{g}}_d(t) = \sum_{n=0}^{\infty} a_n(d)\, t^n$. Then for small t, using the substitution $y = tu$ in the second step,

$$\bar{\mathfrak{f}}^{(d+1)}(t) = \int_0^t y^{(d-2)/2}\, \bar{\mathfrak{g}}_d(y)\, (t-y)^{-1/2}\, \bar{\mathfrak{g}}_1(t-y)\, dy$$

$$= t^{(d-1)/2} \int_0^1 \sum_{m,n=0}^{\infty} a_m(d)\, a_n(1)\, t^{m+n}\, u^{m+(d-2)/2}\, (1-u)^{n-1/2}\, du\,.$$

Now $\int_0^1 u^{m+(d-2)/2}(1-u)^{n-1/2}\, du = B(m+d/2, n+1/2) \le \pi$, where $B(\cdot,\cdot)$ is the beta function, so that we may exchange summation and integration to find $\bar{\mathfrak{g}}_{d+1}$ as proposed. In particular, $\bar{\mathfrak{g}}_{d+1}(0) = \bar{\mathfrak{g}}_d(0)\bar{\mathfrak{g}}_1(0)B(d/2, 1/2) \neq 0$.

This proves (17.19). For the rest, we may assume $d \ge 2$. Again, we write $\mathfrak{g}_d(t) = \sum_{n=0}^{\infty} a_n(d)\, (1-t)^n$. Now we use (17.17), splitting the integral into $\int_{-1}^{1-\varepsilon}$ and $\int_{1-\varepsilon}^{1}$. The first part is analytic near 1, and the second becomes

$$\text{(17.20)} \qquad \sum_{n=0}^{\infty} a_n(d) \int_{1-\varepsilon}^{1} \frac{(1-t)^{n+(d-2)/2}}{1-zt}\, dt\,.$$

(Exchanging integral and summation is justified by dominated convergence when $z \notin [1\,,\infty)$ and ε is sufficiently small.) The integrals appearing in (17.20) are of the form $I_k(z) = \int_{1-\varepsilon}^{1} \big((1-t)^{k+\delta}/(1-zt)\big)\, dt$, where $\delta = 0$ or $1/2$ according to the parity of d, and $k \ge 0$. Now

$$I_0(z) = \begin{cases} \frac{2}{z}\Big(\sqrt{\varepsilon} + \sqrt{\frac{1-z}{z}}\arctan\sqrt{\frac{1-z}{z\varepsilon}}\Big) - \frac{\pi}{z\sqrt{z}}\sqrt{1-z}\,, & \text{if } d \text{ is odd,} \\ \frac{1}{z}\log(1-(1-\varepsilon)z) - \frac{1}{z}\log(1-z)\,, & \text{if } d \text{ is even,} \end{cases}$$

and $I_k(z) = -\frac{1-z}{z}I_{k-1}(z) + \frac{1}{z}\frac{\varepsilon^{k+\delta}}{k+\delta}$. Note that $\sqrt{\frac{1-z}{z}}\arctan\sqrt{\frac{1-z}{z\varepsilon}}$ extends analytically to a neighbourhood of 1. The rest is an elementary exercise: one inserts the solution of the recursion into (17.20) and checks convergence. □

Proof of Theorem 17.13. In the sequel, we always assume $|z| \le \mathtt{r}$. The notation $\mathcal{O}_c\big((r-z)^k\big)$ will indicate that this term, divided by $(r-z)^k$, has a limit when $z \to \mathtt{r}$. Let $L = \lceil \frac{d-2}{2} \rceil$. We shall show that for z near $\mathtt{r}$ with $|z| \le \mathtt{r}$,

$$\text{(17.21)} \quad G(z) = \sum_{k=0}^{L} g_k\, (\mathtt{r}-z)^k + R(z) + \mathcal{O}_c\big((\mathtt{r}-z)^{L+1}\big)\,, \qquad \text{where}$$

$$R(z) = \begin{cases} \big(c_0 + c_1\,(\mathtt{r}-z)\big)(\mathtt{r}-z)^{(d-2)/2}\,, & \text{if } d \text{ is odd,} \\ \big(c_0 + c_1\,(\mathtt{r}-z)\big)(\mathtt{r}-z)^{(d-2)/2}\log(\mathtt{r}-z)\,, & \text{if } d \text{ is even,} \end{cases}$$

unless $d = 6$, in which case there is an additional term $\bar{c}_1(\mathtt{r}-z)^3\log^2(\mathtt{r}-z)$ in $R(z)$. The coefficients depend on d, and $c_0 \neq 0$.

As the random walk has period 2, we have $G(-z) = G(z)$, and besides $\pm\mathtt{r}$, there are no singularities on $\{|z| = \mathtt{r}\}$, by (17.15). We can then apply Darboux's method once more: $R(-z)$ is the term in the expansion near $-\mathtt{r}$ analogous to $R(z)$. The Taylor coefficients at $z = 0$ of $R(z)+R(-z)$ have the asymptotic behaviour asserted for $\mu^{(n)}(o)$. The Riemann–Lebesgue lemma applied to $G(z)-R(z)-R(-z)$ shows that the remainder is $\mathfrak{o}\big(\rho(\mu)^n\, n^{-d/2}\big)$ as $n \to \infty$, concluding the proof.

We are left with having to prove (17.21). It will be convenient to consider the function $B(z) = G(z)/\big(1+G(z)\big)$ instead of $G(z)$, that is, $G(z)+1 = 1/\big(1-B(z)\big)$. In particular, we have $\mathtt{r}B(\mathtt{r}) = 1$ by Lemma 17.14(a).

As $0 < B(\mathtt{r}) < 1$, we see that $G(z)$ will have an expansion as asserted in (17.21), if $B(z)$ has an analogous expansion, with coefficients (say) b_k instead of g_k and d_0, d_1 instead of c_0, c_1. The equation of Lemma 17.14(b) becomes

$$(17.22)\qquad \big(1-B(z)\big)G_0\big(zB(z)\big) = 1/2\,.$$

We now use Proposition 17.16, writing $G_0(z) = \sum_{k=0}^{L-1} a_k(1-z)^k + E_0(z)$, where

$$(17.23)\qquad \begin{aligned} E_0(z) = &\Big(e_0 + e_1(1-z)\Big)\,(1-z)^{(d-2)/2}\log^{\kappa}(1-z)\\ &+ a_L\,(1-z)^L + \mathcal{O}_c\big((1-z)^{L+1}\big)\,, \end{aligned}$$

with $\kappa = 0$ if d is odd, and $\kappa = 1$ if d is even.

We inductively determine $b_0, \ldots, b_{L-1}$ such that for $\ell \le L-1$ one has

$$(17.24)\quad B(z) = \sum_{k=0}^{\ell} b_k\,(\mathtt{r}-z)^k + B_\ell(z)\,,\quad \text{where}\quad B_\ell(z) = \mathfrak{o}\big((\mathtt{r}-z)^\ell\big)\,.$$

For $\ell = 1$ this follows from Lemma 17.14(c), and as $\mathtt{r}B(\mathtt{r}) = 1$, we get that $1 - zB(z) = \mathcal{O}_c(\mathtt{r}-z)$ for $|z| \le \mathtt{r}$. In particular, we obtain that $E_0\big(zB(z)\big) = \mathcal{O}_c\big((\mathtt{r}-z)^{(d-2)/2}\log^{\kappa}(\mathtt{r}-z)\big)$. Now suppose that (17.24) holds for some $\ell < L-1$. Then $1 - zB(z) = (\mathtt{r}-z)P_\ell(z) - z\,B_\ell(z)$ and $\big(1-zB(z)\big)^k = (\mathtt{r}-z)^k P_\ell(z)^k + \mathfrak{o}\big((\mathtt{r}-z)^{\ell+1}\big)$ for $k \ge 2$, where $P_\ell(z) = \sum_{j=0}^{\ell-1}(b_j - \mathtt{r}\,b_{j+1})\,(\mathtt{r}-z)^j + b_\ell\,(\mathtt{r}-z)^\ell$. We substitute this into (17.22) and bring all polynomial terms to the right. This yields the equation

$$(17.25)\quad \begin{aligned} &\alpha_0\,B_\ell(z) + (1-b_0)\,E_0\big(zB(z)\big) + \mathfrak{o}\big((\mathtt{r}-z)^{\ell+1}\big)\\ &= \frac{1}{2} - \left(1 - \sum_{k=0}^{\ell} b_k\,(\mathtt{r}-z)^k\right)\sum_{k=0}^{L-1} a_k\,(\mathtt{r}-z)^k\,P_\ell(z)^k \end{aligned}$$

with $\alpha_0 = -a_0 - a_1(1-b_0)\mathtt{r} = -G_0(1) + G_0'(1)(\mathtt{r}-1)$. Differentiating (17.22) and letting $z \to \mathtt{r}-$, we find that $\alpha_0\, B'(\mathtt{r}) + \big(1 - B(\mathtt{r})\big)B(\mathtt{r})G_0'(1) = 0$. As $0 < B'(\mathtt{r}) < \infty$, it must be that $\alpha_0 < 0$. We note here also that $\beta_0 = P_\ell(\mathtt{r})$ is equal to $B(\mathtt{r}) + \mathtt{r}\, B'(\mathtt{r}) > 0$.

The left hand side of (17.25) is $\mathfrak{o}\big((\mathtt{r}-z)^\ell\big)$. Therefore the polynomial in $(\mathtt{r}-z)$ on the right hand side must start with a power larger than ℓ. Consequently $B_\ell(z) = \mathcal{O}_c\big((\mathtt{r}-z)^{\ell+1}\big)$, and (17.24) also holds for $\ell+1$.

For the rest of the proof, we consider (17.24)–(17.25) with $\ell = L-1$. A better use of (17.23) yields

$$E_0\big(zB(z)\big) = e_0\,\big(\beta_0(\mathtt{r}-z)\big)^{(d-2)/2}\log^\kappa(\mathtt{r}-z) + \mathfrak{o}\Big((\mathtt{r}-z)^{(d-2)/2}\log^\kappa(\mathtt{r}-z)\Big).$$

We substitute this into (17.25) and obtain

$$B_{L-1}(z) = d_0\,(\mathtt{r}-z)^{(d-2)/2}\log^\kappa(\mathtt{r}-z) + C_L(z), \tag{17.26}$$

$d_0 = -(1-b_0)e_0\beta_0^{(d-2)/2}/\alpha_0$ and $C_L(z) = \mathfrak{o}\big((\mathtt{r}-z)^{(d-2)/2}\log^\kappa(\mathtt{r}-z)\big)$.

Next, if d is odd ($\kappa = 0$), then we introduce the new variable $u = \sqrt{r-z}$ and continue by precisely the same method, extending (17.24) stepwise by the next power of u and carrying all polynomial terms to the right hand side of the equation obtained from (17.22), as in (17.25). Comparing the error terms that remain on the left, we first find b_L (the coefficient of $u^{2L} = (r-z)^L$), and then d_1 (the coefficient of $u^{2L+1} = (\mathtt{r}-z)^{d/2}$).

If d is even, then we do not change the variable. The procedure is similar, but more tedious. We first find b_L: via the binomial theorem, (17.26) yields $\big(1 - zB(z)\big)^L = \big((\mathtt{r}-z)\,P_{L-1}(z)\big)^L + \mathcal{O}_c\big((\mathtt{r}-z)^{L+1}\log(\mathtt{r}-z)\big)$ and

$$\begin{aligned}\log\big(1 - zB(z)\big) &= \log(\mathtt{r}-z) + \log\beta_0 \\ &\quad + \mathcal{O}_c\big(\max\{(|\mathtt{r}-z|,\ |(\mathtt{r}-z)^{L-1}\log(\mathtt{r}-z)|\}\big).\end{aligned} \tag{17.27}$$

Consequently

$$E_0\big(zB(z)\big) = e_0\,\beta_0^L(\mathtt{r}-z)^L\log(\mathtt{r}-z) + f_0\,\beta_0^L(\mathtt{r}-z)^L + \mathfrak{o}\big((\mathtt{r}-z)^L\big)$$

with $f_0 = a_L + e_0\log\beta_0$. Once again substituting this into (17.24), we find that $\alpha_0\, C_L(z) + \mathfrak{o}\big((\mathtt{r}-z)^L\big)$ is a polynomial in $(\mathtt{r}-z)$. We see that this must be divisible by $(\mathtt{r}-z)^L$. Therefore $C_L(z) = \mathcal{O}_c\big((\mathtt{r}-z)^L\big)$, and comparing coefficients on left and right, one obtains b_L.

We now have $B(z) = \sum_{k=0}^L b_k(\mathtt{r}-z)^k + D_L(z) + B_L(z)$ with $D_L(z) = d_0(\mathtt{r}-z)^L\log(\mathtt{r}-z)$ and $B_L(z) = \mathfrak{o}\big((\mathtt{r}-z)^L\big)$.

For the rest, the case $L = 2$ ($d = 6$) has to be treated separately because of the "max" in (17.27). We invite the reader to do this as an exercise. So let $L \ge 3$. We restart, writing $1 - zB(z) = (\mathtt{r} - z)P_L(z) - zD_L(z) - zB_L(z)$ and $\big(1 - zB(z)\big)^2 = (\mathtt{r} - z)^2 P_L(z)^2 - 2\beta_0 \mathtt{r}(\mathtt{r} - z)D_L(z) + \mathfrak{o}\big((\mathtt{r} - z)^{L+1}\big)$, while for $k \ge 3$, we find $\big(1 - zB(z)\big)^k = (\mathtt{r} - z)^k P_L(z)^k + \mathfrak{o}\big((\mathtt{r} - z)^{L+2} \log(\mathtt{r} - z)\big)$. We substitute these expansions into (17.22) and find an equation of the form $\big\langle\, \alpha_0 B_L(z)$ + a constant times $(\mathtt{r} - z)^{L+1} \log(\mathtt{r} - z)$ + a remainder term $\mathfrak{o}\big((\mathtt{r} - z)^{L+1}\big)\,\big\rangle = \big\langle$ a polynomial in $(\mathtt{r} - z)\big\rangle$. From this we obtain $B_L(z) = d_1\,(\mathtt{r} - z)^{L+1} \log(\mathtt{r} - z) + \mathcal{O}_c\big((\mathtt{r} - z)^{L+1}\big)$, which concludes the proof. □

18. Intermezzo: Cartesian products

For Cartesian products of Markov chains, as defined in §4.B, it is rather easy to obtain a local limit theorem from the asymptotic behaviour of each of the factors.

(18.1) Proposition. *Let (a_n) and (b_n) be two non-negative sequences such that $a_n \sim C_1\, \rho_1^n\, n^{\lambda_1}$ and $b_n \sim C_2\, \rho_2^n\, n^{\lambda_2}$ when $n \to \infty$, with $C_i, \rho_i > 0$ and $\lambda_i \in \mathbb{R}$. Then for $0 < t < 1$,*

$$c_n = \sum_{k=0}^{n} \binom{n}{k} t^k a_k\, (1-t)^{n-k} b_{n-k} \sim C\, \rho^n\, n^{\lambda_1+\lambda_2}\,,$$

where $\rho = t\,\rho_1 + (1-t)\rho_2$ and $C > 0$.

Proof. We set $\theta = t\rho_1/\rho$. Choose $\varepsilon > 0$, $\varepsilon < \min\{\theta\,,\ 1-\theta\}$. Let n be large enough that $1 - \varepsilon < a_k/\big(C_1\, \rho_1^k\, k^{\lambda_1}\big) < 1 + \varepsilon$, and analogously for b_k, for all k that satisfy $\big|\frac{k}{n} - \theta\big| < \varepsilon$. We write $c_n/\big(\rho^n\, n^{\lambda_1+\lambda_2}\big) = q_n + r_n$, where q_n is the sum over those k, and r_n is the rest. Then

$$\begin{aligned} q_n &\le (1+\varepsilon)^2 C_1 C_2 \sum_{k:|\frac{k}{n}-\theta|<\varepsilon} \binom{n}{k} \theta^k \big(\tfrac{k}{n}\big)^{\lambda_1} (1-\theta)^{n-k} \big(1-\tfrac{k}{n}\big)^{\lambda_2} \\ &\le (1+\varepsilon)^2 C_1 C_2 (\theta + \varepsilon \operatorname{sign} \lambda_1)^{\lambda_1} (1 - \theta + \varepsilon \operatorname{sign} \lambda_2)^{\lambda_2}\, \mathbb{P}\big[\big|\tfrac{1}{n} X_n - \theta\big| < \varepsilon\big]\,, \end{aligned}$$

where X_n is a binomial random variable with parameters n and θ. There is an analogous lower bound, changing sign in front of all occurrences of ε. By Chebyshev's inequality, the probability tends to 1 as $n \to \infty$. So what is left is to show that $r_n \to 0$.

Let $\ell_i = -\min\{0\,,\ \lfloor \lambda_i \rfloor\}$. We split $r_n = r_n' + r_n''$, where r_n' is the part corresponding to $k = 0$ together with $k = n$. It is straightforward to see that $r_n' \to 0$. For the remaining part, we can find a bound M such that

$$r_n'' \le M \sum_{\substack{k=1 \\ |\frac{k}{n}-\theta|\ge\varepsilon}}^{n-1} \binom{n}{k} \theta^k \big(\tfrac{k}{n}\big)^{\lambda_1} (1-\theta)^{n-k} \big(1-\tfrac{k}{n}\big)^{\lambda_2}\,.$$

We have $\left(\frac{k}{n}\right)^{\lambda_1} \le \left(\frac{n}{k}\right)^{\ell_1} \le \frac{n+1}{k+1} \cdots \frac{n+\ell_1}{k+\ell_1}$. In the same way, $\left(\frac{n-k}{n}\right)^{\lambda_2} \le \frac{n+\ell_1+1}{n-k+1} \cdots \frac{n+\ell_1+\ell_2}{n-k+\ell_2}$. Using the formula $\frac{n+1}{k+1}\binom{n}{k} = \binom{n+1}{k+1}$, we obtain

$$\begin{aligned} r''_n &\le M \sum_{\substack{k=1 \\ |\frac{k}{n}-\theta|\ge\varepsilon}}^{n-1} \binom{n+\ell_1+\ell_2}{k+\ell_1} \theta^k (1-\theta)^{n-k} \\ &= \frac{M}{\theta^{\ell_1}(1-\theta)^{\ell_2}} \mathbb{P}\left[\left| \frac{X_{n+\ell_1+\ell_2}}{n} - \frac{\ell_1}{n} - \theta \right| \ge \varepsilon \right], \end{aligned}$$

where $X_{n+\ell_1+\ell_2}$ is binomial with parameters $n+\ell_1+\ell_2$ and θ. Once more by Chebyshev's inequality, this probability tends to 0 as $n \to \infty$. □

This result has an immediate application to random walks. Indeed, if (X_1, P_1) and (X_2, P_2) are two Markov chains, and

$$P = c\,P_1 \otimes I_2 + (1-c)\,I_1 \otimes P_2$$

is a Cartesian product of the two, then

$$(18.2)\quad p^{(n)}(x_1x_2, y_1y_2) = \sum_{k=0}^{n} \binom{n}{k} c^k\, p_1^{(k)}(x_1, y_1)\,(1-c)^{n-k}\, p_2^{(n-k)}(x_2, y_2)\,.$$

This yields the simplest proof of Pólya's classical result that the simple random walk on $\mathbb{Z}^d$ satisfies $p^{(2n)}(\mathbf{0},\mathbf{0}) \sim \mathbf{C}_d\, n^{-d/2}$. For dimension $d = 1$, we know that it is true. The simple random walk in d dimensions is a Cartesian product of d copies of the simple random walk on $\mathbb{Z}$, so that we can apply Proposition 18.1.

Many other examples can be easily constructed. Let us consider one of them, namely the graph $\mathbf{C}_d \times \mathbb{Z}$, where $\mathbf{C}_d$ is the comb lattice in dimension d, as considered in (2.21). We shall use Proposition 18.1 to prove the following.

(18.3) Theorem. *For $d \ge 2$, the simple random walk on $\mathbf{C}_d \times \mathbb{Z}$ is transient.*

We first remark that this is not straightforward by use of the tools that we have at our disposal from Chapter 1. Our graph satisfies IS_2 by Theorem 4.10, but does not satisfy $IS_{2+\varepsilon}$ for any $\varepsilon > 0$. Also note that $\mathbf{C}_d \times \mathbb{Z}$ is a slight "fattening" of $\mathbf{C}_{d+1}$. One (non-trivial) approach is to construct a finite energy flow from o to ∞; see Markvorsen, McGuinness and Thomassen [226].

Here, we shall derive a local limit theorem for the Cartesian product (with $c = 1/2$, say) of the simple random walks on the two factors. As $\mathbf{C}_d$ is not regular, the simple random walk on $\mathbf{C}_d \times \mathbb{Z}$ cannot be obtained in this way. However, the Dirichlet form associated with the latter is equivalent to the one corresponding to the former (i.e., $D_2(\cdot) \ge \varepsilon_1\, D_1(\cdot)$ and vice versa – see Theorems 3.1–3.2), and we can apply Corollary 2.14.

(18.4) Proposition. *The simple random walk on* $\mathbf{C}_d$ *satisfies*

$$p^{(2n)}(\mathbf{0},\mathbf{0}) \sim \frac{2^{-1+2^{1-d}}\, d}{\Gamma(2^{-d})}\, n^{-1+2^{-d}}\,.$$

Proof. We can construct $\mathbf{C}_d$ by taking the two-way-infinite path $\mathbb{Z}$ and countably many copies of $\mathbf{C}_{d-1}$; at each point of $\mathbb{Z}$ we attach a copy of $\mathbf{C}_{d-1}$ by its origin. (That is, $\mathbb{Z}$ is a "spit" running through the centres of all the copies of $\mathbf{C}_{d-1}$.) We write $\mathbf{k} = (k, 0, \ldots, 0)$ for the points in $\mathbf{C}_d$ situated on the "spit" $\mathbb{Z}$. Let $U_d(z) = U_d(\mathbf{0},\mathbf{0}|z)$ and $F_d(z) = F_d(\mathbf{1},\mathbf{0}|z)$ for the generating functions (1.12) of the first return probabilities to the origin $\mathbf{0}$, and (1.11) of the first passage probabilities from $\mathbf{1}$ to $\mathbf{0}$ in $\mathbf{C}_d$, respectively. By Lemma 1.23 and the translation invariance of the random walk along $\mathbb{Z}$, we have $F_d(\mathbf{2},\mathbf{0}|z) = F_d(z)^2$. Therefore, use of Lemma 1.13 yields the following equations.

$$\begin{aligned} F_d(z) &= \frac{z}{2d} + \frac{d-1}{d} U_{d-1}(z) F_d(z) + \frac{z}{2d} F_d(z)^2\,; \\ U_d(z) &= \frac{d-1}{d} U_{d-1}(z) + \frac{z}{d} F_d(z)\,. \end{aligned}$$

In the first equation, the first term on the right corresponds to going from $\mathbf{1}$ to $\mathbf{0}$ at the first step, the second term corresponds to first taking a "detour" into the copy of $\mathbf{C}_{d-1}$ attached at $\mathbf{1}$ until the first return to $\mathbf{1}$ and then restarting once more at $\mathbf{1}$, and the third term corresponds to first going from $\mathbf{1}$ to $\mathbf{2}$ and continuing from there until reaching $\mathbf{0}$. The second equation is obtained in a similar way (staying in the copy of $\mathbf{C}_{d-1}$ attached at $\mathbf{0}$, or first moving to $\pm\mathbf{1}$).

We now eliminate $F_d(z)$ from the two equations and replace $U_d(z)$ with $1 - 1/G_d(z)$, where $G_d(z) = G(\mathbf{0},\mathbf{0}|z)$ on $\mathbf{C}_d$. It is then convenient to introduce the auxiliary function $H_d(z) = d/G_d(z) = d - d\,U_d(z)$. We obtain the recurrence relation $H_d(z)^2 = \big(1 + H_{d-1}(z)\big)^2 - z^2$. Recall that $p^{(n)}(\mathbf{0},\mathbf{0}) = 0$ when n is odd, and that $G_d(z)$ cannot have any singularities on the circle of convergence $\{|z| = 1\}$ besides ± 1. We replace z by $\sqrt{z}$, that is, we consider the functions $\overline{G}_d(z) = \sum_n p^{(2n)}(\mathbf{0},\mathbf{0})\, z^n$ and $\overline{H}_d(z) = d/\overline{G}_d(z)$. Thus we only have to consider the singularity $z = 1$. The recurrence relation becomes

$$\overline{H}_d(z)^2 = \big(1 + \overline{H}_{d-1}(z)\big)^2 - z\,. \tag{18.5}$$

We know from Lemma 1.24 that $\overline{H}_1(z) = \sqrt{1-z}$. Therefore each $\overline{H}_d(z)$ is algebraic, and $z = 1$ is an algebraic singularity. Also, $\overline{H}_d(1) = 0$ by recurrence of the simple random walk. Therefore it admits an expansion as a convergent *Puiseux series*

$$\overline{H}_d(z) = a_d\,(1-z)^{q_d} + \sum_{n=1}^{\infty} a_d(n)\,(1-z)^{q_d(n)}\,,$$

valid in a neighbourhood of $z = 1$ with the exception of $[1\,,\infty)$, where $q_d < q_d(1) < q_d(2) < \cdots$ is a discrete sequence of rational numbers and $a_d \neq 0$ (see e.g. Dimca [97], p.p. 177–179). We have $q_1 = 1/2$ and $a_1 = 1$. Inserting the expansion into (18.5), we can determine recursively q_d and a_d by comparing the lowest exponents on the left and right hand sides. We find $2q_d = q_{d-1}$ and $a_d^2 = 2a_{d-1}$, that is, $q_d = 2^{-d}$ and $a_d = 2^{1-2^{1-d}}$. We now obtain

$$\overline{G}_d(z) = d\,2^{-1+2^{1-d}}(1-z)^{-2^{-d}} + \text{higher order terms},$$

where the latter form again a discrete series in increasing rational powers of $(1-z)$. This is a case where Darboux's theorem applies immediately (singularity analysis might also be used), yielding that $p^{(2n)}(\mathbf{0},\mathbf{0})$ is asymptotically equivalent with the coefficients in the expansion at $z=0$ of $d\,2^{-1+2^{1-d}}(1-z)^{-2^{-d}}$. □

(18.6) Corollary. *If P is defined on $\mathbf{C}_d \times \mathbb{Z}$ as a Cartesian product of the simple random walks on $\mathbf{C}_d$ and $\mathbb{Z}$, respectively, then*

$$p^{(2n)}(0,0) \sim C\,n^{2^{-d}-3/2}\,.$$

In particular, P is transient for $d \geq 2$.

19. Free groups and homogeneous trees

The free group $\mathbb{F}_M$ and the group $\Gamma = \langle a_1,\dots,a_M \mid a_i^2 = o\rangle$ have the homogeneous tree (with degree $2M$ and M, respectively) as their Cayley graphs. Both are free products, and Corollary 17.8(a,b) applies. For the case of nearest neighbour random walks, explicit formulae for $\Phi(t)$ were given in (9.29). Furthermore, on $\mathbb{F}_M$, we obtain the asymptotic behaviour of Corollary 17.8 whenever μ is supported in the set of all integer powers of the free generators and either has some exponential moment or has finite first moment and zero mean over each subgroup $\langle a_i\rangle \simeq \mathbb{Z}$. In this section we study the following questions: what is the local limit theorem when μ does not arise as a free sum? In §B we shall give the answer for the case when μ has arbitrary finite support, and in §C we shall consider random walks whose transition probabilities depend only on the distances between the points. The simple random walk on $\mathbb{T}_M$ is the simplest example of this type. For random walks in $\mathbb{Z}^d$, the local central limit theorem gave us asymptotic estimates that are uniform in space and time over a certain range. In §A we study the analogous question for a variant of simple random walk on the tree.

A. Space-time asymptotics for aperiodic simple random walks on $\mathbb{T}_M$

Instead of the simple random walk, we shall study here the random walk on $\mathbb{T} = \mathbb{T}_M$ whose transition probabilities are

$$p(x,x) = 1/2 \quad \text{and} \quad p(x,y) = 1/(2M) \quad \text{when } y \sim x.$$

The only reason for this choice is that we obtain an aperiodic random walk, so that we do not have to distinguish parities. The modifications needed for obtaining the analogous results for the "pure" simple random walk are straightforward.

The transition probabilities depend only on the distance, so that it is sufficient to consider $p^{(n)}(o,x)$. As usual, we write $|x| = d(x,o)$. With direct calculations, or combining Lemmas 1.24 and 9.2, we get

$$G(o,x|z) = G(z)\,F(z)^{|x|}\,, \quad \text{where} \tag{19.1}$$

$$F(z) = \tfrac{M}{(M-1)z}\Big(\big(1-\tfrac{1}{2}z\big) - \sqrt{(1-z/\mathbf{r})(1-z/\mathbf{s})}\Big) \quad \text{and}$$

$$G(z) = \frac{1}{1-\frac{1}{2}z-\frac{1}{2}zF(z)}\,, \quad \text{with}$$

$$\mathbf{r} = \Big(\tfrac{1}{2}+\tfrac{\sqrt{M-1}}{M}\Big)^{-1} \quad \text{and} \quad \mathbf{s} = \Big(\tfrac{1}{2}-\tfrac{\sqrt{M-1}}{M}\Big)^{-1}.$$

Here, $\mathbf{r} = \rho(P)^{-1}$ is the radius of convergence (principal singularity) of $G(o,x|z)$. We now use Cauchy's integral formula to write

$$\begin{aligned} p^{(n)}(o,x) &= \frac{1}{2\pi i}\int_{\mathcal{C}} \frac{G(z)\,F(z)^{|x|}}{z^{n+1}}\,dz \\ &= \frac{1}{2\pi i}\int_{\mathcal{C}} \frac{G(z)}{z}\,\exp\Big(n\,\big(\alpha\,\log F(z) - \log z\big)\Big)\,dz\,, \end{aligned} \tag{19.2}$$

where $\alpha = |x|/n$ and $\mathcal{C}$ is a positively oriented, simple closed curve in $\mathbb{C}$ which has 0 in its interior and $\mathbf{r}$ in its exterior. We define for $\alpha \in [0\,,\,1]$

$$\varphi(\alpha) = \min\{\alpha\,\log F(z) - \log z : 0 \le z \le \mathbf{r}\}\,. \tag{19.3}$$

The minimum is attained at

$$z(\alpha) = \tfrac{2M}{(M-2)^2}\big(M - w(\alpha)\big)\,, \quad \text{where} \quad w(\alpha) = \sqrt{M^2\alpha^2 + 4(M-1)(1-\alpha^2)}\,.$$

In particular, $z(0) = \mathbf{r}$, $z(1) = 0$, and φ is strictly decreasing, with $\varphi(0) = -\log \mathbf{r}$ and $\varphi(1) = -\log(2M)$.

(19.4) Theorem. *As $n \to \infty$, we have uniformly in $|x| \le (1-c)n$ with $c > 0$*

$$p^{(n)}(o,x) \sim B(|x|/n)\left(1 + \tfrac{M-2}{M}|x|\right)\exp\bigl(n\,\varphi(|x|/n)\bigr)\,n^{-3/2},$$

$$\text{where}\quad B(\alpha) = M\,G\bigl(z(\alpha)\bigr)\Big/\sqrt{2\pi\bigl(M - w(\alpha)\bigr)w(\alpha)}\,.$$

Proof. We subdivide the proof according to the two ranges $a \le \alpha \le 1-c$ and $\alpha \le a$. We start with the first of the two; the choice of a will become evident in the second part.

Case 1. For $\alpha \in [a\,,\,1-c]$, the values $z(\alpha)$ lie in some interval $[\bar{a}\,,\,\mathtt{r} - \bar{c}]$ with $\bar{a}, \bar{c} > 0$. We now choose the integration contour $\mathcal{C} = \{z(\alpha)e^{it} : t \in (-\pi\,,\,\pi]\}$. The function $\psi_\alpha(t) = \alpha\,\log F\bigl(z(\alpha)e^{it}\bigr) - \log\bigl(z(\alpha)e^{it}\bigr)$ has derivatives $\psi_\alpha'(0) = 0$ and $\psi_\alpha''(0) = -z(\alpha)\,w(\alpha)/(2M\alpha^2)$ and Taylor expansion

$$\psi_\alpha(t) = \varphi(\alpha) - \frac{z(\alpha)\,w(\alpha)}{4M\alpha^2}t^2 + R(\alpha,t)\,.$$

The remainder term $R(\alpha,t)/t^2$ is continuous in $[a\,,\,1-c] \times [-\pi\,,\,\pi]$, so that $R(\alpha,t) = \mathfrak{o}(t^2)$ uniformly as $t \to 0$, when $\alpha \in [a\,,\,1-c]$. In particular, we can find $\gamma > 0$ such that $|R(\alpha,t)| \le -\psi_\alpha''(0)t^2/4$ for all $t \in [-\gamma\,,\,\gamma]$. We decompose the integral in (19.2), rewritten in terms of the variable t, into the parts where $|t| \le \gamma$ and $\gamma < |t| \le \pi$, respectively. The second part becomes

$$\frac{1}{2\pi}e^{n\,\varphi(\alpha)}\int_{\gamma<|t|\le\pi}\left(\frac{F\bigl(z(\alpha)e^{it}\bigr)}{F\bigl(z(\alpha)\bigr)}\right)^{n\alpha} e^{-int}G\bigl(z(\alpha)e^{it}\bigr)\,dt\,,$$

which is bounded in absolute value by $e^{n\,\varphi(\alpha)}\,G(\mathtt{r})\,\lambda^n$, where $\lambda = \max\bigl\{\bigl(|F(z)|/F(|z|)\bigr)^a : |z| \in [\bar{a}\,,\,\mathtt{r}-\bar{c}]\,,\ |\arg(z)| \ge \gamma\bigr\}$. Now, in $\{|z| \le \mathtt{r}\}$, the power series defining $F(z)$ has all coefficients strictly positive (for $n \ge 1$), and $|F(z)| < F(|z|)$ strictly, when z is not in $[0\,,\,\mathtt{r}]$. (This is the *clou* of this "saddle point" technique: that $z(\alpha)$ minimizes $|F(z)^\alpha/z|$ over $[0\,,\,\mathtt{r}]$, but maximizes the same function over the circle $\{|z| = z(\alpha)\}$.) Therefore $\lambda < 1$, which will be seen to imply that the integral over $\gamma < |t| \le \pi$ is asymptotically negligible in comparison with the one over $[-\gamma\,,\,\gamma]$. The latter is

$$\frac{1}{2\pi}e^{n\,\varphi(\alpha)}\int_{-\gamma}^{\gamma}\exp\Bigl(-n\tfrac{z(\alpha)w(\alpha)}{4M\alpha^2}t^2 + nR(\alpha,t)\Bigr)G\bigl(z(\alpha)e^{it}\bigr)\,dt\,.$$

We substitute $\theta = t\sqrt{n\frac{z(\alpha)w(\alpha)}{2M\alpha^2}} = t\sqrt{n}b(\alpha)$ to rewrite this as

$$\frac{G\big(z(\alpha)\big)}{2\pi b(\alpha)} e^{n\,\varphi(\alpha)} n^{-1/2}$$
$$\times \int_{-\sqrt{n}b(\alpha)\gamma}^{\sqrt{n}b(\alpha)\gamma} \exp\Big(-\tfrac{1}{2}\theta^2 + nR\big(\alpha, \tfrac{\theta}{\sqrt{n}b(\alpha)}\big)\Big) \frac{G\Big(z(\alpha)\exp\big(i\frac{\theta}{\sqrt{n}b(\alpha)}\big)\Big)}{G\big(z(\alpha)\big)}\, d\theta\,.$$

Due to our choice of γ, we can bound the integrand in absolute value by $\exp(-\frac{\theta^2}{4})$. Thus we can apply Lebesgue's theorem (dominated convergence) to see that the last integral tends to $\int_{\mathbb{R}} e^{-\theta^2/2}\, d\theta$ uniformly in α. We have obtained

$$p^{(n)}(o,x) \sim \frac{G\big(z(\alpha)\big)\alpha}{\sqrt{\pi\, z(\alpha)\, w(\alpha)/M}} e^{n\,\varphi(\alpha)} n^{-1/2}$$

uniformly as $n \to \infty$ and $\alpha = \frac{|x|}{n} \in [a\,,\, 1-c]$. In this range, $1 + \frac{M-2}{M}|x| \sim \frac{M-2}{M}\alpha\, n$, and the result follows.

Case 2. If $\alpha \le a$ and $z(\alpha)$ is close to the singularity $\mathtt{r}$ of $F(z)$, the above technique does not work. The substitution $u = \sqrt{\mathtt{r} - z}$ will make the singularity disappear. We shall integrate along the segment $u = u(\alpha) - it$ for $t \in [-\bar{\gamma}\,,\, \bar{\gamma}]$, where $u(\alpha) = \sqrt{\mathtt{r} - z(\alpha)}$. The number $\bar{\gamma}$ will be chosen below, and does in turn imply the choice of a. In z-coordinates, the line segment transforms into the parabola piece

$$z = z(\alpha, t) = z(\alpha) + t^2 + 2it\, u(\alpha)\,, \quad |t| \le \bar{\gamma}\,.$$

Pick some (small) $\varepsilon_0 > 0$, and choose a such that $|z(\alpha, \bar{\gamma})| \ge \mathtt{r} + \varepsilon_0$ for all $\alpha \le a$ (this is possible because $z(\alpha) \to \mathtt{r}$ as $\alpha \to 0$). Recall that on the other hand, $\mathtt{r} - \bar{c} \le z(\alpha) \le \mathtt{r}$.

Figure 15: the integration contours in Cases 1 and 2 The dashed circle is $|z| = \mathtt{r}$.

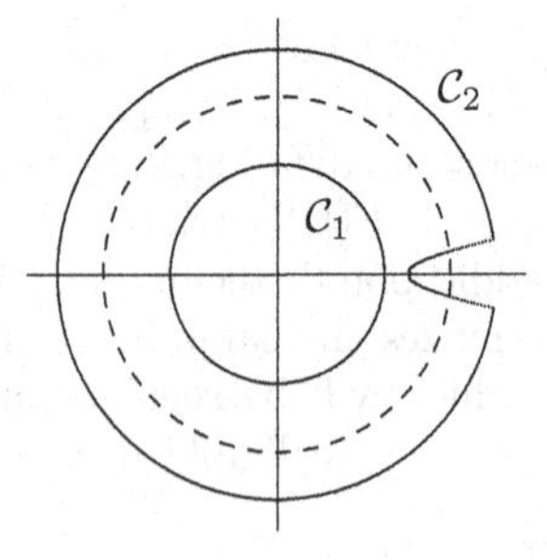

Now observe that $F(z)$ and $G(z)$ are analytic in $\mathbb{C} \setminus [\mathtt{r}\,,\, \mathtt{s}]$. Therefore we may chose as integration contour C in (19.2) the parabola piece together with the arc $\{|z| = |z(\alpha,\bar\gamma)|\,,\ |\arg(z)| \geq \arg\big(z(\alpha,\bar\gamma)\big)\}$; see Figure 15. This contour depends on α, but remains in a bounded region of the plane, where $F(z)$ and $G(z)$ are bounded by some constant $C \geq 1$. Therefore the integral (19.2), taken over the arc only, is bounded above in absolute value by

$$e^{n\,\varphi(\alpha)}\,C\,\bar\lambda^n\,,\quad \text{where}\quad \bar\lambda = \left(\frac{C}{F(\mathtt{r}-\bar c)}\right)^a \frac{\mathtt{r}}{\mathtt{r}+\varepsilon_0}\,.$$

If we choose a small enough so that $\bar\lambda < 1$, this will be asymptotically negligible as compared with the integral over the parabola piece. After our change of variable, the latter is

$$(19.5)\qquad \frac{1}{\pi}\int_{-\bar\gamma}^{\bar\gamma} \exp\big(n\,\overline{\psi}_\alpha(t)\big)\,\frac{G\big(z(\alpha,t)\big)}{z(\alpha,t)}\,\big(u(\alpha)-it\big)\,dt\,,\quad \text{where}$$
$$\overline{\psi}_\alpha(t) = \alpha\,\log F\Big(z(\alpha,t)\Big) - \log\Big(z(\alpha,t)\Big)\,.$$

We compute $u(\alpha)^2 = \mathtt{r} - z(\alpha) = \frac{2M\alpha^2}{2\sqrt{M-1}+w(\alpha)}$ and $\overline{\psi}_\alpha''(0) = -\bar b(\alpha)^2$ with

$$\bar b(\alpha)^2 = \frac{2\big(\mathtt{r}-z(\alpha)\big)w(\alpha)}{M\alpha^2\,z(\alpha)}\,.$$

The latter tends to the non-zero limit $2/\mathtt{r}$ as $\alpha \to 0$. Writing $\overline{\psi}_\alpha(t) = \varphi(\alpha) - \frac{1}{2}\big(\bar b(\alpha)t\big)^2 + \overline{R}(\alpha,t)$, we see as above that $\overline{R}(\alpha,t) = \mathfrak{o}(t^2)_{t\to 0}$ uniformly for $\alpha \in [0\,,\,1-c]$. We finally can choose $\bar\gamma$ such that $|\overline{R}(\alpha,t)| \leq \big(\bar b(\alpha)t\big)^2/4$ for all $t \in [-\bar\gamma\,,\,\bar\gamma]$ and all $\alpha \in [0\,,\,1-c]$. Proceeding as above, we substitute $\theta = \sqrt{n}\,\bar b(\alpha)\,t$. Now, if $\alpha \to 0$, both real and imaginary parts of the last factor in the integral (19.5) tend to 0. This last factor is

$$u(\alpha) - it_n\,,\quad \text{where}\quad t_n = t_n(\theta,\alpha) = \frac{\theta}{\sqrt{n}\,\bar b(\alpha)}\,.$$

Therefore we have to consider two subcases.

Case 2.A. If $\alpha \geq n^{-1/4}$ then $u(\alpha) - it_n \sim u(\alpha)$ uniformly in α for each θ, as $n \to \infty$. Also, $G\big(z(\alpha,t_n)\big)/z(\alpha,t_n) \sim G\big(z(\alpha)\big)/z(\alpha)$. We proceed as in Case 1 and find

$$p^{(n)}(o,x) \sim \frac{\sqrt{2}u(\alpha)G\big(z(\alpha)\big)}{\sqrt{\pi}\,z(\alpha)\,\bar b(\alpha)}\,e^{n\,\varphi(\alpha)}n^{-1/2}\,,$$

which by a few elementary computations transforms into the proposed asymptotics.

Case 2.B. We now assume $\alpha < n^{-1/4}$. First, we refine $\overline{R}(\alpha,t) = \frac{1}{6}\overline{\psi}_\alpha'''(0)t^3 + \mathcal{O}(t^4)$. Also, $\overline{\psi}_\alpha'''(0) = \mathcal{O}(\alpha)$, so that $\overline{R}(\alpha,t) = t^3\big(\mathcal{O}(\alpha) + \mathcal{O}(t)\big)$. Note that, $p^{(n)}(o,x)$ being real, we are only interested in the real part of the integrals (19.2) and (19.5). We decompose $\overline{R}(\alpha,t) = \overline{R}_0(\alpha,t) + i\,\overline{R}_1(\alpha,t)$ (real and imaginary parts). We have $nt_n^3 = t_n\,\theta^2/\overline{b}(\alpha)^2$, whence

$$\cos\big(n\overline{R}_1(\alpha,t_n)\big) \sim 1\,, \quad \sin\big(n\overline{R}_1(\alpha,t_n)\big) \sim n\overline{R}_1(\alpha,t_n)\big) = t_n\big(\mathcal{O}(\alpha)+\mathcal{O}(t_n)\big)\,.$$

Also, by (19.1), we can write $G(z)/z = H(z) + \sqrt{\mathtt{r}-z}K(z)$, with $H(z)$ and $K(z)$ analytic in $\{0 < \mathrm{Re}(z) < \mathtt{s}\}$, $H(\mathtt{r}) = G(\mathtt{r})/\mathtt{r}$ and $K(\mathtt{r}) \neq 0$. Decomposing $H(z) = H_0(z) + i\,H_1(z)$ and $K(z) = K_0(z) + i\,K_1(z)$, both $H_1(z)$ and $K_1(z)$ are $\mathcal{O}(|\mathtt{r}-z|)$ as $z \to \mathtt{r}$, and with our substitution $\sqrt{\mathtt{r}-z} = u(\alpha) - it_n$, this becomes $\mathcal{O}\big(u(\alpha)^2 + t_n^2\big)$.

We now calculate the real part of the integrand of (19.5). It starts with $\exp\big(n\,\varphi(\alpha) - \frac{1}{2}\theta^2 + n\overline{R}_0(\alpha,t_n)\big)$. In the remaining factor,

$$\exp\Big(i\,n\,\overline{R}_1(\alpha,t_n)\Big)\Big(H(z_n) + \big(u(\alpha) - it_n\big)K(z_n)\Big)\Big(u(\alpha) - it_n\Big)\,,$$

with $z_n = z(\alpha,t_n)$, we find that the real part is

$$\begin{aligned} &u(\alpha)\,H_0(z_n) - t_n^2\,K_0(z_n) + \mathfrak{o}(\alpha) + \mathfrak{o}(t_n^2) \\ &\quad \sim \alpha\left(\frac{2M}{2\sqrt{M-1}+w(\alpha)}\right)^{1/2} H\big(z(\alpha)\big) - \theta^2\frac{K\big(z(\alpha)\big)}{n\,\overline{b}(\alpha)^2}\,. \end{aligned}$$

We may as well replace $z(\alpha)$ with its limit $\mathtt{r}$, and in particular, we compute $H(\mathtt{r}) = \frac{2M\sqrt{M-1}}{(M-2)\mathtt{r}^2}$ and $K(\mathtt{r}) = -\frac{2M^2\sqrt{2M\sqrt{M-1}}}{(M-2)^2\mathtt{r}^3}$. Remembering that $\int_{-\infty}^{\infty}\theta^2 e^{-\theta^2/2}\,d\theta = \sqrt{2\pi}$, we use dominated convergence to obtain

$$p^{(n)}(o,x) \sim \frac{e^{n\,\varphi(\alpha)}}{\overline{b}(\alpha)\sqrt{n\pi/2}}\left(\alpha\left(\frac{2M}{2\sqrt{M-1}+w(\alpha)}\right)^{1/2} H\big(z(\alpha)\big) - \frac{K\big(z(\alpha)\big)}{n\,\overline{b}(\alpha)^2}\right).$$

Once more, this transforms into the proposed formula after some elementary calculations. □

In order to compare Theorem 19.4 with the Gaussian estimate (14.12) and in particular with the local limit theorem in the grid, (13.11), let us consider the range $|x|/\sqrt{n} \le C$. Then $\alpha = |x|/n \to 0$. We compute the first and second (right) derivatives of φ at 0 and find the expansion $\varphi(\alpha) = -\log\mathtt{r} - \alpha\,\log\sqrt{M-1} - c_M\,\alpha^2 + \mathcal{O}(\alpha^3)$, where $c_M = \frac{M+2\sqrt{M-1}}{4\sqrt{M-1}}$. Inserting this into the asymptotic estimate, we obtain

$$\begin{aligned} \text{(19.6)}\quad &p^{(n)}(o,x) \\ &\sim B(0)\Big(1 + \tfrac{M-2}{M}|x|\Big)\frac{1}{\sqrt{M-1}^{\,|x|}}\exp\Big(-c_M|x|^2/n\Big)\,\rho(P)^n\,n^{-3/2} \end{aligned}$$

uniformly for $|x|/\sqrt{n}$ bounded. Regarding the full range $0 < |x|/n \le 1$, we have the following simple estimate that will be useful in §28.B.

(19.7) Lemma. $\lim_{n\to\infty} \sup_{|x|\le n} p^{(n)}(o,x)^{1/n} \exp(-\varphi(|x|/n)) = 1$.

Proof. The limit, if it exists, cannot be < 1 (set $x = o$). For showing that it is ≤ 1, we use the simple estimate $p^{(n)}(o,x) \le \binom{n}{|x|} F'(0)^{|x|}$. Indeed, $F'(0) = p(x,y)$ for any edge $[x,y]$, and the random walk has to cross each of the $|x|$ edges on $\pi(o,x)$ on its way from o to x. The binomial coefficient stands for the different choices of the instants when each of these crossings occurs for the first time. We have $\binom{n}{|x|}^{1/n} < e^{\varepsilon/2}$ when $n \to \infty$ and $|x|/n > 1 - \delta_\varepsilon$. Using $F\big(z(\alpha)\big) = \alpha z(\alpha) F'\big(z(\alpha)\big)$, with $\alpha = |x|/n$, we get

$$F'(0)^\alpha \exp\big(-\varphi(\alpha)\big) = \exp\left(-\alpha\log\alpha + (1-\alpha)\log z(\alpha) - \alpha \log \frac{F'\big(z(\alpha)\big)}{F'(0)}\right),$$

which tends to 1 as $\alpha \to 1$. Thus, if $\varepsilon > 0$ then the last expression is $< e^{\varepsilon/2}$ if $|x|/n > 1 - \delta_\varepsilon$ with a suitable choice of δ_ε.

The estimate of Theorem 19.4 yields that for $|x|/n \le 1 - \delta_\varepsilon$, we have also $p^{(n)}(o,x)^{1/n} \exp\big(-\varphi(|x|/n\big) < e^\varepsilon$ when $n \to \infty$. □

Finally, we remark that the method used in Theorem 19.4 works also in the case $M = 2$, that is, on $\mathbb{Z}$. In this case, $\mathtt{r} = 1$ and $G(z) = 1/\sqrt{1-z}$, which causes slight changes in the (easier) computations and the $n^{-1/2}$ in place of $n^{-3/2}$. Without going through all the details again, we state the corresponding result, which can also be derived more combinatorially.

(19.8) Theorem. *Consider the random walk on $\mathbb{Z}$ with $p(k, k \pm 1) = 1/4$ and $p(k,k) = 1/2$. As $n \to \infty$, we have uniformly in $|k| \le (1-c)n$ with $c > 0$*

$$p^{(n)}(0,k) \sim B(|k|/n) \exp\Big(n\,\varphi(|k|/n)\Big)\, n^{-1/2}, \quad \textit{where}$$

$$\varphi(\beta) = -\Big((1-\beta)\log(1-\beta) + (1+\beta)\log(1+\beta)\Big) \quad \textit{and}$$

$$B(\beta) = 1\Big/\sqrt{\pi(1-\beta^2)}\,.$$

Noting that $\varphi(0) = \varphi'(0) = 0$, one recovers the asymptotic estimate of Corollary 13.11 when $|k|/\sqrt{n}$ is bounded. Also, Lemma 19.7 remains valid on $\mathbb{Z}$ without any change.

B. Finite range random walks on free groups

Let $\mathbb{F}_M = \langle a_1, \dots, a_M \mid \rangle$ be the free group on $M \ge 2$ generators. In this subsection we shall prove the following general local limit theorem.

(19.9) Theorem. *Let μ be a finitely supported, irreducible and aperiodic probability measure on $\mathbb{F}_M$. Then as $n \to \infty$*

$$\mu^{(n)}(x) \sim \frac{h(x)}{2\sqrt{\pi\,\rho(P)}}\,\rho(P)^n\,n^{-3/2}\,.$$

We shall need several preliminary steps. We write $d(x,y)$ for the distance in the Cayley graph $\mathbb{T} = \mathbb{T}_{2M}$ of $\mathbb{F}_M$ with respect to $S = \{a_1^{\pm 1}, \dots, a_M^{\pm 1}\}$, and $|x| = d(x,o)$. Also, $\pi(x,y)$ is the geodesic arc in $\mathbb{T}$ between x and y. Let N be the smallest integer such that $\operatorname{supp}\mu$ is contained in $B = B(o,N)$, so that $B(x,N) = xB$.

(19.10) Exercise. *If $w \in \pi(x,y)$ then the random walk starting at x has to pass through wB before reaching y.*

For every $y \in \mathbb{F}_M$ define the stopping time $\mathbf{s}_y (= \mathbf{s}^{yB}) = \inf\{n \ge 0 : Z_n \in yB\}$ and for $x \neq y$ the matrix $\mathcal{H}_{x,y}(z) = \big(H_{x,y}(xa, yb|z)\big)_{a,b\in B}$ with

$$H_{x,y}(xa, yb|z) = \mathbb{E}_{xa}\big(z^{\mathbf{s}_y}\,\delta_{yb}(Z_{\mathbf{s}_y})\big)\,, \quad z \in \mathbb{C}\,.$$

This is a power series with n-th coefficient $h_{x,y}^{(n)}(xa,yb) \le f^{(n)}(xa,yb) \le p^{(n)}(xa,yb)$, so that it certainly converges for $|z| \le \mathbf{r} = 1/\rho(P)$. Recall that the Green function converges at $z = \mathbf{r}$ by Theorem 7.8 – this fact will be important. Let $B^* = B \setminus \{o\}$ and consider the column vectors

$$\mathbf{F}^*(z) = \big(F(b,o|z)\big)_{b\in B^*}\,, \quad \mathbf{F}(z) = \big(F(b,o|z)\big)_{b\in B}\,, \quad \text{and} \quad \mathbf{e}_o = \big(\delta_o(b)\big)_{b\in B}\,.$$

(Recall that $F(o,o|z) \equiv 1$.) Also, we define a column vector and square matrix over B^* by

$$\mathbf{A}(z) = \big(M(b,o|z)\big)_{b\in B^*} \quad \text{and} \quad \mathcal{M}(z) = \big(M(a,b|z)\big)_{a,b\in B^*}\,, \quad \text{where}$$
$$M(a,b|z) = \mu(a^{-1}b)z + \sum_{x\in aB\setminus B} \mu(a^{-1}x)z\,H_{x,o}(x,b|z)\,.$$

As usual, we write $G(z) = G(x,x|z)$, which is independent of x. Using Exercise 19.10, we now obtain the following system of equations for determining $G(o,x|z)$ and the other generating functions involved.

(19.11) Proposition. *If $x,y \in \mathbb{F}_M$ and $\pi(x,y) = [x = x_0, x_1, \dots, x_k = y]$*

then

$$\text{(a)} \qquad G(x,y|z) = F(x,y|z)\,G(z)\,;$$

$$\text{(b)} \qquad G(z) = 1 + z\Big(\mu(o) + \sum_{b\in B^*} \mu(b)\,F(b,o|z)\Big)G(z)\,;$$

$$\text{(c)} \qquad F(x,y|z) = \mathbf{e}_o \cdot \mathcal{H}_{x,y}(z)\mathbf{F}(z)\,;$$

$$\text{(d)} \qquad \mathbf{F}^*(z) = \mathbf{A}(z) + \mathcal{M}(z)\mathbf{F}^*(z)\,;$$

$$\text{(e)} \qquad \mathcal{H}_{x,y}(z) = \prod_{j=1}^{k} \mathcal{H}_{x_{j-1},x_j}(z)\,.$$

$$\text{(f)} \qquad H_{x,y}(xa,yb|z) = \begin{cases} \delta_{yb}(xa)\,, & \textit{if } xa \in yB\,,\\ \sum_{a'\in B} \mu(a')z\,H_{xa,y}(xaa',yb|z)\,, & \textit{otherwise.}\end{cases}$$

Proof. Statements (a) and (b) are part of Lemma 1.13, and (c) follows from Exercise 19.10 and the fact that $F(yb,y|z) = F(b,o|z)$. Also (e) is immediate from Exercise 19.10.

To see (d), we condition on the instant of the *first return* to B, which may occur at the first step (giving rise to the first term of $M(a,b|z)$) or after time 1, so that the first step goes to the exterior of B (giving rise to the second term). Also, $\mathbf{A}(z)$ and $\mathcal{M}(z)$ correspond to the first return taking place at o and in B^*, respectively. Indeed, $M(a,b|z)$ is a power series whose n-th coefficient is the probability that for the random walk starting at $a \in B^*$, the first return to B occurs at time n at the point b.

Similarly, to see (f) in the non-degenerate case, we condition on the first step, which goes from xa to some $xaa' \in xaB$. □

Note that by group invariance $H_{x,y}(xa,yb|z) = H_{o,x^{-1}y}(a,x^{-1}yb|z)$, so that each of the terms $H_{xa}(xaa',yb|z)$ occurring in the right hand side of (f) is by virtue of (e) a matrix element in a product of at most $N+1$ matrices from $\{\mathcal{H}_{o,v}(z) : v \in S\}$. We eliminate the constant ones (with value 0 or 1) among them, and write $H_i(z)$, $i \in \mathcal{I} = \{1,\dots,\kappa\}$, for the non-constant ones among the matrix elements of all the $\mathcal{H}_{o,v}(z)$, $v \in S$.

Consequently, for any fixed $x \in \mathbb{F}_M$, Proposition 19.11 provides a finite system of equations, each of which is of the form ⟨unknown function⟩ = ⟨a polynomial with non-negative coefficients in z and the unknown functions⟩. Given this explicit form of the equations, "elimination theory" of algebraic geometry implies that each function involved is *algebraic,* i.e., satisfies a polynomial equation $Q\big(z,f(z)\big) = 0$, see van der Waerden [333], Section 20. Indeed, the (complicated) algorithm for finding Q works even when considering formal power series in non-commuting variables; see Kuich and Salomaa [210], Section 16.

Every non-constant function occurring in Proposition 19.11 arises as a power series whose radius of convergence must be a singularity that cannot be smaller than $\mathbf{r} = 1/\rho(\mu)$. We shall investigate the nature of these singularities. The main key to our result is the following system of equations.

(19.12) Proposition. *Suppose that $\mu(x) > 0$ for every $x \in S$. Then there are polynomials $\mathcal{Q}_i(z, w_1, \dots, w_\kappa)$, $i = 1, \dots, \kappa$, with positive coefficients such that*

$$H_i(z) = \mathcal{Q}_i\big(z, H_1(z), \dots, H_\kappa(z)\big),$$

and every variable w_j appears as a factor in a non-zero term of $\mathcal{Q}_i$ that also contains the variable w_i. Also, all $\mathcal{Q}_i$ are divisible by z, and some $\mathcal{Q}_i$ must have a non-zero linear term in z.

Proof. Each $i \in \mathcal{I}$ corresponds to a triple (u, a, b), where $u \in S$ and $a, b \in B$, and $H_i(z) = H_{o,u}(a, ub|z)$ is non-constant. (We distinguish functions corresponding to different triples, even if they might actually coincide, as for example in the radial case.)

We apply (f) of Proposition 19.11 with $x = o$ and $y = b$. We iterate this formula N times by conditioning on each of the first $N + 1$ steps instead of the first one only. What we obtain is an equation

$$(19.13) \qquad H_{o,u}(a, ub|z) = \sum_{\substack{x_1, \dots, x_n \\ 1 \le n \le N+1}} \mu(x_1) \cdots \mu(x_n)\, z^n$$
$$+ \sum_{x_1, \dots, x_{N+1}} \mu(x_1) \cdots \mu(x_{N+1})\, z^{N+1}\, H_{a\bar{x}_N, u}(a\bar{x}_N x_{N+1}, ub|z),$$

where all $x_k \in \operatorname{supp} \mu$ and $\bar{x}_N = x_1 \cdots x_N$. In the first of the two sums it must be that $x_1 \cdots x_k \notin uB$ for $k < n$ and $x_1 \cdots x_n \in uB$. In the second, it must be that $x_1 \cdots x_k \notin uB$ for all $k \le N+1$, and we only consider non-zero terms. In particular, $d(\bar{x}_N, u) \le N^2 + 1$, and again, $H_{\bar{x}_N, u}(\bar{x}_N x_{N+1}, ub|z)$ is a matrix element in a product of at most N^2+1 matrices from $\{\mathcal{H}_{o,v}(z) : v \in S\}$. In this way we obtain the required polynomial $\mathcal{Q}_i$, where $i \equiv (u, a, b)$. It certainly has non-negative coefficients and is divisible by z. We now verify its other proposed properties.

It is true that some $\mathcal{Q}_i$ has a non-zero linear term in z, since there must be some point from where one can enter into uB in a single step.

For the other property, let $i \equiv (u, a, b)$ and $j \equiv (v, c, d)$ with $u, v \in S$, $a, b, c, d \in B$ and H_i and H_j non-constant. This implies that $a \notin uB$, and it must be that $a = a_1 \cdots a_N$ with $a_k \in S$ (this and all subsequent representations are tacitly assumed to be reduced) with $a_1 \ne u$. Also, it must be possible to reach ub starting from a before visiting any other element of uB. Now the random walk cannot leave the branch $\mathbb{T}_{o,a_1}$ (see §6.B for notation) before visiting uB, that is, $ub \in \mathbb{T}_{o,a_1}$, and the representation of b must be of the form $u^{-1}a_1 \cdots a_r b_1 \cdots b_\ell$ with $r > 0$ and $r + \ell \le N - 1$.

Analogously $c = c_1 \cdots c_N$ with $c_1 \ne v$ and $d = v^{-1}c_1 \cdots c_s d_1 \cdots d_m$ with $s > 0$, $s + m \le N - 1$.

To see that w_j occurs in $\mathcal{Q}_i$, choose $a_{N+1} \in S \setminus \{a_N^{-1}, v\}$ and let $y = a_{N+1}v^{-1}$. Then ayv $(= aa_{N+1})$, ay, ayc and $ayvd$ lie in the branch $\mathbb{T}_{o,a}$; see Figure 16. The same is true for $x = aycc_N^{-1} = a_1 \cdots a_{N+1}v^{-1}c_1 \cdots c_{N-1}$.

Figure 16

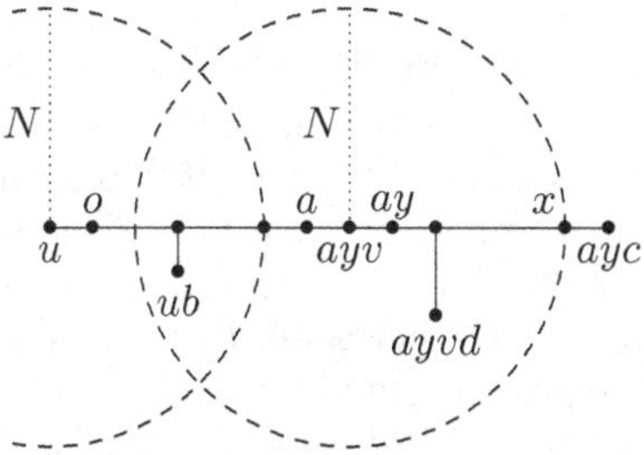

Now $d(x, a) = N + 1$. By assumption $\mu(a_{N+1})$, $\mu(v^{-1})$, $\mu(c_1), \dots, \mu(c_N)$ are all positive. Therefore the right hand side of (19.13) contains $H_{ax,u}(xc_N, ub|z)$ in a non-zero term. We now verify that the latter function is non-constant and contains $H_j(z) = H_{ay,ayv}(ayc, ayvd|z)$, and also $H_i(z)$, in a term of its expansion according to Proposition 19.11(e). Indeed,

$$\mathcal{H}_{ax,u}(z) = \mathcal{H}_{ax,ay}(z)\,\mathcal{H}_{ay,ayv}(z)\,\mathcal{H}_{ayv,o}(z)\,\mathcal{H}_{o,u}(z)\,.$$

Writing $ayc = xc_N$ we see that $H_{x,ay}(ayc, ayc|z) = 1$ is a positive matrix element in the first factor. In the second, $H_{ay,ayv}(ayc, ayvd|z) = H_j(z)$. In the fourth factor, we have the matrix element $H_{o,u}(a, ub|z) = H_i(z)$.

To close the chain, we show finally that $H_{ayv,o}(ayvd, a|z)$ is non-zero. Changing denomination for the sake of convenience, write $ayvd = a_1 \cdots a_L$ (reduced representation by elements of S), so that $N+2 \le L \le 2N+1$, and let $\bar{a}_h = a_1 \cdots a_h$. In this notation, $H_{ayv,o}(ayvd, a|z) = H_{\bar{a}_{N+1},o}(\bar{a}_L, \bar{a}_N|z)$. We have $\bar{a}_L \in \bar{a}_h B$ for all $h \ge L - N$, whence $H_{\bar{a}_{N+1},\bar{a}_{L-N}}(\bar{a}_L, \bar{a}_L|z) = 1$. Finally, $H_{\bar{a}_{L-N},o}(\bar{a}_L, \bar{a}_N|z)$ contains by assumption the non-zero term $\mu(a_L^{-1})\mu(a_{L-1}^{-1}) \cdots \mu(a_{N+1}^{-1})\, z^{L-N}$. This concludes the proof. □

It is much easier to deal with the matrix $\mathcal{M}(z)$.

(19.14) Lemma. *Some matrix element of $\mathcal{M}(z)$ must contain a non-constant function $H_j(z)$. Furthermore, if $\mu(x) > 0$ whenever $|x| \le 2$, then $\mathcal{M}(z)$ is irreducible for every positive z.*

Proof. The first statement is obvious. For the second, let $a, b \in B^*$. If $d(a, b) \le 2$ then $M(a, b|z) > 0$. Otherwise, we can find $a = a_0, a_1, \dots, a_k = b$ in B^* such that $d(a_{i-1}, a_i) \le 2$, whence $M^{(k)}(a, b|z) > 0$. □

Our strategy is now the following. Assuming that $\mu(x) > 0$ whenever $|x| \le 2$, we first prove that all non-constant generating functions occurring in Proposition 19.11 have the same radius of convergence $\mathbf{r}$. We then prove that for each $H_i(z)$, the singularity $z = \mathbf{r}$ is a simple branch point. This will then be carried over to the $F(b,o|z)$, $b \in B^*$, and finally to $G(x,y|z)$.

(19.15) Lemma. *If $\mu(x) > 0$ whenever $|x| \le 2$, then the power series $F(b,o|z)$ and $H_i(z)$ all have radius of convergence $\mathbf{r}$.*

Proof. The generating function $z\Big(\mu(o) + \sum_{b\in B^*} \mu(b)\, F(b,o|z)\Big) = U(o,o|z)$ of the first return probabilities to o is strictly increasing and has radius of convergence at least $\mathbf{r}$. If it is larger than $\mathbf{r}$ then the only possibility for $z = \mathbf{r}$ to be a singularity of $G(z)$ is for it to be a pole. But then the random walk would be ρ-recurrent, in contradiction to Theorem 7.8.

From this we get that some $F(a,o|z)$ with $a \in B^*$ must have radius of convergence equal to $\mathbf{r}$ (and not larger).

Now, if $b \in B^*$ with $d(a,b) \le 2$, then $F(b,o|z) \ge \mu(b^{-1}a)z\,F(a,o|z)$ for all $z > 0$, and the power series $F(b,o|z)$ must diverge for $z > \mathbf{r}$. Therefore also $F(b,0|z)$ has radius of convergence $\mathbf{r}$. Continuing as in the proof of Lemma 19.14, we conclude that all $F(b,o|z)$, $b \in B^*$, have radius of convergence $\mathbf{r}$.

Analogously, using the polynomials of Proposition 19.12, we conclude that all power series $H_i(z)$ must have the same radius of convergence, say $\mathbf{s}$, which is at least $\mathbf{r}$. Now suppose that $\mathbf{s} > \mathbf{r}$.

For small (positive) z, the matrix $I - \mathcal{M}(z)$ is invertible, and

$$\mathbf{F}^*(z) = \big(I - \mathcal{M}(z)\big)^{-1}\,\mathbf{A}(z)\,. \tag{19.16}$$

Increasing z, the entries of $\mathcal{M}(z)$ increase, as do those of $\mathbf{A}(z)$. Thus, (19.16) will define an analytic function until either we reach $\mathbf{s}$ or $\det\big(I - \mathcal{M}(z)\big) = 0$. Now $\mathbf{r}$ *is* a singularity of the left hand side, and having supposed $\mathbf{s} > \mathbf{r}$, we find $\det\big(I - \mathcal{M}(\mathbf{r})\big) = 0$. But then the singularity $\mathbf{r}$ of the left hand side is a pole. As $F(b,o|\mathbf{r}) \le G(b,o|\mathbf{r}) < \infty$, this is impossible. □

In particular, $I - \mathcal{M}(z)$ is invertible for $z \in (o\,,\,\mathbf{r}]$. This fact will be used below. We now know that each of the power series $H_i(z)$, $F(o,x|z)$ (with $x \ne o$) and $G(o,y|z)$ converge at $z = \mathbf{r}$, and that this is an algebraic singularity of each of these functions. Therefore, as in the proof of Proposition 18.4, each of them has an expansion as a *Puiseux series*

$$\begin{cases} H_i(z) = a_0(i) - a_1(i)(\mathbf{r} - z)^{\alpha(i)} + h.o.t.\,, \\ F(o,x|z) = b_0(x) - b_1(x)(\mathbf{r} - z)^{\beta(x)} + h.o.t.\,, \\ G(o,x|z) = c_0(x) - c_1(x)(\mathbf{r} - z)^{\gamma(x)} + h.o.t.\,, \end{cases} \tag{19.17}$$

where *h.o.t.* stands for "higher order terms" which are all of the form $\langle$real coefficient times $(\mathtt{r}-z)^p\rangle$, where $p>0$ is rational. Also $\alpha(i)$, $\beta(x)$ and $\gamma(x)$ are rational and positive, and each series converges in a set of the form $\{|z-\mathtt{r}|<\varepsilon\}\setminus(\mathtt{r}\,,\infty)$. As our functions are strictly increasing in the interval $[0\,,\mathtt{r}]$, in each of the three cases the first two coefficients must be strictly positive.

(19.18) Proposition. *If $\mu(x)>0$ whenever $|x|\le 2$ then $\alpha(i)=1/2$ for all $i\in\mathcal{I}$.*

Proof. We insert the Puiseux series (19.17) of all $H_j(z)$ in the polynomial equation for $H_i(z)$ given by Proposition 19.12. The terms on the right hand side are of the form $\langle$constant times $z^k\rangle$ (where $k\ge 1$), with expansion starting with $\mathtt{r}^k - k\mathtt{r}^{k-1}(\mathtt{r}-z)$, or of the form $\langle$constant times $z^k H_{j_1}(z)\cdots H_{j_\ell}(z)\rangle$ (where $k,\ell\ge 1$), with expansion starting with $C_0 - C_1(\mathtt{r}-z)^{\min\{1,\alpha(j_1),\dots,\alpha(j_\ell)\}}$. Recall that every $H_j(z)$ must occur on the right hand side. Comparison yields $\alpha(i)=\min\{1\,,\alpha(j):j\in\mathcal{I}\}$. Therefore $\alpha(i)=\alpha\le 1$. This was the easier part.

We now consider the Jacobian matrix

$$\mathcal{J}(z)=\left(\frac{\partial Q_i(z,H_1(z),\dots,H_\kappa(z))}{\partial w_j}\right)_{i,j\in\mathcal{I}}.$$

For $z\in(0\,,\mathtt{r}]$, all of its entries are positive by (19.12). We can therefore appeal to the Perron–Frobenius theory of finite non-negative matrices; see Seneta [297]. It says in particular that $\mathcal{J}(z)$ has only one eigenvalue $\lambda(z)$ of largest absolute value, which is positive and a simple root of the characteristic equation. By the implicit function theorem, $\lambda(z)$ is continuous (real-analytic). As all entries of $\mathcal{J}(z)$ increase with z, so does $\lambda(z)$. Also, $\mathcal{J}(0)$ is the zero matrix. We now determine $\lambda(\mathtt{r})$.

As $\lim_{z\to 0+}\lambda(z)=0$, we have $\lambda(z)<1$ for small (positive) z. We claim that $\lambda(z)<1$ for all $z\in(0\,,\mathtt{r})$. Suppose the contrary, that is, $\mathtt{s}<\mathtt{r}$, where by continuity $\mathtt{s}=\min\{z>0:\lambda(z)=1\}$. By the Perron–Frobenius theorem, the left eigenvector $(\mathbf{v}_1,\dots,\mathbf{v}_\kappa)$ of $\mathcal{J}(\mathtt{s})$ corresponding to the eigenvalue $\lambda(\mathtt{s})=1$ has all entries positive. Expand the function

$$\mathcal{F}(z,w_1,\dots,w_\kappa)=\sum_{i=1}^{\kappa}\mathbf{v}_i\Big(w_i-\mathcal{Q}_i(z,w_1,\dots,w_\kappa)\Big)$$

in a Taylor series around the point $\mathfrak{z}=\mathfrak{z}(\mathtt{s})=\big(\mathtt{s},H_1(\mathtt{s}),\dots,H_\kappa(\mathtt{s})\big)$: we have for all i, j

$$\mathcal{F}(\mathfrak{z}) = 0\,, \quad -a = \frac{\partial \mathcal{F}(\mathfrak{z})}{\partial z} < 0\,, \quad \frac{\partial \mathcal{F}(\mathfrak{z})}{\partial w_i} = 0\,,$$
$$-C_0 = \frac{\partial^2 \mathcal{F}(\mathfrak{z})}{\partial z^2} < 0\,, \quad -C_i = \frac{\partial^2 \mathcal{F}(\mathfrak{z})}{\partial z \partial w_i} < 0\,, \quad -C_{i,j} = \frac{\partial^2 \mathcal{F}(\mathfrak{z})}{\partial w_i \partial w_j} < 0\,.$$

Substituting $w_i = H_i(z)$, we find

$$\begin{aligned}(19.19)\quad a(\mathtt{s} - z) = \tfrac{1}{2} C_0 (\mathtt{s} - z)^2 + \tfrac{1}{2}(\mathtt{s} - z) \sum_{i=1}^{\kappa} C_i \big(H_i(\mathtt{s}) - H_i(z)\big) \\ + \tfrac{1}{2} \sum_{i,j=1}^{\kappa} C_{i,j} \big(H_i(\mathtt{s}) - H_i(z)\big)\big(H_j(\mathtt{s}) - H_j(z)\big) + h.o.t.\end{aligned}$$

Now all $H_i(z)$ are analytic and strictly increasing at $z = \mathtt{s}$. We insert their Taylor expansions at $z = \mathtt{s}$ in (19.19) and find an expression of the form

$$a(\mathtt{s} - z) = C(\mathtt{s} - z)^2 + h.o.t.\,,$$

where $a, C > 0$, valid in a neighbourhood of $\mathtt{s}$: a contradiction. Therefore $\lambda(z) < 1$ for all $z < \mathtt{r}$, and by continuity also $\lambda(\mathtt{r}) \leq 1$. Now it cannot be $\lambda(\mathtt{r}) < 1$. Otherwise, $\big(I - \mathcal{J}(\mathtt{r})\big)$ would be invertible, and the implicit function theorem would imply that the $H_i(z)$ were analytic at $z = \mathtt{r}$.

We have found $\lambda(\mathtt{r}) = 1$. We can now repeat the above calculations with $\mathtt{r}$ in place of $\mathtt{s}$, a corresponding left eigenvector of $\mathcal{J}(z)$, and $\mathfrak{z} = \mathfrak{z}(\mathtt{r})$. In the new version of (19.19), we insert the Puiseux expansions (19.17) and find an expression of the form

$$a(\mathtt{s} - z) = D_1(\mathtt{s} - z)^{2\alpha} + D_2(\mathtt{s} - z)^{1+\alpha} + D_3(\mathtt{s} - z)^2 + h.o.t.\,,$$

where $a, D_1, D_2, D_3 > 0$, valid for $z \in (\mathtt{r} - \varepsilon\,, \mathtt{r}]$. Comparing the exponents on the left and right hand sides, and knowing that $\alpha > 0$, we find $\alpha = 1/2$. □

We are now close to the conclusion.

(19.20) Lemma. *If $\mu(u) > 0$ for all u with $|u| \leq 2$ then in (19.17) the exponents are $\beta(x) = \gamma(x) = 1/2$ for all x.*

Proof. Propositions 19.11(e) and 19.18 imply that each non-constant matrix element of each $\mathcal{H}_x(z)$ has a Puiseux expansion of the form $a_0(x) - a_1(x)(\mathtt{r} - z)^{1/2} + h.o.t.$

We have observed that (19.16) holds for all $z \in (0\,,\mathtt{r}]$. The elements of $\mathcal{M}(z)$ are polynomials in z and $w_i = H_i(z)$. Perhaps not all w_i occur, but some do by Lemma 19.14. Write $\mathcal{M}(z, w_1, \dots, w_\kappa)$ for the matrix obtained by replacing the $H_i(z)$ with free variables w_i. It is irreducible for positive z and w_i, and $I - \mathcal{M}(z, w_1, \dots, w_k)$ is invertible in a neighbourhood of $\{(z, H_1(z), \dots, H_\kappa(z))\}$. The inverse $\sum_{n=0}^{\infty} \mathcal{M}(z, w_1, \dots, w_\kappa)^n$ is an analytic function of z and the w_i. It is componentwise strictly increasing for real positive z and all those w_i that occur. Expanding each matrix element at $(\mathtt{r}, H_1(\mathtt{r}), \dots, H_{\mathtt{r}}(z))$ and substituting the Puiseux expansions (19.17), we find $\beta(a^{-1}) = 1/2$ for all $a \in B^*$.

Now (19.11.c) yields $\beta(x) = 1/2$ for all $x \neq o$. Finally, (19.11.b) and the fact that $U(\mathtt{r}) < 1$ imply $\gamma(o) = 1/2$, and (19.11.a) shows that $\gamma(x) = 1/2$ for all x. □

Proof of Theorem 19.9. If μ is irreducible and aperiodic then there is k_0 such that $\mu^{(k)}(u) > 0$ for all $k \geq k_0$ and all $u \in \mathbb{F}_M$ with $|u| \leq 2$.

We can apply the machinery developed above to the random walk with law $\mu^{(k)}$. Its Green function satisfies (19.17) with $\gamma(x) = 1/2$ and $\mathtt{r}^k$ in place of $\mathtt{r}$. We know from Theorem 9.4 that $z = \mathtt{r}^k$ is the only singularity on the circle of convergence. Therefore we can apply Darboux's method as in Theorem 17.3 (even though here we do not know whether the Puiseux series is in integer powers of $\sqrt{\mathtt{r} - z}$ only) and find

$$\mu^{(kn)}(x) \sim \frac{h_k(x)}{2\sqrt{\pi\, \rho(\mu^{(k)})}}\, \rho(\mu^{(k)})^n\, n^{-3/2}\,.$$

Note that $\rho(\mu^{(k)}) = \rho(\mu)^k$. Now $\mu^{(kn+\ell)}(x) = \sum_{|y| \leq \ell N} \mu^{(\ell)}(y)\mu^{(kn)}(y^{-1}x)$, so that

$$\mu^{(n)}(x) \sim \frac{h_{k,\ell}(x)}{2\sqrt{\pi\, \rho(\mu)}}\, \rho(\mu)^n\, n^{-3/2}\,, \quad \text{as } n \to \infty\,,\ n \equiv \ell \text{ modulo } k\,.$$

The same holds with $k+1$ in place of k, and it is now a straightforward exercise to show that $h_{k,\ell}(x)$ is independent of k and ℓ. □

C. Radial random walks on the homogeneous tree

In this final subsection on local limit theorems, we return to techniques of harmonic analysis similar to the use of characteristic functions in Section 13. Here we consider random walks on $\mathbb{T} = \mathbb{T}_M$ ($M \geq 3$) which are *radial,* that is, $p(x,y)$ depends only on the distance between x and y. In accordance with the notation of (12.13), we write $M = q+1$. Let $S_k(x)$ be the sphere centred at x with radius k. Recall that $|S_0| = 1$ and $|S_k| = (q+1)q^{k-1}$ $(k \geq 1)$. If P_k is the transition matrix where $p_k(x,\cdot)$ is equidistribution on $S_k(x)$, the transition matrix P of any radial random walk is of the form

$$(19.21) \qquad P = \sum_{k=0}^{\infty} a_k P_k\,, \quad \text{where} \quad a_k \geq 0\,, \sum_{k=0}^{\infty} a_k = 1\,.$$

It is irreducible if and only if $a_k > 0$ for some odd k and aperiodic if in addition $a_k > 0$ also for some even k. The key to this subsection is the following simple formula.

$$(19.22) \qquad P_1\, P_k = P_k\, P_1 = \tfrac{1}{q+1} P_{k-1} + \tfrac{q}{q+1} P_{k+1} \quad (k \geq 1)\,.$$

This shows that each P_k can be written as $P_k = \widehat{P}_k(P_1)$, where the polynomials $\widehat{P}_k(t)$ are given by

$$(19.23) \qquad \widehat{P}_0(t) = 1\,,\ \widehat{P}_1(t) = t\,,\ t\,\widehat{P}_k(t) = \tfrac{1}{q+1}\widehat{P}_{k-1}(t) + \tfrac{q}{q+1}\widehat{P}_{k+1}(t)\,.$$

$\widehat{P}_k(t)$ is even for even k and odd for odd k. We now write $\rho = \rho(P_1) = \frac{2\sqrt{q}}{q+1}$ and consider these polynomials for $t \in [-\rho\,,\,\rho]$. The linear recurrence is easily solved, and with the substitution $t = \rho\,\cos\theta$, where $\theta \in [0\,,\pi]$,

$$\widehat{P}_k(\rho\,\cos\theta) = \frac{1}{\sqrt{q}^{\,k}}\left(\cos k\theta + \frac{q-1}{q+1}\cos\theta\,\frac{\sin k\theta}{\sin\theta}\right) \quad \text{and}$$

$$\widehat{P}_k'(\rho\,\cos\theta) = \frac{1}{\rho\sqrt{q}^{\,k}}\left(k\frac{\sin k\theta}{\sin\theta} + \frac{q-1}{q+1}\frac{\sin k\theta - k\,\sin\theta\,\cos\theta\,\cos k\theta}{\sin^3\theta}\right).$$

In particular,

$$\widehat{P}_k(\rho) = (-1)^k\widehat{P}_k(-\rho) = \frac{1}{\sqrt{q}^{\,k}}\left(1 + \frac{q-1}{q+1}k\right) \quad \text{and}$$

$$\widehat{P}_k'(\rho) = (-1)^{k-1}\widehat{P}_k'(-\rho) = \frac{1}{3\rho\sqrt{q}^{\,k}}\left(3k^2 + \frac{q-1}{q+1}(2k+k^3)\right),$$

and one easily proves by induction that

$$(19.24) \qquad |\widehat{P}_k(t)| < \widehat{P}_k(\rho) \quad \text{and} \quad |\widehat{P}_k'(t)| < \widehat{P}_k'(\rho) \quad \text{for} \quad |t| < \rho\,.$$

Viewed as a function of $x \in \mathbb{T}$, one calls $\phi_t(x) = \widehat{P}_{|x|}(t)$ a *spherical function.* For a radial transition operator (19.21) we now define its *spherical transform*

$$(19.25) \qquad \widehat{P}(t) = \sum_{k=0}^{\infty} a_k\,\widehat{P}_k(t)\,, \quad t \in [-\rho\,,\,\rho]\,.$$

Then (19.22)–(19.24) imply the following.

(19.26) Lemma. (a) *The series defining $\widehat{P}(t)$ and its termwise derivative converge absolutely and uniformly in $[-\rho\,,\,\rho]$, and summation and differentiation may be exchanged.*

(b) *If P and Q are two radial transition operators then*

$$\widehat{PQ}(t) = \widehat{P}(t)\widehat{Q}(t)\,.$$

(For (b), observe that this holds for $Q = P_1$ and hence for $Q = P_k$, applying (a).)

Now the $\widehat{P}_k$ satisfy a second order linear recurrence relation with constant coefficients and must be orthogonal polynomials with respect to some Borel measure; see the books by Pólya [270], Chihara [73] and Wall [335].

In the present situation, it is easy to find this measure. Consider the Green function associated with the simple random walk P_1, and in particular $G(z) = G(x,x|z)$. Let $\mathfrak{m}$ be the Plancherel measure on the spectrum of P_1, that is, the diagonal element of the spectral resolution. (We have used it previously in (17.15) and (17.17).) Then, setting $S(z) = G(1/z)/z$, we know that $S(z) = \int \frac{1}{z-t}\,d\mathfrak{m}(t)$. We now compute m via the inversion formula

$$\frac{1}{2}\Big(\mathfrak{m}\big((-\infty\,,\,t)\big) + \mathfrak{m}\big((-\infty\,,\,t]\big)\Big) = \frac{1}{\pi}\lim_{y\to 0+}\int_{-\infty}^{t} \operatorname{Im}\big(S(x-iy)\big)\,dx\,.$$

By Lemma 1.24, $S(z) = \Big((q-1)z - \operatorname{sign}\operatorname{Re}(z)\sqrt{(q+1)^2z^2-4q}\Big)/(2-2z^2)$ for $z \in \mathbb{C}\setminus[-\rho\,,\,\rho]$. After a short computation,

$$dm(t) = \frac{q+1}{2\pi}\frac{\sqrt{\rho^2-t^2}}{1-t^2}\,dt\,,\quad t\in[-\rho\,,\,\rho]\,. \tag{19.27}$$

If we expand the integrand in $\int \frac{1}{z-t}\,dm(t)$ in a power series in z and compare with the series defining $G(1/z)/z$, we find $p_1^{(n)}(x,x) = \int t^n\,dm(t)$. Lemma 19.26(b) and the fact that $P_k = \widehat{P}_k(P_1)$ now indeed imply

$$\int_{-\rho}^{\rho} \widehat{P}_k(t)\widehat{P}_\ell(t)\,dm(t) = \sum_{y\in\mathbb{T}} p_k(x,y)p_\ell(y,x) = \delta_{k,\ell}/|S_k|\,. \tag{19.28}$$

Now let P be a radial transition matrix as in (19.21), and write $P^n = \sum_k a_k(n)\,P_k$. Then $p^{(n)}(x,y) = a_k(n)/|S_k|$, where $k = d(x,y)$. Together with (19.28) and Lemma 19.26, this yields the following.

(19.29) Proposition. *If $d(x,y) = k$ then*

$$p^{(n)}(x,y) = \int_{-\rho}^{\rho} \big(\widehat{P}(t)\big)^n\,\widehat{P}_k(t)\,dm(t)\,.$$

We can now prove the local limit theorem for radial random walks on $\mathbb{T}$.

(19.30) Theorem.

$$p^{(n)}(x,y) \sim A\left(1+\frac{q-1}{q+1}d(x,y)\right) q^{-d(x,y)/2}\,\rho(P)^n\,n^{-3/2}\,,$$

where $A = \frac{2q(q+1)}{(q-1)^2\sqrt{2\pi}}\left(\frac{\rho\widehat{P}'(\rho)}{\widehat{P}(\rho)}\right)^{-3/2}$, *and the spectral radius* $\rho(P) = \widehat{P}(\rho)$ *is given by (12.17).*

Proof. Let $k = d(x,y)$. We use once more the method of Laplace and start by substituting $t = \rho\cos\theta$ $(0 \le \theta \le \pi)$ in the integral in (19.29). We expand $\log\widehat{P}(\rho\cos\theta)$ near 0:

$$\log\widehat{P}(\rho\cos\theta) = \log\widehat{P}(\rho) - \tfrac{B}{2}\,\theta^2 + R(\theta)\,,\quad \text{where}\quad B = \tfrac{\rho\widehat{P}'(\rho)}{\widehat{P}(\rho)} > 0\,,$$

and $R(\theta) = \mathfrak{o}(\theta^2)$. Next, we choose $\varepsilon > 0$ such that $|R(\theta)| \le B\,\theta^2/4$ for $0 \le \theta \le \varepsilon$, and subdivide the integral into $\int_0^\varepsilon$ and $\int_\varepsilon^\pi$. By assumption, $a_\ell > 0$ for both some odd and some even ℓ, and (19.24) yields $|\widehat{P}(\rho\cos\theta)|/\widehat{P}(\rho) \le \lambda < 1$ for $\theta \in [\varepsilon\,,\,\pi]$. Then the second integral,

$$\frac{2q}{(q+1)\pi}\rho(P)^n \int_\varepsilon^\pi \left(\frac{\widehat{P}(\rho\cos\theta)}{\widehat{P}(\rho)}\right)^n \widehat{P}_k(\rho\cos\theta)\,\frac{\sin^2\theta}{1-\rho^2\cos^2\theta}\,d\theta\,,$$

is bounded in absolute value by $C_k\,\rho(P)^n\,\lambda^n$ (with $C_k > 0$) and asymptotically negligible with respect to the first one. With the substitution $u = \sqrt{Bn}\,\theta$ the latter transforms into

$$\frac{2q}{(q+1)\pi}\,\widehat{P}_k(\rho)\,\rho(P)^n\,(Bn)^{-3/2}$$

$$\times \int_0^{\sqrt{Bn}\varepsilon} \frac{\widehat{P}_k\Big(\rho\cos\big(\frac{u}{\sqrt{Bn}}\big)\Big)}{\widehat{P}_k(\rho)}\;\frac{\sin^2\big(\frac{u}{\sqrt{Bn}}\big)}{\big(\frac{u}{\sqrt{Bn}}\big)^2}\;\frac{\exp\Big(-\frac{u^2}{2}+nR\big(\frac{u}{\sqrt{Bn}}\big)\Big)u^2}{1-\rho^2\cos^2\big(\frac{u}{\sqrt{Bn}}\big)}\,du\,.$$

Due to our choice of ε, the last integrand is bounded by $\frac{t^2}{1-\rho^2}\,e^{-u^2/4}$, and as $n \to \infty$, the integral converges to $\frac{1}{1-\rho^2}\sqrt{\frac{\pi}{2}}$. □

Note that the above contains a different proof of the formula (12.17) for $\rho(P)$. We also remark that it is straightforward to adapt the proof to the situation when P is not aperiodic, that is, $a_k > 0$ for some odd but no even k. The period is 2 in this case.

Notes and remarks

13. The local central limit theorem on the grid

This material is taken from Ney and Spitzer [246]. See also Rvačeva [283] and Stone [310], in particular for the case when μ is not in the domain of attraction of the normal distribution. In my survey [348], I wrongly stated Theorem 13.12 under the assumption of finite second moment, in which case it is known (in general) only when the drift is 0. Extensions to generalized lattices are due to Guivarc'h [157] and Krámli and Szász [208]. I did not present them here for lack of space.

14. Growth, isoperimetric inequalities, and the asymptotic type of random walk

§A. My principal source here was the very nice to read survey by Pittet and Saloff-Coste [266]. The use of Nash inequalities was introduced to continuous-time Markov chains by Varopoulos [324], and to discrete-time chains by Carlen, Kusuoka and Stroock [51]. See also Coulhon and Saloff-Coste [84], [85]. The present version is based on Coulhon [79]. One should also mention the link with Faber–Krahn-type inequalities, which are not "visible" here but stand behind the arguments. See Coulhon and Grigoryan [82] and the references given there.

§B. The Gaussian upper bound is due to Hebisch and Saloff-Coste [170].

§C. Theorem 14.19 is adapted from Benjamini, Chavel and Feldman [34], who prove the analogue for Brownian motion on manifolds. Lemma 14.21 is due to Carne [52]. For Theorem 14.22 (where again my source was [266]), see Lust-Piquard [217] and Coulhon and Grigoryan [81].

15. The asymptotic type of random walks on amenable groups

§A. This material is adapted from Pittet and Saloff-Coste [268].

§B. The lower bound in Theorem 15.8(b) is due to Alexopoulos [2]. In this reference one can also find a proof of Exercise 15.13.

§C. This is just an exercise following the method of Theorem 15.8(b).

§D. I learned the fancy name "lamplighter groups" for these particular wreath products from Lyons, Pemantle and Peres [221]. Random walks on these groups were first studied by Kaimanovich [182] and Kaimanovich and Vershik [190], who also gave first estimates of the transition probabilities. Theorem 15.15 is taken from Saloff-Coste and Pittet [267]. Theorem 15.16 is a true challenge!

16. Simple random walks on the Sierpiński graphs

§A. The random walk on the Sierpiński graphs is one of the main ingredients in the construction of Brownian motion on the Sierpiński fractal ("gasket") by Barlow and Perkins [24]. The functional equation (16.5) is well known to physicists via what they call "decimation procedure"; see Rammal [274], [275].

§B. These results are due to Grabner and Woess [144]. The method goes back to Odlyzko [248].

In theoretical physics, the asymptotic behaviour of the return probabilities to the origin has been determined via heuristics (plus variants) in the following spirit: Near $z = 1$, there will be an expansion $G(z) \sim (1 - z)^{\eta} H(z)$, with $H(z)$ analytic (which is wrong, as we know). Substituting in (16.5), one finds an equation for η, whence

$p^{(n)}(\mathbf{0},\mathbf{0}) \sim C\,n^{1-\eta}$. See Rammal [274], [275] and Friedberg and Martin [122]. Their non-rigorous arguments fail in detecting the oscillations.

The asymptotic behaviour of Theorem 16.11 is valid for $p^{(n)}(\mathbf{x},\mathbf{y})$ with arbitrary $\mathbf{x},\mathbf{y}$. This can be proved via a *ratio limit theorem* – compare with [144]:

Theorem. (a) *If P is strongly aperiodic then* $\lim_n \big(p^{(n+1)}(x,y)/p^{(n)}(x,y)\big) = \rho(P)$.

(b) *If in addition P is recurrent, then* $\lim_n \big(p^{(n)}(x,y)/p^{(n)}(y,y)\big) = 1$.

Part (a) follows Gerl [132], where this is proved for random walks on groups. This is only the "tip of an iceberg" concerning ratio limit theorems. See my survey [348] for more references.

Regarding off-diagonal estimates, there is a result in the spirit of the Gaussian estimates of Section 14 for the Sierpiński graph in dimension 2, due to Jones [181]:

Theorem. *Set $\delta_s = 2\log 3/\log 5$ and $\delta_w = \log 5/\log 2$. Then there are constants $n_0, c_0, \dots, c_5 > 0$ such that for all $\mathbf{x},\mathbf{y} \in \mathbf{S}_2$,*

$$c_1\,n^{-\delta_s/2}\exp\left(c_2\frac{|\mathbf{x}-\mathbf{y}|^{\delta_w/(\delta_w-1)}}{n^{1/(\delta_w-1)}}\right) \le p^{(n)}(\mathbf{x},\mathbf{y}) \le c_3\,n^{-\delta_s/2}\exp\left(c_4\frac{|\mathbf{x}-\mathbf{y}|^{\delta_w/(\delta_w-1)}}{n^{1/(\delta_w-1)}}\right)$$

for all $n \ge n_0$, and in the lower bound one also needs $n \ge c_0|\mathbf{x}-\mathbf{y}|$.

Here, $|\mathbf{x}-\mathbf{y}|$ is Euclidean distance. The proof makes extensive use of stopping time arguments.

A general rigorous study of random walks on fractal-type graphs is only at the beginning. Recently Barlow and Bass [23], have proved a result similar to that of Jones (but, of course, with different exponents) for random walks on the "graphical Sierpiński carpet". In this case one does not have a nice functional equation such as (16.5). The results are based on methods comprising those presented in Section 14, first applied by the authors to Brownian motion on the Sierpiński carpet itself [22].

Associated with a fractal, there are three characteristic constants, which in the setting of graphs with polynomial growth become the *fractal dimension* δ_f (the growth exponent), the *spectral dimension* δ_s, and the *walk dimension* δ_w. The latter two play the same role as in Jones' theorem above. The expected time until the first exit from the n-ball around the starting point is of order n^{δ_w}; this defines the walk dimension. Obviously, the definitions themselves require certain regularity properties. In typical cases, one has the "Einstein relation" $\delta_s = 2\delta_f/\delta_w$; compare with Telcs [313], [314].

17. Local limit theorems on free products

§**A.** Theorem 17.3 and Corollary 17.8 are from Woess [341]. Lemma 17.9 and Corollary 17.10 are due to Cartwright [57].

§**B.** Theorem 17.13 is also due to Cartwright [56], but my proof here is quite different; [56] elaborates the singular expansion of $G(z)$ only until the first non-polynomial term, which makes the use of Darboux's method slightly problematic. Let me explain in more detail how Darboux's method works in our setting: one looks for the first singular term in the expansion of $G(z)$ near $z = \mathbf{r}$. Typically, one knows an explicit asymptotic equivalent a_n of its Taylor expansion around 0, and wants to conclude that $p^{(n)}(o,o) \sim a_n$. To this end, one wishes to apply the Riemann–Lebesgue lemma to the remainder term, which should yield a decay that is $\mathfrak{o}(a_n)$. If this is not the case, one has to continue the expansion beyond the first singular term until one reaches (hopefully after finitely many steps) a remainder where the Riemann–Lebesgue lemma yields $\mathfrak{o}(a_n)$ Taylor coefficients.

Alternatively to Darboux's method, one might show that the expansion holds in a larger set, so that one can apply singularity analysis (Theorem 16.8) and does not have to determine the additional terms.

18. Intermezzo: Cartesian products

Proposition 18.1 was proved by Cartwright and Soardi [64] in the context of random walks on Cartesian products of groups, also taking into account the periods of P_1 and P_2. Proposition 18.4 was proved by Gerl [138], and later independently by Cassi and Regina [69].

19. Free groups and homogeneous trees

§**A.** This is due to Lalley [211], who proved Theorem 19.4 more generally for random walks on the free group whose law is supported by the identity, the free generators and their inverses. The local limit theorem in this case had been proved by Gerl and Woess [131], [337], [140]. Lalley [211] also finds the uniform asymptotics in the range $(1-c)n \le |x| \le n$. On the other hand, regarding the range of Case 2, in particular 2.A, I could not follow the indications of [211], so that I chose another integration contour.

§**B.** The very significant result of Theorem 19.9 is also due to Lalley [212]. It has a "story": in the mid-80s, Tim Steger had proved the same theorem, at least for symmetric random walks, and has given seminars on it in a few places, including Leoben and Sydney, but has never written it up. Here, I follow Lalley's method. However, there is a mistake in Prop. 2.7 of [212] (which has become Proposition 19.12 here), because the procedure which [212] describes there is not the iteration of the system of equations obtained previously (the one of Proposition 19.11(f)). Of course, this does not compromise Lalley's excellent proof strategy.

I have not gone much into detail regarding the proof of algebraicity of the Green function (after Proposition 19.11). It was proved by Aomoto [8] and Steger [309], [116]. Behind this, there is a certain amount of algebraic geometry. One can also use formal power series and the theory of context-free languages (Woess [343]).

§**C.** Theorem 19.30 is due to Sawyer [290]; see also Picardello [256]. The setting of spherical harmonic analysis on $\mathbb{T}$ was first developed by Cartier [54], [55], and later on, in part independently, by other authors; see the book by Figà-Talamanca and Picardello [115] and the references therein. In particular, the formula (19.27) for the Plancherel measure has been rediscovered many times (with different parametrizations). In the classical context of orthogonal polynomials, it goes back to Geronimus [141]; see also Chihara's book [73].

Finally, note that Corollary 17.8(b) yields a local limit theorem for a class of random walks on free groups that are non-radial and are not required to have finite range.

Further results

In Section 14, I have not presented the main result of Varopoulos' seminal paper [323], which stands at the origin of the material in §14.A:

Theorem. *Let (X,P) be reversible and $d>2$. Then the following are equivalent.*

(a) *There is $\kappa>0$ such that* $\|f\|_{2d/(d-2)} \le \kappa\, D_P(f)$ *for all* $f \in \ell_0(X)$ (Wirtinger inequality)

(b)
$$\sup_{x,y} \frac{p^{(n)}(x,y)}{m(y)} \le C\, n^{-d/2}\,.$$

We have arrived at the estimate (b) in another way. Varopoulos' theorem is well presented in his book with Coulhon and Saloff-Coste [326].

Also, in Section 14 I have not touched the interesting question of off-diagonal lower estimates. A result has been mentioned in the notes to Section 16. The – recent – main general theorem of this type is due to Delmotte [90], [91]; see also the survey by Coulhon [80].

In the case of a *radial* structure, that is, when the random walk admits a factor chain that is nearest neighbour on $\mathbb{N}$, one can use the theory of analytic continued fractions to obtain information on transition probabilities. See Gerl [134] and Woess [339]. The method goes back to Karlin and McGregor [195], although continued fractions are not mentioned there explicitly.

The results of §19.C extend immediately to isotropic random walks on *distance regular* graphs. A graph is said to have this property, when the cardinality $|B(x,k) \cap B(y,\ell)|$ depends only on k, ℓ and $d(x,y)$. The infinite distance regular graphs have been classified by Ivanov [180]; they are precisely the free products $\mathbb{D}_{m,M} = \mathbb{K}_m * \cdots * \mathbb{K}_m$ (M times), where $m, M \geq 2$. When $m = 2$, this is $\mathbb{T}_M$. The corresponding Plancherel measure has again been computed by various authors, including Kuhn and Soardi [209] and Faraut and Picardello [114]. The simplest way is to compute $G(z)$ for the simple random walk (compare with Gerl [133]) and apply the inversion formula.

CHAPTER IV

AN INTRODUCTION TO TOPOLOGICAL BOUNDARY THEORY

This final chapter is devoted to studying the interplay between the spatial behaviour of random walks, harmonic functions, and compactifications of the state space. The word "topological" in the title is thought of as opposed to "measure theoretic". This means that the rich and deep theory of the *Poisson boundary,* related entropy methods and the *weak Liouville property* (constantness of bounded harmonic functions) will be touched only marginally.

Let X be a separable metric space (typically a graph with its integer-valued distance). By a *compactification* $\widehat{X}$ of X we shall always mean a compact Hausdorff space with countable base of the topology, containing X, such that the inclusion $X \hookrightarrow \widehat{X}$ is a homeomorphism, and such that X is open and dense in $\widehat{X}$. Therefore $\widehat{X}$ is metrizable, and the *boundary* $\vartheta X = \widehat{X} \setminus X$ is compact. Our way of thinking is that the compactification has a "natural" description in terms of the geometry of X, prior to considering random walks. We then consider a *transient* random walk Z_n on X with transition matrix P and look for suitable adaptedness conditions which allow us to answer the following questions.

(A) *Convergence to the boundary.* Does there exist a ϑX-valued random variable Z_∞ such that in the $\widehat{X}$-topology, $Z_n \to Z_\infty$ almost surely ($\mathbb{P}_x$) for every x?

(B) *Dirichlet problem at infinity.* Does every continuous function $\vartheta X \to \mathbb{R}$ admit a continuous extension to $\widehat{X}$ which is harmonic on X?

(C) *Identification of the Martin boundary.* Does every positive harmonic function have an integral representation over ϑX?

Question (C) will be formulated more precisely in Section 24. Of course, harmonicity refers to the transition operator P. We start with some general observations regarding (A) and (B).

20. A probabilistic approach to the Dirichlet problem, and a class of compactifications

A. The Dirichlet problem and convergence to the boundary

Let P be the (irreducible) transition matrix of a transient random walk Z_n on the graph X. Suppose that we have a compactification $\widehat{X}$ with boundary ϑX. Set

$$\Omega_\infty = \{\omega \in \Omega : Z_\infty(\omega) = \lim_{n\to\infty} Z_n(\omega) \in \vartheta X \text{ exists in the topology of } \widehat{X}\}.$$

Convergence to the boundary means that $\mathbb{P}_x(\Omega_\infty) = 1$ for every $x \in X$. In this case, Z_∞ is measurable with respect to the Borel σ-algebra of ϑX. Write ν_x for the $\mathbb{P}_x$-distribution of Z_∞, that is, $\nu_x(B) = \mathbb{P}_x[Z_\infty \in B]$, where B is a Borel set in ϑX. (We may also consider ν_x as a Borel measure on $\widehat{X}$ that charges only the boundary.) Factoring with respect to the first step, the Markov property yields

$$\nu_x = \sum_{y\in X} p(x,y)\,\nu_y\,. \tag{20.1}$$

The Borel probability measures family ν_x, $x \in X$, are called the *harmonic measures.* From (20.1) we get $\nu_x \ge p^{(n)}(x,y)\,\nu_y$, and irreducibility yields that all ν_x are mutually absolutely continuous. A function on ϑX which is ν_x-integrable for some x has this property for all $x \in X$. The harmonic measure class can be used for producing harmonic functions: if φ is a ν_x-integrable function on ϑX, then

$$h_\varphi(x) = \int_{\vartheta X} \varphi\, d\nu_x \tag{20.2}$$

defines a harmonic function with respect to P, and (20.2) is an analogue of the classical Poisson integral for harmonic functions in the unit disk. If φ is continuous then h_φ is the right candidate to solve the Dirichlet problem with boundary function φ. What we want is that $\lim_{x\to\xi} h_\varphi(x) = \varphi(\xi)$ for every $\varphi \in C(\vartheta X)$ and every $\xi \in \vartheta X$.

Note that when it exists, then the solution is necessarily unique. Indeed, by continuity and the minimum principle (1.15), any solution h must satisfy $\min_{\vartheta X} \varphi \le h \le \max_{\vartheta X} \varphi$. Now, if h' and h'' are two solutions, then $h' - h''$ solves the Dirichlet problem with all boundary values $= 0$, whence $h' - h'' \equiv 0$.

(20.3) Theorem. *The Dirichlet problem with respect to P and $\widehat{X}$ is solvable if and only if*

(i) *(Z_n) converges to the boundary, and*
(ii) *for the corresponding harmonic measures,*

$$\lim_{x\to\xi} \nu_x = \delta_\xi \quad \textit{weakly for every } \xi \in \vartheta X\,.$$

Proof. The above discussion shows that (i) and (ii) are sufficient conditions. To prove necessity, assume that the Dirichlet problem is solvable.

Step 1. For $\varphi \in C(\vartheta X)$, let h_φ denote the (unique) harmonic function on X which gives the continuous extension of φ to $\widehat{X}$. For $x \in X$, consider the evaluation operator

$$V_x : C(\vartheta X) \to \mathbb{R}, \quad V_x(\varphi) = h_\varphi(x),$$

where $C(\vartheta X)$ is equipped with the max-norm. By uniqueness of the extension, V_x is linear, $V_x(\mathbf{1}) = 1$, and by the minimum (maximum) principle, $\|V_x\| = 1$. Also, if $\varphi \geq 0$ then $V_x(\varphi) \geq 0$. Considering the dual of the Banach space $C(\vartheta X)$, we see that there must be a Borel probability measure ν_x on ϑX such that

$$h_\varphi(x) = V_x(\varphi) = \int_{\vartheta X} \varphi \, d\nu_x$$

for every $\varphi \in C(\vartheta X)$. Solvability of the Dirichlet problem now implies (ii) for the measures ν_x, $x \in X$.

Step 2. To prove (i), we first observe that harmonicity of h_φ for all φ implies that (20.1) holds. By our assumptions on the compactification $\widehat{X}$, there is a countable family $\{\varphi_k : k \in \mathbb{N}\}$ which is dense in $C(\vartheta X)$. For each φ_k, the sequence of random variables

$$h_{\varphi_k}(Z_n) = \int_{\vartheta X} \varphi_k \, d\nu_{Z_n}, \quad n = 0, 1, 2, \ldots,$$

is a bounded martingale. By the martingale limit theorem, it converges almost surely. Hence, using transience, $\mathbb{P}_x(\Omega') = 1$ for every x, where

$$\Omega' = \left\{ \omega \in \Omega : Z_n(\omega) \to \infty, \int_{\vartheta X} \varphi_k \, d\nu_{Z_n(\omega)} \text{ converges for every } k \right\}.$$

(For a sequence (x_n) in X, we write $x_n \to \infty$ if it visits any finite subset of X only finitely many times.) If $\omega \in \Omega'$, then by Helly's principle there is a Borel probability σ_ω on ϑX such that

$$\lim_{n\to\infty} \nu_{Z_n(\omega)} = \sigma_\omega \quad \text{weakly.}$$

By compactness, there must be a sub-sequence (n') and a point $\xi \in \vartheta X$ such that $Z_{n'}(\omega) \to \xi$ in the topology of $\widehat{X}$. By (ii),

$$\lim_{n'\to\infty} \nu_{Z_{n'}(\omega)} = \delta_\xi \quad \text{weakly.}$$

Consequently, $\sigma_\omega = \delta_\xi$. Repeating the argument, we see that every accumulation point η of $\bigl(Z_n(\omega)\bigr)$ must satisfy $\delta_\eta = \sigma_\omega$, and $\eta = \xi$. Therefore $Z_n(\omega) \to \xi \in \vartheta X$, with ξ depending on ω, and we have proved convergence to the boundary.

Step 3. We still have to show that the limit distributions, that is, the $\mathbb{P}_x$-distributions of Z_∞, coincide with the measures ν_x, $x \in X$. What we have to verify is that

$$\mathbb{E}_x\big(\varphi(Z_\infty)\big) = \int_{\vartheta X} \varphi \, d\nu_x \quad \text{for every } \varphi \in C(\vartheta X) \text{ and every } x \in X\,.$$

To see this, let $\varphi \in C(\vartheta X)$. Then $\varphi(Z_\infty) = \lim_n h_\varphi(Z_n)$ almost surely. By dominated convergence, we may exchange expectation and limit. But by harmonicity of h_φ,

$$\mathbb{E}_x\big(h_\varphi(Z_n)\big) = \sum_y p^{(n)}(x,y)h_\varphi(y) = h_\varphi(x) = \int_{\vartheta X} \varphi \, d\nu_x$$

for every n. This concludes the proof. □

The interest of the last theorem lies mainly in the fact that solvability of the Dirichlet problem implies that the random walk converges to the boundary. In the next subsection we shall study a situation where this fact will be useful for proving the latter.

If we have convergence to the boundary, then we can also consider a local version of condition (ii): a point $\xi \in \vartheta X$ is called *regular* if $\lim_{x \to \xi} \nu_x = \delta_\xi$ weakly.

(20.4) Lemma. *Let $\xi \in \vartheta X$ be regular. Then $\xi \in \operatorname{supp} \nu_x$ for every $x \in X$, and if $\operatorname{supp} \nu_x$ has more than one element, then the Green kernel vanishes at ξ, that is,*

$$\lim_{y \to \xi} G(y,x) = 0 \quad \textit{for some } (\Longleftrightarrow \textit{every})\; x\,.$$

Poof. If U is a neighbourhood of ξ in ϑX, then $\lim_{x \to \xi} \nu_x(U) = 1$. Hence $\nu_x(U) > 0$ for some x. As all the ν_x are mutually absolutely continuous, this must hold for all x.

If $|\operatorname{supp} \nu_x| \ge 2$ then we may choose a neighbourhood U of ξ such that $\nu_x(\vartheta X \setminus U) > 0$. Starting at y and stopping the random walk at its first visit to x, we obtain

$$\nu_y(\vartheta X \setminus U) \ge F(y,x)\, \nu_x(\vartheta X \setminus U)\,.$$

If $y \to \xi$ then $\nu_y(\vartheta X \setminus U) \to 0$, so that $F(y,x) \to 0$ and also $G(y,x) = F(y,x)G(x,x) \to 0$. □

In particular, when $|\vartheta X| > 1$, we see that for solvability of the Dirichlet problem, it is necessary that the Green kernel vanishes at infinity (given x, for every $\varepsilon > 0$ one has $G(y,x) < \varepsilon$ for all but finitely many y).

B. Compactifications with "hyperbolic" properties

In this subsection we introduce a class of compactifications which – for the case of transitive random walks – admit a straightforward solution of the Dirichlet problem. We shall then apply this to two classes of graphs and groups in the next two sections.

Let (X, d) be a *proper* metric space, that is, closed balls in X are compact. (In principle, we are only interested in locally finite graphs, but no additional difficulty will arise from this slightly more general level.) Let $\widehat{X}$ be a compactification of X. We say that $\widehat{X}$ is *projective,* if the following holds for sequences (x_n), (y_n) in X:

$$(20.5)\qquad x_n \to \xi \in \vartheta X \quad \text{and} \quad \sup_n d(x_n, y_n) < \infty \quad \text{imply} \quad y_n \to \xi\,.$$

Next, let Γ be a subgroup of $\mathrm{AUT}(X)$, the group of self-isometries of (X, d). Our interest is in compactifications that are compatible with the action of Γ. We call $\widehat{X}$ a *contractive* Γ*-compactification,* if it is projective, every $\gamma \in \Gamma$ extends to a self-homeomorphism of $\widehat{X}$, and the following *convergence property* holds. For any sequence (γ_n) in Γ,

(20.6)

$$\gamma_n x \to \xi \in \vartheta X \quad \text{and} \quad \gamma_n^{-1} x \to \eta \in \vartheta X \quad \text{imply} \quad \gamma_n w \to \xi$$

$$\text{uniformly for } w \in \widehat{X} \text{ outside of every neighbourhood of } \eta.$$

(Here, $x \in X$ is arbitrary by projectivity.) This means that for every choice of neighbourhoods U of ξ and V of η in $\widehat{X}$, there is an index $n(U, V)$ such that $\gamma_n(\widehat{X} \setminus V) \subset U$ for all $n \geq n(U, V)$. The case $\xi = \eta$ is not excluded.

The most typical example of this type of compactification is provided by $X = \mathbb{H}$, the hyperbolic plane (Poincaré disk) with the group of Möbius transformations that leave the unit disk invariant. If we take $\widehat{\mathbb{H}}$ to be the Euclidean closure of the unit disk, then (20.5) and (20.6) hold. More about this will be said in Section 22. We now dedicate a few pages to some purely structure theoretic considerations. The first step is a classification of the elements of Γ, analogous to Möbius transformations of the disk. An element of $\mathrm{AUT}(X)$ is called *elliptic,* if it fixes a compact, non-empty subset of X.

(20.7) Proposition. *Let $\widehat{X}$ be a contractive Γ-compactification of X. If $\gamma \in \Gamma$ is non-elliptic, then one has one of the following:*

(1) γ *is* hyperbolic *– it fixes precisely two elements $\xi, \eta \in \vartheta X$,*

$$\lim_{n\to\infty} \gamma^n x = \xi \quad \textit{and} \quad \lim_{n\to\infty} \gamma^{-n} x = \eta \quad \textit{for every } x \in X\,.$$

(2) γ *is* parabolic *– it fixes precisely one element $\xi \in \vartheta X$,*

$$\lim_{n\to\infty} \gamma^n x = \xi \quad \textit{and} \quad \lim_{n\to\infty} \gamma^{-n} x = \xi \quad \textit{for every } x \in X\,.$$

Proof. Equip AUT(X) with the topology of uniform convergence on compact sets. This is a locally compact group. By the classification of monothetic groups (see e.g. Rudin [281], Th. 2.4.1), the subgroup $\langle\gamma\rangle^-$ is either compact, or discrete and equal to $\langle\gamma\rangle$. As γ is non-elliptic, we are in the second case, and (X being proper) for arbitrary $x \in X$ the sequence $(\gamma^n x)_{n\in\mathbb{Z}}$ has no accumulation points in X. Therefore, using compactness of $\widehat{X}$, we can find a sub-sequence (n') and elements $\xi, \eta \in \vartheta X$ such that $\gamma^{n'}x \to \xi$ and $\gamma^{-n'}x \to \eta$.

We have $d(\gamma^{n'+1}x, \gamma^{n'}x) = d(\gamma x, x)$, and by projectivity, $\gamma\xi = \lim_{n'} \gamma^{n'}\gamma x = \xi$. Analogously, $\gamma\eta = \eta$. By the convergence property, $\gamma^{n'}w \to \xi$ and $\gamma^{-n'}w \to \eta$ for every $w \in \widehat{X}$ different from η and ξ, respectively. In particular, no point besides ξ and η is fixed by γ.

First suppose that $\xi \neq \eta$. Let U, V be disjoint, open neighbourhoods in $\widehat{X}$ of ξ, η, respectively, not containing x. By the convergence property, there is n_0 in (n') such that

$$U \supset \gamma^{n_0}(\widehat{X} \setminus V) \supset \gamma^{n_0}(U^-),$$

and inductively $\gamma^{kn_0}x \in U$ for all $k > 0$. Thus, every accumulation point of the sequence $(\gamma^{kn_0}x)$ must lie in U, and hence – being fixed by γ – must be ξ. In other words, $\gamma^{kn_0}x \to \xi$. By projectivity, also $\gamma^{kn_0+r}x = \gamma^{kn_0}(\gamma^r x) \to \xi$ as $k \to \infty$ with r fixed. Hence

$$\lim_{n\to\infty} \gamma^n x = \xi\,, \quad \text{and analogously} \quad \lim_{n\to\infty} \gamma^{-n} x = \eta\,.$$

Once more by projectivity, this is true for every $x \in X$, and (1) holds.

Now suppose that $\xi = \eta$. By the above, this must be the only accumulation point of $\{\gamma^n x : n \in \mathbb{Z}\}$ in $\widehat{X}$, and (2) holds. □

Note that by the arguments used in the proof, we also get the following: if $U, V \subset \widehat{X}$ are non-empty, open and disjoint and there is $\gamma \in \Gamma$ with

$$\text{(20.8)} \qquad \gamma(\widehat{X} \setminus V) \subset U \quad \text{and} \quad \gamma^{-1}(\widehat{X} \setminus U) \subset V \quad \text{strictly,}$$

then γ is hyperbolic with "forward" fixed point in U and "backward" one in V.

Next, we define the *limit set* $L(\Gamma)$ of Γ. This is the set of accumulation points of an orbit Γx in ϑX. By projectivity, it is independent of $x \in X$. We write $H(\Gamma)$ for the set of points in ϑX which are fixed by some hyperbolic $\gamma \in \Gamma$. We have $H(\Gamma) \subset L(\Gamma)$, and $L(\Gamma)$ is compact and fixed (as a set) by Γ.

(20.9) Proposition. *Let $\widehat{X}$ be a contractive Γ-compactification of X.*

(1) *If $L(\Gamma)$ is non-empty, then it has one, two, or infinitely many elements.*

(2) *If $L(\Gamma)$ has more than one element, then $H(\Gamma)$ is dense in $L(\Gamma)$.*

Proof. We start with (2). Let $\xi, \xi' \in L(\Gamma)$ be distinct, and choose $x \in X$. There are sequences (β_n) and (γ_n) in Γ such that $\beta_n x \to \xi$ and $\gamma_n x \to \xi'$. Without loss of generality, we assume that $\beta_n^{-1} x \to \eta$ and $\gamma_n^{-1} x \to \eta' \in \vartheta X$.

Case 1. $\eta \neq \xi$. Let U, V be disjoint, open neighbourhoods in $\widehat{X}$ of ξ, η, respectively, not containing x. By contractivity, there is n_0 such that $\beta_{n_0}(\widehat{X} \setminus V) \subset U$ and $\beta_{n_0}^{-1}(\widehat{X} \setminus U) \subset V$. By (20.8), β_{n_0} is hyperbolic with fixed points $\xi_1 \in U$ and $\eta_1 \in V$. In particular, $\xi_1 \in H(\Gamma) \cap U$.

Case 2. $\eta' \neq \xi'$. As in Case 1, we can find n_0 such that γ_{n_0} is hyperbolic with fixed points ξ_1 and η_1. One of them, say ξ_1, must be different from η. By (20.6), $\beta_n \xi_1 \to \xi$, and $\beta_n \xi_1$ is a fixed point of the hyperbolic element $\beta_n \gamma_{n_0} \beta_n^{-1}$. Again, $H(\Gamma) \cap U \neq \emptyset$.

Case 3. $\eta = \xi$, $\eta' = \xi'$. We choose open, disjoint $\widehat{X}$-neighbourhoods U of ξ and U' of ξ', not containing x. By (20.6), there is n_0 such that $\beta_{n_0}^{\pm 1}(\widehat{X} \setminus U) \subset U$ and $\gamma_{n_0}^{\pm 1}(\widehat{X} \setminus U') \subset U'$. We set $\gamma = \beta_{n_0} \gamma_{n_0}$. Then

$$\gamma(\widehat{X} \setminus U') \subset \beta_{n_0}(U') \subset U \quad \text{and} \quad \gamma^{-1}(\widehat{X} \setminus U) \subset \gamma_{n_0}^{-1}(U) \subset U'.$$

By (20.8), γ is hyperbolic, and one of its fixed points lies in U. Once more, $H(\Gamma) \cap U \neq \emptyset$.

We now prove (1). Suppose that $|L(\Gamma)| \geq 3$. By (2) there are two hyperbolic elements $\beta, \gamma \in \Gamma$ with $|\operatorname{Fix} \beta \cup \operatorname{Fix} \gamma| \geq 3$. Let $\xi \in \operatorname{Fix} \beta \setminus \operatorname{Fix} \gamma$. Then for every n, $\gamma^n \beta \gamma^{-n}$ is hyperbolic and fixes $\gamma^n \xi$. As γ is hyperbolic, all $\gamma^n \xi$ are different. □

For convergence to the boundary, and the Dirichlet problem, we want the boundary as big as possible: then we can distinguish many different directions of convergence at infinity, that is, there is more information. Therefore we shall only consider the case when $L(\Gamma)$ is infinite. It is then not hard to show that $L(\Gamma)$ is uncountable. Indeed, it is a perfect set.

(20.10) Proposition. *Let $\widehat{X}$ be a contractive Γ-compactification of X with infinite limit set $L(\Gamma)$. Then precisely one of the following holds.*

(a) *Γ contains two hyperbolic elements without common fixed points, and no proper, non-empty compact subset A of $L(\Gamma)$ is fixed by Γ.*

(b) *Γ fixes a unique element ξ_0 of $L(\Gamma)$, and no proper, non-empty compact subset A of $L(\Gamma) \setminus \{\xi_0\}$ is fixed by Γ.*

Proof. (a) Suppose Γ has two hyperbolic elements without common fixed points, and let A be as stated. By Proposition 20.9 there is a hyperbolic element $\gamma \in \Gamma$ with "forward" fixed point $\xi \in \vartheta X \setminus A$. Let η be the "backward" fixed point of γ. If $\zeta \in A \setminus \{\eta\}$ then $\gamma^n \zeta \to \xi$, and γ does not fix A. This proves (a), unless $A = \{\eta\}$. In this case there is another hyperbolic element of Γ whose fixed points are different from ξ and η.

(b) Now assume that any two hyperbolic elements of Γ have a common fixed point. By Proposition 20.9, we can find a hyperbolic element $\gamma \in \Gamma$ with "forward" and "backward" fixed points ξ_0 and η_0. Let β be any other element in Γ. Then $\beta\gamma\beta^{-1}$ is hyperbolic with fixed points $\beta\xi_0$ and $\beta\eta_0$. Therefore $\beta(\{\xi_0, \eta_0\}) \cap \{\xi_0, \eta_0\} \neq \emptyset$.

Now suppose that there are $\beta_1, \beta_2 \in \Gamma$ such that

$$\beta_1(\{\xi_0, \eta_0\}) \cap \{\xi_0, \eta_0\} = \{\xi_0\} \quad \text{and} \quad \beta_2(\{\xi_0, \eta_0\}) \cap \{\xi_0, \eta_0\} = \{\eta_0\}.$$

Then $\gamma_1 = \beta_1\gamma\beta_1^{-1}$ and $\gamma_2 = \beta_2\gamma\beta_2^{-1}$ are hyperbolic and must have a common fixed point. Hence, there is $\zeta \in \vartheta X$ such that $\operatorname{Fix}\gamma_1 = \{\xi_0, \zeta\}$ and $\operatorname{Fix}\gamma_2 = \{\eta_0, \zeta\}$. Let U and V be disjoint $\widehat{X}$-neighbourhoods of ξ_0 and η_0. By hyperbolicity of γ there is an integer n such that $\gamma^n\zeta \in U$ and $\gamma^{-n}\zeta \in V$, so that $\gamma^n\zeta \neq \gamma^{-n}\zeta$. But then $\gamma^n\gamma_1\gamma^{-n}$ and $\gamma^{-n}\gamma_2\gamma^n$ are hyperbolic elements of Γ without common fixed points, a contradiction.

Hence, without loss of generality, we may assume that $\xi_0 \in \gamma(\{\xi_0, \eta_0\})$ for every $\gamma \in \Gamma$. (Otherwise exchange ξ_0 and η_0.) We claim that Γ fixes ξ_0.

Let $\beta \in \Gamma$ with $\beta(\{\xi_0, \eta_0\}) \neq \{\xi_0, \eta_0\}$. Suppose $\beta\xi_0 \notin \{\xi_0, \eta_0\}$, so that $\beta\eta_0 = \xi_0$. Then $\beta^2\eta_0 = \beta\xi_0 \neq \xi_0$ and hence $\beta^2\xi_0 = \xi_0$. This yields $\beta\xi_0 = \beta^{-1}\xi_0 = \eta_0$, a contradiction. Thus, we must have $\beta\xi_0 = \xi_0$ and $\beta\eta_0 \neq \eta_0$. Choose some $\beta_0 \in \Gamma$ with these properties. It must exist, since by assumption, $H(\Gamma)$ is infinite.

Suppose that there is $\beta \in \Gamma$ with $\beta\xi_0 \neq \xi_0$. By the above, it must be that $\beta\xi_0 = \eta_0$ and $\beta\eta_0 = \xi_0$. But then $\beta\beta_0(\{\xi_0, \eta_0\}) = \{\eta_0, \beta\beta_0\eta_0\}$ does not contain ξ_0, again a contradiction.

Finally, we verify the statement regarding invariant subsets. Let A be as given. By Proposition 20.9, there is some hyperbolic $\gamma \in \Gamma$ with "forward" fixed point ξ_0 and the other fixed point outside of A. But then $\gamma^n A \to \xi_0$ by (20.6), and A is not invariant under Γ. □

(20.11) Exercise [Klein's ping-pong Lemma]. *In case (a) of Proposition 20.10, Γ contains a free group with two generators. It is discrete in the topology of uniform convergence on compact sets, and all its elements different from the identity are hyperbolic.*

[Hints: Take hyperbolic $\beta_1, \beta_2 \in \Gamma$ with $\operatorname{Fix}\beta_i = \{\xi_i, \eta_i\}$ (all four different). Take pairwise disjoint neighbourhoods O of the origin, U_i of the ξ_i and V_i of the η_i. Set $\gamma_i = \beta_i^k$, where k is large enough that $\gamma_i(\widehat{X} \setminus V_i) \subset U_i$ and $\gamma_i^{-1}(\widehat{X} \setminus U_i) \subset V_i$, $i = 1, 2$. Show that $\langle \gamma_1, \gamma_2 \rangle$ is a free subgroup, as asserted.

To see this, set $W_i = U_i \cup V_i$. Show that every non-trivial reduced word $\gamma = \gamma_1^{k_1}\gamma_2^{\ell_1}\cdots\gamma_1^{k_r}\gamma_2^{\ell_r}$ with integer exponents (only k_1 and ℓ_r are allowed to be $= 0$) satisfies $\gamma(\widehat{X} \setminus W_i) \subset W_j$ for suitable $i, j \in \{1, 2\}$.]

In the discrete setting, applications of this machinery are twofold: on the one hand, X may be non-discrete and Γ a finitely generated discrete subgroup of $\mathrm{AUT}(X)$ carrying a random walk. On the other hand, X may be a locally finite graph carrying a random walk P, and Γ a closed subgroup of $\mathrm{AUT}(X, P)$ which acts transitively on X. (This includes the case when X is a Cayley graph of the finitely generated group Γ.) Existence of a non-trivial contractive Γ-compactification is of course not automatic.

We shall only consider the second situation. Recall the method described in §8.A of lifting the random walk from the graph X to the group Γ. Let μ be the probability measure on Γ defined in (8.12), and R_n the corresponding right random walk on Γ (8.11). The convolution of a measure μ on Γ with a measure ν on ϑX is defined by

$$\int_{\vartheta X} f(\xi)\, \mu * \nu(d\xi) = \int_{\vartheta X} \int_{\Gamma} f(\gamma\xi)\, \mu(d\gamma)\nu(d\xi)\,.$$

Convergence to the boundary now means that $R_n o \to Z_\infty \in \vartheta X$, and $\alpha R_n o \to \alpha Z_\infty$, as $\alpha R_n o$ models Z_n with starting point αo. Set $\nu = \nu_o$. In view of Proposition 8.13, we get $\nu_{\alpha o} = \delta_\alpha * \nu$. The limit $X_1^{-1} Z_\infty$ of $X_2 \cdots X_n o$ must have the same distribution as Z_∞. Therefore we find that ν is a *stationary measure* for μ, that is, $\mu * \nu = \nu$.

Conversely, if we have a stationary measure ν, then for $\alpha \in \Gamma_o$,

$$\delta_\alpha * \nu = \delta_\alpha * \mu * \nu = \mu * \nu = \nu\,,$$

as μ is Γ_o-invariant on the left by the definition (8.12). Thus, we may define $\nu_x = \delta_{\gamma_x} * \nu$, where (recall) $\gamma_x o = x$, and $\delta_\alpha * \nu = \nu_{\alpha o}$ for arbitrary $\alpha \in \Gamma$. The resulting probability measures ν_x, $x \in X$, are candidates for being the harmonic measures.

We now suppose that our compactification is also contractive, and that ϑX is infinite. By transitivity, $L(\Gamma) = \vartheta X$. We write $L^*(\Gamma) = L(\Gamma)$ in case (a) of Proposition 20.10, and $L^*(\Gamma) = L(\Gamma) \setminus \{\xi_0\}$ in case (b).

(20.12) Lemma. *Suppose that ν is a μ-stationary probability measure on ϑX. Let A be a non-empty, compact proper subset of $L^*(\Gamma)$ such that $\gamma A = A$ or $\gamma A \cap A = \emptyset$ for any $\gamma \in \Gamma$. Then $\nu(A) = 0$.*

Proof. From Proposition 20.10 and its proof, we know that there must be a hyperbolic element $\gamma \in \Gamma$ which does not fix A, and hence $A \cap \mathrm{Fix}\,\gamma = \emptyset$. But then $\gamma^n A$ is contracted to one of the fixed points of γ. Therefore $L^*(\Gamma)$ contains an infinity of mutually disjoint Γ-translates of A.

Let $M = \max\{\nu(\gamma A) : \gamma \in \Gamma\}$. Suppose that $M > 0$. Then there are mutually disjoint sets $A_1 = \gamma_1 A, \dots, A_n = \gamma_n A$ such that

$$\{\gamma A : \nu(\gamma A) = M\} = \{A_1, \dots, A_n\}.$$

Let $B = \bigcup_{i=1}^n A_i$. We claim that $\alpha^{-1}B = B$ for all $\alpha \in \operatorname{supp}\mu$. Suppose the contrary. Then $\alpha^{-1}A_i \cap B = \emptyset$ for some i, and $\nu(\alpha^{-1}A_i) = c < M$. The above discussion shows that then $\nu(\gamma^{-1}A_i) = c$ for all γ in the open neighbourhood $\alpha\Gamma_o$ of α in Γ. Therefore

$$nM = \nu(B) = \int_\Gamma \nu(\gamma^{-1}B)\,\mu(d\gamma) \le (n-1)M + c,$$

a contradiction. Consequently each element of $\operatorname{supp}\mu$ fixes B. Now $\bigcup_{n\in\mathbb{N}}(\operatorname{supp}\mu)^n = \Gamma$. Thus, Γ leaves B invariant. But this is impossible by Proposition 20.10, as B is a compact, proper subset of $L^*(\Gamma)$. Therefore $M = 0$. □

In particular, ν must be *continuous* on $L^*(\Gamma)$, that is, $\nu(\{\xi\}) = 0$ for every $\xi \in L^*(\Gamma)$. We have now collected all the ingredients for a solution of the Dirichlet problem in the transitive case.

(20.13) Theorem. *Suppose that the closed subgroup Γ of* AUT(X, P) *acts transitively on the locally finite graph X. Let $\widehat{X}$ be a contractive Γ-compactification of X whose boundary ϑX is infinite and does not have a point fixed by Γ.*

Then the Dirichlet problem with respect to P and $\widehat{X}$ is solvable.

Proof. For μ given by (8.12), there is at least one stationary probability measure ν on ϑX. Indeed, start with any Borel probability σ on Γ, and consider the sequence of measures $\frac{1}{n}(\sigma + \mu * \sigma + \cdots + \mu^{n-1} * \sigma)$. By compactness (recall that ϑX is metrizable), there must be a weak accumulation point ν in the space of Borel probabilities on ϑX. This is stationary.

Lemma 20.12 implies that ν is continuous. Let $\varphi \in C(\vartheta X)$. We define

$$h_\varphi(x) = \int_{\vartheta X} \varphi(\gamma_x \zeta)\,\nu(d\zeta).$$

Recall the discussion before Lemma 20.12: setting $\nu_x = \delta_{\gamma_x} * \nu$, we have $\sum_y p(x,y)\nu_y = \nu_x$. Therefore h_φ is harmonic. We show that it provides the solution of the Dirichlet problem. Let $\xi \in \vartheta X$. Take a sequence (x_n) in X that converges to ξ, and let $\gamma_n = \gamma_{x_n}$. Thus, $\gamma_n o \to \xi$. By compactness, we may assume that $\gamma_n^{-1} o$ converges to some $\eta \in \vartheta X$. Now $\nu(\{\eta\}) = 0$, while the convergence property guarantees that $\gamma_n \zeta \to \xi$ for all $\zeta \in \vartheta X \setminus \{\eta\}$.

We see that $\varphi(\gamma_n\zeta) \to \varphi(\xi)$ for ν-almost every ζ. Dominated convergence yields $h_\varphi(x_n) \to \varphi(\xi)$. □

A posteriori we see that ν is the unique μ-stationary probability measure on ϑX. The proof does also work in case (b) of Proposition 20.10, if we assume we have a stationary probability measure with $\nu(\{\xi_0\}) = 0$. In practice, this requires that one first proves a.s. convergence of the random walk to $L^*(\Gamma)$.

21. Ends of graphs and the Dirichlet problem

The *end compactification* can be defined for an arbitrary graph, and provides the simplest class of examples where projectivity and the convergence property hold. We have already encountered it in the special case of trees in §6.B.

Let X be a connected, locally finite graph. An *infinite path* or *ray* is a sequence $\pi = [x_0, x_1, \dots]$ of distinct vertices, such that $x_i \sim x_{i-1}$ for all i. (Contrary to trees, this is not the same concept as a *geodesic ray,* where in addition one requires $d(x_i, x_j) = |i-j|$.) If F is a finite set of vertices and/or edges of X, then the (induced) graph $X \setminus F$ has finitely many connected components. Every ray π must have all but finitely many points in precisely one of them, and we say that π *ends up* in that component. Two rays are called *equivalent,* if for any finite $F \subset X \cup E(X)$ they end up in the same component of $X \setminus F$. This is the same as saying that there is a third ray which meets each of the two infinitely often. An *end* of X is an equivalence class of rays. We write ϑX for the space of ends of X, and $\widehat{X} = X \cup \vartheta X$. If C is a component of $X \setminus F$ (F finite), then we write ϑC for the set of those ends whose rays end up in C, and $\widehat{C} = C \cup \vartheta C$ for the completion of C. See Figure 17.

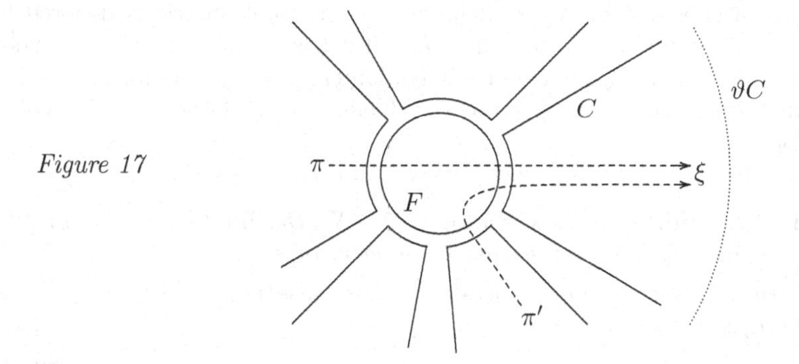

Figure 17

We now explain the topology of ϑX. If $F \subset X \cup E(X)$ is finite, and $w \in \widehat{X} \setminus F$, then there is precisely one component of $X \setminus F$ whose completion contains w. We denote it by $\widehat{C}(w, F)$. Varying F (finite, with $w \notin F$), we obtain a neighbourhood base of w. If $x \in X$, we can take for F the (finite!) set of neighbours of X to see that the topology is discrete on X. It has a countable base and is Hausdorff. If $\xi \in \vartheta X$, then we do not need *all* $\widehat{C}(w, F)$, where F is finite, to obtain a neighbourhood basis. For each end ξ, we can find a *standard* neighbourhood base, that is, one of the form $\widehat{C}(\xi, F_k)$, $k \in \mathbb{N}$, where the finite $F_k \subset X$ are such that

$$F_k \cup \widehat{C}(\xi, F_k) \subset \widehat{C}(\xi, F_{k-1}) \quad \text{for all } k. \tag{21.1}$$

See Figure 18. Another neighbourhood base of ξ is given by $\{\widehat{C}(\xi, B_k) : k \in \mathbb{N}\}$, with the balls $B_k = B(o, k)$ and arbitrary (fixed) root $o \in X$.

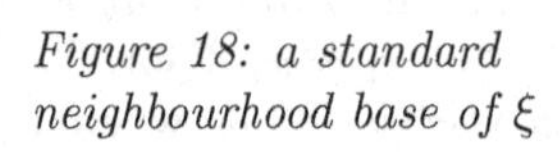

Figure 18: a standard neighbourhood base of ξ

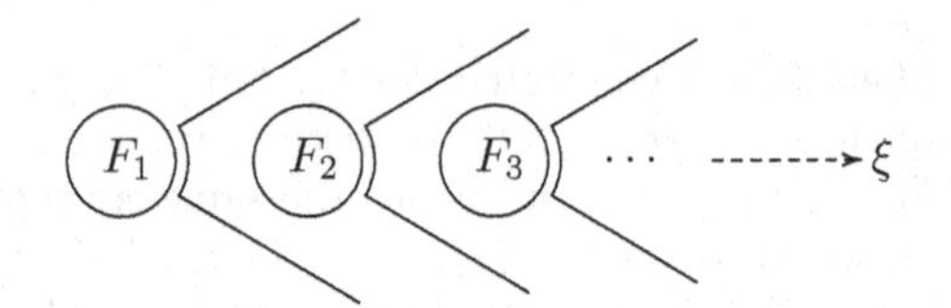

(21.2) Exercise. Show that $\widehat{X}$ is compact. [Hint: Let (w_n) be a sequence of distinct elements in $\widehat{X}$. Then $X \setminus B_1$ has some component C_1 such that $\widehat{C}_1$ contains infinitely many w_n. Among the components of $X \setminus B_2$, there must be C_2 such that $C_2 \subset C_1$ and $\widehat{C}_2$ contains infinitely many w_n. Now continue inductively: $\bigcap_k \widehat{C}_k$ contains precisely one end ξ, and by the obvious diagonal argument, there is a sub-sequence converging to ξ.]

As the number of components of $X \setminus F$ (for finite F) is finite, each $\widehat{C}(w, F)$ is open and compact, and $\widehat{X}$ is totally disconnected.

For a subset A of X, its diameter in the graph metric is denoted by $\operatorname{diam}(A)$. An end ξ of X is called *thin*, if it has a standard neighbourhood base (21.1) with $F_k \subset X$, such that $\operatorname{diam}(F_k) = m < \infty$ for all k. The minimal m with this property is the *diameter* of ξ. Otherwise, ξ is called *thick*.

As usual, we fix a reference vertex o and write $|x| = d(x, o)$.

(21.3) Lemma. (a) *Let $k \geq 1$ and $x, y \in X \setminus B_k$. If $|x| + |y| - d(x, y) > 2k$ then x and y belong to the same component of $X \setminus B_k$.*

(b) *Any sequence (x_n) of vertices satisfying $|x_n| + |x_{n+1}| - d(x_n, x_{n+1}) \to \infty$ converges to an end of X.*

Proof. Let x, y be vertices that belong to different components of $X \setminus B_k$. Any path from x to y must pass through B_k. Hence there is $w \in B_k$ such that $d(x, y) = d(x, w) + d(w, y)$. We obtain $d(x, y) \geq |x| + |y| - 2k$.

If $|x_n| + |x_{n+1}| - d(x_n, x_{n+1}) \to \infty$ then $|x_n| \to \infty$. Suppose that (x_n) has two distinct accumulation points in ϑX. Then there must be k and infinitely many n such that x_n and x_{n+1} lie in distinct components of $X \setminus B_k$, a contradiction. □

For the last preparatory step, recall Definition 3.7, of rough isometries.

(21.4) Lemma. *Let X and X' be roughly isometric, locally finite graphs. Then the rough isometry extends to a continuous mapping from $\widehat{X}$ to $\widehat{X}'$, whose restriction to ϑX is a homeomorphism onto $\vartheta X'$, which preserves thickness and thinness.*

Proof. Let $\varphi : X \to X'$ be our rough isometry, with constants a, b (in place of A, B) as in (3.7). Let $o' = \varphi o$, and write $|x'| = d'(x', o')$ for $x' \in X'$.

Given $\xi \in \vartheta X$, consider a ray $\pi = [x_0, x_1, x_2, \dots]$ representing ξ. Then $|x_n| \to \infty$ and $d(x_n, x_{n+1}) = 1$. Consequently also $|\varphi x_n| + |\varphi x_{n+1}| - d(\varphi x_n, \varphi x_{n+1}) \to \infty$. By Lemma 21.4, (φx_n) converges to an end ξ' of X'. We define $\varphi \xi = \xi'$. In proving continuity, we shall see that this does not depend on the choice of the ray π.

Let (w_n) be a sequence in $\widehat{X}$ converging to ξ. This means that for every k we can find n_k such that each w_n, $n \geq n_k$, is connected to π outside of B_k. That is, there are $x_{r(n)}$ and a path or ray $\pi_n = [y_0, y_1, \dots]$ in $X \setminus B_k$ such that $y_0 = x_{r(n)}$, and π_n terminates in w_n (if $w_n \in X$), or represents w_n (if $w_n \in \vartheta X$), respectively. We find that all points of $\varphi(\pi_n)$ lie outside the ball $B'_{(k-b)/a}$ around o'. Now, $\varphi(\pi_n)$ is not necessarily a path or ray, but $d'(\varphi y_{i+1}, \varphi y_i) \leq a + b$. Therefore we can fill the gap between φy_i and φy_{i+1} by a path in X' of length at most $a + b$. We see that φw_n is connected to $\varphi x_{r(n)}$ by a path or ray (respectively) in $X' \setminus B'_\ell$, where $\ell = (k-b)/a - (a+b)$. If $k \to \infty$ then $\ell \to \infty$, whence $\varphi w_n \to \varphi \xi$.

We now leave it as an exercise to show that φ is one-to-one and onto from ϑX to $\vartheta X'$, and that thickness and thinness of ends are preserved. □

In particular, for Cayley graphs of a finitely generated group, the end compactification is independent of the choice of the finite generating set, and one speaks of the ends of the group itself. The simplest examples of groups with infinitely many ends are free products (with the exception of $\mathbb{Z}_2 * \mathbb{Z}_2$, which has two ends). More generally, the free product of two (or more) rooted graphs has infinitely many ends, unless both have only two elements. On the other hand, every Cartesian product of two infinite, locally finite graphs has one end only.

A. The transitive case

The considerations that follow are not restricted to the transitive case, but this is the context where they will be applied.

(21.5) Proposition. *The end compactification of a locally finite graph X is a contractive* $\mathrm{AUT}(X)$*-compactification.*

Proof. Suppose that (x_n) and (y_n) are sequences in X such that x_n converges to some end ξ, and $d(y_n, x_n) \le M < \infty$. Then $x_n \in \widehat{C}(\xi, B_k)$ and $d(x_n, o) > k + M$ for all $n \ge n_k$. As y_n is connected to x_n by a path of length $\le M$, also $y_n \in \widehat{C}(\xi, B_k)$ for $n \ge n_k$. Therefore $y_n \to \xi$, and $\widehat{X}$ is projective.

If $\gamma \in \mathrm{AUT}(X)$ then it maps rays onto rays, and preserves equivalence between rays. Therefore γ acts bijectively on ϑX. It is straightforward that this action is continuous.

Finally, let (γ_n) be a sequence in $\mathrm{AUT}(X)$ such that $\gamma_n o \to \xi \in \vartheta X$ and $\gamma_n^{-1} o \to \eta \in \vartheta X$. Take $\widehat{X}$-neighbourhoods of ξ and η of the forms $U = \widehat{C}(\xi, B_k)$ and $V = \widehat{C}(\eta, B_k)$. We can find n_0 such that $\gamma_n B_k \subset U$ and $\gamma_n^{-1} B_k \subset V$ for all $n \ge n_0$. Now $\widehat{X} \setminus V$ induces a connected subgraph of X that is disjoint from $\gamma_n^{-1} B_k$ and contains B_k. Therefore $\gamma_n(\widehat{X} \setminus V)$ contains $\gamma_n B_k$ and induces a connected subgraph that must be contained in a component of $X \setminus B_k$. But the component of $X \setminus B_k$ containing $\gamma_n B_k$ is U. Consequently, $\gamma_n(\widehat{X} \setminus V) \subset U$, and the convergence property holds. □

In particular, we see that a quasi-transitive (locally finite, connected) graph has one, two, or infinitely many ends (Proposition 20.9), and in the last case, Proposition 20.10 applies to every quasi-transitive subgroup of $\mathrm{AUT}(X)$. Before stating the obvious corollary of Theorem 20.13, we need more structure theoretic information for a better understanding of case (b) of Proposition 20.10. For this purpose, we give a brief introduction to the powerful theory of cuts and structure trees developed by Dunwoody [108], [109]; see the book by Dicks and Dunwoody [96].

A *cut* of a connected graph X is a set F of edges whose deletion disconnects X. If it disconnects X into precisely two connected components $A = A(F)$ and $A^* = A^*(F) = X \setminus A$, then we call F *tight,* and A, A^* are the *sides* of F.

(21.6) Lemma. *Let X be connected and $e \in E(X)$. For any $k \in \mathbb{N}$ there are only finitely many tight cuts F containing e with $|F| = k$.*

Proof. We use induction on k. For $k = 1$, there is nothing to prove. Let $k > 1$, and suppose that $e = [x, y]$ is contained in some tight cut F with cardinality k. Then the graph $X' = X \setminus \{e\}$ is connected and contains a finite path π from x to y. Therefore $F \setminus \{e\}$ is a tight cut of X' and must contain an edge of π. By the induction hypothesis there are only finitely many tight cuts F' of X' with $|F'| = k - 1$ that contain some edge of π. □

We say that two cuts F, F' *cross,* if all four sets

$$A(F) \cap A(F')\,,\ A(F) \cap A^*(F')\,,\ A^*(F) \cap A(F')\,,\ A^*(F) \cap A^*(F')$$

are non-empty. Dunwoody [109], [96] has proved the following important theorem.

(21.7) Theorem. *Every infinite, connected graph with more than one end has a finite tight cut F with infinite sides, such that F crosses no γF, where $\gamma \in \mathrm{AUT}(X)$.*

A cut with these properties will be called a *D-cut.* Now let F be a D-cut of the locally finite, connected graph X, let Γ be a closed subgroup of $\mathrm{AUT}(X)$, and define

$$\mathcal{E} = \{A(\gamma F), A^*(\gamma F) : \gamma \in \mathrm{AUT}(X)\}\,. \tag{21.8}$$

This collection has the following properties.

(1) All $A \in \mathcal{E}$ are infinite and connected.
(2) If $A \in \mathcal{E}$ then $A^* = X \setminus A \in \mathcal{E}$.
(3) If $A, B \in \mathcal{E}$ and $A \subset B$ then there are only finitely many $C \in \mathcal{E}$ such that $A \subset C \subset B$.
(4) If $A, B \in \mathcal{E}$, then one of $A \subset B$, $A \subset B^*$, $A^* \subset B$ and $A^* \subset B^*$ holds.

Here, (3) follows from Lemma 21.6 and (4) from Theorem 21.7. These properties can now be used to construct a tree $\mathcal{T}$, called the *structure tree* of X with respect to Γ and the D-cut F: think of an unoriented edge of $\mathcal{T}$ as a pair of oriented edges, where the second edge points from the endpoint to the initial point of the first one. Then the edge set of $\mathcal{T}$ is $\mathcal{E}$. That is, if $A \in \mathcal{E}$ then A, A^* constitute such a pair of oppositely oriented edges between the same two vertices. If $A, B \in \mathcal{E}$ and $B \neq A^*$ then the endpoint of A is the initial point of B, if $A \supset B$ and there is no $C \in \mathcal{E}$ such that $A \supset C \supset B$ properly.

In this way, we have defined $\mathcal{T}$ in terms of its edges and their incidence, contrary to our usual approach of defining a graph by starting with its vertices. We now formalize the latter. A vertex of $\mathcal{T}$ is an equivalence class of edges "with the same endpoint", that is, $A, B \in \mathcal{E}$ are *equivalent* in this sense, if $A = B$, or else, if $A \supset B^*$ properly and no $C \in \mathcal{E}$ satisfies $A \supset C \supset B^*$ properly.

We verify that this is indeed an equivalence relation. Symmetry is obvious. To see transitivity, let $A, B, C \in \mathcal{E}$ be distinct and B equivalent to both A and C. Then $A \cap C \supset B^*$. Using (4), of the four possibilities $A \subset C$, $A \supset C$, $A \subset C^*$ and $A \supset C^*$, the first is impossible because otherwise $B^* \subset A \subset C$ properly, the second because otherwise $B^* \subset C \subset A$ properly, and the third because otherwise $B^* \subset C^*$. Therefore $A \supset C^*$. Now suppose $D \in \mathcal{E}$ is such that $A \supset D \supset C^*$ properly. Again using (4), if $D \subset B$

then $C \supset D^* \supset B^*$, whence $D = B$ and $A \supset B \cup B^*$, a contradiction. If $D \supset B$, the same contradiction arises. If $D \subset B^*$ then $C \supset D \supset C^*$, a contradiction. Finally, if $D \supset B^*$ then $A \supset D \supset B^*$, whence $D = B^*$ and $C \supset B^* \supset C^*$, again a contradiction. □

The vertex set is the set of all equivalence classes $[A]$, where $A \in \mathcal{E}$. Neighbourhood in $\mathcal{T}$ is described by $[A] \sim [A^*]$. It is then straightforward to verify that $\mathcal{T}$ is a tree: connectedness follows from (3), and there can be no cycles, as neighbourhood is defined in terms of inclusion of sets (reading around a cycle would give $A \subset A$ properly for some $A \in \mathcal{E}$). The tree is countable by Lemma 21.6, but in general not locally finite. Still, we can define the set $\vartheta\mathcal{T}$ of ends of $\mathcal{T}$ as equivalence classes of rays, as in §6.B, and $\widehat{\mathcal{T}} = \mathcal{T} \cup \vartheta\mathcal{T}$. However, we do not specify a topology. The group Γ acts by automorphisms on $\mathcal{T}$ via $A \mapsto \gamma A$, where $\gamma \in \Gamma$ and $A \in \mathcal{E}$. The action has one or two orbits on $\mathcal{E}$ according to whether $\gamma A(F) = A^*(F)$ for some $\gamma \in \Gamma$ or not. Consequently, Γ acts transitively on $\mathcal{T}$ or else acts transitively on each of the two bipartite classes of $\mathcal{T}$ (that is, the sets of vertices at even/odd distance from a chosen origin). The kernel of the action consists of those $\alpha \in \Gamma$ that leave all Γ-translates of F invariant; in particular, it is compact in Γ. The stabilizer in Γ of a vertex of $\mathcal{T}$ is not necessarily compact. However, this is true for the stabilizer of any (oriented) edge of $\mathcal{T}$. Indeed, if $A \in \mathcal{E}$, then $\gamma A = A$ implies that γ fixes the (finite) D-cut of which A is a side.

The reader is invited to draw a few examples. If $X = \mathbb{T}_M$ then any single edge constitutes a D-cut. If Γ is the whole of $\mathrm{AUT}(\mathbb{T}_M)$, then the structure tree is $\mathbb{T}_M$ again. This is not the case when $\Gamma = \mathbb{Z}_2 * \cdots * \mathbb{Z}_2$ (M times) or the free group. As another example, let $\Gamma = \mathbb{Z}_r * \mathbb{Z}_s = \langle a, b \mid a^r = b^s = o \rangle$ with $rs \geq 6$, and X its Cayley graph with respect to $\{a^{\pm 1}, b^{\pm 1}\}$. In this case, $F = \{[o, a^{\pm 1}]\}$ is a D-cut, and with respect to Γ, the structure tree is the *bi-regular tree,* with degrees r and s for vertices at even and odd distance from the origin, respectively. As an example with thick ends, consider the standard Cayley graph X of the free product $\Gamma = \mathbb{Z}^2 * \mathbb{Z}_2 = \langle a, b, c \mid ab = ba, c^2 = o \rangle$, acted upon by Γ itself. Each copy $\gamma\mathbb{Z}^2$ of the square grid within X, where $\gamma \in \Gamma$, gives rise to a thick end (as an equivalence class of rays that end up in such a copy). The other ends are all thin, they have zero diameter. Let F consist of the single edge $[o, c]$. This is a D-cut, and the structure tree has infinite vertex degrees.

Next, we define the *structure map* $\varphi : \widehat{X} \to \widehat{\mathcal{T}}$. This is easiest by explaining the inverse map. For $A \in \mathcal{E}$, let $\widehat{A}$ be its completion in $\widehat{X}$. Then, given $\bar{w} \in \widehat{\mathcal{T}}$, we say that A points towards $\bar{w}$, if the vertex $[A^*]$ of $\mathcal{T}$ lies on the geodesic from $[A]$ to $\bar{w}$, and

$$\varphi^{-1}\{\bar{w}\} = \bigcap \{\widehat{A} : A \text{ points towards } \bar{w}\}. \tag{21.9}$$

Every element of $\widehat{X}$ is captured in this way: if $x \in X$ then it must be contained in some $A \in \mathcal{E}$, and by Lemma 21.6 there must be a minimal $A \in \mathcal{E}$ with $x \in A$. Then $\varphi x = [A]$. If $\xi \in \vartheta X$, then there are two possibilities: (1) there is a minimal $\widehat{A}$, $A \in \mathcal{E}$, containing ξ, and $\varphi\xi = [A] \in \mathcal{T}$, and (2) there is a (strictly) descending sequence $(A_n)_{n\geq 1}$ in $\mathcal{E}$ such that $\xi \in \bigcap_n \widehat{A}_n$. In this case, the $\widehat{A}_n$ constitute a neighbourhood base of ξ in $\widehat{X}$, while in $\mathcal{T}$ they constitute the edges of a ray (with vertices $[A_n]$) which defines an end $\bar{\xi} \in \vartheta\mathcal{T}$, and $\varphi\xi = \bar{\xi}$. We see that φ is compact-to-one. Note that φ is not necessarily onto, as $\varphi^{-1}\{\bar{w}\}$ may be empty. This is the case, for example, for the structure tree of $\mathbb{Z}_r * \mathbb{Z}_s$ described above: assuming $r \neq s$, the vertices with degree r in $\mathcal{T}$ have empty preimage.

We also note that φ commutes with the action of Γ on $\widehat{X}$ and $\widehat{\mathcal{T}}$, respectively. We can now prove two useful theorems on the structure of (connected) vertex-transitive graphs with infinitely many ends.

(21.10) Theorem. *Let X be a locally finite graph with infinitely many ends, and Γ a closed, transitive subgroup of* $\mathrm{AUT}(X)$.

Then the space of ends of X has a decomposition

$$\vartheta X = \Upsilon_0\,, \quad \vartheta X = \Upsilon_0 \cup \Upsilon_1\,, \quad \textit{or} \quad \vartheta X = \Upsilon_0 \cup \Upsilon_1 \cup \Upsilon_2$$

with the following properties.

- *Each Υ_i is Γ-invariant.*
- *There is $m < \infty$ such that* $\mathrm{diam}(\xi) \leq m$ *for each ξ in Υ_0, and Υ_0 is dense in ϑX.*
- *For $i = 1, 2$, if Υ_i is non-empty then it has a Γ-invariant partition into countably many compact sets.*

Proof. We set $\Upsilon_0 = \varphi^{-1}(\vartheta\mathcal{T})$. This is clearly Γ-invariant. If V is the set of end vertices of the D-cut F, then $\mathrm{diam}(\xi) \leq \mathrm{diam}(V) < \infty$ for every $\xi \in \Upsilon_0$.

To see denseness, given $\xi \in \vartheta X$, let F' be a finite, connected subgraph of X, and $\widehat{C} = \widehat{C}(\xi, F')$. By transitivity, there is $\gamma \in \Gamma$ mapping the D-cut F into (the edge set of) $\widehat{C}$. One of the sides of γF must contain F'. The other side, say A_1, is connected, does not intersect F', but does intersect $\widehat{C}$. Therefore $A_1 \subset \widehat{C}$. Again using transitivity, and property (4) of $\mathcal{E}$, we can now find a strictly descending sequence $A_1 \supset A_2 \supset \cdots$ of elements of $\mathcal{E}$. It gives rise to an element of Υ_0 contained in $\widehat{C}$. (In particular, every vertex of $\mathcal{T}$ has at least two neighbours.)

If Γ acts transitively on $\mathcal{T}$ then we set $\Upsilon_1 = \vartheta X \cap \varphi^{-1}(\mathcal{T})$. This may be empty. Otherwise, $\vartheta X = \Upsilon_0 \cup \Upsilon_1$, and Υ_1 has the Γ-invariant partition $\bigcup_{\bar{x}\in\mathcal{T}} (\vartheta X \cap \varphi^{-1}\bar{x})$ with compact pieces.

If Γ has two orbits $\mathcal{T}_1$ and $\mathcal{T}_2$ on $\mathcal{T}$ then we set $\Upsilon_i = \vartheta X \cap \varphi^{-1}(\mathcal{T}_i)$, $i = 1, 2$. □

The second result regards the "degenerate" case (b) of Proposition 20.10.

(21.11) Theorem. *Let X be a locally finite graph with infinitely many ends, and Γ a closed, transitive subgroup of* AUT(X) *fixing an end ω of X.*

Then the structure tree $\mathcal{T}$ with respect to Γ and a D-cut F is locally finite. The structure map $\varphi : X \to \mathcal{T}$ is a rough isometry, Γ acts transitively on $\mathcal{T}$ and fixes the end $\varphi\omega$ of $\mathcal{T}$. In particular, Γ must be non-unimodular.

Proof. As the action of Γ commutes with φ, it is clear that Γ fixes $\varphi\omega$. We first show that this is an end (and not a vertex) of $\mathcal{T}$. Let F be our D-cut, and let $A = A(F)$ be the one of its sides with $\omega \in \widehat{A}$. Then the open set $\widehat{A}^*$ must contain some end that is fixed by a hyperbolic element γ of Γ, see Proposition 20.9. Now γ must fix ω, and we may assume without loss of generality that this is the "forward" fixed point of γ. As $\widehat{A}$ is an open neighbourhood of ω, and by contractivity, there is k such that $\gamma^k A \subset A$ properly. The sequence $A_n = \gamma^{kn} A \in \mathcal{E}$ is strictly decreasing, and $\omega \in \widehat{A}_n$ for each n. Therefore $\varphi\omega$ cannot be a vertex of $\mathcal{T}$.

Now let $\bar{x} \in \mathcal{T}$, and let $A \in \mathcal{E}$ be the unique oriented edge of $\mathcal{T}$ incident with $\bar{x}$ that points towards $\varphi\omega$, that is, $\bar{x} = [A]$. Let $\bar{y} = [A^*]$ be the neighbour of $\bar{x}$ on the geodesic ray to $\varphi\omega$, and $B \in \mathcal{E}$ the edge incident with $\bar{y}$ pointing towards $\varphi\omega$. Then $B = \gamma A$ or $B = \gamma A^*$ for some $\gamma \in \Gamma$. But $\omega = \gamma\omega \in B \cap \gamma A$, so that it must be that $B = \gamma A$. Consequently $\bar{y} = \gamma\bar{x}$, and (recalling that Γ acts transitively on the bipartite classes) Γ acts transitively on $\mathcal{T}$. In particular, $\varphi X = \mathcal{T}$.

The stabilizer of $\bar{x}$ in Γ also stabilizes A as an edge of $\mathcal{T}$ and hence as a subset of X. Therefore $\Gamma_{\bar{x}}$ is compact. Now let $B_1, B_2, \dots \in \mathcal{E}$ be the oriented edges of $\mathcal{T}$ whose endpoint is $\bar{x}$, and let F_i denote the Γ-translate of F having B_i as one of its sides. Write $\bar{y}_i = [B_i]$, $i \geq 1$, for the initial vertices of B_i in $\mathcal{T}$. For each $i \geq 2$ there must be $\gamma_i \in \Gamma$ such that $\gamma_i\bar{y}_1 = \bar{y}_i$, whence $\gamma_i F_1 = F_i$. On the other hand, each γ_i fixes $\varphi\omega$ and hence also $\bar{x}$, that is, $\gamma_i \in \Gamma_{\bar{x}}$. By compactness, $\Gamma_{\bar{x}} F_1$ is finite, and there are only finitely many $\bar{y}_i$, so that $\mathcal{T}$ is locally finite.

Now $\varphi^{-1}\{\bar{x}\}$ is a *finite* intersection of open and compact sets $\widehat{A}$ with $A \in \mathcal{E}$. It is therefore open. If it contains an end, then by Theorem 21.10 it also contains some element $\xi \in \Upsilon_0$, that is, $\varphi\xi$ is a vertex of $\mathcal{T}$, a contradiction. Therefore $\varphi^{-1}\{\bar{x}\}$ must be a finite subset of X and has finite diameter a. As Γ acts transitively on $\mathcal{T}$, this a is the same for each $\bar{x}$. That is, if $d_X(x, y) \geq a$ then $d_{\mathcal{T}}(\varphi x, \varphi y) \geq 1$. Clearly $d_X(x, y) = 1$ implies $d_{\mathcal{T}}(\varphi x, \varphi y) \leq 1$. Thus, φ is a rough isometry.

We see (compare with Lemma 21.4) that Γ acts on $\mathcal{T}$ continuously in the topology of pointwise convergence, and we know that the kernel Γ^0

of the action is compact. Furthermore, Γ fixes $\varphi\omega$, and $\mathcal{T}$ has infinitely many ends. That is, we have precisely the picture of §12.C, Figure 12. As in (12.15), Lemma 1.29 implies that Γ/Γ^0, and consequently also Γ, are non-unimodular. □

We now see how fruitful a detailed study of structure theory may be for obtaining information on the behaviour of random walks.

(21.12) Corollary. *Let X be a graph with infinitely many ends, and suppose that there is a closed, transitive subgroup Γ of $\mathrm{AUT}(X,P)$. Unless Γ fixes an end of X, the Dirichlet problem for P-harmonic functions is solvable with respect to the end compactification, and the random walk with transition matrix P converges a.s. to a random end which is thin.*

This applies, in particular, to random walks on finitely generated groups with infinitely many ends, and also to the case when X has thick ends.

Proof. If Γ fixes no end, then we can apply Theorem 20.13 to see that the Dirichlet problem is solvable. Convergence to the boundary follows from Theorem 20.3. If ν is the limit distribution on ϑX, then it is μ-invariant, where μ is the probability on Γ defined in (8.12). Now Lemma 20.12 combined with Theorem 21.10 implies that $\nu(\Upsilon_i) = 0$ for $i = 1,2$, whence $\nu(\Upsilon_0) = 1$.

If Γ is discrete, then it is unimodular and cannot fix an end by Theorem 21.11. If X has thick ends, then it cannot be roughly isometric with a tree by Lemma 21.4, and again by Theorem 21.11 there is no end fixed by Γ. □

In the context of finitely generated groups, this result applies to arbitrary (irreducible) random walks on free groups, and more generally, on free products with the sole exception of the infinite dihedral group $\mathbb{Z}_2 * \mathbb{Z}_2$. If $\Gamma = \Gamma_1 * \Gamma_2$, then its end compactification can be described in terms of the two factors as follows. Let $\vartheta\Gamma_i$ be the space of ends of Γ_i (empty, if Γ_i is finite), $i = 1,2$. It is best to have in mind a Cayley graph of Γ of the form $X = X_1 * X_2$, where each X_i is a Cayley graph of Γ_i. In particular, deletion of a single vertex disconnects the graph into two or more infinite components. We set

$$(21.13)\quad \begin{cases} \Upsilon' = \{x_1x_2x_3\cdots : x_j \in \Gamma_{i_j} \setminus \{o\},\ i_j \neq i_{j-1}\} \quad \text{and} \\ \Upsilon'' = \{x_1\cdots x_{k-1}\xi_k : x_j \in \Gamma_{i_j} \setminus \{o\},\ \xi_k \in \vartheta\Gamma_{i_k},\ i_j \neq i_{j-1}\}, \end{cases}$$

where all $i_j \in \{1,2\}$. Each "infinite word" in Υ' is the limit in the end topology of the sequence of its initial pieces $x_1\cdots x_n$ as $n \to \infty$. Two different infinite words represent different ends, which have diameter 0 in X. An element in Υ'' is the limit of the sequence $x_1\cdots x_{k-1}x_k(n)$, where $x_k(n) \in \Gamma_{i_k}$ tends to ξ_k in the topology of $\widehat{\Gamma}_{i_k}$, as $n \to \infty$. We can further decompose $\Upsilon'' = \Upsilon''_1 \cup \Upsilon''_2$, where Υ''_i consists of all elements of Υ'' as in

(21.13) with $i_k = i$. The decomposition $\vartheta\Gamma = \Upsilon' \cup \Upsilon_1'' \cup \Upsilon_2''$ does not in general coincide with that of Theorem 21.10 (as it arises from deleting vertices, not edges), but is easily seen to have the same properties. It will be reconsidered in §26.B.

(21.14) Corollary. *Let $\Gamma = \Gamma_1 * \Gamma_2$, where the Γ_i are finitely generated and we do not have $|\Gamma_1| = |\Gamma_2| = 2$. Then any random walk on Γ converges almost surely to a random infinite word, and the limit distribution is continuous.*

This applies also to free products of more than two groups. Besides free products, there are more general ways for constructing groups with infinitely many ends from given ones, namely, *amalgamated free products* and *HNN-extensions* of groups over finite subgroups. All groups with more than one end are of this type. We do not go into further details here; the interested reader may refer to the books by Stallings [308] and Lyndon and Schupp [218].

B. Geometric adaptedness conditions

When (X, P) is transitive, we have seen that with the exception of the "degenerate" case when an end is fixed under $\mathrm{AUT}(X, P)$, convergence to the boundary and solvability of the Dirichlet problem hold without additional assumptions such as moment conditions, etc. Group invariance alone is so strong that no further adaptedness conditions are needed. If the random walk is not invariant under a "big" group, then we need adaptedness conditions of a geometric nature instead of the algebraic one of transitivity (or at least quasi-transitivity). We first consider the simplest case.

(21.15) Theorem. *Suppose that P has bounded range and is transient. Then the random walk converges to a random end of the graph X.*

If $\operatorname{supp} \nu_x$ has more than one element for some ($\Longleftrightarrow$ every) $x \in X$, then an end $\xi \in \vartheta X$ is regular if and only if the Green kernel vanishes at ξ. In particular, the Dirichlet problem with respect to the end compactification is solvable if and only if the Green kernel vanishes at ∞.

Proof. By local finiteness of X, transience is equivalent with $|Z_n| \to \infty$ a.s. Bounded range yields $|Z_n|+|Z_{n+1}|-d(Z_n, Z_{n+1}) \to \infty$, and convergence to the boundary follows from Lemma 21.3.

Let $K = \max\{d(x, y) : p(x, y) > 0\}$. Then the K-fuzz $X^{(K)}$ is metrically equivalent to X and has the same end compactification, by Lemma 21.4. Therefore we may assume without loss of generality that $K = 1$, that is, P is nearest neighbour.

Assume $|\operatorname{supp} \nu_x| \geq 2$. We know from Lemma 20.4 that the Green kernel vanishes at ξ if $\xi \in \vartheta X$ is regular. Conversely, let $\widehat{C} = \widehat{C}(\xi, F)$, $F \subset X$

finite, be a neighbourhood of ξ such that $\nu_x(\widehat{X} \setminus \widehat{C}) > 0$ for all $x \in X$. If $x \in \widehat{C} \cap X$, then in order to reach $\widehat{X} \setminus \widehat{C}$, the random walk starting at x must pass through F. Therefore, denoting by $\mathbf{s}^F$ the hitting time of F,

$$\nu_x(\widehat{X} \setminus \widehat{C}) = \sum_{y \in F} \mathbb{P}_x[Z_{\mathbf{s}^F} = y]\, \nu_y(\widehat{X} \setminus \widehat{C}) \le \sum_{y \in F} G(x, y).$$

If the Green kernel vanishes at ξ, then $\lim_{x \to \xi} G(x, y) = 0$ for all $y \in F$. Therefore $\lim_{x \to \xi} \nu_x(\widehat{C}) = 1$. This is true for every neighbourhood $\widehat{C}(\xi, F)$, so that $\nu_x \to \delta_\xi$ weakly when $x \to \xi$. □

We remark that it may well be that $\operatorname{supp} \nu_x$ is the whole of ϑX, while there are non-regular points in ϑX. An example may be constructed as follows. Consider the half-line $\mathbb{N}_0$ with its end $+\infty$. To each point $k \in \mathbb{N}$ attach a finite path (a "hair") of length $f(k)$. At the endpoint of each "hair", we attach a copy of the binary tree (by its root). For the tree T obtained in this way (Figure 19), ϑT consists of the ends of all these binary trees (which are all regular) plus the end $+\infty$ of the base line $\mathbb{N}$. The latter is not isolated in ϑT, so that $\operatorname{supp} \nu_x = \vartheta T$. If $f(k)$ increases quickly, e.g. $f(k) \ge k^3$, then one finds that the Green kernel does not vanish at $+\infty$. This can verified by showing that the hitting probabilities $F(k, 1)$ do not tend to 0, when $k \to \infty$. We omit the details, which consist of lengthy computations.

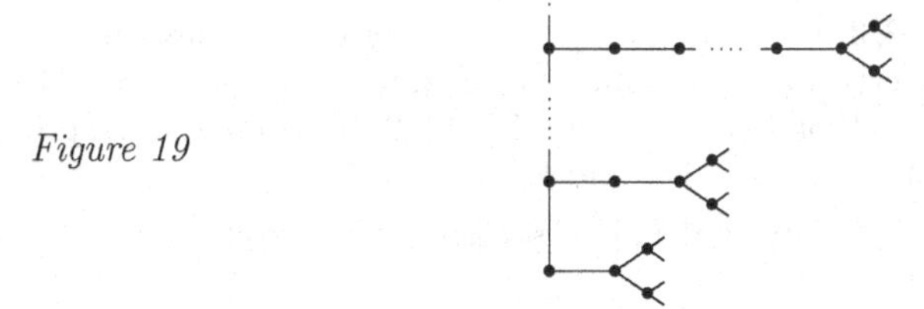

Figure 19

Next, we want to know what can be obtained when P does not have bounded range. We start with convergence to the boundary.

(21.16) Theorem. *If (X, P) is uniformly irreducible and has a uniform first moment, and $\rho(P) < 1$, then Z_n converges almost surely to a random end of X.*

Proof. This is immediate from Lemma 20.4 and the distance estimates of §8.A, as Propositions 8.2 and 8.8(a) imply that $|Z_n| + |Z_{n+1}| - d(Z_n, Z_{n+1}) \to \infty$ a.s. □

Note that via the results of Chapter II, in particular of Section 10, the property $\rho(P) < 1$ is well understood. For convergence to the boundary, we have used those distance estimates of §8.A that yield almost sure convergence. The Dirichlet problem cannot be solved under the same hypotheses as Theorem 21.16. Indeed, consider the nearest neighbour random walk on $\mathbb{Z}$ with $p(k,k+1) = p > 1/2$ and $p(k,k-1) = 1-p$. Then $\vartheta\mathbb{Z} = \{\pm\infty\}$, but $Z_n \to +\infty$ by the law of large numbers. Therefore the limit distribution is not supported by $-\infty$.

Instead of $p^{(n)}(x,y) \le A^{d(x,y)}\,\rho(P)^n$, we shall need the stronger bound of Lemma 8.1(b). We shall now apply those distance estimates of §8.A which give convergence in probability uniformly in $|x|$, plus an additional (decisive) estimate. Recall the constant $\underline{\mathfrak{m}} > 0$ of Proposition 8.2.

(21.17) Lemma. *Suppose (X,P) is uniformly irreducible and has a uniform first moment, and $p^{(n)}(x,y) \le C\,\rho^n$ for all x,y and n, where $\rho < 1$. For $a, \varepsilon > 0$ and $x \in X$, consider the event $\mathcal{A}_x = \mathcal{A}_x(a,\varepsilon)$ in the trajectory space defined by the following properties:*

(1) $$Z_0 = x\,,$$
(2) $$d(Z_n, Z_{n+1}) \le \varepsilon\,|x| \quad \text{for all } n \le a\,|x|\,,$$
(3) $$d(Z_n, Z_{n+1}) \le \varepsilon\, n \quad \text{for all } n \ge a\,|x|\,,$$
(4) $$d(Z_0, Z_n) \ge \underline{\mathfrak{m}}\, n \quad \text{for all } n \ge a\,|x|\,,$$
(5) $$|Z_n| > \varepsilon\,|x| \quad \text{for all } n \ge 0\,.$$

Then there is $\varepsilon_0 > 0$ such that for all $\varepsilon \le \varepsilon_0$ and $a > 0$,

$$\lim_{|x|\to\infty} \mathbb{P}_x(\mathcal{A}_x) = 1\,.$$

Proof. For $i = 2,3,4,5$, write $\mathcal{A}_{x,i}$ for the set of trajectories with $Z_0 = x$ that satisfy property (i). Then $\lim_{|x|\to\infty} \mathbb{P}_x(\mathcal{A}_{x,i}) = 1$ for $i = 2,3$ by Proposition 8.8(b) and for $i = 4$ by Proposition 8.2. To complete the proof, we have to settle $\mathcal{A}_{x,5}$. Let $\varepsilon < 1/2$ and $b > 0$. Then

$$\mathbb{P}_x\big[\exists n : |Z_n| \le \varepsilon\,|x|\big] \le \mathbb{P}_x\big[\exists n \le b\,|x| : |Z_n| \le \varepsilon\,|x|\big] + \mathbb{P}_x\big[\exists n > b\,|x| : |Z_n| \le \varepsilon\,|x|\big]\,.$$

As $|Z_n| \le \varepsilon\,|x|$ implies $d(x,Z_n) \ge (1-\varepsilon)|x| \ge |x|/2$,

$$\mathbb{P}_x\big[\exists n \le b\,|x| : |Z_n| \le \varepsilon\,|x|\big] \le \mathbb{P}_x\left[\frac{1}{b|x|}\sup_{n\le b\,|x|} d(Z_0,Z_n) \ge \frac{1}{2b}\right].$$

If we choose $2b < 1/\overline{\mathfrak{m}}$ then this probability tends to 0 as $|x| \to \infty$ by Proposition 8.8(b). With this choice of b,

$$\mathbb{P}_x\big[\exists n > b\,|x| : |Z_n| \le \varepsilon\,|x|\big] \le \sum_{n>b|x|}\ \sum_{y:|y|\le\varepsilon|x|} p^{(n)}(x,y) \le c\left(\rho^b M^{\varepsilon}\right)^{|x|},$$

where $M \ge 3$ is an upper bound on the vertex degrees and $c = C/(1-\rho)$. If ε is small enough so that $\rho^b M^{\varepsilon} < 1$ then this also tends to 0 as $|x| \to \infty$. □

(21.18) Theorem. *Assume that that the graph X satisfies IS and that P is strongly reversible and uniformly irreducible and has a uniform first moment. Then the Dirichlet problem with respect to the end compactification is solvable.*

Proof. We use once more the method provided by Theorem 20.3. Theorems 10.3 and 10.6 imply that $\rho(P) < 1$, and we have convergence to the boundary. In view of Lemma 8.1(b), we may apply Lemma 21.17.

Let $\xi \in \vartheta X$, and consider a neighbourhood $\widehat{C} = \widehat{C}(\xi, B_k)$, where $B_k = B(o,k)$. Let $x \in \widehat{C} \cap X$, and consider $\mathcal{A}_x = \mathcal{A}_x(\varepsilon, a)$ as in (21.17). If $\omega \in \mathcal{A}_x$ then

$$|Z_n(\omega)| + |Z_{n+1}(\omega)| - d\big(Z_n(\omega), Z_{n+1}(\omega)\big) > \varepsilon\,|x|$$

for all $n \le a\,|x|$, while for $n > a\,|x|$ we get

$$|Z_n(\omega)| + |Z_{n+1}(\omega)| - d\big(Z_n(\omega), Z_{n+1}(\omega)\big) \ge 2(\underline{\mathfrak{m}}\,n - |x|) - \varepsilon\,n \ge b\,|x|\,,$$

where $b = \big((2\underline{\mathfrak{m}} - \varepsilon)a - 2\big)$. Choose $\varepsilon < \min\{\varepsilon_0, 2\underline{\mathfrak{m}}\}$ with ε_0 as in Lemma 21.17, and $a = (\varepsilon + 2)/(2\underline{\mathfrak{m}} - \varepsilon)$. Then

$$|Z_n(\omega)|+|Z_{n+1}(\omega)|-d\big(Z_n(\omega), Z_{n+1}(\omega)\big) \ge \varepsilon\,|x| \quad \text{for all } n \ge 0 \text{ and } \omega \in \mathcal{A}_x\,.$$

Thus, if $\varepsilon\,|x| > 2k$ then Lemma 21.3 implies $Z_n(\omega) \in \widehat{C}$ for all n. As $\lim_n Z_n(\omega) \in \vartheta X$ exists almost surely, we obtain $\nu_x(\widehat{C} \cap \vartheta X) \ge \mathbb{P}_x(\mathcal{A}_x)$, which tends to 1 as $x \to \xi$. $\square$

In the hypotheses of the last theorem, we may replace *IS* and strong reversibility by the condition $p^{(n)}(x,y) \le C\,\rho^n$, with $\rho < 1$.

22. Hyperbolic graphs and groups

Once more, let (X,d) be a proper metric space, and suppose that it is also *geodesic:* for every pair of points $x, y \in X$, there is a *geodesic arc* $\pi(x,y)$ in X, that is, (the image of) an isometric embedding of the real interval $[0\,,\,d(x,y)]$ into X which sends 0 to x and $d(x,y)$ to y. We do not require that geodesics be unique. Any locally finite graph is proper and geodesic, when we consider it as a one-dimensional complex where each edge represents a copy of the unit interval, and the graph metric is extended to the interior points of the edges in the natural way.

A *geodesic triangle* consists of three points u, v, w together with geodesic arcs $\pi(u,v)$, $\pi(v,w)$, $\pi(w,u)$, the *sides.* The triangle is called *δ-thin* (where $\delta \ge 0$), if every point on any one of the sides is at distance at most δ from some point on one of the other two sides; see Figure 20.

(22.1) Definition. One says that X is *hyperbolic,* if there is $\delta \geq 0$ (possibly large) such that every geodesic triangle in X is δ-thin.

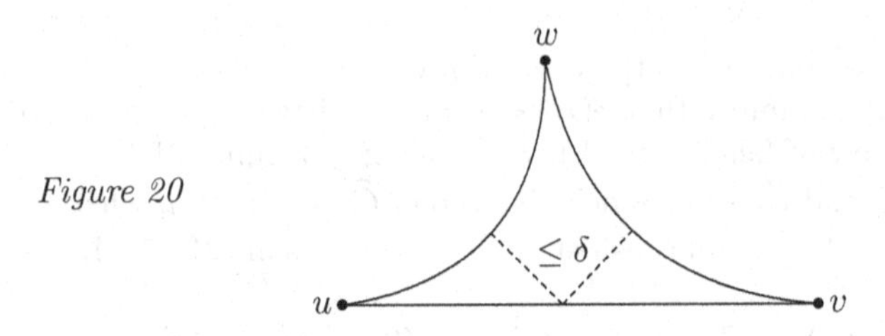

Figure 20

The two most typical examples are trees (where $\delta = 0$) and the hyperbolic plane $\mathbb{H} = \mathbb{H}_2$ (where $\delta = \log(1 + \sqrt{2})$), or more generally, the hyperbolic n-space $\mathbb{H}_n$.

There is a vast literature on hyperbolic spaces, in particular on *hyperbolic groups* (i.e., groups which have a hyperbolic Cayley graph). See the essay by Gromov [151], the notes by Cannon [50], Short (and collaborators) [300], Bowditch [46], and in particular the books by Ghys and de la Harpe [142] and Coornaert, Delzant and Papadopoulos [78].

The proof of the following theorem can be found, e.g., in [142], §5.2.

(22.2) Theorem. *Hyperbolicity of geodesic metric spaces is preserved by rough isometry.*

In particular, we deduce that for a finitely generated group hyperbolicity of one Cayley graph implies that all its Cayley graphs are hyperbolic. Furthermore, every group having a free subgroup with finite index is hyperbolic.

More examples are provided by all finitely generated *Fuchsian groups,* that is, discrete groups of Möbius transformations of $\mathbb{H}$. Nice expositions of their theory are the books by Beardon [28] and Katok [196]. In particular, if G is a co-compact finitely generated Fuchsian group, then it has a compact fundamental domain whose G-translates tessellate $\mathbb{H}$, and the dual graph of the tessellation is a Cayley graph of G which is roughly isometric with $\mathbb{H}$. We see that hyperbolic graphs, though being tree-like in some sense, may have a single, thick end.

Also, many *small cancellation groups* are hyperbolic; see the Appendix in [142].

(22.3) Exercise. Show that the graph in Figure 11 (§10.C) is hyperbolic: it is roughly isometric with the hyperbolic upper half plane. [Hint: every point in $\mathbb{H}$ is at bounded distance from some vertex, and the discrete graph metric is equivalent to the restriction of the metric of $\mathbb{H}$ to the vertex set.]

Returning to (X, d), we choose a reference point o and define for $x, y \in X$

$$|x \wedge y| = \frac{1}{2}\big(|x| + |y| - d(x,y)\big),$$

where $|x| = d(x, o)$. If X is a tree, then this is the length of the confluent, as defined in §6.B, or equivalently, the distance between o and $\pi(x, y)$. For hyperbolic metric spaces, this remains "roughly" true:

(22.4) Lemma. *If X is δ-hyperbolic and $\pi(x, y)$ is a geodesic arc, then*

$$d\big(o, \pi(x,y)\big) - 2\delta \le |x \wedge y| \le d\big(o, \pi(x,y)\big).$$

Proof. The upper bound is straightforward and does not require hyperbolicity.

Choose geodesic arcs $\pi(o, x)$ and $\pi(o, y)$. By continuity, we can find $w \in \pi(x, y)$ such that $d\big(w, \pi(o,x)\big) = d\big(w, \pi(o,y)\big)$. By δ-thinness there must be $u \in \pi(o, x)$ and $v \in \pi(o, y)$ such that $d(u, w) = d(v, w) \le \delta$. Then

$$\begin{aligned} |x| &= |u| + d(u,x) \ge |w| + d(w,x) - 2\delta \quad \text{and} \\ |y| &= |v| + d(v,y) \ge |w| + d(w,y) - 2\delta\,. \end{aligned}$$

Summing the two inequalities, we find the lower bound. □

(22.5) Lemma. *If X is δ-hyperbolic then, for any choice of o and all $u, v, w \in X$,*

$$|u \wedge w| \ge \min\{|u \wedge v|, |v \wedge w|\} - 3\delta\,.$$

Proof. Take geodesic arcs between u, v and w. (The reader is invited to draw a figure.) Let $x \in \pi(u, w)$ be such that $|x| = d\big(o, \pi(u,w)\big)$. By thinness, there is $y \in \pi(u, v) \cup \pi(v, w)$ such that $d(x, y) \le \delta$. Using Lemma 22.4,

$$|u \wedge w| \ge |x| - 2\delta \ge |y| - 3\delta \ge \min\big\{d\big(o, \pi(u,v)\big)\,,\ d\big(o, \pi(v,w)\big)\big\} - 3\delta\,,$$

and the statement follows. □

We remark that the converse of (22.5) also holds: if for any choice of o and all $u, v, w \in X$,

$$|u \wedge w| \ge \min\{|u \wedge v|, |v \wedge w|\} - \delta\,, \tag{22.6}$$

then X is 4δ-hyperbolic in the sense of (22.1). Also, if (22.6) holds for some reference point o, then it holds for every other reference point with 2δ in place of δ. We will not need these facts here, and their proofs are rather simple; see [142] or [78].

Next, we describe the *hyperbolic boundary* ϑX and *compactification* $\widehat{X}$ of X. We assume that X is a locally finite, δ-hyperbolic *graph* (but everything works for any proper, geodesic hyperbolic space). When confusion could arise, we shall write $\vartheta_e X$ for the space of ends and $\vartheta_h X$ for the hyperbolic boundary. Geodesics are determined by their vertices, so that we shall not need the extension of the metric to the edges.

We choose $a > 0$ such that $a' = e^{3\delta a} - 1 < \sqrt{2} - 1$ and define for $x, y \in X$ and a fixed base point o

$$\varrho_a(x,y) = \begin{cases} 0\,, & x = y\,,\\ \exp(-a|x \wedge y|)\,, & x \neq y\,. \end{cases}$$

This is similar to (6.10), but not necessarily a metric unless X is a tree. Lemma 22.5 implies

$$(22.7) \qquad \varrho_a(u,w) \le (1+a')\max\{\varrho_a(u,v), \varrho_a(v,w)\} \quad \text{for all } u,v,w \in X$$

We now define

$$\theta_a(x,y) = \inf\left\{ \sum_{i=1}^{n} \varrho_a(x_{i-1},x_i) : n \ge 1\,,\ x = x_0, x_1, \dots, x_n = y \in X \right\}.$$

Then θ_a is a metric on X.

(22.8) Proposition. $\qquad (1-2a')\,\varrho_a \le \theta_a \le \varrho_a\,.$

Proof. Only the first inequality needs a proof. We show by induction on n that for all $x = x_0, x_1, \dots, x_n = y \in X$,

$$(1-2a')\,\varrho_a(x,y) \le \sum_{i=1}^{n} \varrho_a(x_{i-1},x_i)\,.$$

For $n = 1$ there is nothing to prove. Let $n \ge 2$, and write R for the right hand side of the proposed inequality. If $R \ge 1 - 2a'$ then the latter is certainly true, because $\varrho_a(x,y) \le 1$. Suppose $R < 1 - 2a'$. Let m be the maximal index $\le n$ satisfying $\sum_{i=1}^{m} \varrho_a(x_{i-1},x_i) \le R/2$. Then $m < n$ and also $\sum_{i=m+2}^{n} \varrho_a(x_{i-1},x_i) \le R/2$. By the induction hypothesis,

$$\varrho_a(x,x_m) \le \tfrac{R}{2(1-2a')} \quad \text{and} \quad \varrho_a(x_{m+1},y) \le \tfrac{R}{2(1-2a')}\,,$$

and in addition $\varrho_a(x_m,x_{m+1}) \le R$. Now (22.7) implies

$$\begin{aligned} \varrho_a(x,y) &\le (1+a')^2 \max\{\varrho_a(x,x_m)\,, \varrho_a(x_m,x_{m+1})\,, \varrho_a(x_{m+1},y)\}\\ &\le \tfrac{R}{1-2a'} \max\{\tfrac{1}{2}(1+a')^2\,, (1-2a')(1+a')^2\}\,. \end{aligned}$$

The last "max" is ≤ 1, completing the proof. $\square$

Note that X is discrete in this metric, as $y \neq x$ implies $\varrho_a(x,y) \geq e^{-a|x|}$. We now define $\widehat{X}$ as the completion of X in the metric θ_a. From Proposition 22.8 we see that a sequence (x_n) with $|x_n| \to \infty$ is Cauchy if and only if

$$\lim_{m,n\to\infty} |x_m \wedge x_n| = \infty\,, \tag{22.9}$$

and another Cauchy sequence (y_n) will define the same boundary point if and only if

$$\lim_{n\to\infty} |x_n \wedge y_n| = \infty\,. \tag{22.10}$$

Thus one can also construct the boundary by factoring the set of all sequences in $X^{\mathbb{N}}$ that satisfy (22.9) with respect to the equivalence relation given by (22.10). From (22.8) we also see that the topology of $\widehat{X}$ does not depend on the choice of a, as long as $a' < \sqrt{2}-1$. Furthermore, it is independent of the choice of the base point o.

Similarly to trees, a third, equivalent way is to describe ϑX via equivalence of geodesic rays. Call two rays $\pi = [x_0, x_1, \dots]$ and $\pi' = [y_0, y_1, \dots]$ *equivalent* if

$$\liminf_{n\to\infty} d(y_n, \pi) < \infty\,.$$

The following is true in any connected graph.

(22.11) Lemma. *If $\pi = [x_0, x_1, \dots]$ is a geodesic ray and $u \in X$, then one can find integers $k, \ell \geq 0$ and points $u = u_0, u_1, \dots u_k = x_\ell$ such that $[u, u_1, \dots, u_k, x_{\ell+1}, x_{\ell+2}, \dots]$ is again a geodesic ray.*

Proof. It suffices to show this when $d(u, x_0) = 1$. In this case, $d(x_n, u) \in \{n-1, n, n+1\}$ for all n.

If there is ℓ such that $d(x_\ell, u) = \ell - 1$ then for all $n \geq \ell$

$$n - 1 \leq d(x_n, u) \leq d(x_n, x_k) + d(x_k, u) \leq n - 1\,,$$

and "=" holds in all three inequalities. We set $k = \ell - 1$ and join a geodesic arc $[u = u_0, \dots, u_k = x_\ell]$ with $[x_\ell, x_{\ell+1}, \dots]$.

So now suppose that $d(x_n, u) \geq n$ for all n. If $d(x_n, u) = n+1$ for all n then we just add the initial vertex u to the ray π. Otherwise, we can find ℓ such that $d(x_\ell, u) = \ell$, and as above, we find $d(x_n, u) = n$ for all $n \geq \ell$. Then $k = \ell$, and we proceed exactly as in the first case. $\square$

We now return to hyperbolic graphs.

(22.12) Proposition. (a) *If $\pi' = [y_0, y_1, \dots]$ is equivalent to $\pi = [x_0, x_1, \dots]$ then there is $k \in \mathbb{Z}$ such that $d(y_n, x_{n-k}) \le 2\delta$ for all but finitely many n. If $x_0 = y_0$ then this holds for all n with $k = 0$.*

(b) *Equivalence of geodesic rays is an equivalence relation.*

(c) *Every geodesic ray converges to a point in ϑX. Every boundary point is the limit of some geodesic ray starting at o. Two geodesic rays are equivalent if and only if they converge to the same boundary point.*

Proof. Let π and π' be equivalent. Assume first that $x_0 = y_0$. There are sequences $m(1) < m(2) < \cdots$ and $n(1) < n(2) < \cdots$ such that $d(y_{n(i)}, x_{m(i)}) \le M < \infty$. Given $n \ge 0$, choose i such that $n \le n(i) - M - \delta$. Consider the geodesic subarcs of π and π' from o to $x_{m(i)}$ and to $y_{n(i)}$, respectively. Add an arc from $x_{m(i)}$ to $y_{n(i)}$ to obtain a geodesic triangle. Then y_n lies on one of its sides, and there must be a point on one of the other sides at distance at most δ. By our choice of n and i, this point cannot lie on $\pi(x_{m(i)}, y_{n(i)})$. Therefore it is a point x_ℓ on π. We have $|n - \ell| = \big||y_n| - |x_\ell|\big| \le \delta$, whence $d(y_n, x_n) \le |n - \ell| + d(y_n, x_\ell) \le 2\delta$.

If $y_0 \ne x_0$, then Lemma 22.11 tells us that we can modify a finite initial piece of π' to obtain a geodesic ray starting at x_0. This yields the first statement of (a).

It is clear that equivalence of geodesic rays is symmetric. Transitivity follows immediately from (a), and (b) is verified.

For (c), start with a geodesic ray $\pi = [x_0, x_1, \dots]$. With respect to the origin x_0, we have $|x_m \wedge x_n|_{x_0} = \min\{m, n\} \to \infty$, as $m, n \to \infty$. Convergence to ∞ does not depend on the choice of the origin, so that (x_n) converges to some $\xi \in \vartheta X$ by (22.9). Let $\pi' = [y_0, y_1, \dots]$ be a second geodesic ray. If it is equivalent with π then, by (a), for each $n \ge n_0$ there is $m(n)$ such that $d(y_n, x_{m(n)}) \le \delta$, and $m(n) \to \infty$. Then $|y_n \wedge x_{m(n)}| \to \infty$, and $\theta_a(y_n, x_{m(n)}) \to 0$ in view of Proposition 22.8, whence also y_n tends to ξ. Therefore we have a mapping τ from the equivalence classes of rays to ϑX.

Next, let $\xi \in \vartheta X$, and choose (x_n) in X converging to ξ. Consider the geodesic arcs $\pi_n = \pi(o, x_n)$. By the standard "diagonal" argument, we find a sub-sequence of (π_n) having longer and longer common initial pieces; without loss of generality we assume that this is the whole sequence. That is, there are $y_n \in \bigcap_{k \ge n} \pi_k$ such that $|y_{n-1}| < |y_n| \to \infty$. The y_n lie on a unique ray π starting at o. Let $\pi(y_n, x_n)$ be the terminal piece of π_n. Then $d(o, \pi(y_n, x_n)) = |y_n|$, whence $y_n \to \xi$. Therefore τ is surjective.

To see injectivity, let π and π' as above be two rays converging to $\xi \in \vartheta X$, and such that $x_0 = y_0 = o$. Then $|x_n \wedge y_n| \to \infty$. Take $y_k \in \pi'$. By Lemma 22.4 there is n such that $d(o, \pi(x_n, y_n)) > k + \delta$. On the geodesic triangle with vertices o, x_n, y_n obtained by cutting π and π' at length n and adding the third side $\pi(x_n, y_n)$, the point y_k must have distance at most δ from a point on one of the other two sides. By our choice of n, the latter must lie on π, and $d(y_k, \pi) \le \delta$. Therefore π and π' are equivalent. □

(22.13) Corollary. *If X is a hyperbolic graph then $\widehat{X}$ is compact.*

Proof. Let (x_n) be a sequence in X. If $|x_n|$ does not tend to ∞ then it has a constant sub-sequence by local finiteness. Otherwise, we proceed as in the proof of Proposition 22.12 (surjectivity of τ) to find a sub-sequence that "converges" to a geodesic ray and thus tends to a limit in ϑX.

If (w_n) is an arbitrary sequence in $\widehat{X}$ then there is $x_n \in X$ such that $\theta_a(w_n, x_n) < 1/n$, and if $x_{n'} \to w \in \widehat{X}$ then also $w_{n'} \to w$. □

If X is a hyperbolic graph, then it is easy to understand how its hyperbolic compactification is related to the end compactification: the former is finer, that is, the identity on X extends to a continuous surjection from the hyperbolic to the end compactification which maps $\vartheta_h X$ onto $\vartheta_e X$. This is obvious from the construction of $\vartheta_h X$ via equivalence of rays. (Note that each end is represented by some geodesic ray. Local finiteness is used here.) For trees, the two compactifications are the same. It is easy to construct graphs where this is not true. For example, take a typical Cayley graph of the free product of two one-ended infinite hyperbolic groups (e.g. co-compact Fuchsian groups): each thick end corresponds to a translate of the hyperbolic boundary of one of the free factors.

On the other hand, observe that a graph with infinitely many ends will in general (even when it is transitive) be far from hyperbolic. Examples are provided by the Cayley graphs of $\mathbb{Z}^{d_1} * \mathbb{Z}^{d_2}$, where $d_i \ge 2$.

(22.14) Theorem. *The hyperbolic compactification of a hyperbolic graph is a contractive* $\mathrm{AUT}(X)$*-compactification.*

Proof. Projectivity is obvious. Indeed, if $|x_n| \to \infty$ and $d(y_n, x_n) \le M$ for all n, then $|x_n \wedge y_n| \to \infty$.

Note that the metric θ_a also depends on the base point: $\theta_a = \theta_{a,o}$. If $g \in \mathrm{AUT}(X)$, then g is an isometry $(\widehat{X}, \theta_{a,o}) \to (\widehat{X}, \theta_{a,go})$. Both metrics induce same topology, and $\widehat{X}$ is an $\mathrm{AUT}(X)$-compactification.

The convergence property needs a little more work. Let (g_n) be a sequence in $\mathrm{AUT}(X)$ with $g_n o \to \xi$ and $g_n^{-1} o = \eta \in \vartheta X$. Denote by $B_\theta(w, r)$ the open θ_a-balls, where $w \in \widehat{X}$ and $r > 0$. Take open $\widehat{X}$-neighbourhoods U, V of ξ and η, respectively. We can find $0 < \varepsilon < 1$ and n_0 such that

$$\{g_n o : n \ge n_0\} \cup \{\xi\} \subset B_\theta(g_{n_0} o, \varepsilon/3) \subset B_\theta(g_{n_0} o, \varepsilon) \subset U \quad \text{and}$$
$$\{g_n^{-1} o : n \ge n_0\} \cup \{\eta\} \subset B_\theta(g_{n_0}^{-1} o, \varepsilon/3) \subset B_\theta(g_{n_0}^{-1} o, \varepsilon) \subset V\,.$$

Let $x \in X$ with $\theta_a(g_{n_0}^{-1}o, x) \geq 2\varepsilon/3$. Choose $n \geq n_0$. Then $\theta_a(g_n^{-1}o, x) \geq \varepsilon/3$ and hence $|g_n^{-1}o \wedge x| \leq C = -\frac{1}{a}\log\frac{\varepsilon}{3}$. But then

$$|g_n x \wedge g_n o| = |g_n o| - |g_n^{-1}o \wedge x| \geq |g_n o| - C \to \infty \quad \text{as } n \to \infty .$$

Hence there is n_1 such that $\theta_a(g_n x, g_n o) \leq \exp(-a(|g_n o| - C) < \varepsilon/3$ for all $n \geq n_1$ and all x as above. As every $\zeta \in \vartheta X \setminus V$ can be approximated by elements of $X \setminus B_\theta(g_{n_0}^{-1}o, 2\varepsilon/3)$, the convergence property follows. □

We now want to study the Dirichlet problem in the transitive case. Again, we first derive information on the "degenerate" case (b) of Proposition 20.10. This requires a preliminary step.

A two-way infinite path $[\ldots, x_{-1}, x_0, x_1, \ldots]$ in a graph X such that $d(x_k, x_\ell) = |k-\ell|$ is called a *geodesic line.* If X is hyperbolic then $[x_0, x_1, \ldots]$ and $[x_0, x_{-1}, \ldots]$ are rays defining two different elements $\xi, \eta \in \vartheta X$, and we say that the line is a geodesic from η to ξ, denoted by $\pi(\eta, \xi)$.

(22.15) Lemma. *If X is a hyperbolic graph and $\xi, \eta \in \vartheta X$ are distinct, then there is a geodesic between the two.*

Proof. Consider rays $[x_0, x_1, \ldots]$ and $[y_0, y_1, \ldots]$ converging to ξ and η, respectively. By (22.10), the sequence $|x_n \wedge y_n|$ must be bounded, as otherwise (x_n) and (y_n) would have sub-sequences converging to the same boundary point. Let $k = \max\{|x_n \wedge y_n| : n \geq 0\}$ and ℓ such that $|x_\ell \wedge y_\ell| = k$. Thus, $d(x_n, y_n) \geq 2n - k$, and $d(x_\ell, y_\ell) = 2\ell - k$. If $n \geq \ell$ then $d(x_n, y_n) \leq 2(n-\ell) + d(x_\ell, y_\ell) = 2n - k$. Therefore $d(x_n, y_n) = 2n - k$ for all $n \geq \ell$.

Now insert a geodesic arc $\pi(y_\ell, x_\ell)$ between the rays $[\ldots, y_{\ell+2}, y_{\ell+1}, y_\ell]$ and $[x_\ell, x_{\ell+1}, x_{\ell+2}, \ldots]$. This gives $\pi(\eta, \xi)$ as required. □

(22.16) Proposition. *Let X be a hyperbolic graph with $|\vartheta X| > 2$ and G a closed, transitive subgroup of* AUT(X). *If G fixes a point on the boundary, then G is non-unimodular.*

Proof. Let M be the constant vertex degree of X, and let ξ be the unique point in ϑX fixed by G. By Proposition 20.9(1), ϑX is infinite. For every $N > 0$ we can find distinct points $\eta_0, \ldots, \eta_N \in \vartheta X \setminus \{\xi\}$. Choose geodesics $\pi(\eta_i, \xi) = [x_n^i, n \in \mathbb{Z}]$. Proposition 22.12(a) allows us to label the sequences in such a way that

$$d(x_n^i, x_n^0) \leq 2\delta \quad \text{for all } i = 1, \ldots, N \text{ and } n \geq 0 .$$

By transitivity, we may assume that $x_0^0 = o$.

If K is sufficiently large, the points $y_i \in \{x_n^i : n < o\}$ with $|y_i| = K$ $(i = 0, \ldots, N)$ must be distinct. In particular, $y_0 = x_{-K}^0$ and $y_i = x_{-\ell(i)}^i$ with $|K - \ell(i)| \leq |x_0^i| \leq 2\delta$.

For every $i = 1, \ldots, N$, there is $g_i \in G$ with $g_i y_0 = y_i$. As g_i fixes ξ, the ray $[g_i x^0_{n-K}, n \in \mathbb{N}_0]$ is a geodesic from y_i to ξ. The same is true for $[x^i_{n-\ell(i))}, n \in \mathbb{N}_0]$. Hence, by Proposition 22.12(a),

$$|g_i o| \le d(g_i o, x^i_{K-\ell i}) + d(x^i_{K-\ell i}, x^i_0) + |x^i_0| \le 6\delta\,.$$

Now there are at most $M^{6\delta}$ elements $x \in X$ with $|x| \le 6\delta$. Hence for one of these x we must have $|I| \ge N/M^{6\delta}$, where $I = \{i \in \{1, \ldots, N\} : g_i o = x\}$ (the pigeonhole principle). Choose and fix $i \in I$. Then $g_i^{-1} g_j \in G_o$ for all $j \in I$, and the points $g_i^{-1} g_j y_0$ are all distinct. Thus $|G_o y_0| \ge N/M^{6\delta}$.

On the other hand, let $h \in G_{y_0}$. Then $[h x^0_{n-K}, n \in \mathbb{N}_0]$ is a ray from y_0 to ξ, and (22.12.a) yields $|ho| \le 2\delta$. Therefore $|G_{y_0} o| \le M^{2\delta}$.

For $g \in G$ with $go = y_0$ we get from Lemma 1.29 that $\Delta(g) = |G_o y_0|/|G_{y_0} o| \ge N/M^{8\delta}$. As N may be chosen arbitrarily large, we find $g \in G$ with $\Delta(g) > 1$. □

As in the case of ends, we can now apply Theorems 20.13 and 20.3:

(22.17) Corollary. *Let X be a hyperbolic graph with infinite boundary, and suppose that there is a closed, transitive subgroup Γ of $\mathrm{AUT}(X, P)$. Unless Γ fixes an element of X, the Dirichlet problem for P-harmonic functions is solvable with respect to the hyperbolic compactification, and the random walk with transition matrix P converges a.s. to a random point of ϑX.*

This applies, in particular, to random walks on finitely generated hyperbolic groups with infinite boundary.

For the following, combine Corollary 12.12 with Exercise 20.11 and Proposition 22.16.

(22.18) Exercise. *Every vertex-transitive hyperbolic graph with infinite hyperbolic boundary satisfies IS.*

Again, we conclude this section with results where group-invariance is replaced by geometric adaptedness conditions. The following is the analogue of Theorem 21.16.

(22.19) Theorem. *Suppose that X is a hyperbolic graph. If (X, P) is uniformly irreducible and has a uniform first moment, and $\rho(P) < 1$, then Z_n converges almost surely to a random point of ϑX.*

Proof. Propositions 8.2 and 8.8(a) imply that $\liminf \frac{1}{n}|Z_n \wedge Z_{n+1}| > \underline{\mathfrak{m}} > 0$ almost surely. Consequently, if $m, n \in \mathbb{N}$ and $n \ge m$ then

$$\theta_a(Z_m, Z_n) \le \sum_{k=m}^{\infty} \exp\bigl(-a|Z_k \wedge Z_{k+1}|\bigr)\,,$$

which tends to 0 almost surely. Therefore, (Z_n) is a Cauchy sequence and converges almost surely in $\widehat{X}$. Transience of (Z_n) and local finiteness of X imply that the limit must be a boundary point. □

A hyperbolic analogue of Theorem 21.15, with the additional condition $\rho(P) < 1$, will be considered later, in Exercise 27.17. The following is the hyperbolic analogue of Theorem 21.18.

(22.20) Theorem. *Assume that the hyperbolic graph X satisfies IS and that P is strongly reversible and uniformly irreducible and has a uniform first moment. Then the Dirichlet problem with respect to the hyperbolic compactification is solvable.*

Proof. As Theorems 10.3 and 10.6 imply $\rho(P) < 1$, we have convergence to the boundary (Theorem 22.19) with harmonic measures ν_x, $x \in X$, on ϑX. Consider $\xi \in \vartheta X$ and a neighbourhood $B_\theta(\xi, r)$, where $r > 0$.

We apply Lemma 21.17 and take $\mathcal{A}_x = \mathcal{A}_x(a, \varepsilon)$ with $\varepsilon < \min\{\varepsilon_0, \underline{\mathfrak{m}}/2\}$ and $a = 2/\underline{\mathfrak{m}}$.

Let $\omega \in \mathcal{A}_x$. We claim that $\theta_a\big(x, Z_n(\omega)\big) < r/2$ when x is sufficiently large. Thus, if in addition $x \in B_\theta(\xi, r/2)$, then $\lim Z_n(\omega) \in B_\theta(\xi, r)$, and

$$\nu_x\big(\vartheta X \cap B_\theta(\xi, r)\big) \ge \mathbb{P}_x(\mathcal{A}_x) \to 1\,, \quad \text{as } x \to \xi\,.$$

To prove the claim, first observe that for $n \ge 2|x|/\underline{\mathfrak{m}}$, (21.17.4) implies $|Z_n(\omega)| \ge \underline{\mathfrak{m}}\, n - |x| \ge \underline{\mathfrak{m}}\, n/2$. In combination with (21.17.3), our choice of ε implies

$$|Z_n(\omega) \wedge Z_{n+1}(\omega)| \ge \underline{\mathfrak{m}}\, n/4 \quad \text{for all } n \ge 2|x|/\underline{\mathfrak{m}}\,.$$

On the other hand, (21.17.2) and (21.17.5) yield

$$|Z_n(\omega) \wedge Z_{n+1}(\omega)| > \varepsilon\, |x|/2 \quad \text{for all } n \le 2|x|/\underline{\mathfrak{m}}\,.$$

Therefore

$$\begin{aligned}\theta_a\big(x, Z_n(\omega)\big) &\le \sum_{k=0}^{\infty} \exp\big(-a|Z_k(\omega) \wedge Z_{k+1}(\omega)|\big)\\ &\le \frac{2|x|}{\underline{\mathfrak{m}}} \exp\Big(-a\frac{\varepsilon|x|}{2}\Big) + \sum_{k \ge 2|x|/\underline{\mathfrak{m}}} \exp\Big(-a\frac{\underline{\mathfrak{m}}\, k}{4}\Big)\,,\end{aligned}$$

which tends to 0 as $|x| \to \infty$. □

Again, in this theorem *IS* and strong reversibility may be replaced by the condition $p^{(n)}(x, y) \le C\, \rho^n$, with $\rho < 1$.

23. The Dirichlet problem for circle packing graphs

In this section we continue the considerations regarding simple random walk on the graphs studied in §6.D. Recall in particular Theorems 6.29 and 6.34. If X is a disk triangulation graph which is CP-hyperbolic, then we realize it as the contacts graph of a circle packing whose carrier is the open unit disk $\mathbb{D}$. Thus, the set of accumulation points of (this isomorphic copy of) X coincides with the unit circle $\vartheta\mathbb{D}$. In this way we obtain a compactification of X.

(23.1) Theorem. *Let X be the contacts graph of a circle packing $\mathcal{P}$ of the unit disk $\mathbb{D}$, and suppose that X has bounded geometry. Then the Dirichlet problem with respect to the boundary $\vartheta\mathbb{D}$ is solvable for the simple random walk on X.*

We now collect the tools for proving the theorem. For an arbitrary Markov chain (X, P), consider the first hitting time $\mathbf{s}^A$ of $A \subset X$, and define the transition matrix P^A by

$$p^A(x,y) = \mathbb{P}_x[\mathbf{s}^A < \infty\,,\ Z_{\mathbf{s}^A} = y]\,,$$

the probability that the first visit to A occurs in y. We start with a lemma on functions with finite Dirichlet sum for reversible chains. Recall the relevant notation (Section 2).

(23.2) Lemma. *Let (X, P) be reversible and f a function on X with $D(f) < \infty$. If $A \subset X$ is finite then $\min\{D(h) : h \equiv f \text{ on } X \setminus A\}$ is assumed for the function $h = P^{X\setminus A}f$.*

Proof. Analogously to (2.11), the minimum is attained for a unique function h.

For arbitrary functions h with $h \equiv f$ on $X \setminus A$, consider $D(h)$ as a function of the real variables $h(x)$, $x \in A$. The function h that gives the minimum must be a stationary point. But $\partial D(h)/\partial h(x) = 0$ just means that h is harmonic at x, and this holds for every $x \in A$.

As A is finite and $X \setminus A \neq \emptyset$, the matrix $P^{X\setminus A}$ is stochastic, and by factoring through the first step of the random walk, one finds that $P^{X\setminus A}f$ is harmonic on A, while it coincides with f on $X \setminus A$. Therefore $h - P^{X\setminus A}f$ is harmonic on A and $\equiv 0$ on $X \setminus A$. By the maximum (minimum) principle, we must have $h - P^{X\setminus A}f = 0$ on A. □

We return to the setting of Theorem 23.1. Note that the assumptions (which we shall use throughout this section) contain the fact that the centre of each circle C_x of $\mathcal{P}$ is $z(x) = x$. As usual, M denotes the maximum vertex degree in X. We shall use the last lemma in the following, crucial, distance estimate.

(23.3) Proposition. *Let $x_0 \in X$, $\delta = \inf\{|x_0 - \xi| : \xi \in \vartheta\mathbb{D}\}$ and $t > 1$. There is a constant c depending only on M such that*

$$\mathbb{P}_{x_0}[\exists\, n : |Z_n - x_0| > t\delta] \le 4c/\log t\,.$$

The proof requires another auxiliary lemma. Let $r, s \in \mathbb{R}$, $r < s$. We define the projection $\psi = \psi_{r,s} : \overline{\mathbb{R}} \to [r, s]$, to wit, $\psi(u) = \max\{r, \min\{s, u\}\}$ is the element of $[r\,,\, s]$ closest to $u \in \overline{\mathbb{R}}$. We then define $f = f_{r,s} : \mathbb{C} \to \mathbb{R}$ by

$$f(z) = \psi\Big(\log \frac{|z - x_0|}{\delta}\Big)$$

and consider its restriction $g = g_{r,s}$ to X.

(23.4) Lemma. *There is a constant c depending only on M such that $D(g) \le 4c\,(s - r)$.*

Proof. The natural logarithm maps $\mathbb{C} \setminus \{0\}$ onto the strip $\mathbb{R} \times (-\pi\,,\,\pi]$. In addition to ψ and f, we define the functions $\Psi(u + iv) = \psi(u) + iv$ and

$$F(z) = F_{r,s}(z) = \Psi\Big(\log \frac{z - x_0}{\delta}\Big)\,.$$

Recall the ring lemma 6.33 and the argument that led to (6.35), with x_0 in the place of $o = 0$: if $x \ne x_0$ then $\mathfrak{r}(x) \le a\,\mathfrak{s}(x)$, where $\mathfrak{s}(x)$ is the smallest radius of a circle tangent with x, and

$$\begin{aligned}\max\{|z - x_0| : z \in C_x\} &\le 2a\,\mathfrak{s}(x) + \min\{|z - x_0| : z \in C_x\}\\ &\le (2a + 1)\,\min\{|z - x_0| : z \in C_x\}\,.\end{aligned}$$

Let U be any closed disk contained in the annulus $\{\delta\, e^r \le |z - x_0| \le \delta\, e^s\}$, and such that $\max\{|z - x_0| : z \in U\} \le (2a + 1)\,\min\{|z - x_0| : z \in U\}$. Then $|F'(z_1)/F'(z_2)| \le 2a + 1$ for all $z_1, z_2 \in U$. By the chain rule, $\text{area}\big(F(U)\big) \ge \text{area}(U)\,\min_{z \in U}|F'(z)^2|$ and $\text{diam}\big(F(U)\big) \le \text{diam}(U)\,\max_{z \in U}|F'(z)|$. Therefore $\text{diam}\big(F(U)\big)^2 \le c_1\,\text{area}\big(F(U)\big)$, where $c_1 = (4a + 2)^2/\pi$ depends only on a (and thus only on M). This implies in turn

$$c_1\,\text{area}\big(F(U)\big) \ge \Big(\max\{f(z) : z \in U\} - \min\{f(z) : z \in U\}\Big)^2\,.$$

Now, if $x \in X$ is such that C_x intersects $\{\delta\, e^r \le |z - x_0| \le \delta\, e^s\}$, then it contains a disk U such that $\{|z - x_0| : z \in U\} = \{|z - x_0| : z \in C_x\} \cap [\delta\, e^r\,,\, \delta\, e^s]$. Therefore

$$c_1\,\text{area}\big(F(C_x)\big) \ge \Big(\max\{f(z) : z \in C_x\} - \min\{f(z) : z \in C_x\}\Big)^2\,. \tag{23.5}$$

If C_x does not intersect the annulus, then both sides of (23.5) are 0. Therefore (23.5) holds for all $x \in X$.

Take two neighbours $x, y \in X$. Then $g(x) - g(y) = \big(f(x) - f(z)\big) + \big(f(z) - f(y)\big)$, where $\{z\} = C_x \cap C_y$, and (23.5) yields

$$\big(g(x) - g(y)\big)^2 \le 2c_1\Big(\text{area}\big(F(C_x)\big) + \text{area}\big(F(C_y)\big)\Big).$$

We sum over all edges of X. Since the circles of $\mathcal{P}$ have disjoint interiors, we find

$$D(g) \le 2c_1 M\, \text{area}\big(F(\mathbb{C} \setminus \{0\})\big) = 4\pi c_1 M\,(s - r). \qquad \square$$

Proof of Proposition 23.3. Let $B = \{x \in X : |x - x_0| \le t\delta\}$. We assume that $X \setminus B \ne \emptyset$, as otherwise we have nothing to prove.

We set $r = \log\sqrt{t}$, $s = 2r$, and $g = g_{r,s}$. Next, let $A_1 \subset A_2 \subset \cdots$ be a sequence of finite subsets of B with $\bigcup_n A_n = B$, and define $g_n = P^{X\setminus A_n} g$. We shall show that

$$\big(g_n(x_0) - r\big)r \le \sqrt{c\, r\, D(g_n)}\,. \tag{23.6}$$

If this holds, (23.2) and (23.4) imply $g_n(x_0) \le 2c + r$. On the other hand,

$$\begin{aligned} g_n(x_0) &= \sum_{x \in X\setminus B} p^{X\setminus A_n}(x_0, x)\, f(x) + \sum_{x \in B} p^{X\setminus A_n}(x_0, x)\, f(x) \\ &\ge 2r\, p^{X\setminus A_n}(x_0, X \setminus B) + r\, p^{X\setminus A_n}(x_0, B)\,. \end{aligned}$$

We let $n \to \infty$ and take the lim sup of the last expression. Then we get

$$2r\,\mathbb{P}_{x_0}[\exists\, n : |Z_n - x_0| > t\delta] + r\,\mathbb{P}_{x_0}[\,|Z_n - x_0| \le t\delta \ \ \forall n] \le 2c + r\,,$$

and $\mathbb{P}_{x_0}[\exists\, n : |Z_n - x_0| > t\delta] \le 2c/r = 4c/\log t$, as required.

In order to prove (23.6), let A be a finite subset of B, and set $g_A = P^{X\setminus A} g$. Choose $\tau \in (1\,, \sqrt{t})$. Let X_τ be the set of all $x \in X$ such that C_x intersects the circle $\{|z - x_0| = \delta\,\tau\}$. We claim that there is $x_\tau \in X_\tau$ satisfying $g_A(x_\tau) \ge g_A(x_0)$.

If $x_0 \notin A$ then $g_A(x_0) = r \le g_A(x)$ for all x. So suppose that $x_0 \in A$. Let A' be the component of $A \setminus X_\tau$ that contains x_0, and dA' the set of vertices in $X \setminus A'$ having a neighbour in A'. As g_A is harmonic in A', the maximum principle implies that there is $y \in dA'$ such that $g_A(y) \ge g_A(x_0)$. If $y \in X_\tau$ then we are done. Otherwise, $y \in X \setminus A$ and $|y - x_0| < \delta\,\tau$. Therefore $g_A(y) = g(y) = r$. Note that X_τ is infinite. Consequently there must be $x \in X_\tau \setminus A$, and $g_A(x) = g(x) \ge r = g_A(x_0)$.

Now start at a point of C_{x_τ} that lies on the circle $\{|z - x_0| = \delta\,\tau\}$ and move along that circle in (say) the clockwise direction towards $\vartheta\mathbb{D}$. Any two consecutive disks of the packing that are met on the way must be such that their centres are neighbours in X (because X is a triangulation graph). Thus, we obtain an infinite path starting at x_τ, all of whose vertices are in X_τ. Also, $\mathfrak{r}(x) \to 0$ as $x \to \vartheta\mathbb{D}$. In particular, there must be a vertex $y_\tau \in X \setminus A$ on this path such that $|y_\tau - x_0| < \delta\sqrt{t}$, and $g_A(y_\tau) = r$. We write $\pi_\tau = [x_\tau = y_0, y_1, \dots, y_n = y_\tau]$ for the truncation of the path at y_τ. Then

$$g_A(x_0) - r \le g_A(x_\tau) - g_A(y_\tau) \le \sum_{i=1}^n |g_A(y_i) - g_A(y_i)| \le \sum_{e \in E_\tau} |\nabla g_A(e)|,$$

where E_τ is the set of edges of X having both endpoints in X_τ. As a function of τ, the last term is piecewise constant and has finite support by finiteness of A. Therefore, dividing by τ and integrating over $(1, \sqrt{t})$,

$$(23.7) \qquad \big(g_A(x_0) - r\big) r \le \int_1^{\sqrt{t}} \sum_{e \in E_\tau} \frac{|\nabla g_A(e)|}{\tau}\, d\tau = \sum_{e \in E(X)} |\nabla g_A(e)| \int_1^{\sqrt{t}} \frac{\mathbf{1}_{E_\tau}(e)}{\tau}\, d\tau\,.$$

Now $e \in E_\tau$ implies $C_x \cap \{|z - x_0| = \delta\,\tau\} \neq \emptyset$, where $x = e^-$, and

$$\int_1^{\sqrt{t}} \frac{\mathbf{1}_{E_\tau}(e)}{\tau}\, d\tau \le \max\{f_{0,r}(z) : z \in C_x\} - \min\{f_{0,r}(z) : z \in C_x\} \le \sqrt{c_1\, \mathrm{area}\big(F_{0,r}(C_x)\big)}$$

by (23.5). As in the proof of Lemma 23.4, we conclude that

$$\sum_{e \in E(X)} \left(\int_1^{\sqrt{t}} \frac{\mathbf{1}_{E_\tau}(e)}{\tau}\, d\tau \right)^2 \le c\,r\,.$$

The Cauchy–Schwarz inequality applied to (23.7) now yields (23.6). □

Proof of Theorem 23.1. We use Theorem 20.3 and first show convergence to the boundary. Let $\Omega_0 = \{\omega \in \Omega : d\big(Z_n(\omega), Z_0(\omega)\big) \to \infty\}$, where $d(\cdot,\cdot)$ is the graph metric. By transience, $\mathbb{P}_x(\Omega_0) = 1$ for every $x \in X$. For $\omega \in \Omega_0$, the set $L(\omega)$ of all accumulation points of $\big(Z_n(\omega)\big)$ is contained in $\vartheta\mathbb{D}$. Choose $\varepsilon \in (0\,,\,1)$, and let $\Omega_\varepsilon = \{\omega \in \Omega_0 : \mathrm{diam}\big(L(\omega)\big) \ge \varepsilon\}$.

Now take $\delta = \varepsilon^2/4$, $t = 1/\varepsilon$ and consider the set $A = \{x \in X : |x - \xi| \ge \delta$ for all $\xi \in \vartheta\mathbb{D}\}$. If $y \in X \setminus A$ then by Proposition 23.3

$$\mathbb{P}_y(\Omega_\varepsilon) \le \mathbb{P}_y[\exists n : |Z_n - y| > \varepsilon/4] \le 4c/\log(1/\varepsilon)\,.$$

If $x \in X$ is arbitrary then by transience

$$\mathbb{P}_x(\Omega_\varepsilon) = \sum_{y \in X \setminus A} p^{X \setminus A}(x, y)\, \mathbb{P}_y(\Omega_\varepsilon) \le 4c/\log(1/\varepsilon)\,.$$

If we let $\varepsilon \to 0$ then we see that $\mathbb{P}_x\big[\text{diam}\big(L(\cdot)\big)\big] = 0$, and Z_n converges $\mathbb{P}_x$-almost surely.

Now let $\xi \in \vartheta\mathbb{D}$ and $\varepsilon > 0$. Consider $U = \{z \in \overline{\mathbb{D}} : |z - \xi| < \varepsilon\}$. With $\delta = \varepsilon^2/4$ and $t = 1/\varepsilon$ as above, take $x \in X$ with $|x - \xi| < \delta$. Then

$$\nu_x(U \cap \vartheta\mathbb{D}) \geq \mathbb{P}_x[\,|Z_n - x| \leq \varepsilon/4 \;\; \forall n] \geq 1 - 4c/\log(1/\varepsilon)\,.$$

Thus, ν_x tends to the point mass at ξ when $x \to \xi$. □

24. The construction of the Martin boundary

In this section we briefly introduce the essential features of the construction of the compactification associated with the cone of positive harmonic functions for an irreducible Markov chain (X, P).

Recall the definitions (§7.A) of t-harmonic and t-superharmonic functions and the respective positive cones $\mathcal{H}^+(P,t)$ and $\mathcal{S}^+(P,t)$ with the base $\mathcal{B}(P,t)$ of the latter. We start with $t = 1$, and omit t (as well as P, when no confusion can arise) in this case.

Suppose that (X, P) is *transient*, so that $0 < G(x,y) < \infty$ for all $x, y \in X$. For a function $f : X \to \mathbb{R}$, its *potential* is the function $g = Gf$, where $Gf(x) = \sum_y G(x,y)f(y)$. Here, the last sum is assumed to converge absolutely (f is *G-integrable*).

(24.1) Exercise. *If g is the potential of f, then $f = (I-P)g$. Furthermore, $P^n g \to 0$ pointwise, as $n \to \infty$. If $f \geq 0$ then $g \in \mathcal{S}^+$, and g is harmonic (that is, $Pg(x) = g(x)$) in every $x \in X \setminus \text{supp}\, f$.*

We deduce the discrete version of a basic theorem of potential theory.

(24.2) Riesz decomposition theorem. *If $u \in \mathcal{S}^+$ then there are a unique potential $g = Gf$ and harmonic function $h \geq 0$ such that $u = g + h$. In particular, if u is bounded above by a potential, then $h \equiv 0$.*

Proof. We have $P^n u \geq P^{n+1}u \geq 0$. Define $h(x) = \lim_n P^n u(x)$. As $Pu(x)$ is finite, we can use monotone convergence to obtain

$$Ph(x) = P\left(\lim_{n\to\infty} P^n u\right)(x) = \lim_{n\to\infty} P^{n+1}u(x) = h(x)\,,$$

whence $h \in \mathcal{H}^+$. The function $f = u - Pu$ is P^k-integrable, and

$$u - P^{n+1}u = \sum_{k=0}^{n}(P^k u - P^{k+1}u) = \sum_{k=0}^{n} P^k f\,.$$

Letting $n \to \infty$, we find $u - h = Gf = g$.

To see uniqueness, suppose that $u = g_1 + h_1$ is another decomposition. Then $P^n u = P^n g_1 + h_1$ for each n. Letting $n \to \infty$, Exercise 24.1 yields $P^n g_1 \to 0$, so that $h_1 = h$ and consequently $g_1 = g$.

Finally, if $u \le \bar{g} = G\bar{f}$, then $h \le \lim_n P^n \bar{g} = 0$. □

We now want to gain some information about the compact convex set $\mathcal{B} = \{u \in \mathcal{S}^+ : u(o) = 1\}$, where o is our reference point in X. If $y \in X$, then $K(\cdot, y) \in \mathcal{B}$, where $K(x,y) = G(x,y)/G(o,y) = F(x,y)/F(o,y)$.

(24.3) Definition. (1) The *Martin kernel* is $K(x,y) = F(x,y)/F(o,y)$.

(2) A *minimal harmonic function* is a function $h \in \mathcal{H}^+$ such that $h(o) = 1$ and, whenever $h \ge h_1$ on X for $h_1 \in \mathcal{H}^+$, the function h_1/h is constant.

(24.4) Proposition. *The extremal elements of $\mathcal{B}$ are precisely the Martin kernels $K(\cdot, y)$, where $y \in X$, and the minimal harmonic functions.*

Proof. Let $u \in \mathcal{B}$ be extremal. Then one of the two pieces in the Riesz decomposition must vanish. Suppose $u = Gf$, where $f \ge 0$, $f \ne \mathbf{0}$. Then $u = \sum_y c_y\, K(\cdot, y)$, where $c_y > 0$ for $y \in \operatorname{supp} f$, a convex combination. By transience, all $K(\cdot, y)$ are different (strictly superharmonic at y, harmonic elsewhere). Extremality yields that $|\operatorname{supp} f| = 1$, and $u = K(\cdot, y)$ for some $y \in X$. Otherwise, u is harmonic, and obviously minimal.

Conversely, suppose that $K(\cdot, y) = c \cdot u_1 + (1-c) \cdot u_2$ with $u_i \in \mathcal{B}$. By (24.2), each u_i is a potential Gf_i. Along with $K(\cdot, y)$, u_i must also be harmonic in $X \setminus \{y\}$. Therefore $\operatorname{supp} f_i = \{y\}$, and $u_i = K(\cdot, y)$. Finally, if h is minimal harmonic and $h = c \cdot u_1 + (1-c) \cdot u_2$ with $u_i \in \mathcal{B}$ then $u_i \in \mathcal{H}^+$, and minimality implies $u_i = h$. □

Now note that, by transience, we can embed X into the set of extremal elements of $\mathcal{B}$ via the map $y \mapsto K(\cdot, y)$, and identify X with its image. It will turn out that the other extremal elements, namely the minimal harmonic functions, all sit in the closure of X in $\mathcal{B}$, that is, they are pointwise limits of sequences $K(\cdot, y_n)$. One possibility is now to define the Martin compactification just as this closure. This coincides with the original approach of Doob [101]. However, when P does not have finite range, it may happen that X is not discrete in $\mathcal{B}$, while this is one of our requirements for a compactification of a discrete set.

(24.5) Definition. The *Martin compactification* $\widehat{X}(P)$ is the unique smallest compactification of the discrete set X to which all kernels $K(x, \cdot)$, $x \in X$, extend continuously. The *Martin boundary* is $\mathcal{M}(P) = \mathcal{M}(X, P) = \widehat{X}(P) \setminus X$.

The term "smallest" refers to the partial order on compactifications, where id_X extends to a continuous surjection from the "bigger" onto the "smaller" compactification; "equal" means homeomorphic.

We can think of the Martin boundary in terms of equivalence classes of sequences: a sequence $(y_n) \in X^{\mathbb{N}}$ converges to the Martin boundary, if it tends to ∞ (i.e., leaves every finite subset of X), and $K(\cdot, y_n)$ converges pointwise. Two sequences are equivalent, if the limits coincide at each point of X. The extended kernels are also called Martin kernels and written $K(x,\xi)$, where $x \in X$ and $\xi \in \widehat{X}(P)$. By construction, the kernels separate the points of the boundary: if $\xi, \eta \in \mathcal{M}$ are distinct then there is $x \in X$ such that $K(x,\xi) \neq K(x,\eta)$.

Another way of constructing the Martin boundary is the following: note that $F(o,y) \geq F(o,x)F(x,y)$, so that $K(x,y) \leq 1/F(o,x)$. Choose positive weights w_x $(x \in X)$ such that $\sum_X \big(w_x/F(o,x)\big) < \infty$, and define the metric

$$\theta(y_1,y_2) = \sum_{x \in X} w_x \Big(|K(x,y_1) - K(x,y_2)| + |\delta_x(y_1) - \delta_x(y_2)| \Big)$$

on X. The completion of (X,θ) is (homeomorphic with) the Martin compactification. We now prepare for the integral representation theorem.

(24.6) Approximation theorem. *If $h \in \mathcal{H}^+$ then there is a sequence of potentials $g_n = Gf_n$ with $f_n \geq 0$ such that $g_n \to h$ pointwise from below.*

Proof. We use the method of "balayage". Let $A \subset X$ be finite. The *reduced* function of h on A is

$$R^A[h](x) = \inf\{u(x) : u \in \mathcal{S}^+ ,\ u \geq h \text{ on } A\}.$$

This is a superharmonic function, $R^A[h] \leq h$ on X and $= h$ on A. Let $f_A = h \cdot \mathbf{1}_A$. Then ($A$ being finite) Gf_A is finite and $Gf_A \geq h$ on A. Therefore $R^A[h] \leq Gf_A$, and by Theorem 24.2, $R^A[h]$ is a potential. Furthermore, if $A \subset B$, then $R^A[h] \leq R^B[h]$.

We now choose an increasing sequence (A_n) of finite subsets of X with union X. Then the sequence of potentials $R^{A_n}[h]$ tends to h from below. □

The following is one of the two main theorems in Martin boundary theory.

(24.7) Poisson–Martin representation theorem. *For every $h \in \mathcal{H}^+$ there is a positive Borel measure ν^h on $\mathcal{M} = \mathcal{M}(P)$ such that*

$$h(x) = \int_{\mathcal{M}} K(x,\cdot)\, d\nu^h .$$

Proof. By the approximation Theorem, we have $h = \lim_n Gf_n$ pointwise, where $f_n \geq 0$. We can write

$$Gf_n(x) = \int_{\widehat{X}} K(x,\cdot)\, d\nu_n , \quad \text{where} \quad \nu_n = \sum_{y \in X} \frac{f_n(y)}{K(o,y)}\, \delta_y .$$

We consider the ν_n as Borel measures on the compact metric space $\widehat{X} = \widehat{X}(P)$. For all but finitely many n, their total mass is $h(o)$. By Helly's principle for measures on compact metric spaces (see e.g. Parthasarathy [253]), some sub-sequence of the ν_n converges weakly to a measure ν^h on $\widehat{X}$. All $K(x,\cdot)$ are continuous, whence $h(x) = \int_{\widehat{X}} K(x,\cdot)\, d\nu^h$.

The integral defines a superharmonic function. If we had $\nu^h(y) > 0$ for some $y \in X$ then h would be strictly superharmonic at y. Therefore ν^h must be supported by $\mathcal{M}$. □

We can now show that every minimal harmonic function corresponds to some point of $\mathcal{M}(P)$.

(24.8) Theorem. *If h is minimal harmonic then the representing measure ν^h is unique and a point mass. Thus, $h = K(\cdot,\xi)$ for some $\xi \in \mathcal{M}(P)$.*

Proof. Note that $\nu^h(\mathcal{M}) = h(o) = 1$. Let $B \in \mathcal{M}$ be a Borel set with $0 < \nu^h(B) < 1$ (we shall see that this is impossible). Then we can write $h = \nu^h(B) \cdot h_B + \big(1 - \nu^h(B)\big) \cdot h_{\mathcal{M}\setminus B}$, where

$$h_B(x) = \frac{1}{\nu^h(B)} \int_B K(x,\cdot)\, d\nu^h .$$

By minimality of h we must have $h = h_B = h_{\mathcal{M}\setminus B}$, whence

$$\int_B h(x)\, d\nu^h(\xi) = \nu^h(B)\, h_B(x) = \int_B K(x,\xi)\, d\nu^h(\xi)$$

for every $x \in X$ and every Borel set $B \subset \mathcal{M}$. (If $\nu^h(B) = 0$ or $= 1$ then this is trivially true.) Therefore $K(x,\cdot) \equiv h(x)$ holds ν^h-almost surely for every x, and, X being countable, also $\nu^h(A) = 1$, where

$$A = \{\xi \in \mathcal{M} : h(x) = K(x,\xi)\ \forall\, x \in X\} .$$

We infer that $h = K(\cdot,\xi)$ for some $\xi \in \mathcal{M}$. This ξ must be unique, as by construction $K(\cdot,\xi) \neq K(\cdot,\eta)$ when $\xi,\eta \in \mathcal{M}$ are distinct. □

We shall write $\mathcal{M}_{\min} = \mathcal{M}_{\min}(P)$ for the *minimal Martin boundary,* i.e., the set of all $\xi \in \mathcal{M}$ for which $K(\cdot,\xi)$ is minimal harmonic. We now state further important results on the Martin compactification without proofs.

(24.9) Uniqueness theorem. *$\mathcal{M}_{\min}$ is a Borel subset of $\mathcal{M}$. Given $h \in \mathcal{H}^+$, there is a unique measure ν^h such that $\nu^h(\mathcal{M} \setminus \mathcal{M}_{\min}) = 0$ and $h(x) = \int_{\mathcal{M}} K(x,\cdot)\, d\nu^h$.*

Henceforth, ν^h will always denote this particular measure supported by $\mathcal{M}_{\min}(P)$. The most important theorem of general Martin boundary theory is the following.

(24.10) Convergence theorem. *For transient (X,P), the sequence (Z_n) converges $\mathbb{P}_x$-almost surely for every x to an $\mathcal{M}_{\min}$-valued random variable Z_∞. The hitting distributions ν_x, $x \in X$, satisfy*

$$\nu_x(B) = \int_B K(x,\cdot)\, d\nu_o .$$

In particular, we see that $\nu_o = \nu^{\mathbf{1}}$, the measure on $\mathcal{M}_{\min}$ that represents the constant harmonic function $\mathbf{1}$. Also, the Radon–Nikodym derivative of ν_x with respect to ν_o is $K(x,\cdot)$. The measure space $(\mathcal{M}, \nu_o)$ is called the *Poisson boundary.* (More precisely, we should use the whole family of measures $(\nu_x)_{x\in X}$.) If φ is a ν_o-integrable function on $\mathcal{M}$, then

$$h(x) = \int_{\mathcal{M}} K(x,\cdot)\,\varphi\,d\nu_o = \int_{\mathcal{M}} \varphi\,d\nu_x \tag{24.11}$$

is a harmonic function. It is called the *Poisson integral* of φ. If φ is ν_o-a.e. bounded then $h \in \mathcal{H}^\infty = \mathcal{H}^\infty(X,P)$, the space of bounded harmonic functions. The converse is also true, that is, (24.11) defines a Banach space isomorphism between the spaces $\mathcal{H}^\infty(X,P)$ and $L^\infty(\mathcal{M},\nu_o)$:

(24.12) Theorem. *If $h \in \mathcal{H}^\infty$ then there is $\varphi \in L^\infty(\mathcal{M},\nu_o)$ such that $h(x) = \int_{\mathcal{M}} K(x,\cdot)\,\varphi\,d\nu_o$. The function φ is ν_o-almost everywhere unique.*

Proof. First suppose that $0 \le h \le 1$. Using uniqueness of ν^h and $\nu^{\mathbf{1}-h}$ (both supported on $\mathcal{M}_{\min}$), we see that $\nu^h + \nu^{\mathbf{1}-h} = \nu^{\mathbf{1}}$. Therefore ν^h is absolutely continuous with respect to $\nu^{\mathbf{1}} = \nu_o$, and has bounded Radon–Nikodym derivative φ. Uniqueness of φ follows from uniqueness of ν^h. If $h \in \mathcal{H}^\infty$ is arbitrary, then we can find $\bar{h} \in \mathcal{H}^\infty$ and numbers a, b such that $0 \le \bar{h} \le 1$ and $h = a\cdot\bar{h} + b$. We get $\varphi_h = a\cdot\varphi_{\bar{h}} + b$. □

The last theorem is the starting point of measure theoretic boundary theory: as a model for the set of limit points at infinity of the Markov chain, and for describing the space of bounded harmonic functions, one is primarily interested in $(\mathcal{M},\nu_o)$ as a measure space, and not so much as a topological space. That is, different models of the Poisson boundary will be identified when they are isomorphic as measure spaces, while different models of the Martin compactification are considered the same when they are homeomorphic, and the homeomorphism extends id_X. In particular, the Poisson boundary is called *trivial,* if ν_o is a point mass.

(24.13) Corollary. *For transient (X,P), the following statements are equivalent.*

(a) *The Poisson boundary is trivial.*

(b) *All bounded harmonic functions are constant (*the weak Liouville property*).*

(c) *The constant function* $\mathbf{1}$ *is minimal harmonic.*

(d) *There is $\xi \in \mathcal{M}$ such that $Z_n \to \xi$ $\mathbb{P}_x$-almost surely for every x.*

The following can be deduced from the martingale convergence theorem.

(24.14) Probabilistic Fatou theorem. *If φ is ν_o-integrable on $\mathcal{M}$ and h its Poisson integral, then $h(Z_n) \to \varphi(Z_\infty)$ $\mathbb{P}_o$-almost surely.*

For the unproven results stated above, the reader is referred to Kemeny, Snell and Knapp [197], Revuz [276] and – best – Dynkin [110].

We shall use Definition 24.5 also when (X, P) is recurrent. In this case, $K(x, y) = 1$ for all $x, y \in X$, and the Martin compactification is the one-point-compactification. Theorems 24.9 and 24.10 remain trivially true. The Poisson boundary is trivial (and of course (24.13.d) is no longer valid).

Next, we consider *t-harmonic* functions, that is, the elements of $\mathcal{H}^+(P,t)$. From Lemma 7.2, we know that they can exist only when $t \geq \rho(P)$. Let us assume that $G(x, y|1/t) < \infty$, that is, either $t > \rho(P)$, or $t = \rho(P)$ and (X, P) is ρ-transient. Then $G(\cdot, y|1/t)$ is a function in $\mathcal{S}^+(P,t)$ for each $y \in X$. We can proceed as in the case $t = 1$ and define the Martin kernel

$$K(x, y|t) = \frac{F(x, y|1/t)}{F(o, y|1/t)}. \tag{24.15}$$

The associated Martin compactification is $\widehat{X}(t) = \widehat{X}(P,t)$. It is the unique minimal compactification of X to which all functions $K(x, \cdot|t)$ extend continuously. The corresponding boundary is $\mathcal{M}(t) = \mathcal{M}(P,t)$.

(24.16) Lemma. *If P has finite range at $x \in X$ (that is, $\{y : p(x, y) > 0\}$ is finite) then $K(\cdot, \xi|t)$ is t-harmonic at x for every $\xi \in \mathcal{M}(t)$.*

We leave this as an exercise; compare with Lemma 7.6. In general, assume that $\mathcal{H}^+(P,t)$ is non-empty, and let $h \in \mathcal{H}^+(P,t)$. By the minimum principle, $h(x) > 0$ for all x. The *h-process* is the Markov chain (X, P_h) with transition probabilities

$$p_h(x, y) = \frac{p(x, y)\, h(y)}{t\, h(x)}, \tag{24.17}$$

see (7.5). The associated Green and Martin kernels are

$$G_h(x, y) = \frac{G(x, y|1/t)\, h(y)}{h(x)} \quad \text{and} \quad K_h(x, y) = \frac{h(o)}{h(x)}\, K(x, y|t).$$

We see that

$$u \in \mathcal{S}(P,t) \iff u/h \in \mathcal{S}(P_h, 1) \quad \text{and} \quad u \in \mathcal{H}(P,t) \iff u/h \in \mathcal{H}(P_h, 1).$$

In the special case when $t = \rho(P)$ and (X, P) is ρ-recurrent, h is unique up to a constant factor (Proposition 7.4), the h-process is recurrent, and we see that $F(x, y|1/\rho) = h(x)/h(y)$ and $K(x, y|\rho) = h(x)/h(o)$. Otherwise, the h-process is transient, and we can apply all the above results. Minimal t-harmonic functions are defined in the obvious way, and $\mathcal{M}_{\min}(P,t) = \mathcal{M}_{\min}(t)$ is the corresponding Borel subset of $\mathcal{M}(P,t)$.

(24.18) Corollary. *For each $h \in \mathcal{H}^+(P,t)$, there is a unique Borel measure ν^h on $\mathcal{M}(t)$ such that $\nu^h\big(\mathcal{M}(t) \setminus \mathcal{M}_{\min}(t)\big) = 0$ and*

$$h(x) = \int_{\mathcal{M}_{\min}(t)} K(x, \cdot|t)\, d\nu^h .$$

When the h-process is transient, $\frac{1}{h(o)}\nu^h$ is its limit distribution on $\mathcal{M}(t)$. If $h(o) = 1$ then h is minimal if and only if the h-process has trivial Poisson boundary.

The remaining sections of this book will be devoted to the following questions: given particular classes of graphs or groups and random walks adapted to their structure, can we describe the set of minimal t-harmonic functions and the Martin compactification (which so far is only defined abstractly) in terms of the underlying geometry? The Poisson boundary will be considered only marginally, in those cases when it is directly linked with our study of the Martin boundary. Note that when one knows the minimal harmonic functions, then one can describe all positive and hence also all bounded harmonic functions. In this sense, determining the Poisson boundary is "easier" than determining the minimal harmonic functions. On the other hand, knowing the whole Martin compactification will tell us in which "directions" one has to go to infinity in X in order to find the extended Martin kernels, and in particular, the minimal harmonic functions. Thus, it is "harder" to determine the whole compactification than to determine the minimal harmonic functions. Therefore it will be no surprise that the cases where the Martin compactification is known explicitly are fewer than those where the minimal harmonic functions have been determined, and many more examples are available where the Poisson boundary is completely understood.

25. Generalized lattices, Abelian and nilpotent groups, and graphs with polynomial growth

A. Exponentials and extended exponentials

In this subsection we study the minimal t-harmonic functions on almost transitive graphs with polynomial growth, in particular on generalized lattices. Abelian and nilpotent groups will arise as special cases. On several occasions, we shall apply the following inequality, whose proof is an illustrative exercise left to the reader.

(25.1) Harnack inequality. *Let (X,P) be uniformly irreducible and $t \geq \rho(P)$. Then there is a constant $C_t \geq 1$ such that $u(x) \leq C_t^{d(x,y)}\, u(y)$ for every $u \in \mathcal{S}^+(P,t)$.*

An automorphism α of a graph X is called *bounded,* if $\sup_X d(\alpha x, x) < \infty$. The following is our first application of the Harnack inequality.

(25.2) Lemma. *Suppose that (X, P) is uniformly irreducible and $\alpha \in \mathrm{AUT}(X, P)$ is bounded. If $h \in \mathcal{H}^+(P, t)$ is minimal, then $h(\alpha x)/h(x) = h(\alpha o)$ for every $x \in X$.*

Proof. Along with h, the translate $h_\alpha(x) = h(\alpha x)$ is also t-harmonic. Applying (25.1) to h, we get $h(\alpha x) \leq C_t^r\, h(x)$ for all x, where $r = \max_X d(\alpha x, x)$. Minimality implies $h_\alpha = c \cdot h$, and we must have $c = h(\alpha o)$. □

Here are two cases where this applies.

(25.3) Proposition. *Suppose that Γ is a quasi-transitive subgroup of $\mathrm{AUT}(X, P)$.*

(1) *Let $\mathfrak{Z}$ be the centre of Γ. If $h \in \mathcal{H}^+(P, t)$ is minimal then*

$$\frac{h(\alpha\gamma x)}{h(x)} = \frac{h(\alpha x)}{h(x)} \frac{h(\gamma x)}{h(x)} \quad \text{for all } x \in X\,,\ \alpha \in \mathfrak{Z} \text{ and } \gamma \in \Gamma\,.$$

(2) *Let $\mathfrak{K}$ be a compact normal subgroup of Γ. Then every function in $\mathcal{H}^+(P, t)$ is constant on each $\mathfrak{K}$-orbit.*

Proof. (1) Let o_i, $i \in \mathcal{I}$ (finite), be representatives of the Γ-orbits on X, and $o = o_1$ the reference point. Take $\alpha \in \mathfrak{Z}$ and define $r = \max\{d(\alpha o_i, o_i) : i \in \mathcal{I}\}$. If $x \in X$ then $x = \gamma o_i$ for some $i \in \mathcal{I}$ and $\gamma \in \Gamma$. As $\alpha\gamma = \gamma\alpha$, we have $d(\alpha x, x) = d(\gamma\alpha o_i, \gamma o_i) \leq r$. Lemma 25.2 applies, and $h(\alpha\gamma x) = h(\alpha o) h(\gamma x) = h(\alpha x) h(\gamma x)/h(x)$.

(2) In view of the integral representation over $\mathcal{M}_{\min}(t)$ (Theorem 24.9), it suffices to show this for minimal $h \in \mathcal{H}^+(P, t)$. This time, let $r = \max\{d(\alpha o_i, o_i) : i \in \mathcal{I},\ \alpha \in \mathfrak{K}\}$. If $\alpha \in \mathfrak{K}$ and $x \in X$ then $x = \gamma o_i$ for some $i \in \mathcal{I}$ and $\gamma \in \Gamma$. We have $\beta = \gamma^{-1}\alpha\gamma \in \mathfrak{K}$. Therefore $d(\alpha x, x) = d(\beta o_i, o_i) \leq r$. Lemma 25.2 implies $h_\alpha = c \cdot h$, where $c = h(\alpha o)$. If $K = \mathfrak{K} o$, which is a finite set, then $\sum_K h_\alpha(x) = \sum_K h(x)$. Therefore $c = 1$, proving the claim. □

We now want to determine the minimal t-harmonic functions for random walks on generalized lattices. We take up the material developed in §6.A and §8.B. The notation is that of (6.2). Recall the function $\varphi_P(\mathbf{c})$ and the extended exponentials $g_{\mathbf{c}}(i\mathbf{k})$, $\mathbf{c} \in \mathbb{R}^d$, defined in (8.18).

(25.4) Theorem. *If X is a generalized lattice, P satisfies (6.2) and $t \geq \rho(P)$, then the minimal t-harmonic functions are precisely the extended exponentials $g_{\mathbf{c}}$ with $\varphi_P(\mathbf{c}) = t$.*

In particular, all bounded harmonic functions are constant.

Proof. *Step 1.* Let $h \in \mathcal{H}^+(P,t)$ be minimal. Proposition 25.3(1), with $\Gamma = \mathfrak{Z}(\Gamma) = \mathbb{Z}^d$, implies that there are $\mathbf{c}_i \in \mathbb{R}^d$ such that

$$h(i\mathbf{k}) = h(i\mathbf{0})\, e^{\mathbf{c}_i \cdot \mathbf{k}} \quad \text{for all } \mathbf{k} \in \mathbb{Z}^d \text{ and } i \in \mathcal{I}\,.$$

We now show that the vectors $\mathbf{c}_i$, $i \in \mathcal{I}$, must coincide: in the graph structure of our generalized lattice X, we have $\max\{d(i\mathbf{k}, j\mathbf{k}) : i,j \in \mathcal{I}\,,\ \mathbf{k} \in \mathbb{Z}^d\} = r < \infty$. The Harnack inequality (25.1) implies

$$e^{(\mathbf{c}_i - \mathbf{c}_j)\cdot \mathbf{k}} \le C_t^r\, h(j\mathbf{0})/h(i\mathbf{0}) \quad \text{for all } i,j \in \mathcal{I}\,,\ \mathbf{k} \in \mathbb{Z}^d\,.$$

This yields that $\mathbf{c}_i = \mathbf{c}$ is independent of i. We find that $h(i\mathbf{k}) = h(i\mathbf{0})\, e^{\mathbf{c}\cdot\mathbf{k}}$. Using the notation of (8.16), the relation $Ph = t \cdot h$ translates into

$$t\, h(i\mathbf{0}) = \sum_{j \in \mathcal{I}} \widetilde{p}_{\mathbf{c}}(i,j)\, h(j\mathbf{0})\,.$$

Thus, $\big(h(i\mathbf{0})\big)_{i\in\mathcal{I}}$ is a positive right eigenvector of $\widetilde{P}_{\mathbf{c}}$ with eigenvalue t. It must be the Perron–Frobenius eigenvector $\widetilde{g}_{\mathbf{c}}$, and $t = \varphi_P(\mathbf{c})$. This proves that $h = g_{\mathbf{c}}$, the extended exponential defined in (8.18).

Step 2. Now consider the case $t = 1$. We want to show that the constant function $\mathbf{1} = g_{\mathbf{0}}$ is also minimal. We know from the first step that

$$\{K(\cdot,\xi) : \xi \in \mathcal{M}_{\min}(P)\} \subset \{g_{\mathbf{c}} : \varphi_P(\mathbf{c}) = 1\}\,.$$

On the latter set, the topology of pointwise convergence coincides with the Euclidean topology on $\mathcal{C}_1 = \{\mathbf{c} : \varphi_P(\mathbf{c}) = 1\}$. Thus, we can identify $\mathcal{M}_{\min}(P)$ with a Borel subset of B of $\mathcal{C}_1$, and the uniqueness theorem (24.9) yields existence of a probability measure ν on $\mathcal{C}_1$ with $\operatorname{supp}\nu \subset B$ such that

$$1 = \int_{\mathcal{C}_1} g_{\mathbf{c}}(\mathbf{k})\, d\nu(\mathbf{c}) \quad \text{for all } \mathbf{k} \in \mathbb{Z}^d\,.$$

If $\operatorname{supp}\nu \ne \{\mathbf{0}\}$ then the integral is not bounded in $\mathbf{k}$. Therefore $\nu = \delta_{\mathbf{0}}$, and $\mathbf{0} \in B$. This shows that $\mathbf{1}$ is minimal, and the Poisson boundary is trivial.

Step 3. It is now easy to deduce from Step 2 and the last statement of Corollary 24.18 that each $g_{\mathbf{c}}$ is minimal t-harmonic, where $t = \varphi_P(\mathbf{c})$. Indeed, $g_{\mathbf{c}}$ being an extended exponential, the $g_{\mathbf{c}}$-process is also $\mathbb{Z}^d$-invariant and irreducible, so that it must have trivial Poisson boundary. □

If P has finite range, φ_P is finite on all of $\mathbb{R}^d$. From Proposition 8.20 we see that for any $t > \rho(P)$, the set $\mathcal{C}_t = \{\mathbf{c} \in \mathbb{R}^d : \varphi_P(\mathbf{c}) = t\}$ is the boundary of the convex open set $\{\varphi_P < t\}$. Thus $\mathcal{C}_t$ is homeomorphic with the unit sphere in $\mathbb{R}^d$. The topology of $\mathcal{M}_{\min}(t)$ (i.e., pointwise convergence of the extended exponentials $g_{\mathbf{c}}$) coincides with the Euclidean topology of $\mathcal{C}_t$. At the "bottom", $\mathcal{C}_{\rho(P)}$ consists of a single point. In particular, every positive $\rho(P)$-harmonic function is a multiple of the unique $\rho(P)$-harmonic extended exponential; by Exercise 8.24, this is true without any moment condition. If $\mathfrak{m}(P) = \mathbf{0}$ then $\rho(P) = 1$ by Corollary 8.15, and again, only existence of the first moment suffices to have this.

(25.5) Corollary. *Every $h \in \mathcal{H}^+(P,t)$ has a unique integral representation $h = \int_{\mathcal{C}_t} g_{\mathbf{c}}\, d\nu^h(\mathbf{c})$, where ν^h is a Borel measure on $\mathcal{C}_t$.*

If $\mathfrak{m}(P) = \mathbf{0}$, then all positive harmonic functions are constant (the strong Liouville property).

As mentioned above, there is a unique $\rho(P)$-harmonic extended exponential even when P does not have finite support. But then the shape and size of $\mathcal{C}_t$ may vary considerably with t. For example, take $X = \mathbb{Z}$ and the random walk with law μ of the form

$$\mu(0) = 0\,, \quad \mu(-n) = c_1\, e^{-a_1 n}/n^{b_1}\,, \quad \mu(n) = c_2\, e^{-a_2 n}/n^{b_2} \tag{25.6}$$

($n \in \mathbb{N}$), with suitable $a_i, b_i \geq 0$ and $c_i > 0$. Then the function φ_μ is finite in the interval between $-a_1$ and a_2, open or closed at the endpoints according to the choice of the six parameters. Thus, one can find the following behaviour: $\mathcal{C}_t$ has cardinality 1 for $t = \rho(P)$, and increasing t, the cardinality is first 2, then 1 and finally 0 (e.g. choosing $a_1 = a_2 > 0$ and $b_1 > b_2 > 1$).

(25.7) Exercise. Construct examples where $\rho(P) < 1$, and yet the strong Liouville property holds.

An *exponential* on a group Γ is a positive function satisfying $g(\beta\gamma) = g(\beta)g(\gamma)$ for all $\beta, \gamma \in \Gamma$. If Γ is Abelian and μ an irreducible probability measure on Γ, then the minimal elements of $\mathcal{H}^+(\mu,t)$ for $t \geq \rho(\mu)$ are precisely the t-harmonic exponentials on Γ. This is seen by an obvious modification of the proof of Theorem 25.4. (Consider the single orbit case, and replace $\mathbb{Z}^d$ by Γ.)

We shall see that the results regarding the Liouville property extend to quasi-transitive random walks on graphs with polynomial growth. We first consider an intermediate case.

(25.8) Theorem. *Suppose that $\mathrm{AUT}(X,P)$ contains a discrete, nilpotent subgroup $\mathfrak{N}$ which acts quasi-transitively. If $h \in \mathcal{H}^+(P,t)$ is minimal, then for every $x \in X$, the function $\gamma \mapsto h(\gamma x)/h(x)$ is an exponential on $\mathfrak{N}$.*

Proof. *Step 1.* Denote by $\mathfrak{Z}$ the centre of $\mathfrak{N}$. We know from Proposition 25.3(1) that $h(\alpha x) = h(\alpha o)h(x)$ for every $\alpha \in \mathfrak{Z}$ and $x \in X$.

Step 2. Let $\mathfrak{P}$ be the subgroup of $\mathfrak{N}$ generated by all commutators $[\beta,\gamma] = \beta\gamma\beta^{-1}\gamma^{-1}$ (with $\beta,\gamma \in \mathfrak{N}$) that lie in $\mathfrak{Z}$. We claim that $h(\alpha x) = h(x)$ for all $x \in X$ and $\alpha \in \mathfrak{P}$.

To prove this, fix $\beta \in \mathfrak{N}$, and consider the commutators $[\beta,\gamma]$ that lie in $\mathfrak{Z}$, where $\gamma \in \mathfrak{N}$. If $[\beta,\gamma] \in \mathfrak{Z}$ and $\delta \in \mathfrak{N}$ then $[\beta,\delta\gamma] = \beta\delta\beta^{-1}[\beta,\gamma]\delta^{-1} = [\beta,\delta][\beta,\gamma]$. In particular, $[\beta,\gamma^n] = [\beta,\gamma]^n \in \mathfrak{Z}$ for all $n \in \mathbb{Z}$.

Now consider the the set of functions $h_n(x) = h(\beta^n x)/h(\beta^n o)$, where $n \in \mathbb{Z}$. They are all in $\mathcal{H}^+(P,t)$. Let $\mathcal{B}_\beta$ be the closure of this set with respect to pointwise convergence, a compact set. Every function $u \in \mathcal{B}_\beta$ is in $\mathcal{S}^+(P,t)$ and (by Step 1) satisfies $u(\alpha x) = h(\alpha o)u(x)$ for all $x \in X$ and $\alpha \in \mathfrak{Z}$.

Define a transformation $T_\beta : \mathcal{B}_\beta \to \mathcal{B}_\beta$ by $T_\beta u(x) = u(\beta x)/u(\beta o)$. It is continuous, and by the fixed point theorem of Schauder–Tychonoff (see Dunford and Schwartz [107], I, pp. 468 and 470), there is $u \in \mathcal{B}_\beta$ such that $T_\beta u = u$, that is, $u(\beta x)/u(\beta o) = u(x)$ for all $x \in X$. Fix this u. For all $\gamma \in \mathfrak{N}$, we have $d(\gamma\beta o, \gamma o) = d(\beta o, o) = r_\beta < \infty$. By the Harnack inequality (25.1), we have $u(\gamma o) \le C\, u(\gamma\beta o)$ for all γ, where $C = C_t^{r_\beta}$ Now assume that $[\beta,\gamma] \in \mathfrak{Z}$. Then $\beta\gamma^n = [\beta,\gamma]^n\gamma^n\beta$, and

$$u(\beta o)u(\gamma^n o) = u(\beta\gamma^n o) = h([\beta,\gamma]o)^n\, u(\gamma^n\beta o) \ge C^{-1}\, h([\beta,\gamma]o)^n\, u(\gamma^n o)\,.$$

We find $h([\beta,\gamma]o)^n \le C\, u(\beta o)$ for all $n \in \mathbb{Z}$, which implies $h([\beta,\gamma]o) = 1$.

Every element of $\mathfrak{P}$ is of the form $\alpha = [\beta_1,\gamma_1]\cdots[\beta_k,\gamma_k]$ with $[\beta_i,\gamma_i] \in \mathfrak{Z}$. Therefore, using Step 1 once more, $h(\alpha o) = 1$.

None of these arguments depends on the choice of the base point o. If we choose $x \in X$ as the base point, then we have to renormalize, replacing h with $\frac{1}{h(x)}h$. Therefore $h(\alpha x) = h(x)$.

Step 3. We have not yet used nilpotency of our group. Let

$$\mathfrak{N} = \mathfrak{N}_0 \trianglerighteq \mathfrak{N}_1 \trianglerighteq \cdots \trianglerighteq \mathfrak{N}_{r-1} \trianglerighteq \mathfrak{N}_r = \{\iota\}\,,$$

where $\mathfrak{N}_i = [\mathfrak{N},\mathfrak{N}_{i-1}]$. We use induction on the degree of nilpotency r. If $r = 1$, the group is Abelian and the proposed result is true. Suppose it is true for $r-1$. As $\mathfrak{N}_{r-1}$ is contained in $\mathfrak{Z}$, it is also a subgroup of $\mathfrak{P}$.

Now consider the factor graph $\widetilde{X} = \mathfrak{N}_{r-1}\backslash X$. It is locally finite, and $\widetilde{\mathfrak{N}} = \mathfrak{N}/\mathfrak{N}_{r-1}$ acts quasi-transitively on it. We write π for both the natural projections $X \to \widetilde{X}$ and $\Gamma \to \widetilde{\Gamma}$. The transition probabilities $\widetilde{p}(\pi(x),\pi(y)) = \sum_{w\in\mathfrak{N}_{r-1}y} p(x,w)$ of the factor chain $\widetilde{P}$ are $\widetilde{\mathfrak{N}}$-invariant. From Step 2 we know that h is constant on the orbits of $\mathfrak{P}$, and *a fortiori* also on the orbits of $\mathfrak{N}_{r-1}$ in $\mathfrak{N}$. Therefore we can define the function $\widetilde{h}$ on $\widetilde{X}$ by $\widetilde{h}(\pi(x)) = h(x)$. Then $\widetilde{h}$ is t-harmonic on the factor graph with respect to the factor chain. Minimality of h implies minimality of $\widetilde{h}$, because every function in $\mathcal{H}^+(\widetilde{P},t)$ lifts via π to a function in $\mathcal{H}^+(\mu,t)$.

The degree of nilpotency of $\widetilde{\mathfrak{N}}$ is $r-1$, and we can use the induction hypothesis to obtain

$$\frac{h(\beta\gamma x)}{h(x)} = \frac{\widetilde{h}\big(\pi(\beta\gamma x)\big)}{\widetilde{h}\big(\pi(x)\big)} = \frac{\widetilde{h}\big(\pi(\beta)\pi(x)\big)}{\widetilde{h}\big(\pi(x)\big)} \frac{\widetilde{h}\big(\pi(\gamma)\pi(x)\big)}{\widetilde{h}\big(\pi(x)\big)} = \frac{h(\beta x)}{h(x)} \frac{h(\gamma x)}{h(x)} . \quad \square$$

The last two theorems have several interesting consequences.

(25.9) Corollary. *Let $\mathfrak{N}$ be a finitely generated nilpotent group and μ an irreducible probability measure on $\mathfrak{N}$. Then*

(a) *every function in $\mathcal{H}^+(\mu, t)$ is constant on each coset of the commutator subgroup $\mathfrak{N}'$, and*

(b) *the minimal elements of $\mathcal{H}^+(\mu, t)$ are precisely the t-harmonic exponentials on $\mathfrak{N}$.*

(c) *If, in addition, μ has finite support, then*

$$\rho(\mu) = \min\{ \textstyle\int_{\mathfrak{N}} g \, d\mu : g \text{ an exponential on } \mathfrak{N} \} .$$

Proof. (a) Applying Theorem 25.8, with X a Cayley graph of $\mathfrak{N}$, we see that every minimal t-harmonic function is a (pure) exponential, and hence constant on the cosets of $\mathfrak{N}'$. By Theorem 24.9, this must hold for every element of $\mathcal{H}^+(\mu, t)$.

(b) Let g be a t-harmonic exponential. We have to show that g is minimal. Suppose $g \geq h$, where $h \in \mathcal{H}^+(\mu, t)$. By (a), h is constant on each coset of $\mathfrak{N}'$. We pass to the factor graph and factor chain with respect to this normal subgroup. This is the random walk on the Abelian group Γ/Γ' whose law is given by $\widetilde{\mu}\big(\pi(x)\big) = \mu(x\Gamma')$; compare with Step 3 of the proof of Theorem 25.8. We see that $\widetilde{h}$ as well as $\widetilde{g}$ is in $\mathcal{H}^+(\widetilde{\mu}, t)$, and $\widetilde{g}$ is minimal. Therefore $\widetilde{g}/\widetilde{h}$ and, equivalently, g/h are constant.

(c) We know that finite range implies $\rho(\mu) = \min\{t : \mathcal{H}^+(\mu, t) \neq \emptyset\}$; see Lemma 24.16. Now (b) tells us that $\rho(\mu)$ is the minimal t for which there is a t-harmomic exponential. If g is any exponential, then it is t-harmonic with $t = \int_{\mathfrak{N}} g \, d\mu$, if this number is finite. $\square$

When $\rho(\mu) = 1$ then $\rho(\widetilde{\mu}) = 1$, and we infer from Exercise 8.24 (see the observations before Corollary 25.5) that the constant function $\mathbf{1}$ is the unique harmonic exponential, even without assuming finite range. Hence, again, the strong Liouville property holds. This holds in particular when μ is symmetric. It is quite easy to generalize this to random walks on quasi-transitive graphs with polynomial growth:

(25.10) Corollary. *Suppose that the random walk (X, P) is quasi-transitive, and that the graph X has polynomial growth. Then all bounded harmonic functions are constant. If in addition $\rho(P) = 1$ then all positive harmonic functions are constant. This holds, in particular, for strongly reversible random walks.*

Proof. Recall Theorem 5.11 and its proof: there is a compact normal subgroup $\mathfrak{K}$ of Γ such that $\Gamma/\mathfrak{K}$ is a finitely generated group with polynomial growth acting quasi-transitively and with finite vertex stabilizers on the factor graph $\mathfrak{K}\backslash X$. Proposition 25.3(2) tells us that every function in $\mathcal{H}^+(P, t)$ is constant on each $\mathfrak{K}$-orbit.

As above, this allows us to reduce all considerations to the factor chain on the graph $\mathfrak{K}\backslash X$. (The spectral radius of the factor chain is $\limsup_n p^{(n)}(o, \mathfrak{K}o)^{1/n} = \rho(P)$ by finiteness of $\mathfrak{K}o$.) We find a nilpotent subgroup with finite index $\mathfrak{N}$ in $\Gamma/\mathfrak{K}$ (Theorem 3.17), which acts with finite vertex stabilizers on $\mathfrak{K}\backslash X$.

Thus, we may assume without loss of generality that we are in the situation of Theorem 25.8. Then we know that every minimal harmonic function is constant on each orbit of the commutator group $\mathfrak{N}'$. Again, by the integral representation theorem this is true for every function in $\mathcal{H}^+(P, t)$. Now consider the factor chain P' on $X' = \mathfrak{N}'\backslash X$. The factor group is finitely generated and Abelian, and (X', P') is a random walk on a generalized lattice. Every bounded harmonic function (which may be assumed positive without loss of generality) is constant (Theorem 25.4), and the same must hold for (X, P). We have $\rho(P) \le \rho(P')$. Suppose $\rho(P) = 1$. Then $\rho(P') = 1$. Therefore (X', P') and consequently also (X, P) have the strong Liouville property. Finally, strong reversibility implies $\rho(P) = 1$ (Corollary 12.12.) □

In the context of the last corollary, one can also describe the whole set of minimal t-harmonic functions. They are obtained by lifting the exponentials $g_{\mathbf{c}} \in \mathcal{H}^+(P', t)$ from the generalized lattice X' first to the graph $\mathfrak{K}\backslash X$ (constant on the $\mathfrak{N}'$-orbits) and then to X itself (constant on the $\mathfrak{K}$-orbits).

B. The Martin compactification of random walks on the grid

Having found the minimal harmonic functions, we now determine the whole Martin compactification in the special case of random walks on $\mathbb{Z}^d$. In view of Corollary 25.5, we expect that under suitable conditions, the Martin boundary will consist of a single point, when the mean vector is 0, and otherwise will be homeomorphic with the unit sphere in $\mathbb{R}^d$. The first of these cases is considerably simpler than the second, and is based on the following Green kernel estimate.

(25.11) Theorem. *Let μ be an (irreducible) probability measure on $\mathbb{Z}^d$, $d \ge 3$, with finite moment of even order $2r > \max\{2, d-2\}$. If $\mathfrak{m}(\mu) = \mathbf{0}$*

and Σ is the covariance matrix, then

$$G(\mathbf{0},\mathbf{k}) \sim \frac{\Gamma(\frac{d-2}{2})}{2\pi\sqrt{\det\Sigma}}\Big(\pi\,\Sigma^{-1}[\mathbf{k}]\Big)^{-(d-2)/2} \quad \textit{as } |\mathbf{k}|\to\infty\,.$$

Proof. We first assume that μ is aperiodic, and use Theorem 13.10. To start, consider $\widetilde{G}(\mathbf{0},\mathbf{k}) = \sum_{n=1}^{\infty} \alpha_n(\mathbf{k})$, where

$$\alpha_n(\mathbf{k}) = (\det\Sigma)^{-1/2}\,(2n\pi)^{-d/2}\,\exp\Big(-\tfrac{1}{2n}\Sigma^{-1}[\mathbf{k}]\Big)\,.$$

If we substitute $t_n = n/\Sigma^{-1}[\mathbf{k}]$ (recall that Σ^{-1} is positive definite) then $\Delta t_n = t_n - t_{n-1} = 1/\Sigma^{-1}[\mathbf{k}]$ tends to 0 as $|\mathbf{k}|\to\infty$, and

$$\sqrt{\det\Sigma}\,\sqrt{2\pi}^{\,d}\,\Big(\Sigma^{-1}[\mathbf{k}]\Big)^{(d-2)/2}\,\widetilde{G}(0,\mathbf{k}) = \sum_{n=1}^{\infty} t_n^{-d/2}\,\exp(-1/(2t_n))\,\Delta t_n$$

is a Riemannian sum of $\int_0^\infty t^{-d/2}\,\exp(-\frac{1}{2t})\,dt = 2^{(d-2)/2}\Gamma(\frac{d-2}{2})$. Therefore $\widetilde{G}(\mathbf{0},\mathbf{k})$ has the asymptotic behaviour that we have asserted for $G(\mathbf{0},\mathbf{k})$. We now show that $\lim_{|\mathbf{k}|\to\infty}\big(G(\mathbf{0},\mathbf{k})/\widetilde{G}(\mathbf{0},\mathbf{k})\big) = 1$, or equivalently, that $G(\mathbf{0},\mathbf{k}) - \widetilde{G}(\mathbf{0},\mathbf{k}) = \mathfrak{o}\big(|\mathbf{k}|^{-(d-2)}\big)$. Set

$$\beta_n = \sup_{\mathbf{k}\in\mathbb{Z}^d}\,(2n\pi)^{d/2}\,\max\{1,\,\big(|\mathbf{k}|/\sqrt{n}\big)^{d-1}\}\,\big|\mu^{(n)}(\mathbf{k}) - \alpha_n(\mathbf{k})\big|\,.$$

Then $\beta_n \to 0$ as $n\to\infty$ by Theorem 13.10. Given $\varepsilon > 0$, we have $\beta_n < \varepsilon$ for all $n > n_\varepsilon$. Then

$$|\mathbf{k}|^{d-2}\,\big|G(\mathbf{0},\mathbf{k}) - \widetilde{G}(\mathbf{0},\mathbf{k})\big|$$
$$\le \frac{1}{|\mathbf{k}|(2\pi)^{d/2}}\sum_{n=1}^{n_\varepsilon}\frac{\beta_n}{n^{1/2}} + \frac{1}{|\mathbf{k}|(2\pi)^{d/2}}\sum_{n=n_\varepsilon}^{|\mathbf{k}|^2}\frac{\beta_n}{n^{1/2}} + \frac{|\mathbf{k}|^{d-2}}{(2\pi)^{d/2}}\sum_{n>|\mathbf{k}|^2}^{\infty}\frac{\beta_n}{n^{d/2}}\,.$$

The first of the three terms on the right hand side tends to 0 when $|\mathbf{k}|\to\infty$. The second is bounded by

$$\frac{\varepsilon}{|\mathbf{k}|(2\pi)^{d/2}}\int_0^{|\mathbf{k}|^2} x^{-1/2}\,dx = \frac{2\varepsilon}{(2\pi)^{d/2}}\,,$$

and the third is bounded by

$$\frac{|\mathbf{k}|^{d-2}}{(2\pi)^{d/2}}\int_{|\mathbf{k}|^2}^{\infty} x^{-d/2}\,dx = \frac{2\varepsilon}{(2\pi)^{d/2}(d-2)}\,.$$

Therefore the error term is bounded by a constant times ε, as required.

If μ is periodic, then we replace it with $\bar{\mu} = \frac{1}{2}(\delta_0+\mu)$. The corresponding Green kernels and covariance matrices satisfy $G_{\bar{\mu}}(\mathbf{0},\mathbf{k}) = 2G_\mu(\mathbf{0},\mathbf{k})$ and $\Sigma_{\bar{\mu}} = \frac{1}{2}\Sigma_\mu$. It is now immediate that the asymptotic formula for $G_{\bar{\mu}}$ carries over to G_μ. □

In particular, $\lim_{|\mathbf{k}|\to\infty}\big(G(\mathbf{m},\mathbf{k})/G(\mathbf{0},\mathbf{k})\big) = 1$ for each $\mathbf{m}\in\mathbb{Z}^d$.

(25.12) Corollary. *Under the conditions of Theorem 25.11, the Martin compactification of $(\mathbb{Z}^d, \mu)$ is the one-point compactification.*

Next, we study the – harder – case when μ has mean vector $\mathfrak{m} \neq \mathbf{0}$. We assume that $d \geq 2$ and that $\operatorname{supp}\mu$ is finite; in dimension $d = 1$ the result will follow from Corollary 26.14(a) below. We know that the minimal harmonic functions are the exponentials $g_{\mathbf{c}}(\mathbf{k}) = e^{\mathbf{c}\cdot\mathbf{k}}$ with $\mathbf{c} \in C_1 = \varphi_\mu^{-1}\{1\}$, where $\varphi_\mu(\mathbf{c}) = \sum_{\mathbf{k}} e^{\mathbf{c}\cdot\mathbf{k}}\,\mu(\mathbf{k})$.

We shall prove that the Martin compactification is obtained by embedding $\mathbb{Z}^d$ into the unit ball in $\mathbb{R}^d$ via $\mathbf{k} \mapsto \frac{1}{1+|\mathbf{k}|}\mathbf{k}$ and taking the closure $\widehat{\mathbb{Z}^d}$. In this compactification, the boundary is the unit sphere $\mathbb{S}_{d-1}$ in $\mathbb{R}^d$, and a sequence $(\mathbf{k}_n)$ converges to $\mathbf{u} \in \mathbb{S}_{d-1}$ if and only if $|\mathbf{k}_n| \to \infty$ and $\frac{1}{|\mathbf{k}_n|}\mathbf{k}_n \to \mathbf{u}$. See Figure 21.

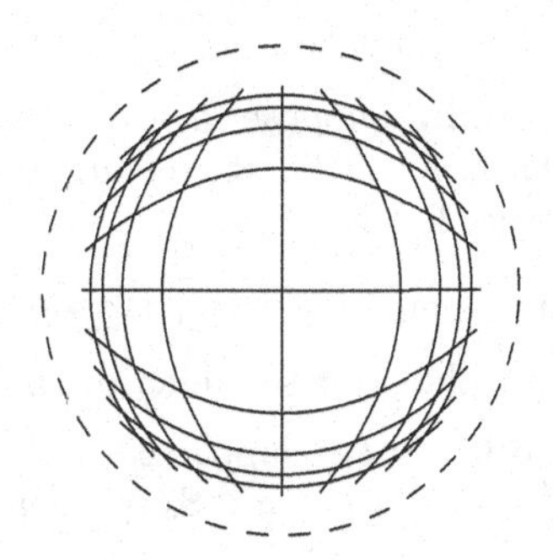

Figure 21: the compactification of $\mathbb{Z}^2$

Take $\mathbf{c} \in C_1$. Then the $g_{\mathbf{c}}$-process is the random walk with law $\mu_{\mathbf{c}}(\mathbf{k}) = e^{\mathbf{c}\cdot\mathbf{k}}\,\mu(\mathbf{k})$. Its mean vector $\mathfrak{m}_{\mathbf{c}} = \operatorname{grad}\varphi_\mu(\mathbf{c}) \neq \mathbf{0}$ depends continuously on $\mathbf{c}$. The set C_1 is the smooth boundary of a strictly convex set. Hence, given $\mathbf{x} \in \mathbb{R}^d \setminus \{\mathbf{0}\}$, there is a unique $\mathbf{c}(\mathbf{x}) \in C_1$ such that $\mathfrak{m}_{\mathbf{c}(\mathbf{x})}$ is a positive multiple of $\mathbf{x}$. It is determined by the equation

$$\frac{\operatorname{grad}\varphi_\mu(\mathbf{c})}{|\operatorname{grad}\varphi_\mu(\mathbf{c})|} = \frac{\mathbf{x}}{|\mathbf{x}|}. \tag{25.13}$$

For each $\mathbf{u} \in \mathbb{S}_{d-1}$, let $R_{\mathbf{u}}$ be the rotation that sends $\mathbf{u}$ to the first coordinate vector $\mathbf{e}_1$ and leaves the orthogonal complement of $\{\mathbf{u}, \mathbf{e}_1\}$ invariant. For a vector $\mathbf{z} = (z_1, \dots, z_n) \in \mathbb{C}^d$, we write $\mathbf{z}' = (z_2, \dots, z_n)$. We shall need the $(d-1) \times (d-1)$ matrix

$$Q_{\mathbf{u}} = \big(q_{\mathbf{u}}(i,j)\big)_{i,j=2}^d \quad \text{with} \quad q_{\mathbf{u}}(i,j) = \sum_{\mathbf{k}\in\mathbb{Z}^d} [R_{\mathbf{u}}\mathbf{k}]_i\,[R_{\mathbf{u}}\mathbf{k}]_j\,\mu(\mathbf{k}).$$

The associated quadratic form is $Q_{\mathbf{u}}[\mathbf{z}'] = \mathbf{z}' \cdot Q_{\mathbf{u}}\mathbf{z}'$, where $\mathbf{z}' \in \mathbb{C}^{d-1}$.

(25.14) Lemma. *There is $\lambda \ge 1$ such that for all $\mathbf{c} \in \mathcal{C}_1$, $\mathbf{u} \in \mathbb{S}_{d-1}$ and $\mathbf{x}' \in \mathbb{R}^{d-1}$*

$$\lambda^{-1} \le |\mathfrak{m}_{\mathbf{c}}| \le \lambda \quad \text{and} \quad \lambda^{-1}\,|\mathbf{x}'|^2 \le Q_{\mathbf{u}}[\mathbf{x}'] \le \lambda\,|\mathbf{x}'|^2 .$$

Proof. The first statement is a consequence of the fact that $\mathbf{c} \mapsto \mathfrak{m}_{\mathbf{c}}$ is continuous and $\mathfrak{m}_{\mathbf{c}} \ne \mathbf{0}$ for all $\mathbf{c} \in \mathcal{C}_1$.

Let $\mathcal{Q}_{\mathbf{c}} = \sum_{\mathbf{k}} \mathbf{k}\,\mathbf{k}^t\,\mu_{\mathbf{c}}(\mathbf{k})$ be the $d \times d$ matrix of second order moments (without any rotation) of μ, where $\mathbf{c} = \mathbf{c}(\mathbf{u})$. Then $\mathcal{Q}_{\mathbf{c}}$ is positive definite by irreducibility of μ; compare with Lemma 13.1. Therefore $R_{\mathbf{u}}\mathcal{Q}_{\mathbf{c}}R_{\mathbf{u}}^t$ is also positive definite. From this matrix we obtain $Q_{\mathbf{u}}$ by deleting the first row and column. Thus $Q_{\mathbf{u}}$ is positive definite as well. Again, the mapping $\mathbf{u} \mapsto Q_{\mathbf{u}}$ is continuous, whence there are a positive lower and a finite upper bound on the eigenvalues of the $Q_{\mathbf{u}}$, and the second statement follows. □

We shall prove the following.

(25.15) Theorem. *Suppose that $d \ge 2$ and μ has finite support and non-zero mean. Then, as $|\mathbf{m}| \to \infty$, setting $\mathbf{u} = \mathbf{u}(\mathbf{m}) = \mathbf{m}/|\mathbf{m}|$ and $\mathbf{c} = \mathbf{c}(\mathbf{m})$, we have for every $\mathbf{k} \in \mathbb{Z}^d$*

$$G(\mathbf{k},\mathbf{m}) \sim |\mathfrak{m}_{\mathbf{c}}|^{-1}\,\sqrt{\det Q_{\mathbf{u}}}\,\bigl(2\pi|\mathbf{m}|\bigr)^{-(d-1)/2} \exp\bigl(\mathbf{c}\cdot(\mathbf{k}-\mathbf{m})\bigr).$$

This leads to the proposed description of the Martin boundary:

(25.16) Corollary. *Let $\operatorname{supp}\mu$ be finite and $\mathfrak{m}(\mu) \ne \mathbf{0}$. If $\mathbf{m}$ tends to infinity in $\mathbb{Z}^d$ and $\frac{1}{|\mathbf{m}|}\mathbf{m} \to \mathbf{u} \in \mathbb{S}_{d-1}$, then for each $\mathbf{k} \in \mathbb{Z}^d$*

$$K(\mathbf{k},\mathbf{m}) \to e^{\mathbf{c}\cdot\mathbf{k}} \quad \text{where} \quad \mathbf{c} = \mathbf{c}(\mathbf{u}) .$$

In order to prove Theorem 25.15, we shall work with the Green function $G_{\mathbf{c}}(\mathbf{k},\mathbf{m}) = G(\mathbf{k},\mathbf{m})\,g_{\mathbf{c}}(\mathbf{m}-\mathbf{k})$ of the $g_{\mathbf{c}}$-process. In other words, we take the exponential from the right to the left hand side of the asymptotic formula. Note that $\mathbf{c} = \mathbf{c}(\mathbf{m})$ varies with $\mathbf{m}$, so that we shall have to take care that all our subsequent estimates are uniform in $\mathbf{c}$ and $\mathbf{u}$, respectively. The characteristic function of $\mu_{\mathbf{c}}$ is $\phi_{\mathbf{c}}(\mathbf{v}) = \sum_{\mathbf{k}} \mu_{\mathbf{c}}(\mathbf{k})\,e^{i\,\mathbf{k}\cdot\mathbf{v}}$, where $\mathbf{v} \in \mathbb{W}_d$.

(25.17) Lemma. *For $\mathbf{m} \in \mathbb{Z}^d$,*

$$G_{\mathbf{c}}(\mathbf{0},\mathbf{m}) = \frac{1}{(2\pi)^d} \int_{\mathbb{W}_d} \frac{e^{-i\,\mathbf{m}\cdot\mathbf{v}}}{1-\phi_{\mathbf{c}}(\mathbf{v})}\,d\mathbf{v}$$

Proof. By Spitzer [307], P. 7.5, the only point where $\phi_{\mathbf{c}}(\mathbf{v}) = 1$ is $\mathbf{v} = \mathbf{0}$; compare with Lemma 13.1. (The "aperiodicity" of [307] means that the group generated by $\operatorname{supp}\mu$ is the whole of $\mathbb{Z}^d$.) By Fourier inversion,

$$\sum_{n=0}^{N} \mu_{\mathbf{c}}^{(n)}(\mathbf{m}) = \frac{1}{(2\pi)^d} \int_{\mathbb{W}_d} \frac{1-\phi_{\mathbf{c}}(\mathbf{v})^{N+1}}{1-\phi_{\mathbf{c}}(\mathbf{v})}\,e^{-i\,\mathbf{m}\cdot\mathbf{v}}\,d\mathbf{v} .$$

We show that $|1-\phi_{\mathbf{c}}(\mathbf{v})|^{-1}$ is integrable on $\mathbb{W}_d$ (or equivalently, on some neighbourhood of $\mathbf{0}$), so that we can exchange the integral with the limit when $N \to \infty$ by dominated convergence.

Since $1-\phi_{\mathbf{c}}(\mathbf{v}) = -i\mathfrak{m}_{\mathbf{c}}\cdot\mathbf{v}+\frac{1}{2}\mathcal{Q}_{\mathbf{c}}[\mathbf{v}]+\mathcal{O}(|\mathbf{v}|^3)_{\mathbf{v}\to\mathbf{0}}$, we have $|1-\phi_{\mathbf{c}}(\mathbf{v})| \geq \varepsilon_1|-i\mathfrak{m}_{\mathbf{c}}\cdot\mathbf{v}+\frac{1}{2}\mathcal{Q}_{\mathbf{c}}[\mathbf{v}]|$ in $\mathbb{W}_d$, where $\varepsilon_1 > 0$. Up to a change of variables that transfers a suitable neighbourhood U of $\mathbf{0}$ into another neighbourhood of $\mathbf{0}$, we may assume for our purpose that $\frac{1}{2}\mathcal{Q}_{\mathbf{c}}[\mathbf{v}] = |\mathbf{v}|^2$ and $\mathfrak{m}_{\mathbf{c}} = \varepsilon_2\mathbf{e}_1$, where $\varepsilon_2 > 0$. As above, we write $\mathbf{v} = (v_1, \mathbf{v}')$, where $\mathbf{v}' \in \mathbb{R}^{d-1}$. Then

$$\left|-i\,\mathfrak{m}_{\mathbf{c}}\cdot\mathbf{v}+\tfrac{1}{2}\mathcal{Q}_{\mathbf{c}}[\mathbf{v}]\right| \geq \left|-i\,\varepsilon_2 v_1+|\mathbf{v}'|^2\right| = \left((\varepsilon_2 v_1)^2+|\mathbf{v}'|^4\right)^{1/2} \geq C|v_1|^{2/3}|\mathbf{v}'|^{2/3}$$

by Hölder's inequality, where $C > 0$. To conclude, we observe that $|v_1|^{-2/3}|\mathbf{v}'|^{-2/3}$ is integrable near $\mathbf{0}$. □

For our asymptotic estimate, we can restrict the integral in Lemma 25.17 to any suitable neighbourhood of $\mathbf{0}$. To see this, given $\varepsilon \in (0\,,\pi)$, let $f = f_\varepsilon \in C^\infty(\mathbb{W}_d)$ be a radial function (i.e., invariant under rotations around the origin) such that $0 \leq f \leq 1$, $f(\mathbf{v}) = 1$ for $|\mathbf{v}| \leq \varepsilon/3$ and $f(\mathbf{v}) = 0$ for $|\mathbf{v}| \geq 2\varepsilon/3$, $\mathbf{v} \in \mathbb{W}_d$. Then the periodic continuation of $(1-f)/(1-\phi_{\mathbf{c}})$ to $\mathbb{R}^d$ is a C^∞-function whose derivatives are uniformly bounded with respect to $\mathbf{c}$, because they depend continuously on $\mathbf{c} \in \mathcal{C}_1$ and $\mathcal{C}_1$ is compact. Therefore the Riemann–Lebesgue lemma implies

$$(25.18)\quad G_{\mathbf{c}}(\mathbf{k},\mathbf{m}) = \frac{1}{(2\pi)^d}\int_{\mathbb{W}_d} e^{-i(\mathbf{m}-\mathbf{k})\cdot\mathbf{v}}\frac{f(\mathbf{v})}{1-\phi_{\mathbf{c}}(\mathbf{v})}\,d\mathbf{v} + \mathfrak{o}\left(|\mathbf{m}|^{-(d-1)/2}\right),$$

and the $\mathfrak{o}(\cdot)$ is uniform in $\mathbf{c}$.

Our principal tool is the following extended version of the function φ_μ. For $\mathbf{u} \in \mathbb{S}_{d-1}$ and $\mathbf{c} = \mathbf{c}(\mathbf{u})$, we define the "rotated" Laplace transform

$$(25.19)\qquad \mathfrak{L}_{\mathbf{u}}(\mathbf{z}) = \sum_{\mathbf{k}\in\mathbb{Z}^d} \exp(\mathbf{z}\cdot R_{\mathbf{u}}\mathbf{k})\,\mu_{\mathbf{c}}(\mathbf{k}), \quad \mathbf{z} \in \mathbb{C}^d.$$

It is analytic on the whole of $\mathbb{C}^d$ for each $\mathbf{u}$.

(25.20) Proposition. *There is $\varepsilon > 0$ such that in $\mathcal{U}_\varepsilon = \{\mathbf{z} \in \mathbb{C}^d : |\mathbf{z}| < \varepsilon\}$ one can decompose*

$$1-\mathfrak{L}_{\mathbf{u}}(\mathbf{z}) = \left(z_1 - \mathcal{A}_{\mathbf{u}}(\mathbf{z}')\right)\mathcal{B}_{\mathbf{u}}(\mathbf{z}).$$

The functions $\mathcal{A}_{\mathbf{u}}$ and $\mathcal{B}_{\mathbf{u}}$ are analytic in $\mathcal{U}'_\varepsilon$ and $\mathcal{U}_\varepsilon$, respectively, and depend continuously on $\mathbf{u} \in \mathbb{S}_{d-1}$ as well as all their derivatives. Furthermore,

$$\mathcal{A}_{\mathbf{u}}(\mathbf{z}') = -\tfrac{1}{2}Q_{\mathbf{u}}[\mathbf{z}'] + \mathcal{O}(|\mathbf{z}'|^3),$$

and the $\mathcal{O}(\cdot)$ is uniform in $\mathbf{u}$ when $\mathbf{z}' \to \mathbf{0}$.

Proof. Using (25.13), we compute $\operatorname{grad}\mathfrak{L}_{\mathbf{u}}(\mathbf{0}) = \sum_{\mathbf{k}} \mu_{\mathbf{c}}(\mathbf{k})\, R_{\mathbf{u}}\mathbf{k} = R_{\mathbf{u}}\mathfrak{m}_{\mathbf{c}} = |\mathfrak{m}_{\mathbf{c}}|\, \mathbf{e}_1$. Also, $\mathfrak{L}_{\mathbf{u}}(\mathbf{0}) = 1$. Therefore, for fixed $\mathbf{u}$, the proposed decomposition of $1 - \mathfrak{L}_{\mathbf{u}}$ in a suitable neighbourhood of the origin follows from the Weierstrass preparation theorem; see Hörmander [175], Thm. 7.5.1. To see that the neighbourhood can be chosen independently of $\mathbf{u}$, we have to look at the first three lines of the proof of Thm. 7.5.1 in [175] in order to see how the neighbourhood is constructed.

First, we need $r > 0$ such that $\mathfrak{L}_{\mathbf{u}}(z_1, \mathbf{0}') \neq 1$ when $0 < |z_1| < 2r$. Now, the mapping $(\mathbf{u}, \mathbf{z}) \mapsto \frac{\partial}{\partial z_1}\mathfrak{L}_{\mathbf{u}}(\mathbf{z})$ is continuous and equal to $|\mathfrak{m}_{\mathbf{c}}| \geq \lambda^{-1} > 0$ at $(\mathbf{u}, \mathbf{0})$ for each $\mathbf{u} \in \mathbb{S}_{d-1}$. Therefore it has real part $> \lambda^{-1}/2$ in a neighbourhood $U_{\mathbf{u}} \times V_{\mathbf{u}}$ of $(\mathbf{u}, \mathbf{0})$. The sphere is covered by finitely many $U_{\mathbf{u}(j)}$, and $\bigcap_j V_{\mathbf{u}(j)}$ is a neighbourhood of $\mathbf{0}$ in $\mathbb{C}^d$ where $\operatorname{Re}\big(\frac{\partial}{\partial z_1}\mathfrak{L}_{\mathbf{u}}\big) > \lambda^{-1}/2$ for all $\mathbf{u}$. Therefore r can be chosen independently of $\mathbf{u}$.

Next we need $\delta > 0$ such that $\mathfrak{L}_{\mathbf{u}}(\mathbf{z}) \neq 1$ when $|z_1| = r$ and $|\mathbf{z}'| < \delta$. The same type of argument as above shows that δ can be chosen independently of $\mathbf{u}$ (exercise !).

According to [175], the decomposition is then valid in $\{|z_1| < r\,,\ |\mathbf{z}'| < \delta\}$.

The function $\mathcal{A}_{\mathbf{u}}$ satisfies $\mathfrak{L}_{\mathbf{u}}\big(\mathcal{A}_{\mathbf{u}}(\mathbf{z}'), \mathbf{z}'\big) = 1$ for all $\mathbf{z}' \in \mathcal{U}'_\varepsilon$ and $\mathbf{u} \in \mathbb{S}_{d-1}$. This yields that $\mathcal{A}_{\mathbf{u}}$ and its derivatives depend continuously on $\mathbf{u}$. Consequently, the same must hold for $\mathcal{B}_{\mathbf{u}}$. Differentiating the implicit equation, we find $\operatorname{grad}\mathcal{A}_{\mathbf{u}}(\mathbf{0}') = \mathbf{0}'$ and the matrix of second order derivatives of $\mathcal{A}_{\mathbf{u}}$ at $\mathbf{0}'$ is $-Q_{\mathbf{u}}$. Near the origin, each derivative of $\mathcal{A}_{\mathbf{u}}$ is uniformly bounded, so that the $\mathcal{O}(|\mathbf{z}'|^3)$ in its expansion is uniform. □

Proof of Theorem 25.15. Let $\mathbf{k} \in \mathbb{Z}^d$ be fixed. Let ε be as in Proposition 25.20 or smaller, $\mathbf{u} = \mathbf{u}(\mathbf{m})$ and $\mathbf{c} = \mathbf{c}(\mathbf{m})$. The characteristic function of $\mu_{\mathbf{c}}$ is $\phi_{\mathbf{c}}(\mathbf{v}) = \mathfrak{L}_{\mathbf{u}}(i\, R_{\mathbf{u}}\mathbf{v})$, as $R_{\mathbf{u}}$ leaves the inner product invariant. From (25.18) we get that $G_{\mathbf{c}}(\mathbf{k}, \mathbf{m}) = G^\varepsilon_{\mathbf{c}}(\mathbf{k}, \mathbf{m}) + \mathfrak{o}\big(|\mathbf{m}|^{-(d-1)/2}\big)$, where

$$G^\varepsilon_{\mathbf{c}}(\mathbf{k}, \mathbf{m}) = \frac{1}{(2\pi)^d} \int_{|\mathbf{v}|<\varepsilon} e^{-i(\mathbf{m}-\mathbf{k})\cdot\mathbf{v}} \frac{f(\mathbf{v})}{1 - \mathfrak{L}_{\mathbf{u}}(i\, R_{\mathbf{u}}\mathbf{v})}\, d\mathbf{v}\,.$$

Now remember that $R_{\mathbf{u}}\mathbf{m} = |\mathbf{m}|\, \mathbf{e}_1$. Substituting $R_{\mathbf{u}}\mathbf{v} = \mathbf{w}$, we obtain

$$G^\varepsilon_{\mathbf{c}}(\mathbf{k}, \mathbf{m}) = -\frac{1}{(2\pi)^d} \int_{|\mathbf{w}|<\varepsilon} e^{-i\,|\mathbf{m}|w_1} \frac{f_{\mathbf{u},\mathbf{k}}(\mathbf{w})}{i\,w_1 - \mathcal{A}_{\mathbf{u}}(i\,\mathbf{w}')}\, d\mathbf{w}\,, \quad \text{where}$$
$$f_{\mathbf{u},\mathbf{k}}(\mathbf{w}) = -\exp(i\, R_{\mathbf{u}}\mathbf{k}\cdot\mathbf{w})\, f(\mathbf{w})/\mathcal{B}_{\mathbf{u}}(i\,\mathbf{w})\,.$$

We extend $f_{\mathbf{u},\mathbf{k}}$ to a function in $C^\infty(\mathbb{R}^d)$ by setting it equal to 0 outside of $\{\mathbf{w} : |\mathbf{w}| < \varepsilon\}$. It has compact support, and all its derivatives depend continuously on $\mathbf{u} \in \mathbb{S}_{d-1}$ and hence are bounded uniformly in $\mathbf{u}$. For fixed

$\mathbf{w}' \in \mathbb{R}^{d-1}$, $|\mathbf{w}'| \le \varepsilon$, we consider its Fourier transform in the first variable:

$$\widehat{f}_{\mathbf{u},\mathbf{k}}(t,\mathbf{w}') = \int_{-\infty}^{\infty} e^{i\,tw_1} f_{\mathbf{u},\mathbf{k}}(w_1,\mathbf{w}')\,dw_1\,.$$

It is continuous in $(t,\mathbf{w}')$ and once more by the Riemmann–Lebesgue lemma, $\widehat{f}_{\mathbf{u},\mathbf{k}}(t,\mathbf{w}') = \mathfrak{o}\big(|t|^{-(d+1)/2}\big)$ uniformly in $\mathbf{u}$ and $\mathbf{w}'$, as $t \to \infty$.

We now choose ε small enough that in the expansion $\mathcal{A}_{\mathbf{u}}(i\,\mathbf{w}') = \frac{1}{2}Q_{\mathbf{u}}[\mathbf{w}'] + \mathcal{O}(|\mathbf{w}'|^3)$, the $\mathcal{O}(|\mathbf{w}'|^3)$ is bounded in absolute value by $\frac{1}{4}Q_{\mathbf{u}}[\mathbf{w}']$, when $|\mathbf{w}'| < \varepsilon$ (see Proposition 25.20). Then $\mathrm{Re}\big(i\,w_1 - \mathcal{A}_{\mathbf{u}}(i\,\mathbf{w}')\big) < 0$, and

$$\frac{1}{i\,w_1 - \mathcal{A}_{\mathbf{u}}(i\,\mathbf{w}')} = -\int_0^{\infty} \exp\big(i\,w_1 s - \mathcal{A}_{\mathbf{u}}(i\,\mathbf{w}')s\big)\,ds\,.$$

We obtain

$$\begin{aligned} G^{\varepsilon}_{\mathbf{c}}(\mathbf{k},\mathbf{m}) &= \frac{1}{(2\pi)^d}\int_{|\mathbf{w}'|<\varepsilon}\int_0^{\infty} e^{-s\,\mathcal{A}_{\mathbf{u}}(i\,\mathbf{w}')}\int_{-\varepsilon}^{\varepsilon} e^{i(s-|\mathbf{m}|)w_1}\, f_{\mathbf{u},\mathbf{k}}(\mathbf{w})\,dw_1\,ds\,d\mathbf{w}' \\ &= \frac{1}{(2\pi)^d}\int_{|\mathbf{w}'|<\varepsilon}\int_{-|\mathbf{m}|}^{\infty} e^{-(t+|\mathbf{m}|)\mathcal{A}_{\mathbf{u}}(i\,\mathbf{w}')}\,\widehat{f}_{\mathbf{u},\mathbf{k}}(t,\mathbf{w}')\,dt\,d\mathbf{w}'\,. \end{aligned}$$

We decompose the integral into two parts, where the first is over $t \in [-|\mathbf{m}|\,,\,-|\mathbf{m}|/2)$ and $|\mathbf{w}'| < \varepsilon$, and the second over $t \in [-|\mathbf{m}|/2\,,\,\infty)$ and $|\mathbf{w}'| < \varepsilon$. Since the exponential is bounded by 1 in absolute value and $\widehat{f}_{\mathbf{u},\mathbf{k}}(t,\mathbf{w}') = \mathfrak{o}\big(|t|^{-(d+1)/2}\big)$, the first part is $\mathfrak{o}\big(|\mathbf{m}|^{-(d-1)/2}\big)$ as $|\mathbf{m}| \to \infty$. In the second part, we substitute $\sqrt{t+|\mathbf{m}|}\,\mathbf{w}' = \mathbf{x}'$. We obtain

$$G_{\mathbf{c}}(\mathbf{k},\mathbf{m}) = \mathfrak{o}(|\mathbf{m}|^{-(d-1)/2}) + \frac{|\mathbf{m}|^{-(d-1)/2}}{(2\pi)^d}\int_{-|\mathbf{m}|/2}^{\infty}\frac{1}{(1+\frac{t}{|\mathbf{m}|})^{(d-1)/2}}$$

$$\times\int_{|\mathbf{x}'|<\varepsilon\sqrt{t+|\mathbf{m}|}} \exp\Big(-\tfrac{1}{2}Q_{\mathbf{u}}[\mathbf{x}'] + \mathcal{O}\big(\tfrac{1}{\sqrt{t+|\mathbf{m}|}}|\mathbf{x}'|^3\big)\Big)\,\widehat{f}_{\mathbf{u},\mathbf{k}}\big(t,\tfrac{1}{\sqrt{t+|\mathbf{m}|}}\mathbf{x}'\big)\,d\mathbf{x}'\,dt\,.$$

Recall that the $\mathcal{O}(\cdot)$ is uniform in $\mathbf{u}$ when the argument tends to 0. Therefore (and due to our choice of ε), we may use dominated convergence to see that

$$\begin{aligned} G_{\mathbf{c}}(\mathbf{k},\mathbf{m}) &\sim \frac{|\mathbf{m}|^{-(d-1)/2}}{(2\pi)^d}\int_{-|\mathbf{m}|/2}^{\infty}\int_{\mathbb{R}^{d-1}} \exp(-\tfrac{1}{2}Q_{\mathbf{u}}[\mathbf{x}'])\,\widehat{f}_{\mathbf{u},\mathbf{k}}(t,\mathbf{0}')\,d\mathbf{x}'\,dt \\ &= \sqrt{\det Q_{\mathbf{u}}}\,\big(2\pi|\mathbf{m}|\big)^{-(d-1)/2}\,\frac{1}{2\pi}\int_{-|\mathbf{m}|/2}^{\infty}\widehat{f}_{\mathbf{u},\mathbf{k}}(t,\mathbf{0}')\,dt\,, \end{aligned}$$

when $|\mathbf{m}| \to \infty$. To conclude, we observe that

$$\frac{1}{2\pi}\int_{-\infty}^{\infty}\widehat{f}_{\mathbf{u},\mathbf{k}}(t,\mathbf{0}')\,dt = f_{\mathbf{u},\mathbf{k}}(\mathbf{0}) = -1/\mathcal{B}_{\mathbf{u}}(\mathbf{0}) = 1/|\mathfrak{m}_{\mathbf{c}}|\,. \qquad \square$$

(25.21) Exercise. *Let μ have finite support on $\mathbb{Z}^d$, $d \geq 2$. When $t = \rho(\mu)$, the Martin boundary $\mathcal{M}(\mu, t)$ consists of a single point.*

When $t > \rho(\mu)$, the boundary is homeomorphic with the unit sphere; if $|\mathbf{m}| \to \infty$ and $\frac{1}{|\mathbf{m}|} \to \mathbf{u} \in \mathbb{S}_{d-1}$ then $K(\mathbf{k}, \mathbf{m}|t) \to e^{\mathbf{c}\cdot\mathbf{k}}$ for each $\mathbf{k} \in \mathbb{Z}^d$, where $\mathbf{c}$ is the unique point with $\varphi_\mu(\mathbf{c}) = t$ that satisfies (25.13).

26. Trees, ends, and free products

The theme of this section is the relation between the Martin and end compactifications of a graph. In comparing different compactifications of X, by a *natural surjection* of one onto the other we mean a continuous mapping that extends id_X and maps one boundary onto the other.

We shall start with a general result. Modifying the definition (1.20), we say that (X, P) is *weakly uniformly irreducible* if there is $K < \infty$ such that $x \sim y$ implies $p^{(k)}(x, y) > 0$ for some $k \leq K$. As usual, X is supposed to be a connected, locally finite graph.

If $A \subset X$ and $x, w \in X$ then we shall write

$$F^A(x, w) = \mathbb{P}_x[Z_{\mathbf{s}^A} = w]$$

for the probability that the random walk starting at x first hits the set A at the point w. For a function $f : X \to \mathbb{R}$ (or $A \to \mathbb{R}$), we write $F^A f(x) = \sum_{w \in A} F^A(x, w) f(w)$, as usual when dealing with kernels. If $y \in X$ and the random walk starting at x has to pass through A with probability 1, given that it reaches y (that is, $\mathbf{s}^A \leq \mathbf{s}^y$ almost surely, if $Z_0 = x$), then

$$F(x, y) = \sum_{w \in A} F^A(x, w) F(w, y) . \tag{26.1}$$

This identity will be crucial when dealing with ends.

(26.2) Theorem. *If (X, P) has bounded range, is weakly uniformly irreducible and transient, then the Martin compactification $\widehat{X}(P)$ surjects naturally onto the end compactification $\widehat{X}_e$.*

Proof. As P has bounded range, it is nearest neighbour on some k-fuzz $X^{(k)}$. The latter has the same end compactification as X by Lemma 21.4. Therefore we may assume that P is nearest neighbour on X.

Let (x_n) be a sequence in X converging in the Martin topology to a boundary point. We write $h = \lim_n K(\cdot, x_n)$. We have to show that (x_n) converges in the end compactification.

Suppose the contrary. Then (x_n) must have two different accumulation points $\xi, \eta \in \vartheta_e X$. There is some finite ball $B \subset X$ such that

$d(\widehat{C}(\xi,B),\widehat{C}(\eta,B)) > K$, where K is the bound giving weak uniform irreducibility. We write U for the set of neighbours of B in $\widehat{C}(\xi,B)$, and analogously $V = \{x \in \widehat{C}(\eta,B) : d(x,B) = 1\}$. There are sub-sequences $(x_{n'})$ and $(x_{n''})$ lying entirely in $C(\xi,B)$ and $C(\eta,B)$, respectively. If $x \in U$ then the random walk starting at x has to pass through V in order to reach any of the $x_{n''}$. We apply (26.1) and divide by $F(o,x_{n''})$, so that

$$K(x,x_{n''}) = \sum_{y\in V} F^V(x,y)K(y,x_{n''})\,.$$

In the same way, $K(y,x_{n'}) = \sum_{x\in U} F^U(y,x)K(x,x_{n'})$ for every $y \in V$. These sums are finite, and letting $n',n'' \to \infty$, we find $h(x) = F^V h(x)$ on U and $h(y) = F^U h(y)$ on V.

Next, let $V^* = \{y \in V : F^V(x,y) > 0\}$ be the set of points where the first visit to V may occur with positive probability, when $x \in U$. This set is independent of $x \in U$. Indeed, if $x, w \in U$ then there is a path $x = x_0, x_1, \ldots, x_\ell = w$ that lies entirely in $\widehat{C}(B,\xi)$, whence $d(x_i,V) > K$ for all i. Weak uniform irreducibility implies $p^{(k_i)}(x_{i-1},x_i) > 0$ for some $k_i \le K$. Being nearest neighbour, in K or fewer steps the random walk cannot get beyond distance K from the starting point x_{i-1} and, in particular, cannot hit V. This implies

$$F^V(x,y) \ge p^{(k_1)}(x,x_1)\,p^{(k_2)}(x_1,x_2)\cdots p^{(k_\ell)}(x_{\ell-1},w)\,F^V(w,y)\,.$$

Analogously, $U^* = \{x \in U : F^U(y,x) > 0\}$ is independent of $y \in V$.

Now let f be the restriction of h to U^* and Q the matrix indexed by U^* with entries $q(x,w) = \sum_{y\in V^*} F^V(x,y)\,F^U(y,w)$. The matrix is substochastic, and irreducible, as all its elements are > 0. Suppose that there is $x_0 \in U^*$ such that $\sum_{w\in U^*} q(x_0,w) < 1$. Then, for arbitrary $x \in U^*$,

$$\sum_{w\in U^*} q^{(2)}(x,w) = \sum_{u\in U^*} q(x,u) \sum_{w\in U^*} q(u,w) < \sum_{u\in U^*} q(x,u) \le 1\,,$$

and $\rho(Q^2) < 1$. But $Q^2 f = f$, and the Perron–Frobenius eigenvalue of Q^2 must be $\rho(Q^2) = 1$. We conclude that

$$\sum_{w\in U^*}\sum_{y\in V^*} F^V(x,y)F^U(y,w) = 1 \quad \text{for all } x \in U^*\,.$$

But this contradicts transience, as it implies that for any starting point $x \in U^*$, the random walk returns to U^* with probability 1. □

(26.3) Exercise. Verify that the theorem also holds with $\widehat{X}(P,t)$ in the place of the "ordinary" Martin compactification, whenever $t > \rho(P)$ or $t = \rho(P)$ and the random walk is ρ-transient. [Hint: Use a suitable h-process.]

A. Thin ends and trees

Under the conditions of Theorem 26.2, we have a natural surjection $\sigma : \widehat{X}(P) \to \widehat{X}_e$, and in this subsection we shall study the following two questions. (1) When is σ a homeomorphism, so that the Martin boundary is realized by the space of ends of X? (2) Given an end ξ, under what conditions is $\sigma^{-1}(\xi)$ a single point in $\mathcal{M}(P)$?

We shall also give answers for $\widehat{X}(P,t)$ in place of $\widehat{X}(P)$. To simplify, we shall write $t \gtrsim \rho(P)$, when either $t > \rho(P)$, or $t = \rho(P)$ and the random walk is ρ-transient.

We start with the simpest case, namely that of nearest neighbour random walks on trees. Let T be a tree and o the root. Recall that the *confluent* $v \wedge w$ of two elements $v, w \in \widehat{T}$ is the last common element on the geodesics $\pi(o,v)$ and $\pi(o,w)$.

(26.4) Theorem. *Let P be an irreducible nearest neighbour random walk on the locally finite tree T. If $t \gtrsim \rho(P)$, then the Martin compactification $\widehat{X}(P,t)$ coincides with the end compactification. The extended Martin kernel is given by*

$$K(x,\xi|t) = K(x,c|t)\,, \quad \text{where } \xi \in \vartheta T \quad \text{and} \quad c = x \wedge \xi\,.$$

For every $\xi \in \vartheta T$, the function $K(\cdot,\xi|t)$ is minimal t-harmonic.

Proof. In view of Theorem 26.2, we only have to show the following to prove that $\widehat{X}(P,t) = \widehat{X}_e$: if (x_n) converges to the end ξ of T, then $K(x,x_n|t)$ converges to $K(x,c|t)$, where $c = x \wedge \xi$.

To see this, let y be the point immediately after c on $\pi(o,\xi)$. Then the branches $T_{o,y}$ and $T_{x,y}$ (see §6.B) coincide. All but finitely many x_n lie in $T_{o,y}$, and in this case, Lemma 1.23 tells us that $F(o,x_n|1/t) = F(o,c|1/t)F(c,x_n|1/t)$ and $F(x,x_n|1/t) = F(x,c|1/t)F(c,x_n|1/t)$. Therefore $K(x,x_n|t) = K(x,c|t)$ for all those n, and consequently $\lim_n K(x,x_n|t) = K(x,c|t)$.

For minimality, suppose that $K(\cdot,\xi|t) = a \cdot h_1 + (1-a) \cdot h_2$, where the h_i are t-superharmonic and $h_i(o) = 1$. Finite range implies that $K(\cdot,\xi|t)$ is t-harmonic (Lemma 24.16). Therefore the h_i are also t-harmonic.

Now let $x \in T$, and y a point on the geodesic ray $\pi(x,\xi)$. For the h_i-process (24.17) we know that $F_{h_i}(x,y) \leq 1$. Rewriting this in terms of P, we find $F(x,y|1/t)\,h_i(y) \le h_i(x)$. On the other hand, $K(x,\xi|t) = F(x,y|1/t)\,K(y,\xi|t)$ by the above. Therefore

$$\begin{aligned} K(x,\xi|t) &= a \cdot h_1(x) + (1-a)\cdot h_2(x) \\ &\ge a \cdot F(x,y|1/t)\,h_1(y) + (1-a)\cdot F(x,y|1/t)\,h_2(y) \\ &= F(x,y|1/t)\,K(y,\xi|t) = K(x,\xi|t)\,. \end{aligned}$$

The inequalities used cannot be strict, and $h_i(x) = F(x,y|1/t)\,h_i(y)$. In particular, with $y = x \wedge \xi$, we find $h_i(x) = F(x,y|1/t)\,h_i(y)$ and $1 = h_i(o) = F(o,y|1/t)\,h_i(y)$. Dividing, we see that $h_i(x) = K(x,y|t) = K(x,\xi|t)$. □

As a matter of fact, we see that the extended Martin kernels are *locally constant:* for each $x \in T$ and $\xi \in \vartheta T$, the kernel $K(x,\cdot|t)$ is constant on the neighbourhood $\{v \in \widehat{T} : x \wedge v = x \wedge \xi\}$ of ξ.

In the transient case, we can also compute the hitting distributions ν_x on ϑT, $x \in T$, in terms of the $F(x,y)$. The sets $\vartheta T_{x,y}$, $y \in T \setminus \{x\}$, generate the Borel σ-algebra of ϑT. First, if x, y are arbitrary ($x \ne y$) and w is the neighbour of y on $\pi(x,y)$, then a variation of Lemma 1.23 yields $\nu_x(\vartheta T_{x,y}) = F(x,w)\,\nu_w(\vartheta T_{w,y})$, since the random walk must reach w if it "ends up" in $\widehat{T}_{x,y} = \widehat{T}_{w,y}$. Second,

$$\nu_w(\vartheta T_{w,y}) = F(w,y)\,\nu_y(\vartheta T_{w,y}) = F(w,y)\bigl(1 - \nu_y(\vartheta T_{y,w})\bigr)\,,$$

since $\vartheta T_{w,y} = \vartheta T \setminus \vartheta T_{y,w}$. Exchanging y and w, we find a second inequality in the two unkowns $\nu_w(\vartheta T_{w,y})$ and $\nu_y(\vartheta T_{y,w})$. We can solve the system of equations, and find

$$\nu_x(\vartheta T_{x,y}) = F(x,y)\frac{1 - F(y,w)}{1 - F(w,y)F(y,w)}\,, \tag{26.5}$$

where $w \in \pi(x,y)$, $w \sim y$. Note that $F(w,y)F(y,w) \le U(w,w) < 1$ by transience. If $h \in \mathcal{H}^+(P,t)$, where (as usual) $t \gtrsim \rho(P)$, then we can apply (26.5) to the h-process in order to find the (unique) representing measure ν^h on the boundary, see (24.18).

(26.6) Example. Consider the simple random walk on the homogeneous tree $\mathbb{T} = \mathbb{T}_{q+1}$. We know that $F(x,y|z) = F(z)^{d(x,y)}$, where $F(z)$ has been computed in the proof of Lemma 1.24. Thus, if $x \in \mathbb{T}$, $\xi \in \vartheta\mathbb{T}$ and $c = x \wedge \xi$ then

$$K(x,\xi|t) = F(1/t)^{d(x,c)-d(o,c)} = F(1/t)^{\mathfrak{h}(x,\xi)}\,,$$

where $\mathfrak{h}(x,\xi)$ is the height (horocycle number) of x with respect to the end ξ; see (12.13), where ω replaces ξ. In particular, when $t = 1$, $F(1) = 1/q$ and $K(x,\xi) = q^{-\mathfrak{h}(x,\xi)}$. The hitting distribution ν_x on the boundary is invariant under the stabilizer of x in $\mathrm{AUT}(\mathbb{T})$. We find $\nu_x(\vartheta\mathbb{T}_{x,y}) = 1/|S(x,k)|$, where $k = d(x,y)$ and $S(x,k)$ is the sphere of radius k centred at x. □

One of the next questions will be what happens when instead of nearest neighbour, we consider a bounded range random walk on a locally finite tree. The tool for answering this will also give an answer to question (2) posed at the beginning of this subsection. Recall that $F^V(x,y)$ is the probability that y is the first point of $V \subset X$ reached by the random walk starting in x. With arbitrary finite subsets U, V of X we associate the matrix $\mathfrak{F}(U,V) = \big(F^V(x,y)\big)_{x\in U, y\in V}$. If $V = \{y\}$ then we write $\mathbf{f}(U,y)$ (a column vector) instead of $\mathfrak{F}(U,V)$. With any non-negative matrix $Q = \big(q(x,y)\big)_{x\in U, y\in V}$ we associate the coefficient

$$\lambda(Q) = \max\{c : q(x,y) \ge c\, q(x',y) \text{ for all } x, x' \in U\,,\ y \in V\}\,.$$

Finally, recall the definition (21.1) of a standard neighbourhood base of an end of X.

(26.7) Theorem. *Suppose that (X,P) is nearest neighbour and transient. Let ξ be an end of X. If there is a standard neighbourhood base $\{\widehat{C}(\xi, V_n) : n \in \mathbb{N}\}$, such that*

$$\lambda\big(\mathfrak{F}(V_n, V_{n+1})\big) > 0 \quad \textit{for all } n \quad \textit{and} \quad \sum_{n=1}^{\infty} \lambda\big(\mathfrak{F}(V_n, V_{n+1})\big) = \infty\,,$$

then ξ corresponds to a unique point in $\mathcal{M}(P)$ under the natural surjection σ, and $\sigma^{-1}(\xi)$ is minimal.

Proof. Suppose that $U \cap \widehat{C}(\xi, V_k) = \emptyset$ and that $y \in \widehat{C}(\xi, V_n)$, where $n \ge k$. Then the random walk starting at any $x \in U$ has to pass through $V_k, \dots, V_n$ (in this order) if it visits y. By (26.1), this implies

$$\mathbf{f}(U,y) = \mathfrak{F}(U,V_k)\,\mathfrak{F}(V_k,V_{k+1}) \cdots \mathfrak{F}(V_{n-1},V_n)\,\mathbf{f}(V_n,y)\,. \tag{26.8}$$

If $U = \{o, x\}$, the Martin kernel $K(x,y)$ is the quotient of the two entries of $\mathbf{f}(U,y)$. If $y \to \xi$ then $n \to \infty$, and as we want to show that $K(x,y)$ converges, we need a result on convergence of matrix products such as those occurring in (26.8). We interrupt the proof in order to set up the necessary machinery.

Denote by $\mathcal{C}_d$ the non-negative cone in $\mathbb{R}^d$, by $\mathcal{C}_d^o$ its interior (the strictly positive cone), by $\mathcal{S}_d$ the standard simplex in $\mathcal{C}_d$ (the vectors summing up to 1), and by $\mathcal{S}_d^o$ its intersection with $\mathcal{C}_d^o$. Furthermore, $\mathcal{P}_d$ is the central projection from the origin of $\mathcal{C}_d \setminus \{\mathbf{0}\}$ onto $\mathcal{S}_d$. On $\mathcal{S}_d^o$, we introduce the hyperbolic metric

$$\theta_d(\mathbf{a},\mathbf{b}) = \log \max_{u,u'} \frac{a(u)b(u')}{b(u)a(u')}\,,$$

where $\mathbf{a} = \big(a(u)\big)_{u\in U}$ and $\mathbf{b} = \big(b(u)\big)_{u\in U}$ with $|U| = d$. This metric induces the Euclidean topology in $\mathcal{S}^o_d$. If no confusion can arise, we shall omit the subscript d. A non-negative, non-zero matrix $Q = (q(u,v))_{u\in U, v\in V}$ is said to have its *zeros disposed in columns,* if $\lambda(Q) > 0$, that is, in any column, either all elements are > 0 or all are $= 0$. Its *Birkhoff contraction coefficient* is

$$\tau(Q) = \inf\left\{\frac{\theta(\mathcal{P}\,Q\mathbf{a}, \mathcal{P}\,Q\mathbf{b})}{\theta(\mathbf{a},\mathbf{b})} : \mathbf{a},\mathbf{b} \in \mathcal{S}^o_{|V|}\,, \mathbf{a} \neq \mathbf{b}\right\}.$$

(26.9) Birkhoff's contraction lemma.

$$\tau(Q) = \frac{1-\sqrt{\phi(Q)}}{1+\sqrt{\phi(Q)}}\,, \quad \textit{with} \quad \phi(Q) = \min \frac{q(u,v)q(u',v')}{q(u',v)q(u,v')}\,,$$

where the minimum ranges over all non-negative entries of Q.

For a proof, see e.g. Seneta [297], Chap. 3, where only square matrices are considered; the present case requires no changes. A sequence $(\mathbf{c}_n)$ of non-zero vectors in $\mathcal{C}_d$ is said to *converge in direction,* if $\mathcal{P}\mathbf{c}_n$ converges to a point in $\mathcal{S}_d$.

(26.10) Proposition. *For $n = 1, 2, \ldots$, let Q_n be a non-negative, non-zero $d(n-1) \times d(n)$ matrix whose zeros are disposed in columns. If*

$$\tau(Q_1 \cdots Q_n) \to 0\,, \quad \textit{and in particular, if} \quad \sum_{n=1}^{\infty} \sqrt{\phi(Q_n)} = \infty\,,$$

then there is a unique $\mathbf{c} \in \mathcal{S}^o_{d(0)}$ such that for any choice of vectors $\mathbf{a}_n \in \mathcal{C}^o_{d(n)}$, the sequence $Q_1 \cdots Q_n\mathbf{a}_n$ converges in direction to $\mathbf{c}$.

Proof. Set $R_n = Q_1 \cdots Q_n$. Since $\tau(R_n) \le \tau(Q_1)\cdots\tau(Q_n)$, it is clear from Lemma 26.9 that divergence of $\sum_n \sqrt{\phi(Q_n)}$ implies $\tau(R_n) \to 0$.

We have $\mathcal{P}Q_n\mathcal{S}^o_{d(n)} \subset \mathcal{S}^o_{d(n-1)}$. Thus, the sequence of sets $D_n = \mathcal{P}Q_1 \cdots Q_n\mathcal{S}^o_{d(n)}$ $(n = 1, 2, \ldots)$ is decreasing. We prove that their intersection contains a single point $\mathbf{c} \in \mathcal{S}^o_{d(0)}$. It is sufficient to find a sub-sequence whose intersection is $\{\mathbf{c}\}$.

Write $R_n = \big(r_n(u,v)\big)_{u\in U, v\in V_n}$, that is, the rows of Q_1 are indexed by the set U and the columns of Q_n and rows of Q_{n+1} by the set V_n. We have $\lambda(R_n) \ge \lambda > 0$ for all n, where $\lambda = \lambda(Q_1)$.

Now choose for each n a non-zero column $\mathbf{r}_n(v_n) = \big(r_n(u,v_n)\big)_{u\in U}$ of R_n. By compactness, there must be a sub-sequence $\big(n(k)\big)$ and a vector $\mathbf{c} = \big(c(u)\big)_{u\in U}$ in $\mathcal{S}_{d(0)}$ such that $\mathcal{P}\mathbf{r}_{n(k)}(v_{n(k)}) \to \mathbf{c}$. We must have $\lambda\, c(u) \le c(u') \le \lambda^{-1}c(u)$ for all u, u', whence $\mathbf{c} \in \mathcal{S}^o_{d(0)}$.

If $\tau(R_n) \to 0$ then $\phi(R_n) \to 1$, which implies that

$$\frac{r_n(u,v_n)}{r_n(u',v_n)} \Big/ \frac{r_n(u,v'_n)}{r_n(u',v'_n)} \to 1$$

uniformly in $u, u' \in U$ and $v'_n \in V_n$ corresponding to non-zero columns of R_n, as $n \to \infty$. We infer that

$$\frac{r_{n(k)}(u,v'_{n(k)})}{r_{n(k)}(u',v'_{n(k)})} \to \frac{c(u)}{c(u')} \quad \text{and consequently} \quad \mathcal{P}\mathbf{r}_{n(k)}(v'_{n(k)}) \to \mathbf{c}$$

uniformly in $v'_{n(k)} \in V_{n(k)}$ corresponding to non-zero columns of $R_{n(k)}$. Now $D_{n(k)}$ is the convex hull of the projections onto $\mathcal{S}_{d(0)}$ of the non-zero column vectors of $R_{n(k)}$. Therefore $\bigcap_k D_{n(k)} = \{\mathbf{c}\}$, as proposed. □

Proof of Theorem 26.7 (concluded). We return to the identity (26.8). By assumption, each matrix $\mathfrak{F}(V_k, V_{k+1})$ has its zeros disposed in columns. Also, $\sqrt{\phi(\mathfrak{F}(V_k, V_{k+1}))} \geq \lambda(\mathfrak{F}(V_k, V_{k+1}))$.

Given x, irreducibility yields that there are $x_0 = o, x_1, \ldots, x_\ell = x$ such that $x_i \neq x$ for $i < \ell$ and $p(x_i, x_{i+1}) > 0$. Analogously, there are $y_0 = x, y_1, \ldots, y_m = o$ such that $y_j \neq o$ for $j < m$ and $p(y_j, y_{j+1}) > 0$. For $U = \{o, x\}$, we choose $\kappa = \kappa(U)$ large enough that V_κ does not contain any of the x_i and y_j. Then also $\mathfrak{F}(U, V_\kappa)$ has its zeros disposed in columns, since for $y \in V_\kappa$

$$F^{V_\kappa}(o,y) \geq p(x_0,x_1)\cdots p(x_{\ell-1},x_\ell)\, F^{V_\kappa}(x,y) \quad \text{and}$$
$$F^{V_\kappa}(x,y) \geq p(y_0,y_1)\cdots p(y_{m-1},y_m)\, F^{V_\kappa}(o,y)\,.$$

If we set $Q_1 = \mathfrak{F}(U, V_\kappa)$, $Q_2 = \mathfrak{F}(V_\kappa, V_{\kappa+1})$, $\ldots$, then $\sum \sqrt{\phi(Q_n)} = \infty$. Proposition 26.10 and (26.8) imply that there is a vector $\mathbf{f}(U,\xi) \in \mathcal{S}^{o}_{|U|}$ such that

$$\mathcal{P}\,\mathbf{f}(U, y_n) \to \mathbf{f}(U,\xi)$$

for every sequence (y_n) in X that converges to the end ξ. We note that the same argument works for arbitrary finite $U \subset X$, and that

$$\mathbf{f}(U,\xi) = \mathcal{P}\,\mathfrak{F}(U,V_k)\mathbf{f}(V_k,\xi) \quad \text{for every} \quad k \geq \kappa(U)\,. \tag{26.11}$$

Returning to $U = \{o, x\}$, we see that $K(x, y_n)$ converges to the quotient of the two entries of $\mathbf{f}(U,\xi)$. This proves that $\sigma^{-1}(\xi)$ is a single point of the Martin boundary, and we identify this point with the end ξ.

We are left with proving minimality of $K(\cdot,\xi)$, and proceed as for Theorem 26.4. Suppose that $K(\cdot,\xi) = a \cdot h_1 + (1-a) \cdot h_2$, where the h_i are harmonic and $h_i(o) = 1$. For $A \subset X$, we write $\mathbf{h}_i(A) = \big(h(x)\big)_{x\in A}$ and $\mathbf{k}(A,\xi) = \big(K(x,\xi)\big)_{x\in A}$. The h_i-process satisfies $\sum_{y\in A} F^A_{h_i}(x,y) \le 1$, and $F^A_{h_i}(x,y) = F^A(x,y)h(y)/h(x)$. Thus $h_i(x) \ge \sum_{y\in A} F^A(x,y)h_i(y)$. With $U = \{o,x\}$ and $\kappa = \kappa(U)$ as above, we get

$$\mathbf{h}_i(U) \ge \mathfrak{F}(U,V_\kappa)\,\mathfrak{F}(V_\kappa,V_{\kappa+1})\cdots\mathfrak{F}(V_{n-1},V_n)\,\mathbf{h}_i(V_n) \tag{26.12}$$

elementwise for all $n \ge \kappa$. On the other hand, (26.11) implies

$$\mathbf{k}(U,\xi) = \mathfrak{F}(U,V_\kappa)\mathfrak{F}(V_\kappa,V_{\kappa+1})\cdots\mathfrak{F}(V_{n-1},V_n)\,\mathbf{k}(V_n,\xi)$$

for all $n \ge \kappa$. As in the proof of Theorem 26.4, we infer that equality must hold in (26.12). Letting $n \to \infty$, we find $\mathcal{P}\,\mathbf{h}_i(U) = \mathbf{f}(U,\xi)$. Therefore $h_i(x) = h_i(x)/h_i(o) = K(x,\xi)$. □

We now explain a few applications of Theorem 26.7. Usually, there are obvious lower bounds on the numbers $\lambda\big(\mathfrak{F}(V_n,V_{n+1})\big)$. As above, when $\sigma^{-1}(\xi)$ is a single element of the Martin boundary, we identify the end ξ with the latter. Thin ends have been defined in Section 21.

(26.13) Corollary. *Suppose that (X,P) is uniformly irreducible and has bounded range. If $t \gtrsim \rho(P)$, then every thin end is a point of $\mathcal{M}_{\min}(P,t)$.*

Proof. Again, we may suppose without loss of generality that the random walk is nearest neighbour (if an end of X is thin then it is also thin as an end of any k-fuzz of X). If h is any function in $\mathcal{H}^+(P,t)$, which exists by Lemma 24.16, then the h-process is transient and uniformly irreducible because of the Harnack inequality (25.1). Also $\widehat{X}(P,t) = \widehat{X}(P_h,1)$. Thus, we may also assume without loss of generality that $t = 1$ and (X,P) is transient.

Let K and ε_0 be the constants of uniform irreducibility in (1.20). Take a thin end ξ of X. We can find a standard neighbourhood base $\{C(\xi,V_n) : n \in \mathbb{N}\}$ with $\mathrm{diam}(V_n) = m < \infty$ and $d(V_n,V_{n+1}) > K + m/2$ for all n. We show that the corresponding numbers $\lambda\big(\mathfrak{F}(V_n,V_{n+1})\big)$ are bounded below by a fixed constant, so that Theorem 26.7 applies.

Let $x,x' \in V_n$. Then there is a path $[x_0 = x, x_1, \dots, x_r = x']$ in X with length $r \le m$. We have $d(x_i,V_n) \le m/2$ for each i. Given i, there is $k_i \le K$ such that $p^{(k_i)}(x_{i-1},x_i) \ge \varepsilon_0$. Every element on a path of length $\le K$ starting at x_{i-1} is at distance at most K from x_{i-1}, and hence at distance at most $K + m/2$ from V_n. None of its elements lies in V_{n+1}. Therefore, as the function $x \mapsto F^{V_{n+1}}(x,y)$ (where $y \in V_{n+1}$) is harmonic outside of V_{n+1}, it is also P^{k_i}-harmonic at x_{i-1}. This yields

$$F^{V_{n+1}}(x_{i-1},y) \ge p^{(k_i)}(x_{i-1},x_i)F^{V_{n+1}}(x_i,y) \quad \text{for all } y \in V_{n+1}\,,$$

and $\lambda\big(\mathfrak{F}(V_n,V_{n+1})\big) \ge \varepsilon_0^m$. □

(26.14) Corollary. (a) *Let T be a tree. Suppose that (T,P) is uniformly irreducible and has bounded range. Then for every $t \gtrsim \rho(P)$, the Martin compactification $\widehat{X}(P,t)$ coincides with the end compactification, and all ends are in $\mathcal{M}_{\min}(P,t)$.*

(b) *In particular, let Γ be a group with a non-cyclic free subgroup of finite index, and μ a finitely supported irreducible probability on Γ. Then for every $t \geq \rho(\mu)$, the Martin compactification $\widehat{X}(\mu,t)$ coincides with the end compactification of Γ.*

For (b), recall that the random walk must be ρ-transient by Theorem 7.8. The same as (a) holds, more generally, for graphs having no thick ends, although such graphs are not necessarily roughly isometric with a tree. The following is one of our very few excursions into measure theoretic boundary theory; it combines various results from previous sections with Corollary 26.13 and applies also in the presence of thick ends.

(26.15) Corollary. *Let X be a graph with infinitely many ends, and suppose that P has finite range and $\mathrm{AUT}(X,P)$ acts transitively on X. Let ν be the limit distribution on $\vartheta_e X$ of the random walk starting at o. Then the pair $(\vartheta_e X, \nu)$ is a model for the Poisson boundary of (X,P).*

Proof. For the existence of ν, that is, convergence of Z_n to a random end Z_∞, see Theorem 21.16; we know that $\rho(P) < 1$ from Theorem 10.10. By "model" we mean of course that the measure space $(\vartheta_e X, \nu)$ is isomorphic with the measure space $(\mathcal{M}, \widehat{\nu})$, where $\widehat{\nu}$ is distribution of the limit $\widehat{Z}_\infty$ of Z_n in the Martin compactification, when $Z_0 = o$.

The isomorphism is the natural surjection σ. Indeed, by continuity, $\sigma(\widehat{Z}_\infty) = Z_\infty$, so that ν is the image of $\widehat{\nu}$ under σ. Now consider the decomposition $\vartheta_e X = \Upsilon_0 \cup \Upsilon_1 \cup \Upsilon_2$ of Theorem 21.10. If μ is the probability on $\Gamma = \mathrm{AUT}(X,P)$ defined in (8.12), then ν is μ-invariant, and Lemma 20.12 implies that $\nu(\Upsilon_i) = 0$ for $i = 1,2$. On the other hand, each end ξ in Υ_0 is thin, so that $|\sigma^{-1}(\xi)| = 1$. Thus, σ is one-to-one on a Borel subset of $\mathcal{M}(P)$ whose complement has measure 0. □

B. Free products

We take up the material developed in §9.C, and consider a free product $(X,o) = \mathop{*}_{i \in \mathcal{I}} (X_i, o_i)$ of connected, locally finite graphs X_i. We exclude the case $|\mathcal{I}| = |X_1| = |X_2| = 2$, as then X has precisely two ends, which are thin. Think of the sets xX_i as "leaves" of the free product, where $i \in \mathcal{I}$ and $x \in X_i^{\top}$. Suppose that we have a compactification $\widehat{X}_i$ with boundary ϑX_i of each X_i. (If some X_i is finite, then $\widehat{X}_i = X_i$.) Then there is a canonical way to construct a compactification $\widehat{X}$ of X, which will be denoted by $\widehat{\mathop{*}_{i \in \mathcal{I}}} \widehat{X}_i$.

It corresponds to $|\mathcal{I}|+1$ different ways in which a sequence can "move to infinity" in X.

The first piece of the boundary in our compactification is the set of *infinite words*. It was denoted Υ' in (21.13) for the free product of two groups; here we shall write

$$X_\infty = \{y_1 y_2 y_3 \cdots : y_n \in \bigcup_i X_i',\ \mathfrak{i}(y_{n+1}) \neq \mathfrak{i}(y_n)\ \forall n \geq 1\}, \tag{26.16}$$

while X consists of the analogous finite words as in (9.16). The other pieces of the boundary are the sets

$$x\vartheta X_i = \{x\xi : \xi \in \vartheta X_i\}, \quad \text{where } i \in \mathcal{I} \text{ and } x \in X_i^{\top}, \tag{26.17}$$

also thought of as words whose last letter is a boundary point of X_i. The boundary ϑX in the compactification $\widehat{X} = \widehat{*}_{i\in\mathcal{I}} \widehat{X}_i$ is the union of all these parts.

As with trees, if $u, v \in \widehat{X}$ are different, then we define their *confluent* $u \wedge v$ to be the longest common initial word of v and w. (If $u = v$ then $u \wedge v = u$.) We write $|x|_w$ for the *word length* of $x \in X$ (representation as in (9.16)), and $|\xi|_w = \infty$ if $\xi \in \vartheta X$. We explain the topology of the compactification in terms of convergence of sequences (v_n) in $\widehat{X}$.

(1) If $\eta \in X_\infty$ then $v_n \to \eta$ if $|v_n \wedge \eta|_w \to \infty$.

(2) If $\zeta \in \vartheta X_i$ and $x \in X_i^{\top}$ then $v_n \to x\zeta$ if for all but finitely many n, the representation (9.16), (26.16) or (26.17) of v_n, respectively, starts with xw_n, where $w_n \in \widehat{X}_i$ and $w_n \to \zeta$ in the topology of $\widehat{X}_i$.

Thus, if $x \in X_i^{\top}$ then in our compactification of X, the closure of xX_i is $x\widehat{X}_i$, which is homeomorphic with $\widehat{X}_i$ via the correspondence $xv \leftrightarrow v$.

Now let $P = \sum_{i\in\mathcal{I}} \alpha_i \overline{P}_i$ be a weighted "free sum" of transition operators P_i on X_i, as defined in (9.16). We assume that each (X_i, P_i) has finite range and is transitive, so that (X, P) is also transitive and, by Theorem 7.8, ρ-transient. For $t \geq \rho(P)$, we want to describe the Martin compactification $\widehat{X}(P,t)$, at least in the transitive case. With t, we associate the numbers $t_i = 1/\zeta_i(1/t)$. By Proposition 9.18 and Lemma 17.1, we have $t_i \gtrsim \rho(P_i)$. In these circumstances, we have the following.

(26.18) Theorem. $\qquad \widehat{X}(P,t) = \widehat{*}_{i\in\mathcal{I}} \widehat{X}_i(P_i, t_i)\,.$

Proof. Transitivity implies that $F(ux, uy|z) = F_i\big(x, y|\zeta_i(z)\big)$ for *all* $u \in X_i^{\top}$ and $x, y \in X_i$ (and not only when $u = o$ as in the general case – see Lemma 17.1(b)). Let $x \in X$, and let $\big(y(n)\big)$ be a sequence in X that converges to a point in $\widehat{*}_{i\in\mathcal{I}} \widehat{X}_i(P_i, t_i)$.

Case 1. Suppose that $y(n) \to \eta \in X_\infty$. Consider the confluent $u = x \wedge \eta$ with word representation $u = u_1 \cdots u_k$. Then $x = u_1 \cdots u_k x_{k+1} \cdots x_\ell$ with $\ell \geq 0$ and $\eta = u_1 \cdots u_k y_{k+1} y_{k+2} \cdots$ with $y_{k+1} \neq x_{k+1}$. There is n_u such that the word representation of $y(n)$ will start with uy_{k+1} for all $n \geq n_u$. The random walk starting at x must pass through uy_{k+1} before reaching $y(n)$, whence

$$F\big(x, y(n)|z\big) = F\big(x, uy_{k+1}|z\big)\, F\big(uy_{k+1}, y(n)|z\big)\,.$$

The same holds with o in place of x. Therefore, for all $n \geq n_u$,

$$K\big(x, y(n)|t\big) = \frac{F(x, uy_{k+1}|1/t)}{F(o, uy_{k+1}|1/t)}\,, \tag{26.19}$$

which is the limit of $K\big(x, y(n)|t\big)$ and defines $K(x, \eta|1/t)$.

Case 2. Now suppose that $y(n) \to y\zeta$, where $y \in X_i^\top$ and $\zeta \in \mathcal{M}(P_i, t_i)$. Given $x \in X$, the "cactus-like" structure implies that there is a point $v_i = v_{y,i}(x) \in X_i$ such that the first visit to yX_i of the random walk starting at x must occur at yv_i. Also, for all but finitely many n, the word representation of $y(n)$ is of the form $y(n) = yy_i(n)w(n)$, where $w(n) \in X_i^\perp$ and $y_i(n) \in X_i$ converges to ζ in the topology of $\widehat{X}_i(P_i, t_i)$. There is n_x such that $y_i(n) \neq v_i$ for all $n \geq n_x$. But then the random walk starting at x must pass through yv_i in order to reach $y(n)$, whence

$$F\big(x, y(n)|z\big) = F\big(x, yv_i|z\big)\, F\big(yv_i, yy_i(n)|z\big)\, F\big(yy_i(n), y(n)|z\big)\,.$$

The same relation holds when we replace x and v_i with o. Therefore, for all $n \geq n_x$,

$$\begin{aligned} K\big(x, y(n)|t\big) &= \frac{F(x, yv_i|1/t)}{F(o, y|1/t)}\, \frac{F\big(yv_i, yy_i(n)|1/t\big)}{F\big(y, yy_i(n)|1/t\big)} \\ &= \frac{F(x, yv_i|1/t)}{F(o, y|1/t)}\, K_i\big(v_i, y_i(n)|t_i\big) \\ &\to \frac{F(x, yv_i|1/t)}{F(o, y|1/t)}\, K_i(v_i, \zeta|t_i) \end{aligned} \tag{26.20}$$

when $n \to \infty$. The last line defines $K(x, y\zeta|t)$.

To make sure that we have determined the Martin compactification, we have to show that the functions $K(\cdot, \xi|t)$ are different for different ξ in the boundary of our compactification. Recall Theorem 26.2, which implies that sequences in X converging to different ends cannot converge to the same

Martin kernel. The end compactification of X is $\widehat{\ast}_{i\in\mathcal{I}} \widehat{X}_i$, where each $\widehat{X}_i$ is the end compactification of X_i. Suppose that $\xi, \xi' \in \widehat{\ast}_{i\in\mathcal{I}} \widehat{X}_i(P_i, t_i)$ and $K(\cdot, \xi|t) \equiv K(\cdot, \xi'|t)$. If $\xi \in X_\infty$, then it is a thin end, and $\xi' = \xi$. If $\xi = y\zeta \in y\mathcal{M}(P_i, t_i)$ then we also must have $\xi' = y\zeta' \in y\mathcal{M}(P_i, t_i)$, as otherwise ξ and ξ' correspond to different ends of X. But then (26.20) implies that $K_i(\cdot, \zeta|t_i) = K_i(\cdot, \zeta'|t_i)$ on X_i, and $\zeta = \zeta'$ as required. □

Under the same assumptions, we can also determine the minimal Martin boundary.

(26.21) Proposition.

$$\mathcal{M}_{\min}(P, t) = X_\infty \cup \bigcup_{i\in\mathcal{I}} \{y\zeta : y \in X_i^\top,\ \zeta \in \mathcal{M}_{\min}(P_i, t_i)\}.$$

Proof. We know from Corollary 26.13 that every infinite word is in $\mathcal{M}_{\min}(P, t)$.

Let $\zeta \in \mathcal{M}_{\min}(P_i, t_i)$ and $y \in X_i^\top$, and suppose that $K(\cdot, y\eta|t) = a \cdot g + (1-a) \cdot h$, where $g, h \in \mathcal{H}^+(P, t)$ and $g(o) = h(o) = 1$.

Step 1. Let $x \in X$ be such that $v_{y,i}(x) = o_i$. Then, as in Theorem 26.4,

$$\begin{aligned} K(x, y\xi|t) &= a \cdot g(x) + (1-a) \cdot h(x) \\ &\geq \big(a \cdot F(x, y|1/t)\, g(y) + (1-a) \cdot F(x, y|1/t)\, h(y)\big) \\ &= F(x, y|1/t)\, K(y, y\xi|t) = K(x, y\xi|t) \end{aligned}$$

by (26.20), and equality must hold. This implies $g(x) = F(x, y|1/t)\, g(y)$ and $h(x) = F(x, y|1/t)\, h(y)$. In particular, $g(y) = h(y) = 1/F(o, y|1/t)$.

Step 2. The reader is now invited to reconsider the proof of Proposition 9.18, in particular the stopping time $\mathbf{s}(1) = \min\{n > 0 : Z_n \in yX_i$ and $\mathfrak{i}(Z_n) = i\}$, where $Z_0 = yu$ with $u \in X_i$. For arbitrary $v \in X_i$, the expression $\mathbb{E}_{yu}\big(z^{\mathbf{s}(1)} \mathbf{1}_{[Z_{\mathbf{s}(1)} = yv]}\big)$ coincides with $w(v) = \zeta_i(z)\, p_i(u, v) = p_i(u, v)/t_i$, when $z = 1/t$. (Note that transitivity is used here when $u \neq o_i$.) We have

$$h(yu) \geq \sum_{v\in X_i} \mathbb{E}_{yu}\big(t^{-\mathbf{s}(1)} \mathbf{1}_{[Z_{\mathbf{s}(1)} = yv]}\big)\, h(yv),$$

because the right hand side divided by $h(yu)$ is the probability that the h-process ever satisfies $Z_n \in yX_i$ and $\mathfrak{i}(Z_n) = i$. The same relation holds with g in place of h.

Now, for $u \in X_i$, define $g_i(u) = g(yu)\, F(o, y|1/t)$ and $h_i(u) = h(yu)\, F(o, y|1/t)$. We have $v_{y,i}(yu) = u$, and (26.20) implies together with the above that

$$\begin{aligned}K_i(u,\zeta|t_i) &= F(o,y|1/t)\,K(yu,x\zeta|t) = F(o,y|1/t)\big(a\cdot g(yu)+(1-a)\cdot h(yu)\big)\\ &\ge F(o,y|1/t)\sum_{v\in X_i}\frac{p_i(u,v)}{t_i}\big(a\cdot g(yv)+(1-a)\cdot h(yv)\big)\\ &= F(o,y|1/t)\sum_{v\in X_i}\frac{p_i(u,v)}{t_i}\,K(yv,y\zeta|t)\\ &= \sum_{v\in X_i}\frac{p_i(u,v)}{t_i}\,K_i(v,\zeta|t_i) = K_i(u,\zeta|t_i)\,.\end{aligned}$$

Therefore the inequality cannot be strict, g_i and h_i lie in $\mathcal{H}^+(P_i,t_i)$, $g_i(o_i) = h_i(o_i) = 1$ and $K_i(\cdot,\zeta|t_i) = a\cdot g_i + (1-a)\cdot h_i$. Minimality of ζ implies $g_i = h_i = K_i(\cdot,\zeta|t_i)$.

Step 3. We know from Step 2 that $g(x) = h(x) = K(x,y\zeta|t)$ also for those x which are of the form $x = yu$ with $u \in X_i$. To conclude, we must prove this relation for all x of the form $x = yuw$ with $u \in X_i'$ and $w \in X_i^{\perp}$. But then $v_{y,i}(x) = u$, and we get from (26.20) that $K(x,y\zeta|t) = F(x,yu|1/t)\,K(yu,y\zeta|t)$. Furthermore

$$g(x) \ge F(x,yu|1/t)\,g(yu) = F(x,yu|1/t)\,K(yu,y\zeta|t) = K(x,y\zeta|t)\,,$$

and the same holds for $h(x)$. Once again, the inequalities cannot be strict. □

(26.22) Exercise. (a) As a matter of fact, we have only proved that the set on the right in Proposition 26.21 is contained in $\mathcal{M}_{\min}(P,t)$. Verify the other inclusion.

(b) Check that the finite range assumption is not really needed for Theorem 26.18 and Proposition 26.21.

(26.23) Examples. (1) Consider the simple random walk P on $X = \mathbb{Z}^d * \mathbb{Z}^d$. If P_0 is the simple random walk on $\mathbb{Z}^d$, then $\mathcal{M}(P_0,t) = \mathcal{M}_{\min}(P_0,t) \cong \mathbb{S}_{d-1}$ for $t > \rho(P_0) = 1$, while $\mathcal{M}(P_0,1) = \{\infty\}$ is a single point. If $t > \rho(P)$ then $\widehat{X}(P,t)$ is obtained from the end topology by replacing each thick end, which is the end of a subgraph $y\mathbb{Z}^d$ isomorphic with $\mathbb{Z}^d$, with the sphere $y\mathbb{S}_{d-1}$. The latter is the boundary of $y\mathbb{Z}^d$ in the corresponding topology. When $d \le 4$, the same is also true for $t = \rho(P)$.

However, when $d \ge 5$ and $t = \rho(P)$, then we know from (9.25.3) that $t_0 = 1/\zeta(1/\rho(P))$ has value $t_0 = 1$, and the corresponding boundary of each subgraph $y\mathbb{Z}^d$ is its end $y\infty$. In this case, the Martin compactification coincides with the end compactification.

(2) On the other hand, we know from Corollary 17.10 that there is another symmetric, nearest neighbour random walk $\bar{P}$ on $X = \mathbb{Z}^d * \mathbb{Z}^d$ for

which $\bar{t}_0 = 1/\varsigma(1/\rho(\bar{P})) > 1 = \rho(\bar{P}_0)$. In this case, the Martin boundary $\mathcal{M}(\bar{P}, t)$ can be described as above, but it does not collapse to the space of ends when t reaches the value $\rho(\bar{P})$. The Martin compactification $\widehat{X}(\bar{P}, t)$ is topologically the same for all $t \geq \rho(\bar{P})$. □

27. The Martin boundary of hyperbolic graphs

Hyperbolic graphs are in some sense generalizations of trees. For a tree, the end compactification coincides with the hyperbolic compactification. Therefore we may expect that under suitable conditions, the results concerning Martin boundaries of random walks on trees (Theorem 26.4 and Corollary 26.14(a)) have analogues in the hyperbolic context. We shall prove the following result.

(27.1) Theorem. *Let X be a hyperbolic graph, and suppose that P is uniformly irreducible with bounded range. Then for every $t > \rho(P)$ the Martin compactification $\widehat{X}(P, t)$ coincides with the hyperbolic compactification, and $\mathcal{M}_{\min}(P, t) = \mathcal{M}(P, t) = \vartheta X$, the hyperbolic boundary.*

The proof needs careful preparation with a potential theoretic flavour. We refine the definition of $F^A(x, y)$, setting $f^{(A,n)}(x, y) = \mathbb{P}_x[\mathbf{s}^A = n\,,\ Z_n = y]$ and dually, $l^{(A,n)}(x, y) = \mathbb{P}_x[Z_n = y\,,\ Z_0 \in A\,,\ Z_i \notin A \text{ for } 1 \leq i \leq n]$. For $z \in \mathbb{C}$, let

$$F^A(x, y|z) = \sum_{n=0}^{\infty} f^{(A,n)}(x, y)\, z^n \quad \text{and} \quad L^A(x, y|z) = \sum_{n=0}^{\infty} l^{(A,n)}(x, y)\, z^n\,.$$

Both power series converge for $|z| < \mathrm{r} = 1/\rho(P)$. If $A = \{x\}$ then we write $L^A(x, y|z) = L(x, y|z)$. Also, we omit z when it is equal to 1. We have $G(x, y|z) = G(x, x|z)L(x, y|z)$; compare with Lemma 1.13(b). If we take an invariant measure ν (that is, $\nu P = \nu$; its existence follows from finite range by reasoning analogous to Lemma 7.6), then we can define the corresponding *dual chain* or *ν-process* P^ν by $p^\nu(x, y) = \nu(y)\, p(y, x)/\nu(x)$. Then $L^A_{[P]}(x, y|z) = \nu(y)\, F^A_{[P^\nu]}(y, x|z)/\nu(x)$, and $G_{[P]}(x, y|z) = \nu(y)\, G_{[P^\nu]}(y, x|z)/\nu(x)$. We have the identity

$$\begin{aligned} (27.2) \qquad G(x, y|z) &= G_{X\setminus A}(x, y|z) + \sum_{w \in A} F^A(x, w|z)\, G(w, y|z) \\ &= G_{X\setminus A}(x, y|z) + \sum_{w \in A} G(x, w|z)\, L^A(w, y|z)\,, \end{aligned}$$

where $G_{X\setminus A}$ denotes the Green function of the random walk restricted to $X \setminus A$, as in (2.6). This is seen as follows. The first term corresponds to going from x to y without visiting A. In the remainder, we can condition either on the first or on the last visit to A, giving the two different decompositions. In particular, when the random walk has to pass through A on the way from x to y, we have

$$G(x,y|z) = \sum_{w \in A} G(x,w|z)\, L^A(w,y|z)\,. \tag{27.3}$$

Obviously, $F^A(x,y|z) = 0$ when $y \notin A$ and $L^A(x,y|z) = 0$ when $x \notin A$. For nearest neighbour random walk, we shall use the following.

(27.4) If $x \in A$ and $1 \le z \le \mathbf{r}$ then $L^A(x,y|z) \ge z^{d(x,y)}\, L^A(x,y)$,

since it needs at least $d(x,y)$ steps to reach y. From the proof of the approximation theorem (24.6), recall the definition of the reduced function $R^A[h](x) = \inf\{g(x) : g \in \mathcal{S}^+(P)\,,\ g \ge h \text{ on } A\}$ of $h \in \mathcal{S}^+(P)$.

(27.5) Lemma. $R^A[h](x) = \sum_{y \in A} F^A(x,y)h(y)\,.$

Proof. Set $u(x) = \sum_{y\in A} F^A(x,y)h(y)$. Then $u \in \mathcal{S}^+(P)$ and $u = h$ on A. Therefore $u \ge R^A[h]$ on X.

Let $g \in \mathcal{S}^+(P)$, $g \ge h$ on A. We can define the g-process as in (24.17) (with $t = 1$). The matrix P^g is not necessarily stochastic, but it is substochastic; at points where g is strictly superharmonic, there is a positive probability that the g-process "vanishes". The sum $\sum_{y\in A} F^A_{[P^g]}(x,y)$ is the probability that the g-process starting at x ever hits A, and is ≤ 1. On the other hand, $F^A_{[P^g]}(x,y) = F^A(x,y)g(y)/g(x)$. Therefore $\sum_{y\in A} F^A(x,y)g(y) \le g(x)$. As $g(y) \ge h(y)$ for every $y \in A$, we obtain $g \ge u$ on X. Thus $u \le R^A[h]$. $\square$

A measure ν on X is called *t-excessive,* if $\nu P \le t \cdot \nu$. The corresponding positive cone is denoted by $\mathcal{S}^+_*(P,t)$. It is non-empty if and only if $t \ge \rho(P)$; compare with §7.A. The Harnack inequality (25.1) is also valid for t-excessive measures, with the same constant C_t. Finally, we need the following tool.

(27.6) Resolvent equation. *For $|z|, |z'| < \mathbf{r}(P)$ and $x, y \in X$,*

$$z\, G(x,y|z) - z'\, G(x,y|z') = (z - z') \sum_{w \in X} G(x,w|z)G(w,y|z')\,.$$

The proof is straightforward, using the relation $(I - zP)\mathcal{G}(z) = I$, where $\mathcal{G}(z) = \big(G(x,y|z)\big)_{x,y\in X}$.

We now return to our hyperbolic graph X. Recall the key lemma (1.23) for trees: our main tool will be a weaker version of that. Since hyperbolicity is preserved by rough isometries, we can assume that P is nearest neighbour (replacing X with a suitable k-fuzz). Also, it is enough to assume that $t = 1$ and $\rho(P) < 1$. The general result will follow, because $\widehat{X}(P, t) = \widehat{X}(P_h, 1)$, where $h \in \mathcal{H}^+(P, t)$ is arbitrary (its existence is guaranteed by Lemma 7.6) and $\rho(P_h) = \rho(P)/t$. Without loss of generality, we may assume that the parameter δ of hyperbolicity (Definition 22.1) is a positive integer.

For $x \in X$, we define $U_{o,x} = \{w \in X : |w \wedge x| \geq |x| - 7\delta\}$ and $V_{x,o} = X \setminus U_{o,x}$. Obviously, $U_{o,x} \neq X$ if and only if $|x| > 7\delta$. More generally, we define $U_{a,x}$ and $V_{x,a}$ in the same way, replacing the reference point o with $a \in X$. We start with a structure theoretic lemma.

(27.7) Lemma. *Let $x, y \in X$ be such that $|y| = |x| + d(x, y)$ and $d(x, y) \geq 21\delta$. Then $U_{o,y} \subset U_{o,x}$,*

(1) *if $w \in U_{o,y}$ and $d(w, y) \geq r$ then $B(w, r) \subset U_{o,x}$, and*

(2) *if $v \in V_{x,o}$ and $d(v, x) \geq 2r + d(x, y) + 1$ then $B(v, r) \subset V_{y,o}$.*

Proof. We have $w \in U_{o,y}$ if and only if $|y| + d(y, w) \leq |w| + 14\delta$. As $|y| - |x| = d(x, y)$, the triangle inequality implies $|y| + d(y, w) \geq |x| + d(x, w)$, and $w \in U_{o,x}$. Now suppose $d(x, y) \geq 21\delta$. If $w \in U_{o,y}$ then by Lemma 22.5

$$|w \wedge x| \geq \min\{|w \wedge y|, |y \wedge x|\} - 3\delta \geq \min\{|y| - 7\delta, |x|\} - 3\delta = |x| - 3\delta.$$

Next, if $v \in dV_{x,o}$ then, as $|v \wedge x|$ is a multiple of $1/2$,

$$|x| - 7\delta - \tfrac{1}{2} \geq |v \wedge x| \geq \min\{|v \wedge w|, |w \wedge x|\} - 3\delta \geq \min\{|v \wedge w| - 3\delta, |x| - 6\delta\}.$$

Therefore $|v \wedge w| \leq |x| - 4\delta - \frac{1}{2}$, that is, $d(v, w) \geq |v| + |w| - 2|x| + 8\delta + 1$. As v has a neighbour in $U_{o,x}$, we find $|v \wedge x| \geq |x| - 7\delta - 1$, that is, $|v| \geq d(v, x) + |x| - 14\delta - 2$. Recall that $|w| \geq d(w, y) + |y| - 14\delta$. Consequently, for all $v \in dV_{x,o}$, $w \in U_{o,y}$,

$$d(v, w) \geq d(v, x) + d(w, y) + |y| - |x| - 20\delta - 1 \geq d(v, x) + d(w, y).$$

Suppose in addition that $d(w, y) \geq r$. If $B(w, r)$ is not contained in $U_{o,x}$ then it contains some $v \in dV_{x,o}$. But then $d(v, w) \geq 1 + r$, a contradiction. This proves (1). To see (2), let r be such that $B(v, r)$ is not contained in $V_{y,o}$, where $v \in V_{x,o}$. Then there is some $w \in B(v, r) \cap U_{o,y}$, and on some geodesic segment from v to w, we find $v' \in B(v, r) \cap dV_{x,o}$. This implies $r \geq d(v', w) \geq d(w, y) \geq d(v, x) - d(x, y) - d(v, w) \geq d(v, x) - d(x, y) - r$. This is impossible when $d(v, x) \geq 2r + d(x, y) + 1$. □

Obviously, the lemma holds with any vertex a in place of o. We now choose and fix t in the interval $\big(\rho(P), 1\big)$ and set $z = 1/t$.

(27.8) Proposition. *There is a constant $C' = C'(z)$ such that*

$$G(x,w) \le C'\,F(x,v)\,G(v,w|z) \quad \text{for all } w \in U_{x,v} \cup V_{v,y}$$

whenever $x, y \in X$ and v lies on a geodesic segment from x to y.

Proof. We shall only use properties that do not depend on the base point. Therefore we may assume that $x = o$. Throughout this proof, we let $\ell = 21\delta$. In our geodesic segment $\pi(o,y)$, consider the initial piece $\pi(o,v)$. Let m be the integer part of $|v|/\ell$, and consider the points $v_0, \dots, v_m \in \pi(o,v)$ where $d(v_k, v) = (m-k)\ell$. We then consider the sets $W_k = U_{o,v_k} \cup V_{v_k,y}$. By Lemma 27.7

(i) $v_k \in W_k \subset W_{k-1}$ and $d(v_k, dW_k) < \ell$, and

(ii) if $w \in W_k$ with $d(w, v_k) \ge 2r + \ell + 1$ then $B(w,r) \subset W_{k-1}$.

We shall work with a fixed choice of r. Let $C_1 \ge 1$ be the constant appearing in the Harnack inequality (25.1), valid for $\mathcal{S}^+(P)$ and $\mathcal{S}^+_*(P)$. Since $\mathcal{S}^+(P,t) \subset \mathcal{S}^+(P)$ (as $t < 1$), it is also valid for $\mathcal{S}^+(P,t)$ and analogously for $\mathcal{S}^+_*(P,t)$. Now $r \ge \ell$ is chosen such that $t^r\, C_1^{2\ell} \le 1$. The constant of our proposition is going to be $C' = C_1^{4r+2\ell}$.

We shall use induction on k to show that

$$\text{(27.9)} \qquad G(o,w) \le C'\,F(o,v_k)\,G(v_k,w|z) \quad \text{for all } w \in W_k\,.$$

Let $k = 0$. Then $|v_0| < \ell$, and the function $G(\cdot, w)$ is superharmonic for any w. By Harnack, $G(o,w) \le C_1^\ell\, G(v_0,w) \le C_1^\ell\, G(v_0,w|z)$, since $z > 1$. Also $G(\cdot, v_0)$ is superharmonic, whence $G(v_0,v_0) \le C_1^\ell\, G(o,v_0)$, that is, $C_1^\ell\, F(o,v_0) \ge 1$. Putting things together, we find that (27.9) holds for $k = 0$ and *all* $w \in X$.

On replacing v_{k-1} with v_k, the distance from o increases, and we shall compensate this by proving the validity of the inequality on a smaller set, namely W_k. So suppose that (27.9) holds for $k-1$. We use the Harnack inequality once more to deduce that $G(v_{k-1},w|z) \le C_1^\ell\, G(v_k,w|z)$ and $C_1^\ell\, F(v_{k-1},v_k) \ge 1$. Therefore

$$\begin{aligned} \text{(27.10)} \quad G(o,w) &\le C'\,C_1^{2\ell}\,F(o,v_{k-1})\,F(v_{k-1},v_k)\,G(v_k,w|z) \\ &\le C'\,C_1^{2\ell}\,F(o,v_k)\,G(v_k,w|z) \qquad \text{for all } w \in W_{k-1}\,. \end{aligned}$$

Let $w \in W_k$ with $d(w,v_k) \ge 2r + \ell + 1$. Set $A = \{a \in X : d(a,w) = r\}$. Then $A \subset W_{k-1}$ by (ii), and (27.10) holds for all $a \in A$ in place of w. The random walk is nearest neighbour and has to pass through A on the way from o to w. (Indeed, note that $o \notin W_1$. If $k-1 = 0$ then $|w| \ge$

$d(w,v_1) - |v_1| \ge 2r - \ell + 2 > r$. In both cases, $o \notin B(w,r)$.) Therefore, using (27.3), (27.4) and our choice of r,

$$G(o,w) = \sum_{a \in A} G(o,a)\, L^A(a,w) \le C' C_1^{2\ell} \sum_{a \in A} F(o,v_k)\, G(v_k,a|z)\, t^r\, L^A(a,w|z)$$
$$\le C'\, F(o,v_k) \sum_{a \in A} G(v_k,a|z)\, L^A(a,w|z) = C'\, F(o,v_k)\, G(v_k,w|z)\,,$$

and the inequality (27.9) holds for w.

On the other hand, let $w \in W_k$ with $d(w,v_k) \le 2r + \ell$. The measures $G(o,\cdot)$ and $G(v_k,\cdot|z)$ are excessive, and the Harnack inequality implies

$$G(o,w) \le C_1^{2r+\ell}\, G(o,v_k) \le C_1^{2r+\ell}\, F(o,v_k)\, G(v_k,v_k|z)$$
$$\le C_1^{4r+2\ell}\, F(o,v_k)\, G(v_k,w|z)\,.$$

Again, (27.9) is verified, and the proof is complete. □

(27.11) Corollary. *There is a constant $C'' = C''(z)$ such that*

$$G(w,y) \le C''\, G(w,v|z)\, L(v,y) \quad \textit{for all } w \in V_{v,x} \cup U_{y,v}$$

whenever $x,y \in X$ and v lies on a geodesic segment from x to y.

Proof. Consider a dual chain P^ν, where $\nu P = \nu$. It is also nearest neighbour and uniformly irreducible. Therefore we can apply Proposition 27.8 to P^ν, with a suitable constant C'' in place of C'. Also, we exchange the roles of x and y. When we rewrite the resulting inequality in terms of the original chain, we obtain the proposed statement. □

We can now provide the main tool for determining the Martin boundary.

(27.12) Theorem. *Suppose that X is a hyperbolic graph, and that P is uniformly irreducible with bounded range and $\rho(P) < 1$. Then for each $r \ge 0$ there is a constant $C(r) \ge 1$ such that*

$$F(x,v)F(v,y) \le F(x,y) \le C(r)\, F(x,v)F(v,y)$$

whenever $x,y \in X$ and w is at distance at most r from a geodesic segment bewtween x and y.

Proof. The lower bound is trivial. For the upper bound, we may suppose once more that P is nearest neighbour, and use the same numbers t and $z = 1/t$ and Harnack constant C_1 as in the proof of Proposition 27.8. We first consider the case $r = 0$, that is, v lies on a geodesic segment from x to y.

If $d(x,v) \le 7\delta$ then Harnack applied to $F(\cdot\,,y)$ implies $F(x,y) \le C_1^{7\delta}\,F(v,y)$, and applied to $G(\cdot\,,v)$ implies $1 \le C_1^{7\delta}\,F(x,v)$. Therefore $F(x,y) \le C_1^{14\delta}\,F(x,v)F(v,y)$.

So now suppose that $d(x,v) > 7\delta$. Then $x \notin U_{x,v}$, while $y \in U_{x,v}$. Being nearest neighbour, the random walk has to pass through the set $A = dU_{x,v}$ on the way from x to y. Combining this fact with Proposition 27.8, we get

$$(27.13)\quad G(x,y) = \sum_{a\in A} G(x,a)L^A(a,y) \le C'\,F(x,v)\sum_{a\in A} G(v,a|z)L^A(a,y)\,.$$

Every point $a \in A$ is at distance 1 from some point a' in $V_{v,x}$. The inequality of Corollary 27.11 is valid for a' in place of w. The usual reasoning with the Harnack inequality implies that it also holds for a, if we replace the constant C'' with C_1^2C''. Therefore the superharmonic function $C_1^2C''\,G(\cdot\,,v|z)L(v,y)$ dominates the superharmonic function $G(\cdot\,,y)$ on the set A. By Lemma 27.5, it dominates the reduced function $R^A[G(\cdot\,,y)] = \sum_{a\in A} F^A(\cdot\,,a)G(a,y) = \sum_{a\in A} G(\cdot,a)L^A(a,y)$ on the whole of X (we have used (27.2) here), that is

$$(27.14)\quad \sum_{a\in A} G(w,a)L^A(a,y) \le C_1^2C''\,G(w,v|z)L(v,y) \quad \text{for all } w \in X\,.$$

We now use the resolvent equation. With $z' = 1$ and $z = 1/t$ $(t < 1)$, it becomes $G(v,a|z) = \sum_w \nu(w)\,G(w,a)$, where $\nu = t\cdot\delta_v + (1-t)\cdot G(v,\,\cdot\,|z)$. Combining (27.13) and (27.14), we now find

$$\begin{aligned} G(x,y) &\le C'\,F(x,v) \sum_{w\in X, a\in A} \nu(w)G(w,a)L^A(a,y) \\ &\le C_1^2C'C''F(x,v)\left(\sum_{w\in X} \nu(w)G(w,v|z)\right)L(v,y)\,. \end{aligned}$$

Set $z' = (z+\mathrm{r})/2$. We compute, using (27.6),

$$\begin{aligned} \sum_{w\in X} \nu(w)G(w,v|z) &= \tfrac{1}{z}G(v,v|z) + (1-\tfrac{1}{z})\sum_{w\in X} G(v,w|z)G(w,v|z) \\ &\le \tfrac{1}{z}G(v,v|z) + (1-\tfrac{1}{z})\sum_{w\in X} G(v,w|z)G(w,v|z') \\ &\le \tfrac{1}{z}G(v,v|z) + (1-\tfrac{1}{z})\tfrac{z'}{z'-z}G(v,v|z') \le D\,, \end{aligned}$$

where $D = 2\mathrm{r}(\mathrm{r}+z)/(\mathrm{r}-z)^2$. For the upper bound, note that Lemma 1.9 implies $G(v,v|z) \le \mathrm{r}/(\mathrm{r}-z)$ for $0 \le z < \mathrm{r}$. On the other hand,

$G(v,v) \geq 1$. Therefore, setting $C(0) = \max\{C_1^{14\delta}, C_1^2 C'C''D\}$, we get $G(x,y) \leq C(0)\, F(x,v)G(v,v)L(v,y) = C(0)\, F(x,v)G(v,y)$. The proof for the case $r = 0$ is complete after dividing both sides by $G(y,y)$.

If $r \geq 0$ is arbitrary, then we can find v' on a geodesic from x to y with $d(v,v') \leq r$. Then $F(x,y) \leq C(0)\, F(x,v')F(v',y)$, and the usual Harnack machinery (compare with the beginning of the proof) implies $F(x,y) \leq C(0)C_1^{2r}\, F(x,v)F(v,y)$. □

We have completed our preparations for the main result of this section.

Proof of Theorem 27.1[4]**.** We assume again that $\rho(P) < 1$ and that δ is a positive integer.

Step 1. We first show that the hyperbolic compactification projects naturally onto the Martin compactification. This is based on the following.

Claim. If (y_n) and (y_n') are two sequences in X converging to the same point $\xi \in \vartheta X$ then, setting $\varepsilon_1 = 1/C(2\delta)^2$ (with $C(\cdot)$ as in Theorem 27.12),

$$\liminf_{n\to\infty} \frac{K(x,y_n)}{K(x,y_n')} \geq \varepsilon_1 \quad \text{for all } x \in X\,. \tag{27.15}$$

Proof. Let $\pi(o,\xi)$ and $\pi(x,\xi)$ be geodesic rays from o to ξ and x to ξ, respectively. By Proposition 22.12(a), there are $v \in \pi(o,\xi)$ and $\bar v \in \pi(x,\xi)$ with $d(v,\bar v) \leq \delta$. Now consider geodesics $\pi(o,y_n)$ and $\pi(x,y_n)$. Let x_n be the point on $\pi(o,\xi)$ with $|x_n| = |y_n|$. Add a segment $\pi(x_n,y_n)$ to obtain a geodesic triangle with vertices o, x_n, y_n. Lemma 22.4 and (22.10) imply that $d\big(o,\pi(x_n,y_n)\big) \to \infty$. If n is sufficiently large then we will have $d\big(o,\pi(x_n,y_n)\big) > |v| + \delta$. Given v, there must be a point w_n on one of the other two sides with $d(v,w_n) \leq \delta$. Our choice of n implies that it must lie on $\pi(o,y_n)$.

In the same way, replacing o with x, we find $\bar w_n$ on $\pi(x,y_n)$ with $d(\bar v,\bar w_n) \leq \delta$, and $d(v,\bar w_n) \leq 2\delta$. Also, we repeat the same argument with geodesic segments $\pi(o,y_n')$ and $\pi(x,y_n')$.

Therefore, if n is sufficiently large, then v is at distance at most 2δ from each of $\pi(o,y_n)$, $\pi(x,y_n)$, $\pi(o,y_n')$ and $\pi(x,y_n')$. We can now apply Theorem 27.12, and obtain

$$\begin{aligned}\frac{K(x,y_n)}{K(x,y_n')} &= \frac{F(x,y_n)F(o,y_n')}{F(o,y_n)F(x,y_n')}\\ &\geq \frac{F(x,v)F(v,y_n)\,F(o,v)F(v,y_n')}{C(2\delta)F(o,v)F(v,y_n)\,C(2\delta)F(x,v)F(v,y_n')} = \frac{1}{C(2\delta)^2}\,.\end{aligned}$$

[4] See comments on page 317

Having proved (27.15), we now let L_ξ be the set of all limit points in the Martin boundary $\mathcal{M}(P)$ of sequences in X which converge to ξ in the hyperbolic topology. Bounded range implies that $K(\cdot,\alpha) \in \mathcal{H}^+(P)$ for every $\alpha \in L_\xi$. By (27.15), $K(\cdot,\alpha) \geq \varepsilon_1 \cdot K(\cdot,\beta)$ for all $\alpha,\beta \in L_\xi$.

We next show in Step 2 that there is $\alpha \in L_\xi$ such that $K(\cdot,\alpha)$ is minimal harmonic. Then the last inequality will imply that $K(\cdot,\beta) = K(\cdot,\alpha)$ for all $\beta \in L_\xi$, that is, L_ξ consists of the single point α. This is the natural image of ξ, completing Step 1.

Step 2. Let $\pi(o,\xi)$ be a geodesic from o to ξ. There must be a sequence (x_n) of points on $\pi(o,\xi)$ such that $|x_{n+1}| > |x_n|$ and $x_n \to \alpha \in L_\xi$ in the Martin topology. Consider the set $\mathcal{H}_\alpha = \{h \in \mathcal{H}^+ : \sup_x h(x)/K(x,\alpha) = 1\}$. If we can show that $\mathcal{H}_\alpha = \{K(\cdot,\alpha)\}$ then minimality of $K(\cdot,\alpha)$ follows.

Setting $\varepsilon = 1/C(0)$, Theorem 27.12 yields $K(x_k,x_n) \geq \varepsilon/F(o,x_k)$ whenever $0 \leq k \leq n$. Therefore

$$F(x,x_k)K(x_k,\alpha) \geq \varepsilon\, K(x,x_k) \quad \text{for all } x \in X\,.$$

If $h \in \mathcal{H}^+$ is arbitrary then for all x

$$h(x) \geq F(x,x_k)\, h(x_k) \geq \varepsilon\, K(x,x_k)\, \frac{h(x_k)}{K(x_k,\alpha)}\,. \tag{27.16}$$

Now let $h \in \mathcal{H}_\alpha$, and apply (27.16) to $h' = K(\cdot,\alpha) - h$. Then

$$h'(x) \geq \varepsilon\, K(x,\alpha) \limsup_{k\to\infty} \frac{h'(x_k)}{K(x_k,\alpha)}\,.$$

As $\inf_X \big(h'/K(\cdot,\xi)\big) = 0$, we must have $\lim_k \big(h(x_k)/K(x_k,\alpha)\big) = 1$. We use this fact, and apply (27.16) to our $h \in \mathcal{H}_\alpha$. Letting $k \to \infty$, we infer $h \geq \varepsilon \cdot K(\cdot,\alpha)$. This holds for every $h \in \mathcal{H}_\alpha$.

Set $c_n = \varepsilon\big(1 + (1-\varepsilon) + \cdots + (1-\varepsilon)^n\big)$. We show inductively that $h \geq c_n \cdot K(\cdot,\alpha)$ for all $n \geq 0$. This is true for $n = 0$. Suppose it holds for $n-1$. Then the function $\frac{1}{1-c_{n-1}}\big(h - c_{n-1}\cdot K(\cdot,\alpha)\big)$ is also an element of $\mathcal{H}_\alpha$ and $\geq \varepsilon \cdot K(\cdot,\alpha)$. This yields $h \geq \big(c_{n-1} + \varepsilon(1-c_{n-1})\big)K(\cdot,\alpha) = c_n \cdot K(\cdot,\alpha)$. Letting $n \to \infty$, we get $h \geq K(\cdot,\alpha)$. Therefore $h = K(\cdot,\alpha)$ for every $h \in \mathcal{H}_\alpha$. This concludes the proof of minimality of $K(\cdot,\alpha)$, and completes Step 2 and thus also Step 1.

Step 3. In view of steps 1 and 2, we can define $K(x,\xi) = \lim_n K(x,y_n)$, where (y_n) is an arbitrary sequence in X converging to $\xi \in \vartheta X$ in the hyperbolic topology. To conclude, we now prove that the natural projection obtained via Step 1 is one-to-one: let ξ,η be distinct points in ϑX; we have to show that $K(\cdot,\xi) \neq K(\cdot,\eta)$.

By Lemma 22.15, there is a geodesic $[\ldots, y_{-1}, y_0, y_1, \ldots]$ with $y_n \to \xi$ and $y_{-n} \to \eta$. From Harnack we get $C_1^{|y_0|} F(y_0, \cdot) \geq F(o, \cdot) \geq C_1^{-|y_0|} F(y_0, \cdot)$. Fix $k \geq 0$. Then Theorem 27.12 implies for $n \geq k$

$$\begin{aligned}\frac{K(y_k, y_n)}{K(y_k, y_{-n})} &\geq \frac{F(y_k, y_n)}{C_1^{|y_0|} C(0)\, F(y_0, y_k) F(y_k, y_n)} \frac{F(y_0, y_{-n})}{C_1^{|y_0|} C(0)\, F(y_k, y_0) F(y_0, y_{-n})} \\ &= \frac{1}{C_1^{2|y_0|} C(0)^2\, F(y_0, y_k) F(y_k, y_0)} .\end{aligned}$$

Letting $n \to \infty$, we find

$$\frac{K(y_k, \xi)}{K(y_k, \eta)} \geq \frac{1}{C_1^{2|y_0|} C(0)^2\, G(y_0, y_k) G(y_k, y_0)} \qquad \text{for all } k \geq 0 .$$

Via the resolvent equation, the hypothesis $\rho(P) < 1$ implies that $\sum_w G(y_0, w) G(w, y_0) < \infty$; compare with the last computation in the proof of Theorem 27.12. Therefore $G(y_0, y_k) G(y_k, y_0) \to 0$ as $k \to \infty$, and there must be k such that $K(y_k, \xi) > K(y_k, \eta)$. □

We return briefly to the solution of the Dirichlet problem given in Theorem 22.20. If we replace the uniform first moment condition by the stronger bounded range assumption, then we can weaken the other hypotheses:

(27.17) Exercise. *Suppose that X is a hyperbolic graph with $|\vartheta X| \geq 2$ and P is uniformly irreducible with bounded range and $\rho(P) < 1$. Then a point $\xi \in \vartheta X$ is regular for the Dirichlet problem with respect to the hyperbolic compactification if and only if the Green kernel vanishes at ξ.*

[Hint: In view of Lemma 20.4, what one has to prove is the "if". This is done by showing that vanishing of the Green kernel at ξ implies that $\frac{d\nu_x}{d\nu_o}(\eta) = K(x, \eta) \to 0$ when $x \to \xi$, uniformly for η outside of any neighbourhood U of ξ in $\widehat{X}$. To see this, observe that there is $k = k(U)$ such that $\limsup_{x \to \xi} |x \wedge y| \leq k$ whenever $y \in X \setminus U$, and apply Theorem 27.12.]

Theorem 27.1 applies, in particular, to random walks on finitely generated *Fuchsian groups,* that is, discrete groups of Möbius transformations of the open unit disk $\mathbb{D}$ onto itself, or equivalently, orientation-preserving automorphisms with respect to the hyperbolic (Poincaré) metric. If Γ is such a group, then its limit set $L(\Gamma)$ is defined (as in §20.B) as the set of accumulation points in the unit circle of an orbit Γx, where $x \in \mathcal{D}$. If $|L(\Gamma)| = 2$ then γ is called *elementary* (it is infinite and cyclic in this case). We suppose that Γ is non-elementary. Then it is non-amenable: this is the case corresponding to Proposition 20.10(a), and there is a free subgroup with two generators. In particular, $\rho(\mu) < 1$ for the law μ of any random walk on Γ (Corollary 12.5(a)). Cayley graphs of Fuchsian groups are one of the most typical classes of hyperbolic graphs; see e.g. Ghys and de la Harpe [142]. Thus, the Martin compactification $\widehat{\Gamma}(\mu)$ coincides with the hyperbolic compactification. While the hyperbolic boundary $\vartheta\Gamma$ depends only on the abstract group, the limit set $L(\Gamma)$ depends on its concrete realization as a Möbius group. The relation between the two is as follows: there is a natural continuous surjection of $\widehat{\Gamma}$ onto $\Gamma \cup L(\Gamma)$ which is one-to-one except possibly at countably many boundary points, where it is two-to-one. (These exceptional points arise when the standard fundamental domain of the group has cusps, i.e., it comes from a punctured surface. One can enlarge the punctures slightly to become small holes such that the fundamental groups remain isomorphic. For the new fundamental group, the limit set is homeomorphic with its hyperbolic boundary.) For more details on Fuchsian groups, see Beardon [28] or Katok [196]. Regarding Exercise 27.17, on Fuchsian groups this is of course true without any range or moment condition (Corollary 22.17).

28. Cartesian products

In this last section, we want to determine the Martin boundary of a Cartesian product of two random walks in terms of the boundaries of the factors. We first find a general description of the minimal t-harmonic functions. Then we shall consider a final example where we determine the whole Martin compactification.

A. Minimal harmonic functions on Cartesian products

Given a Markov chain (X, P) and a reference point $o \in X$, we shall write $\mathcal{E}(P, t)$ for the set of minimal t-harmonic functions, where $t \geq \rho(P)$ (the letter $\mathcal{E}$ stands for extremal). We know that $\mathcal{E}(P, t)$ is non-empty for every $t \geq \rho(P)$, when X is infinite and P has finite range (Lemma 7.6). When the range is not finite, this does not remain true, and when X is finite, there are no t-harmonic functions for any $t > 1 = \rho(P)$. Every $h \in \mathcal{E}(P, t)$ is of the form $h = K(\cdot, \xi | t)$ with $\xi \in \mathcal{M}_{\min}(P, t)$.

We start with two irreducible Markov chains (X_1, P_1) and (X_2, P_2). For simplicity, we also write P_1 for the (non-irreducible) transition operator $P_1 \otimes I_2$ on $X = X_1 \times X_2$, and analogously P_2 for $I_1 \otimes P_2$. On X, we consider a Cartesian product $P = c\,P_1 + (1-c)\,P_2$. Note that P_1 and P_2 commute as transition operators on X. Also recall (Exercise 7.3) that

$$\rho(P) = c\,\rho(P_1) + (1-c)\,\rho(P_2)\,.$$

For functions $f_i : X_i \to \mathbb{R}$ $(i = 1, 2)$, we define $f = f_1 \otimes f_2$ by $f(x_1 x_2) = f(x_1) f(x_2)$. The following key lemma is straightforward.

(28.1) Lemma. *If $h_1 \in \mathcal{H}(P_1,t_1)$ and $h_2 \in \mathcal{H}(P_2,t_2)$ then $h_1 \otimes h_2 \in \mathcal{H}(P,t)$, where $t = c\,t_1 + (1-c)\,t_2$.*

Given $t \geq \rho(P)$, we consider the segment

$$I(t) = \{(t_1,t_2) : t_i \geq \rho(P_i)\,,\ c\,t_1 + (1-c)\,t_2 = t\}\,.$$

It consists of a single point, when $t = \rho(P)$. In each X_i, we choose a reference point o_i, and our reference point in X is $o = o_1o_2$.

(28.2) Proposition. *If $h \in \mathcal{E}(P,t)$, $t \geq \rho(P)$, then there are $(t_1,t_2) \in I(t)$, $h_1 \in \mathcal{E}(P_1,t_1)$ and $h_2 \in \mathcal{E}(P_2,t_2)$ such that $h = h_1 \otimes h_2$.*

Proof. *Claim 1.* There is $(t_1,t_2) \in I(t)$ such that $P_1h = t_1 \cdot h$ and $P_2h = t_2 \cdot h$ on X.

Proof. First, as P and P_1 commute, $PP_1h = P_1Ph = t \cdot P_1h$, and $P_1h \in \mathcal{H}^+(P,t)$. Second, $h = \frac{1}{t} \cdot Ph \geq \frac{c}{t} \cdot P_1h$. As h is minimal, we must have $P_1h = t_1 \cdot h$ for some $t_1 > 0$. In the same way, $P_2h = t_2 \cdot h$ for some $t_2 > 0$. It is clear that $(t_1,t_2) \in I(t)$, and Claim 1 is proved.

Therefore $h(\cdot\,x_2) \in \mathcal{H}^+(P_1,t_1)$ for every $x_2 \in X_2$. By Theorem 24.9, for every $x_2 \in X_2$ there is a unique Borel measure ν^{x_2} on $\mathcal{M}(P_1,t_1)$ with no mass outside of $\mathcal{M}_{\min}(P_1,t_1)$, such that

$$h(x_1x_2) = \int_{\mathcal{M}(X_1,t_1)} K_1(x_1,\cdot\,|t_1)\,d\nu^{x_2} \quad \text{for all } x_1x_2 \in X\,. \tag{28.3}$$

(Obviously, $K_i(\cdot,\cdot|t_i)$ denotes the Martin kernel of P_i on $\widehat{X}_i(P_i,t_i)$.) By Claim 1,

$$\begin{aligned}\int_{\mathcal{M}(X_1,t_1)} K_1(x_1,\cdot\,|t_1)\,t_2\,d\nu^{x_2} &= t_2 \cdot h(x_1x_2) = P_2h(x_1x_2)\\ &= \int_{\mathcal{M}(X_1,t_1)} K_1(x_1,\cdot\,|t_1) \sum_{y_2 \in X_2} p_2(x_2,y_2)\,d\nu^{y_2}.\end{aligned}$$

By the uniqueness of the representing measure,

$$\sum_{y_2 \in X_2} p_2(x_2,y_2)\,\nu^{y_2} = t_2 \cdot \nu^{x_2}\,. \tag{28.4}$$

In particular, all the ν^{x_2}, $x_2 \in X_2$, are mutually absolutely continuous and have the same support $S = \operatorname{supp}\nu^{x_2} \subset \mathcal{M}(P_1,t_1)$. (Irreducibility of P_2 is used here.)

Claim 2. S has only one point.

Proof. Suppose the contrary. Then there are two closed disjoint subsets $A, B \subset S$ such that $\nu^{x_2}(A) > 0$ and $\nu^{x_2}(B) > 0$ for some and hence all $x_2 \in X_2$. Consider

$$h_A(x_1x_2) = \int_A K_1(x_1, \cdot\,|t_1)\, d\nu^{x_2} \quad \text{and} \quad h_B(x_1x_2) = \int_B K_1(x_1, \cdot\,|t_1)\, d\nu^{x_2} .$$

Then $h_A, h_B \in \mathcal{H}^+(P,t)$ by (28.4), and $h_A + h_B \le h$. By minimality of h there are constants $c_A, c_B > 0$ such that $h = c_A \cdot h_A = c_B \cdot h_B$. Again by uniqueness of the representing measure, $\nu^{x_2} = c_A \cdot \nu^{x_2}\big|_A = c_B \cdot \nu^{x_2}\big|_B$. This contradicts disjointness of A and B, and Claim 2 is proved.

Thus, $S = \{\xi_1\}$ with $\xi_1 \in \mathcal{M}_{\min}(P_1,t_1)$, and each ν^{x_2} is a multiple of the point mass at ξ_1, that is, $\nu^{x_2} = h_2(x_2) \cdot \delta_{\xi_1}$, where $h_2(x_2) > 0$. By (28.4), $h_2 \in \mathcal{H}^+(P_2,t_2)$. If we set $h_1 = K_1(\cdot, \xi_1|t_1)$, then (28.3) yields $h = h_1 \otimes h_2$.

In conclusion, it is immediate that h_2 must be minimal in $\mathcal{H}^+(P_i,t_i)$, as otherwise h also would not be minimal in $\mathcal{H}^+(P,t)$. □

The main result of this subsection is that tensor products of minimal harmonic functions for P_1 and P_2 are minimal for P, and conversely.

(28.5) Theorem. $$\mathcal{E}(P,t) = \bigcup_{(t_1,t_2)\in I(t)} \mathcal{E}(P_1,t_1) \otimes \mathcal{E}(P_2,t_2) .$$

Proof. Proposition 28.2 says that $\mathcal{E}(P,t)$ is contained in the set on the right. Using this fact, we prove the converse inclusion. Let $h = h_1 \otimes h_2$, where $h_i \in \mathcal{E}(P_i,t_i)$ and $(t_1,t_2) \in I(t)$. Then

$$h(x_1x_2) = \int_{\mathcal{M}(P,t)} K(x_1x_2, \cdot\,|t)\, d\nu$$

for a unique probability measure ν which has no mass outside of $\mathcal{M}_{\min}(P,t)$.

Consider the set $\mathcal{E}_{t_1,t_2} = \mathcal{E}(P_1,t_1) \otimes \mathcal{E}(P_2,t_2)$. Note that we do not yet know whether this is a Borel set in the topology of pointwise convergence of functions in $\mathcal{S}^+(P,t)$. Recall that the Martin boundary $\mathcal{M}(P,t)$ embeds homeomorphically into $\mathcal{S}^+(P,t)$ via the map $\xi \mapsto K(\cdot,\xi|t)$. Write $\mathcal{M}_{t_1,t_2}$ for the preimage of $\mathcal{E}_{t_1,t_2}$ under this embedding. By $\mathcal{M}^-_{t_1,t_2}$ we denote its closure in the topology of the Martin boundary. This is a Borel set, and its image in $\mathcal{S}^+(P,t)$ is contained in the closure of $\mathcal{E}_{t_1,t_2}$ with respect to pointwise convergence.

Claim. $\operatorname{supp} \nu \subset \mathcal{M}^-_{t_1,t_2}$.

Proof. Let $\varepsilon > 0$. Consider the preimages A_ε and B_ε in $\mathcal{M}(P,t)$ of the sets

$$\{u \in \mathcal{S}^+(P,t) : P_1 u \le (t_1 - \varepsilon) \cdot u\} \cap \mathcal{E}(P,t) \quad \text{and}$$
$$\{u \in \mathcal{S}^+(P,t) : P_2 u \le (t_2 - \varepsilon) \cdot u\} \cap \mathcal{E}(P,t) ,$$

respectively. Then A_ε and B_ε are Borel sets and contained in $\mathcal{M}_{\min}(P,t)$. Let

$$h'(x_1y_1) = \int_{A_\varepsilon} K(x_1x_2,\cdot\,|t)\,d\nu\,.$$

By Proposition 28.2, for $\zeta \in A_\varepsilon$ one has

$$K(\cdot,\zeta|t) \in \mathcal{E}(P_1,s_1)\otimes\mathcal{E}(P_2,s_2)\,,$$

where $(s_1,s_2)\in I(t)$ and $s_1 \le t_1-\varepsilon$, whence $s_2 = (t-c\,s_1)/(1-c) \ge s_\varepsilon = (t-c(t_1-\varepsilon))/(1-c)$, and $s_\varepsilon > t_2$. Hence

$$t_2^n \cdot h = P_2^n h \ge P_2^n h' \ge s_\varepsilon^n \cdot h' \quad \text{for all } n\in\mathbb{N}\,.$$

This yields $h'\equiv 0$. Thus $\nu(A_\varepsilon)=0$, and in the same way $\nu(B_\varepsilon)=0$.

The set $\mathcal{E}^-_{t_1,t_2}$ has empty intersection with $\mathcal{E}_{s_1,s_2}$ whenever $(s_1,s_2)\in I(t)$ is different from (t_1,t_2), because pointwise convergence preserves superharmonicity with respect to P, P_1 and P_2. Therefore, again using Proposition 28.2,

$$\mathcal{M}_{\min}(P,t)\setminus\mathcal{M}^-_{t_1,t_2} = \bigcup_{n=1}^{\infty}\big(A_{1/n}\cup B_{1/n}\big)\,,$$

which is a ν-null set. This proves the claim.

Next, set $U=\mathcal{M}_{\min}(P,t)\cap\mathcal{M}^-_{t_1,t_2}$. Then $\nu\big(\mathcal{M}(P,t)\setminus U\big)=0$, and

$$h(x_1x_2) = \int_U K(x_1x_2,\cdot\,|t)\,d\nu\,.$$

Suppose that $\operatorname{supp}\nu$ has more than one element. Then we can find a Borel set $U'\subset U$ such that both U' and $U\setminus U'$ have positive ν-mass. We set $h'(x_1x_2) = \int_{U'} K(x_1x_2,\cdot\,|t)\,d\nu$. Then $h_1(x_1)h_2(x_2) = h(x_1x_2) > h'(x_1x_2)$. Once more by Proposition 28.2, $K(\cdot,\xi|t)\in\mathcal{E}(P_1,t_1)\otimes\mathcal{E}(P_2,t_2)$ for every $\xi\in U$. This implies that $P_1h' = t_1\cdot h'$. We deduce that h_1 dominates the function $h'(\cdot\,x_2)/h(x_2)$, which is in $\mathcal{H}^+(P_1,t_1)$. By minimality of h_1 we must have

$$h'(x_1x_2)/h_2(x_2) = c(x_2)h_1(x_1) \quad \text{for all } x_1x_2\in X\,,$$

where $c(x_2)>0$. In the same way, minimality of h_2 yields

$$h'(x_1x_2)/h_1(x_1) = d(x_1)h_2(x_2) \quad \text{for all } x_1x_2\in X\,,$$

where $d(x_1)>0$. Comparing these identities, we find that $c(x_2)=d(x_1)$ for all $x_1x_2\in X$, so that they coincide with their values at o_1 and o_2, respectively, which are equal to $\nu(U')$. Therefore $h' = \nu(U') \cdot h$. But this contradicts the uniqueness of the representing measure (Theorem 24.9), and ν must be a point mass. □

In several known examples of finite range random walks (X, P), the Martin compactifications $\widehat{X}(P, t)$ with respect to different eigenvalues $t \geq \rho(P)$ are naturally homeomorphic, so that $\mathcal{M}(P, t) = \mathcal{M}(P)$ is independent of t. If in addition the Martin kernel $K(x, \xi|t)$ is jointly continuous in $\xi \in \mathcal{M}(P)$ and $t \geq \rho(P)$, and all kernels $K(\cdot, \xi|t)$ are minimal t-harmonic functions, then we say that (X, P) has *strictly stable* Martin boundary.

If (X_1, P_1) and (X_2, P_2) both have strictly stable boundaries and P is a Cartesian product of the two, then it follows from Theorem 28.5 that $\mathcal{M}_{\min}(P, t)$ with the Martin topology is homeomorphic with $\mathcal{M}(P_1) \times \mathcal{M}(P_2) \times I(t)$. This applies, for example, if the P_i are ρ-transient nearest neighbour random walks on trees T_i, $i = 1, 2$. Note that $I\big(\rho(P)\big) = \big\{\big(\rho(P_1), \rho(P_2)\big)\big\}$ is a single point, so that $\mathcal{M}_{\min}\big(P, \rho(P)\big) \cong \mathcal{M}(P_1) \times \mathcal{M}(P_2)$ is considerably smaller than $\mathcal{M}_{\min}(P, t)$ for $t > \rho(P)$.

In various cases, the above requirements of stability are satisfied only for eigenvalues $t > \rho(P)$, while the Martin boundary at $\rho(P)$ is smaller in the sense that $\widehat{X}(P, t)$ surjects naturally onto $\widehat{X}\big(P, \rho(P)\big)$. In this case we speak of a *stable* Martin boundary. A finite range random walk on $\mathbb{Z}^d$ is such an example. It is an instructive exercise to work out how for an eigenvalue $t > 1$, the minimal Martin boundary of a Cartesian product of two simple random walks on $\mathbb{Z}$ becomes homeomorphic with a circle, when it is described in terms of Theorem 28.5.

In the examples that we have encountered so far, the full Martin boundary has coincided with the minimal boundary. Typical examples where this is not the case are provided by Cartesian products. One such example will be considered next.

B. The Martin compactification of $\mathbb{T} \times \mathbb{Z}$

Our final example is the computation of the full Martin compactification of the simple random walk on $\mathbb{T} \times \mathbb{Z}$, where $\mathbb{T} = \mathbb{T}_M$ with $M \geq 3$. For computational convenience, we shall work with a Cartesian product $P = c\,P_1 \otimes I_2 + (1-c)\,I_1 \otimes P_2$, where P_1 and P_2 are the aperiodic simple random walks on $\mathbb{T}$ and $\mathbb{Z}$ considered in §19.A (with $p_i(x, x) = 1/2$). This is the same as determining the Martin compactification for a "pure" simple random walk. Indeed, setting $c = 1/2$, one can decompose $P = \frac{1}{2}(I + \overline{P})$, where $\overline{P}$ is "pure"; by Lemma 9.2, $K_{\overline{P}}(\cdot, \cdot|t) = K_P(\cdot, \cdot|\frac{1+t}{2})$ and $\widehat{X}(\overline{P}, t) = \widehat{X}(P, \frac{1+t}{2})$.

It will be convenient to view $\mathbb{T}$ as a Cayley graph of the free product $\mathbb{Z}_2 * \cdots * \mathbb{Z}_2$ (M times), so that x^{-1} will denote the inverse and xy the product of $x, y \in \mathbb{T}$ in this group. The respective roots (group identities) are denoted by o_1 and o_2 $(= 0)$.

We shall use two different methods, one of a more algebraic nature (when $t = \rho(P)$, the bottom of the spectrum), and the other much more analytic (when $t > \rho(P)$). We start with $t = \rho$, where $\rho = \rho(P)$. We have $\rho_1 = \rho(P_1) = \frac{1}{2} + \frac{\sqrt{M-1}}{M}$ and $\rho_2 = \rho(P_2) = 1$. The Martin compactification $\widehat{\mathbb{Z}}(P_2, 1)$ is the one-point compactification $\mathbb{Z}_\infty = \mathbb{Z} \cup \{\infty\}$; the Martin kernel has constant value 1. Regarding $\mathbb{T}$, recall the spherical function $\phi(x) = \left(1 + \frac{M-2}{M}|x|\right) M^{-|x|/2}$ on $\mathbb{T}$. This is the unique function in $\mathcal{H}^+(P_1, \rho_1)$ which is radial (i.e., it depends only on $|x| = d(x, o_1)$ in $\mathbb{T}$). We shall need the normalized translate

$$\phi(x, y) = \phi(x^{-1}y)/\phi(y), \quad x, y \in \mathbb{T}. \tag{28.6}$$

Another prerequisite is the following characterisation of $K_1(\cdot, \xi)$, where $\xi \in \vartheta\mathbb{T}$. See (12.13) for the definition of horocycles.

(28.7) Lemma. *Suppose that $h \in \mathcal{H}^+(P_1, \rho_1)$ is constant on each horocycle $H_k = H_k(\xi)$ with respect to $\xi \in \vartheta\mathbb{T}$. Then $h = h(o_1)\, K_1(\cdot, \xi|\rho_1)$.*

Proof. Write $\tilde{h}(k)$ for the constant value of h on H_k. Then

$$\frac{1}{2M}\tilde{h}(k-1) + \frac{1}{2}\tilde{h}(k) + \frac{M-1}{2M}\tilde{h}(k+1) = \rho_1\, \tilde{h}(k).$$

Let $g(k) = (M-1)^{k/2}\, \tilde{h}(k)$. Then g is positive and $g(k-1) + g(k+1) = 2g(k)$. Hence g is constant and $h(x) = h(o_1)\,(M-1)^{-\mathfrak{h}(x,\xi)/2} = h(o_1)\, K_1(x, \xi|\rho_1)$; compare with Example 26.6 and Lemma 1.24. □

Reconsider Theorem 28.5. For $t = \rho$, we have $I(t) = \{(\rho_1, \rho_2)\} = \{(\rho_1, 1)\}$. Since $\mathcal{E}(P_2, 1) = \{\mathbf{1}\}$, the minimal ρ-harmonic functions for P are precisely the functions $x_1x_2 \mapsto K_1(x_1, \xi|\rho_1)$, where $\xi \in \vartheta\mathbb{T}$. Therefore (Corollary 24.18) every function $h \in \mathcal{H}^+(P, \rho)$ is of the form $h(x_1x_2) = h_1(x_1)$, where $h_1 \in \mathcal{H}^+(P_1, \rho_1)$.

(28.8) Theorem. *The Martin compactification $\widehat{\mathbb{T}\times\mathbb{Z}}(P, \rho)$ is $\widehat{\mathbb{T}} \times \mathbb{Z}_\infty$ with the product topology, where $\widehat{\mathbb{T}}$ is the end compactification of $\mathbb{T}$ and $\mathbb{Z}_\infty$ is the one-point compactification of $\mathbb{Z}$. The directions of convergence of the Martin kernel are as follows.*

(1) *If $y_1(n) \to \xi \in \vartheta\mathbb{T}$ and $y_2(n)$ is an arbitrary sequence in $\mathbb{Z}$ then*

$$K\big(x_1x_2, y_1(n)y_2(n)\big) \to K_1(x_1, \xi|\rho_1) \quad \text{for all } x_1x_2 \in \mathbb{T} \times \mathbb{Z}.$$

(2) *If $y_1 \in \mathbb{T}$ and $y_2(n) \in \mathbb{Z}$, $|y_2(n)| \to \infty$, then*

$$K\big(x_1x_2, y_1y_2(n)\big) \to \phi(x_1, y_1) \quad \text{for all } x_1x_2 \in \mathbb{T} \times \mathbb{Z}.$$

Proof. What we have to prove are statements (1) and (2). We start with (1). By compactness, we may assume without loss of generality that

$y_1(n)y_2(n)$ is such that $K(x_1x_2, y_1(n)y_2(n)|\rho)$ converges pointwise to a limit $h(x_1x_2)$. Then $h(x_1x_2) = h_1(x_1)$, where $h_1 \in \mathcal{H}^+(P_1, \rho_1)$. Suppose that $x_1, x_1' \in \mathbb{T}$ lie in the same horocycle with respect to ξ. Choose n large enough that $x_1 \curlywedge y_1(n)$ lies on the geodesic ray from $x_1 \curlywedge x_1'$ to ξ, where $\curlywedge$ is the confluent operation with respect to ξ; see (12.13). Then there is an automorphism γ of $\mathbb{T}$ that fixes this geodesic ray, and also $y_1(n)$, and exchanges x_1 with x_1'. The element $(\gamma, \iota) \in \mathrm{AUT}(\mathbb{T} \times \mathbb{Z})$ leaves P invariant. Therefore we see that $K(x_1x_2, y_1(n)y_2(n)|\rho) = K(x_1'x_2, y_1(n)y_2(n)|\rho)$ for all but finitely many n, and $h(x_1) = h(x_1')$. Lemma 28.7 implies that $h_1 = K_1(\cdot, \xi|\rho)$, as asserted.

(2) Again, we may assume that $K(x_1x_2, y_1y_2(n)|\rho) \to h_1(x_1)$ for all $x_1x_2 \in \mathbb{T} \times \mathbb{Z}$, where $h_1 \in \mathcal{H}^+(P_1, \rho_1)$. We now use the fact that P_1 is invariant under the stabilizer of y_1 in $\mathrm{AUT}(\mathbb{T})$. An argument completely analogous to that for (1) (replacing ξ with y_1) shows that $h_1(x_1) = h_1(x_1')$ whenever $d(x_1, y_1) = d(x_1', y_1)$. Therefore h_1 is radial around y_1, and as $h_1(o_1) = 1$, we must have $h_1(x_1) = \phi(x_1, y_1)$. □

For $t > \rho$, the computations are harder. We shall first derive a "renewal theorem", that is, the spatial asymptotic behaviour of the Green kernel $G(o_1o_2, y_1y_2|1/t)$, when $|y_1| + |y_2| \to \infty$ (lengths in $\mathbb{T}$ and $\mathbb{Z}$, respectively). We use the method of Laplace; not all computations will be laid out in detail. We have

$$(28.9) \qquad G(o_1o_2, y_1y_2|1/t) = \sum_{\substack{k \ge |y_1| \\ m \ge |y_2|}} s_{k,m}(y_1y_2)\,, \quad \text{where}$$

$$s_{k,m}(y_1y_2) = \frac{1}{t^{k+m}} \binom{k+m}{k} c^k\, p_1^{(k)}(o_1, y_1)\, (1-c)^m\, p_2^{(m)}(o_2, y_2)\,.$$

Set $\lambda = |y_1|/|y_2|$, $\alpha_k = |y_1|/k$ and $\beta_m = |y_1|/m$. Suppose for the moment that $k, m \to \infty$, $|y_1| \le (1-\varepsilon)k$ and $0 < |y_2| \le (1-\varepsilon)m$, We may then use Stirling's formula to approximate the binomial coefficients, and replace the transition probabilities by the expressions of Theorems 19.4 and 19.8; we write B_1, φ_1 for the functions of Theorem 19.4 (for $\mathbb{T}$) and B_2, φ_2 for those of Theorem 19.8 (for $\mathbb{Z}$). Then

$$(28.10) \qquad s_{k,m}(y_1y_2) \sim \frac{1}{\sqrt{2\pi}} \left(1 + \tfrac{M-2}{M}|y_1|\right) |y_1|^{-5/2}\, B_1(\alpha_k)\, B_2(\lambda\beta_m) \cdot \\ \times \alpha_k^{3/2}\, \beta_m^{1/2} \sqrt{\alpha_k + \beta_m}\, \exp\big(|y_1|\, \Phi_{\lambda,t}(\alpha_k, \beta_m)\big)\,,$$

where $\qquad \Phi_{\lambda,t}(\alpha, \beta) =$

$$\frac{1}{\alpha}\left(\log\left(1 + \frac{\alpha}{\beta}\right) + \log\frac{c}{t} + \varphi_1(\alpha)\right) + \frac{1}{\beta}\left(\log\left(1 + \frac{\beta}{\alpha}\right) + \log\frac{1-c}{t} + \varphi_2(\lambda\beta)\right)$$

with $0 < \alpha \le 1$ and $0 < \beta \le 1/\lambda$. We recall that $\varphi_1(\alpha) = \alpha \log F_1\big(1/t_1(\alpha)\big) + \log t_1(\alpha)$, where F_1 is the function F of (19.1), and $t_1(\alpha) = 1/z(\alpha)$. The function φ_2 is obtained as in (19.3), with $F_2(z) = \frac{2}{z}\big((1-\frac{1}{2}z) - \sqrt{1-z}\big)$ replacing the $F(z)$ there, and $\varphi_2(\beta) = \beta \log F_2\big(1/t_2(\beta)\big) + \log t_2(\beta)$ with $t_2(\beta) = 1/(1-\beta^2)$. In particular, $t_1(\alpha)$ and $t_2(\beta)$ are determined by the equations

$$(28.11)\qquad \alpha\, t_1 = \sqrt{(t_1 - \tfrac{1}{2})^2 - \tfrac{M-1}{M^2}} \quad\text{and}\quad \beta\, t_2 = \sqrt{(t_2 - \tfrac{1}{2})^2 - \tfrac{1}{4}}\,.$$

Our strategy will be to show that the principal contribution to the sum in (28.10) comes from the point where $\Phi_{\lambda,t}$ is maximal. We compute and study the gradient and find that the absolute maximum is attained at the unique stationary point $(\alpha,\beta) = \big(\alpha(\lambda),\beta(\lambda)\big)$, which is given by

$$(28.12)\qquad t_1(\alpha)\left(1+\frac{\alpha}{\beta}\right)\frac{c}{t} = 1 \quad\text{and}\quad t_2(\lambda\beta)\left(1+\frac{\beta}{\alpha}\right)\frac{1-c}{t} = 1\,,$$

Via (28.11), this transforms into the system of equations

$$(28.13)\qquad \begin{cases} c\,t_1 + (1-c)\,t_2 = t \quad\text{and} \\ \sqrt{(t_2-\tfrac{1}{2})^2 - \tfrac{1}{4}} = \tfrac{\lambda c}{1-c}\sqrt{(t_1-\tfrac{1}{2})^2 - \tfrac{M-1}{M^2}}\,, \end{cases}$$

which has a unique solution $(t_1,t_2) = \big(t_1(\alpha), t_2(\lambda\beta)\big)$ in $[\rho_1\,,\infty)\times[1\,,\infty)$. Again via (28.11), we can now compute $\big(\alpha(\lambda),\beta(\lambda)\big)$, which lies in the interior of the domain of $\Phi_{\lambda,t}$ for every real $\lambda \ge 0$. In particular, when $\lambda = 0$, we have $t_2 = 1$ and $t_1 = \big(t-(1-c)\big)/c$. Using (28.12) and the definition of the φ_i, we compute the value of the maximum of $\Phi_{\lambda,t}$,

$$M(\lambda) = \log F_1(1/t_1) + \lambda \log F_2(1/t_2)\,,$$

where (t_1,t_2) is the solution of (28.13). The next step (whose details are omitted) is to compute the Hessian matrix $H_{\lambda,t}$ of $\Phi_{\lambda,t}$ at $\big(\alpha(\lambda),\beta(\lambda)\big)$, and to ensure that it is negative definite for each $\lambda \in [0\,,\infty)$ and continuous in λ. We shall write $H_{\lambda,t}[\alpha,\beta] = (\alpha,\beta)\,H_{\lambda,t}\,(\alpha,\beta)^t$ for the associated quadratic form. To conclude our preparations, we remark that (28.13) also makes sense when $\lambda = +\infty$: we take $1/\lambda = 0$ to the left hand side of the second equation and find $t_1 = \rho_1$, $t_2 = (t - c\rho_1)/(1-c)$.

(28.14) Theorem. *Suppose that $t > \rho$, $|y_1| + |y_2| \to \infty$ and $\lambda = |y_2|/|y_1| \to \eta \in [0\,,\infty]$. Let (t_1,t_2) be the solution of (28.13), depending on λ. Then*

$$G(o_1o_2, y_1y_2|1/t) \sim F_1(1/t_1)^{|y_1|}F_2(1/t_2)^{|y_2|}\left(1+\tfrac{M-2}{M}|y_1|\right)C(y_1y_2)\,,$$

where $C(y_1y_2) = |y_1|^{-3/2} D(|y_2|/|y_1|) = |y_2|^{-3/2} D^*(|y_1|/|y_2|)$ *with* $D(\lambda)$ *and* $D^*(\lambda)$ *continuous and strictly positive for* $\lambda \in [0, \infty)$.

(The function D is used when $|y_2|/|y_1| \to \eta < \infty$), while D^* is used when $|y_1|/|y_2| \to \eta^* < \infty$).)

Proof. We assume first that $|y_2|/|y_1| \to \eta < \infty$, in particular $|y_1| \to \infty$. We split the sum of (28.9) into two parts, $G(o_1o_2, y_1y_2|1/t) = S_1(y_1y_2) + S_2(y_1y_2)$, where S_1 is the sum over all k, m for which $|\alpha_k - \alpha(\lambda)| \le \delta$ and $|\beta_m - \beta(\lambda)| < \delta$, and S_2 is the rest. Here, $\delta > 0$ is chosen such that when λ is close to η, (i) the points (α_k, β_m) of S_1 are in the interior of the domain of $\Phi_{\eta,t}$ and at bounded distance from its boundary, and (ii) for some $\varepsilon_0 > 0$ depending on η, we have $\Phi_{\lambda,t}(\alpha, \beta) - M(\lambda) < \exp(-\varepsilon_0(\alpha^2 + \beta^2))$ whenever $|\alpha - \alpha(\lambda)| < \delta$, $|b - \beta(\lambda)| < \delta$.

To see that this is possible when $|y_1|$ is sufficiently large, note that (1) the point $\big(a(\eta), b(\eta)\big)$ lies in the interior of the domain, (2) the dependence of $\big(\alpha(\lambda), \beta(\lambda)\big)$ on λ is continuous, and (3) when $\lambda \to \eta < \infty$ we have $\Phi_{\lambda,t}(\alpha, \beta) \to \Phi_{\eta,t}(\alpha, \beta)$ uniformly for $\alpha \in (0, 1]$ and β in any bounded subinterval of $(0, \min\{1/\lambda, 1/\eta\}]$. Uniform convergence obtains because

$$\text{(28.15)} \qquad \begin{aligned} \Phi_{\lambda,t}(\alpha, \beta) - \Phi_{\eta,t}(\alpha, \beta) &= \big(\varphi_2(\lambda\beta) - \varphi_2(\eta\beta)\big)/\beta \\ &= (\lambda - \eta)\, F_2\big(1/t_2(\tilde{\lambda}\beta)\big) \end{aligned}$$

with $\tilde{\lambda}$ between η and λ.

We first analyse S_1. As $k \ge |y_1|/\big(\alpha(\lambda) + \delta\big)$ and $m \ge |y_1|/\big(\beta(\lambda) + \delta\big)$, we may use the approximation of (28.10). We gather the factor $\exp\big(|y_1| M(\lambda)\big)$ outside of the sum, and observe that

$$\Phi_{\lambda,t}(\alpha, \beta) - M(\lambda) = \tfrac{1}{2} H_{\lambda,t}[a - \alpha(\lambda), b - \beta(\lambda)] + \mathfrak{o}\Big(\big(\alpha - \alpha(\lambda)\big)^2 + \big(\beta - \beta(\lambda)\big)^2\Big),$$

and the $\mathfrak{o}(\cdot)$ for the argument tending to 0 is uniform in λ, as $\lambda \to \eta$. Next, we perform a "change of variables", substituting $\sigma_k = \big(\alpha_k - \alpha(\lambda)\big)\sqrt{|y_1|}$ and $\tau_m = \big(\beta_m - \beta(\lambda)\big)\sqrt{|y_1|}$. Set $\Delta\sigma_k = \sigma_k - \sigma_{k+1}$ and $\Delta\tau_m = \tau_m - \tau_{m+1}$. When $|y_1| \to \infty$, we have in the range of summation of S_1

$$\Delta\sigma_k \sim \frac{|y_1|^{3/2}}{k^2} = \frac{\alpha_k^2}{\sqrt{|y_1|}} \quad \text{and} \quad \Delta\tau_m \sim \frac{|y_1|^{3/2}}{m^2} = \frac{\beta_m^2}{\sqrt{|y_1|}},$$

which tend to 0 uniformly. We obtain

$$S_1(y_1y_2) \sim \frac{1}{\sqrt{2\pi}} \exp\big(|y_1| M(\lambda)\big) \Big(1 + \tfrac{M-2}{M}|y_1|\Big) |y_1|^{-3/2} \sum_{|\sigma_k|, |\tau_m| \le \delta} \bar{s}_{k,m},$$

where

$$\bar{s}_{k,m} = B_1\Big(\alpha(\lambda) + \tfrac{\sigma_k}{\sqrt{|y_1|}}\Big) B_2\Big(\lambda\beta(\lambda) + \lambda\tfrac{\tau_m}{\sqrt{|y_1|}}\Big)\Big(\beta(\lambda) + \tfrac{\tau_m}{\sqrt{|y_1|}}\Big)^{-1}$$
$$\times \sqrt{\frac{1}{\alpha(\lambda)+\frac{\sigma_k}{\sqrt{|y_1|}}} + \frac{1}{\beta(\lambda)+\frac{\tau_m}{\sqrt{|y_1|}}}}\, \exp\Big(\tfrac{1}{2}H_{\lambda,t}[\sigma_k,\tau_m] + \mathfrak{o}(\sigma_k^2+\tau_m^2)\Big)\Delta\sigma_k\,\Delta\tau_m\,.$$

The $\mathfrak{o}(\sigma^2+\tau^2)$ tends to 0 as $(\sigma^2+\tau^2)/|y_1| \to 0$, uniformly with respect to $\lambda \sim \eta$. Due to our choice of δ, the non-exponential terms in $\bar{s}_{k,m}$ are bounded by some constant, and as $\lambda \sim \eta$, the exponential term is bounded by $\exp\big(-\varepsilon_0(\sigma_k^2+\tau_m^2)\big)$. Therefore we may use dominated convergence to conclude that

$$\sum_{|\sigma_k|,|\tau_m|\le\delta} \bar{s}_{k,m} \sim \iint_{\mathbb{R}^2} \exp\big(\tfrac{1}{2}H_{\lambda,t}[\sigma,\tau]\big)\,d\sigma d\tau = D(\lambda)\,,$$

as $\lambda \to \eta < \infty$, and D is continuous and positive. Note that $\exp\big(|y_1|M(\lambda)\big) = F_1(1/t_1)^{|y_1|}\,F_2(1/t_2)^{|y_2|}$ with $t_1 = t_1\big(\alpha(\lambda)\big)$ and $t_2 = t_2\big(\lambda\beta(\lambda)\big)$. Therefore $S_1(y_1y_2)$ has the asympotic behaviour asserted by the theorem.

Next, we show that $S_2(y_1y_2)/S_1(y_1y_2) \to 0$, that is,

$$S_2(y_1y_2)\,\exp\big(-|y_1|M(\lambda)\big) = \mathfrak{o}\big(|y_1|^{-1/2}\big)\,, \tag{28.16}$$

as $|y_1| \to \infty$, $\lambda = |y_2|/|y_1| \to \eta < \infty$ and either $|y_2| \to \infty$ or y_2 remains fixed. Since $\varphi_i(\cdot) \le \log\rho_i$, we have

$$\Phi_{\lambda,t}(\alpha,\beta) \le \frac{1}{\alpha}\log\left(\frac{c\rho_1}{t}\Big/\frac{\beta}{\alpha+\beta}\right) + \frac{1}{\beta}\log\left(\frac{1-c}{t}\Big/\frac{\alpha}{\alpha+\beta}\right) \le -\varepsilon_1\Big(\frac{1}{\alpha}+\frac{1}{\beta}\Big),$$

where $\varepsilon_1 = -\log\frac{c\rho_1+(1-c)}{t} > 0$. This is seen by using the inequality $a\log(u/a)+(1-a)\log(v/(1-a)) \le \log(u+v)$, where $u,v>0$ and $0\le a\le 1$.

Define $a_k(y_1) = p_1^{(k)}(o_1,y_1)\exp\big(-k\varphi_1(|y_1|/k)\big)$, and analogously $b_m(y_2)$. For all $k,m\ge 1$, Stirling's formula yields $\binom{k+m}{k} \le c_1(1+\frac{m}{k})^k(1+\frac{k}{m})^m$, where $c_1>0$. For arbitrary $\varepsilon>0$, Lemma 19.7 tells us that there are k_ε, m_ε such that $a_k(y_1)\le e^{\varepsilon k}$ and $b_m(y_2)\le e^{\varepsilon m}$ for all $k\ge k_\varepsilon$, $m\ge m_\varepsilon$. We may assume that $|y_1|\ge k_\varepsilon$. Let $c_\varepsilon = \max\{1, b_m(y_2) : |y_2|\le m < m_\varepsilon\}$. Then $b_m(y_2)\le c_\varepsilon e^{\varepsilon m}$ for all $m\ge|y_2|$. Therefore

$$s_{k,m}(y_1,y_2) \le c_1\,c_\varepsilon\,\exp\Big(|y_1|\Phi_{\lambda,t}(\alpha_k,\beta_m) + \varepsilon(k+m)\Big)$$

for all $k \geq |y_1| \to \infty$, $m \geq |y_2| \geq 0$.

We choose $\varepsilon < \varepsilon_1/2$, and subdivide S_2 into four parts. The first one, denoted by $T_1(y_1 y_2)$, is over the range $k, m \geq r|y_1|$, where r will be chosen below, and we find

$$\begin{aligned} T_1(y_1 y_2)\exp(-|y_1|M(\lambda)) &\leq c_1\, c_\varepsilon \sum_{k,m \geq r|y_1|} \exp\bigl((k+m)(\varepsilon - \varepsilon_1) - |y_1|M(\lambda)\bigr) \\ &\leq \frac{c_1\, c_\varepsilon}{(1-e^{-\varepsilon_1/2})^2} \exp\Bigl(-|y_1|\bigl(r\varepsilon_1 + M(\lambda)\bigr)\Bigr), \end{aligned}$$

which tends to 0 faster than $|y_1|^{-1/2}$, if r is such that $r\frac{\varepsilon_1}{2} + M(\eta) \geq 1$.

The second part, denoted by $T_2(y_1 y_2)$, is over the range $k \geq r|y_1|$, $m < r|y_1|$. We use the same upper bound as for T_1, and

$$T_2(y_1 y_2)\exp(-|y_1|M(\lambda)) \leq \frac{c_1\, c_\varepsilon\, r|y_1|}{1-e^{-\varepsilon_1/2}} \exp\Bigl(-|y_1|\bigl(r\varepsilon_1 + 2M(\lambda)\bigr)/2\Bigr),$$

which also tends to 0 exponentially fast with the above choice of r.

The third part, $T_3(y_1 y_2)$, over the range $k < r|y_1|$, $m \geq r|y_1|$, is treated precisely like T_2.

The fourth part, $T_4(y_1 y_2)$, is over the range $k, m < r|y_1|$, excluding those (k, m) that appear in $S_1(y_1 y_2)$. By (28.15) there is $\varepsilon_2 > 0$ such that $\Phi_{\lambda,t}(\alpha,\beta) - M(\lambda) < -\varepsilon_2$ for all (α,β) with $|\alpha - \alpha(\lambda)| > \delta$ or $|\beta - \beta(\lambda)| > \delta$. Therefore

$$\begin{aligned} T_4(y_1 y_2)\exp(-|y_1|M(\lambda)) &\leq c_1\, c_\varepsilon \sum_{k,m < r|y_1|} \exp\bigl((k+m)\varepsilon - |y_1|\varepsilon_2\bigr) \\ &\leq r^2 |y_1|^2 \exp\bigl(-|y_1|(\varepsilon_2 - 2r\varepsilon)\bigr), \end{aligned}$$

which tends to 0 faster than $|y_1|^{-1/2}$, if – with the r given above – we choose $\varepsilon < \varepsilon_2/(2r)$.

This concludes the proof for $\eta \in [0\,,\infty)$. For $\eta \in (0\,,\infty]$, one has to exchange the roles of the first and second factors in the Cartesian product. In particular, one works with $\lambda^* = |y_1|/|y_2| \to \eta^* \in [0\,,\infty)$. The procedure is precisely the same, and the reader will certainly appreciate not having to read through similar details once more. □

We can now determine the Martin compactification of $\mathbb{T} \times \mathbb{Z}$ for $t > \rho(P)$. Recall that for $t_2 > 1$, we have $\widehat{\mathbb{Z}}(P_2, t_2) = \mathbb{Z} \cup \{-\infty, +\infty\}$, with $K_2(x_2, \pm\infty | t_2) = 1/F_2(1/t_2)^{\pm x_2}$. By an abuse of the notation used above, we write $\bigl(t_1(\lambda), t_2(\lambda)\bigr)$ for the solution of the equations (28.13), where $\lambda \in [0\,,\infty]$. This depends continuously on λ, $\bigl(t_1(0), t_2(0)\bigr) = \bigl(\frac{t-(1-c)}{c}, 1\bigr)$ and $\bigl(t_1(\infty), t_2(\infty)\bigr) = \bigl(\rho_1, \frac{t - c\rho_1}{1-c}\bigr)$.

(28.17) Theorem. *For each $t > \rho$, the Martin boundary is $\mathcal{M}(P,t) = (\vartheta\mathbb{T}\times[-\infty\,,\,+\infty])\cup(\mathbb{T}\times\{-\infty,+\infty\})$. The topology of the Martin compactification and directions of convergence of the Martin kernel are as follows:*

(a) *If $y_1 \to \xi \in \vartheta\mathbb{T}$ and $y_2/|y_1| \to \eta \in [-\infty\,,\,\infty]$ then $y_1y_2 \to (\xi,\eta)$. Let $x_1x_2 \in \mathbb{T}\times\mathbb{Z}$.*

If $\eta > 0$ then $K(x_1x_2,y_1y_2|t) \to K_1\big(x_1,\xi\big|t_1(\eta)\big)\,K_2\big(x_2,+\infty\big|t_2(\eta)\big)$.

If $\eta < 0$ then $K(x_1x_2,y_1y_2|t) \to K_1\big(x_1,\xi\big|t_1(|\eta|)\big)\,K_2\big(x_2,-\infty\big|t_2(|\eta|)\big)$.

If $\eta = 0$ then $K(x_1x_2,y_1y_2|t) \to K_1\big(x_1,\xi\big|t_1(0)\big)$.

(b) *If y_1 is fixed and $y_2 \to \pm\infty$ then $y_1y_2 \to (y_1,\pm\infty)$, and*

$$K(x_1x_2,y_1y_2|t) \to \phi(x_1,y_1)\,K_2\big(x_2,\pm\infty\big|t_2(\infty)\big).$$

Proof. (a) We assume $\eta \geq 0$, that is, y_2 is fixed or tends to $+\infty$. The case $\eta \leq 0$ will follow by symmetry. Set $\lambda = |y_1|/|y_2|$ and $\bar\lambda = |x_2^{-1}y_2|/|x_1^{-1}y_1|$. (Recall that we write $\mathbb{Z}$ multiplicatively here.) Then both λ and $\bar\lambda$ tend to η. Theorem 28.14 yields

$$\begin{aligned} K(x_1x_2,y_1y_2|t) \sim & F_1\big(1/t_1(\bar\lambda)\big)^{\mathfrak{h}(x_1,y_1)}\,F_2\big(1/t_2(\bar\lambda)\big)^{\mathfrak{h}(x_2,y_2)} \\ & \times \left(\frac{F_1\big(1/t_1(\bar\lambda)\big)}{F_1\big(1/t_1(\lambda)\big)}\right)^{|y_1|}\left(\frac{F_2\big(1/t_2(\bar\lambda)\big)}{F_2\big(1/t_2(\lambda)\big)}\right)^{|y_2|}, \end{aligned}$$

where $\mathfrak{h}(x_i,y_i) = |x_i^{-1}y_i| - |y_i|$. As $y_1 \to \xi$, the exponent $\mathfrak{h}(x_1,y_1)$ stabilizes at $\mathfrak{h}(x_1,\xi)$, and the first factor converges to $K_1\big(x_1,\xi\big|t_1(\eta)\big)$. When $\eta = 0$, the second factor converges to 1, because $t_2(\lambda) \to 1$. Otherwise, $y_2 \to +\infty$, the exponent $\mathfrak{h}(x_2,y_2)$ stabilizes at $-x_2$, and the second factor converges to $K_2\big(x_2,+\infty\big|t_2(\eta)\big)$. What is left is to prove that $\left(\frac{F_1(1/t_1(\bar\lambda))}{F_1(1/t_1(\lambda))}\right)^{|y_1|}\left(\frac{F_2(1/t_2(\bar\lambda))}{F_2(1/t_2(\lambda))}\right)^{|y_2|} \to 1$. Rewrite the logarithm of this term as

$$|y_1|\big(A(\bar\lambda)-A(\lambda)\big) - |y_1|(\bar\lambda-\lambda)\log F_2\big(1/t_2(\bar\lambda)\big)\,,$$

where $A(\lambda) = \log F_1\big(1/t_1(\bar\lambda)\big) + \lambda\log F_2\big(1/t_2(\bar\lambda)\big)$. We have $A'(\lambda) = \log F_2\big(1/t_2(\bar\lambda)\big)$ by straightforward computations using (28.13) and the formulae for the functions $F_i(\cdot)$. Thus $A(\bar\lambda)-A(\lambda) = (\bar\lambda-\lambda)\log F_2\big(1/t_2(\widetilde\lambda)\big)$ with $\widetilde\lambda$ between λ and $\bar\lambda$. Therefore we have to show that

$$|y_1|(\bar\lambda-\lambda)\Big(\log F_2\big(1/t_2(\widetilde\lambda)\big) - \log F_2\big(1/t_2(\bar\lambda)\big)\Big) \to 0\,.$$

When $\eta < \infty$, this is obvious, since the last factor tends to 0 (as $\bar\lambda$ and $\widetilde\lambda$ tend to η), while

$$|y_1|(\bar\lambda - \lambda) = \frac{|y_1|}{|x_1^{-1}y_1|}\mathfrak{h}(x_2,y_2) + \frac{|y_2|}{|x_1^{-1}y_1|}\mathfrak{h}(x_1,y_1) \sim \mathfrak{h}(x_2,y_2) + \eta\,\mathfrak{h}(x_1,\xi)\,,$$

which remains bounded. When $\eta = \infty$ and both $|y_1|, y_2 \to \infty$, we have to repeat the same computations after exchanging the roles of y_1 and y_2. In particular, λ is replaced with $\lambda^* = 1/\lambda$; both sides of the second equation in (28.13) have to be multiplied by λ^* to lead to the "new" solution $\big(t_1^*(\lambda^*), t_2^*(\lambda^*)\big)$. This is completely analogous to what we have just done.

(b) We assume without loss of generality that $y_2 \to +\infty$. Using Theorem 28.14 in the form where $C(y_1y_2) = |y_2|^{-3/2}D^*(|y_1|/|y_2|)$, we obtain

$$K(x_1x_2, y_1y_2|t) \sim \frac{1+\frac{M-2}{M}|x_1^{-1}y_1|}{1+\frac{M-2}{M}|y_1|}\,F_1(1/\rho_1)^{\mathfrak{h}(x_1,y_1)}\,F_2\big(1/t_2(\bar\lambda)\big)^{\mathfrak{h}(x_2,y_2)}$$
$$\times\left(\frac{F_1\big(1/t_1(\bar\lambda)\big)}{F_1\big(1/t_1(\lambda)\big)}\right)^{|y_1|}\left(\frac{F_2\big(2/t_2(\bar\lambda)\big)}{F_2\big(1/t_2(\lambda)\big)}\right)^{|y_2|},$$

since $|x_1^{-1}y_1|$ and $|y_1|$ remain bounded. The first line of the right hand side has the proposed asymptotic behaviour. The "remainder" in the second line is treated as in (a) (again replacing λ with its inverse λ^*) and tends to 1. □

Thus, we may think of the Martin compactification $\widehat{\mathbb{T}\times\mathbb{Z}}(P,t)$ as a two-way infinite cylinder with base $\widehat{\mathbb{T}}$, side $\vartheta\mathbb{T}\times[-\infty\,,+\infty]$ and the two caps $\mathbb{T}\times\{+\infty\}$ and $\mathbb{T}\times\{-\infty\}$; the graph $\mathbb{T}\times\mathbb{Z}$ sits inside the cylinder.

Notes and remarks

Here is an incomplete list of references regarding measure theoretic boundary theory: Avez [9], [11], Furstenberg [126], [127], Kaimanovich and Vershik [190], Ledrappier [214], Kaimanovich [183], [185], [187], [189]. One of the most outstanding experts in this branch of the field is V. A. Kaimanovich. I hope that he will conclude writing his own book on this subject soon.

Besides measure theoretic boundary theory (that is, the study of all bounded harmonic functions), another topic that – alas – is missing here for lack of space is harmonic functions with finite Dirichlet sum for reversible Markov chains. A lot of material is contained in the book by Soardi [304], but since then, various important contributions have appeared; see in particular Benjamini and Schramm [37], [38], and their joint paper with Lyons and Peres [35]. The forthcoming book by Lyons and Peres can be expected to become a valuable source.

20. A probabilistic approach to the Dirichlet problem, and a class of compactifications

§**A.** In potential theory, the "only if" of Theorem 20.3 should be known to the experts. In the context of denumerable Markov chains, part of the proof is implicit in Furstenberg [126], but to my knowledge had not been written down in a "swept out" way before I made this effort in [349].

§**B.** This material is taken from Woess [347]. As presented here, it can be traced back to indications hidden between the lines of Furstenberg [126] and [127], who defines the notions of *proximal* and *mean proximal* spaces associated with group actions and random walks: with a pair (Γ, μ), where Γ is a group and μ a probability on Γ, he associates a locally compact Γ-space M together with a μ-invariant Borel probability ν. Proximality of (M, ν) is then a measure theoretic predecessor of the convergence property. There is a natural topology on $\Gamma \cup M$, and the right random walk on Γ with law μ starting at the identity converges almost surely to an M-valued random variable with distribution ν. In general, $\Gamma \cup M$ is not compact. Furstenberg's indications have been elaborated by Cartwright and Soardi [65] to prove convergence to ends for random walks on non-amenable groups of automorphisms of trees; see also Woess [345] for arbitrary vertex-transitive graphs. A closer look shows that what one uses is precisely the two axioms (20.5) and (20.6), and this is formalized in [347]. I am not sure to what extent I deserve credit for having introduced "contractive Γ-compactifications". Several years earlier, Gehring and Martin [129] had introduced the concept of *convergence groups* in the theory of (quasi-)conformal mappings of the complex plane. Besides the restriction to actions on $\widehat{\mathbb{C}}$, the main difference is that in [129] they do not compactify, i.e., they do not distinguish between the action of Γ on some set and on its boundary. In 1992 G. Mess pointed out to me that the main results of [347] are basically contained in [129] (while the referee of [347] seemed not to share this opinion).

It should be remarked that the "contractive" formalism only leads to the most basic structures where the Dirichlet problem can be solved, namely ends and Gromov-hyperbolic spaces. In Ballmann [19] and [20], one finds a solution of the Dirichlet problem with respect to the visibility boundary for discrete groups acting quasi-transitively on non-negatively curved spaces under certain additional assumptions (rank 1).

21. Ends of graphs and the Dirichlet problem

The definition and study of ends of topological spaces go back to Hopf [174] and to Freudenthal [121], who was the first to construct the end compactification of a locally finite graph, without using the term "graph". Later, and independently, Halin [162] re-introduced the concept of ends in graph theory and studied their properties extensively in a noteworthy series of papers [163], [164]. For ends of groups as well as graphs, basic references are the books by Stallings [308] and Dicks and Dunwoody [96]. Lemma 21.4 was already proved by Freudenthal [121] in the case when φ is a metric equivalence.

§**A.** Thomassen's lemma (21.6) [317] is a clever simplification of previous methods. The fundamental result here is Dunwoody's theorem (21.7). Theorem 21.10 and Corollary 21.12 are due to Woess [345]. Theorem 21.11 was proved by Möller [242]. In the case when Γ fixes an end ω, one also has a detailed understanding of the behaviour of the random walk; see Cartwright, Kaimanovich and Woess [59]: let Δ be the group module, and suppose that $\mathfrak{m} = \int_\Gamma \log \Delta(\gamma)\, d\mu(\gamma)$ is finite, where μ is as in (8.12).

Theorem. (a) *If* $\mathfrak{m} > 0$ *and* $M_1(P) < \infty$ *then the random walk converges a.s. to a random end, and the Dirichlet problem is solvable.*

(b) *If* $\mathfrak{m} \geq 0$ *then the Dirichlet problem with respect to the end compactification is not solvable. If* $\mathfrak{m} > 0$ *then the random walk converges to* ω *a.s.*

If $\mathfrak{m} = 0$ then a.s. convergence to ω is known only under an exponential moment condition.

§**B.** Theorem 21.15 was proved by Cartwright, Soardi and Woess [66] and (for trees) Benjamini and Peres [36]. Among the predecessors, there is Derriennic's result for finite range random walks on free groups [93]. The other results are due to Kaimanovich and Woess [191]. The example of Figure 19 is due to Benjamini and Peres [36] and

Kaimanovich and Woess [191]; recently, Amghibech [3] has given a general criterion of Dirichlet-regularity of ends of a tree that applies here.

22. Hyperbolic graphs and groups

Hyperbolic metric spaces were introduced and described in a very influential essay of Gromov [151]. Gromov attributes the "invention" of hyperbolic groups to I. Rips. The task of explaining Gromov's thoughts to a wider audience and working out the details has been performed simultaneously by several work groups [142], [78], [300], [46], and one usually has to switch back and forth between these references for finding the facts that one needs. I have provided the proofs of those basic features that are used here.

The first proof of Theorem 22.14 appeared in Woess [347]. Proposition 22.16 is also from [347], as well as Corollary 22.17. Previously, Series [298] had considered the special case of finite range random walks on Fuchsian groups, later generalized by Ancona [5]. See also below (Section 27). The "geometric" results (22.19) and (22.20) are due to Kaimanovich and Woess [191].

23. The Dirichlet problem for circle packing graphs

The results of this section are due to Benjamini and Schramm [37].

More recently [38], they have developed another, very interesting, method of constructing a boundary for simple random walks on planar graphs (not only triangulations). This is done via *square tilings* of a cylinder, where each edge of the graph corresponds to a square. The side length of each square (possibly $= 0$) is the value on the corresponding edge of the minimum energy unit flow from a chosen root o to ∞. With the exception of an atypical case, the limit set of the tiling is a circle, the limit distribution of the simple random walk starting at o is normalized Lebesgue measure, and the corresponding Dirichlet problem is solvable.

A remote predecessor of these square tilings is Gerl's *volume* of a tree; see [135] (also Woess [340]). Another predecessor, regarding circle boundaries, is Northshield [247].

24. The construction of the Martin boundary

In the classical continuous setting of potential theory, the Martin boundary was introduced by Martin [227] in 1941. In a seminal paper (one of many), Doob [101] explained how to contruct the Martin compactification for denumerable Markov chains. In principle, we follow here Doob's approach, via the approximation theorem and Helly's principle. The uniqueness and convergence theorems are then deduced from the representation theorem (this is not explained too clearly in [101]). However, Doob does not require that X be discrete in the compactification, that is, he just takes the closure of the embedding $y \mapsto K(\cdot, y)$ of X in $\mathcal{S}^+(P)$. Shortly after Doob, Hunt [178] presented a different, more probabilistic, approach, where one first uses a martingale argument to prove the convergence theorem and then deduces the representation theorem (including uniqueness) from the latter. In the context of Markov chains, the standard sources are the books by Kemeny, Snell and Knapp [197] and Revuz [276], but I prefer the presentation of Dynkin [110] because of its clarity, simplicity and convenient notation; [110] follows Hunt's approach. Based on [110], I have made an effort to give a careful exposition in my lecture notes [351]. A recent exposition that follows Doob's ideas more closely is due to Sawyer [291]. Here, I do not consider *recurrent* Martin boundary theory. See the corresponding notes in my survey [348].

25. Generalized lattices, Abelian and nilpotent groups, and graphs with polynomial growth

§**A.** Blackwell [45] was the first to prove that all bounded harmonic functions for random walks on $\mathbb{Z}^d$ are constant. Theorem 25.4 was proved by Choquet and Deny [74] (for random walks on locally compact Abelian groups) and obviously independently by Doob, Snell and Williamson [102] (for random walks on $\mathbb{Z}^d$). Both seem to have been unaware of Blackwell's theorem. Steps 1 and 3 of the proof of Theorem 25.4 given here, as well as Lemma 25.2, are adapted from [102]. Step 2 follows Sawyer [291]. The convex set $\mathcal{C}_t$ appearing in Corollary 25.5 was studied carefully by Hennequin [171] (for random walks on $\mathbb{Z}^d$), see also Babillot [14], [15] (for $\mathbb{R}^d$-invariant random walks).

Theorem 25.8 is adapted from Margulis [225], who proved it for discrete nilpotent groups; see also the minor corrections in Kesten's review [202] of [225]. For Corollary 25.10, compare with Erickson [113].

§**B.** The results in the drift-free case are due to Spitzer [307], P. 26.1, who considers only dimension $d = 3$. (The extension to $d > 3$ is obvious.) If one relaxes the moment condition, then the Martin boundary may have more than one point, even though the minimal Martin boundary is a singleton; see also Cartwright and Sawyer [61] (for $d = 1$). Complete results regarding this general question are recent and due to Uchiyama [320].

The results when the mean vector is non-zero (Theorem 25.15 and Corollary 25.16) are originally due to Ney and Spitzer [246] and constituted for me the hardest challenge in writing this book. Ney and Spitzer used the local central limit theorem (13.10); the disadvantage of Fourier analysis is here that it gives an additive error term instead of asymptotic equivalence. The consequence is that when one recomposes the local estimates to obtain a uniform equivalent of the Green function, one has to control numerous different error terms, which was done by Ney and Spitzer in a heroic effort. The other, perhaps more natural, approach is to work directly with the Fourier transform of the Green function via Lemma 25.17. This was first done in the 80s by Babillot [14], [15], who studied random walks on $\mathbb{R}^d$; the tool is comparison with the Green kernel of the Gaussian process with the same first and second moments. Via this approach, Babillot also treats the continuous-space analogue of generalized lattices. However, its adaptation to $\mathbb{Z}^d$ is not as straightforward as one might think at first glance.

A particularly warm acknowledgment goes to Martine Babillot for having patiently resisted the assault of my email questions, until at last she convinced me that the – much more recent – method elaborated here does work. It follows the indications given in the appendix of her "Thèse d'Habilitation" [16].

Theorem 25.15 and Corollary 25.16 remain obviously valid when instead of finite support, one starts with a suitable exponential moment condition which guarantees that equation (25.13) has a solution for each $\mathbf{x} \neq \mathbf{0}$, and that the Laplace transform (25.19) is analytic in a neighbourhood of $\mathbf{0}$ for each $\mathbf{u}$. On the other hand, I have no precise estimate of the amount of work needed to carry this method over to the case of generalized lattices.

In dimension 1, see the renewal theorem of Blackwell [44] (Spitzer [307], P. 24. 6) for random walks with non-zero mean. In the case of finite range, we can use Corollary 26.14 to show that the Martin boundary is $\{-\infty, +\infty\}$.

26. Trees, ends, and free products

The comparison between Martin boundaries and the ends of general locally finite graphs was studied systematically by Picardello and Woess [258], [259]. Theorem 26.2 is from [259]. This was preceded by a result in the setting of abstract potential theory of Taylor [311].

§**A.** Theorem 26.4 and Example 26.6 are from the influential paper of Cartier [53]. For nearest neighbour random walks on free groups, the result had been proved previously by Dynkin and Malyutov [111]. Theorem 26.7 is due to Picardello and Woess [258], [259], based on previous work of Derriennic [93], who introduced this method for determining the Martin boundary of finite range random walks on free groups. Corollary 26.13 is from [259], the special case of Corollary 26.14(a) from [258]. Regarding Corollary 26.14(b), one might expect to obtain a more general result by requiring that Γ has only thin ends. But it was proved by Woess [344] that such a group must be virtually free; this applies, in particular, to groups whose Cayley graphs are roughly isometric with a tree. (The terminology "rough isometry" is not used in [344].) It is known as an important Theorem of Gromov [151] that a group roughly isometric with a free group is virtually free.

Corollary 26.15, which uses partial knowledge of the Martin boundary to obtain complete knowledge of the Poisson boundary, is the main result in Woess [345]. Previously, Kaimanovich [183] had considered free products of groups and stated that the Poisson boundary coincides with the set of infinite words (as in (21.13) and (26.16); this corresponds to the set Υ_0 appearing in Theorem 21.9) whenever the law of the random walk has finite first moment. However, his proof only worked for free products of finite groups, which are virtually free. Methods that Kaimanovich has elaborated more recently [189] can be used to show that Corollary 26.15 applies more generally when P is assumed to have finite first moment, and also applies to other types of groups.

§**B.** This is from Woess [342].

27. The Martin boundary of hyperbolic graphs

The – very strong – results of this section are due to Ancona [5], [4]. For random walks on Fuchsian groups, Theorem 27.1 and the result of Exercise 27.17 had been proved previously by Series [298] via an extension of the methods of Derriennic [93]. Ancona's method is completely different, and the main tool (Theorem 27.12) is formulated in a more general way than presented here, for so-called Φ-chains. In the context of random walks, I do not know any further example where Φ-chains arise, so that I restricted the exposition to hyperbolic graphs. The conclusion of the proof of Theorem 27.1 given here (before Exercise 27.17) is due to Kaimanovich [188], with the exception of minimality, which is extrapolated from Ancona [4].

Theorem 27.1 has a non-empty intersection with the results on ends of §26.A, namely Corollary 26.14 for $t > \rho(P)$. As a matter of fact, in this context (trees) it is not hard to prove the result of Theorem 27.12 more directly, and also for $t = \rho(P)$.

28. Cartesian products

§**A.** This material is taken from Picardello and Woess [262]. Much earlier, Molchanov [241] had considered the *direct* product $P_1 \otimes P_2$ of two Markov chains and proved the analogue of Proposition 28.2 under a condition of (strong) aperiodicity; [241] uses the associated space-time chain, whose Green kernel splits into the product of the Green kernels of the space-time chains of the factors (this method does not work for Cartesian products). The paper [262] was preceded slightly by analogous results for products of manifolds, where one takes the sum of the Laplacians. (This is the continuous analogue of Cartesian products of Markov chains, or rather, the latter are the discrete analogue of the former.) See Taylor [312] and Freire [120], also Ancona [6].

A more thorough discussion of stability of Martin boundaries can be found in Picardello and Woess [261]; this has direct applications to the Martin boundary of Cartesian products [262].

§**B.** Theorem 28.8 is modelled after Guivarc'h and Taylor [160], who considered the Martin boundary of the product of two or more hyperbolic disks at the bottom of the spectrum. Theorem 28.14 is due to Crotti [87], whose computations follow the similar case of the Cartesian product of two homogeneous trees of degree ≥ 3 (Picardello and Woess [263]); recurrence of $\mathbb{T}_2 = \mathbb{Z}$ causes some changes here).

Further results

We have already mentioned various additional results regarding the Dirichlet problem.

Rough isometry has somehow disappeared from the considerations of Chapter IV, with exception of the fact that ends and hyperbolic boundaries are rough-isometry-invariant. The point is that the Liouville properties for bounded or positive harmonic functions are not-rough-isometry invariant. T. Lyons [223] has given an example which shows (when interpreted in this way) that there are two roughly isometric graphs such that for the respective simple random walks, one of them has non-constant bounded harmonic functions, while the other has all positive harmonic functions constant. Benjamini [33] and Benjamini and Schramm [37] have provided further examples.

For studying Martin boundaries, homogeneous trees are the most rewarding structures. In Section 26, we have only seen results for bounded range random walks. Sawyer [290] has shown that for radial random walks with finite second moment on $\mathbb{T} = \mathbb{T}_{q+1}$ $(q \geq 2)$, the Martin compactification is the end compactification, and the Martin kernels $K(\cdot, \xi)$, $\xi \in \vartheta\mathbb{T}$, coincide with those of the simple random walk, given in Example 26.6 (with $t = 1$). Cartwright and Sawyer [61] have extended this beyond the finite second moment assumption, and have also shown that for arbitrary an radial random walk, the positive harmonic functions are precisely those of the simple random walk. Returning to finite range, Picardello and Woess [260] have studied the following random walk P on $\mathbb{T}$: assign a positive integer radius $r(x)$ to each $x \in \mathbb{T}$. Then $Pf(x)$ is the arithmetic average of the values of f on the ball with radius $r(x)$ centred at x. If the radius function $r(\cdot)$ satisfies a suitable logarithmic Lipschitz inequality, then the Martin boundary is the space of ends, and the Martin kernels coincide with those of the simple random walk.

Ballmann and Ledrappier [21] have refined the discretization method of Brownian motion on manifolds that was developed by Furstenberg [126], Lyons and Sullivan [224] and Kaimanovich [184]. In particular, they provide an example of a random walk on the free group whose Martin boundary is (homeomorphic with) the unit circle. It is based on the realization of the free group as a lattice in $SL(2, \mathbb{R})$. The law of this random walk does not have finite support, and is not known explicitly.

For nearest neighbour random walks on the free group, Lalley [211] has used his uniform local limit theorem to describe the space-time Martin boundary. (The references given here do not aim at completeness regarding boundaries of non-irreducible Markov chains such as space-time chains.)

Recently, Denker and Sato [92] have constructed a non-irreducible random walk on a suitable graph, whose Martin boundary is (homeomorphic with) the Sierpiński fractal. Previously, Benakli [31] had constructed hyperbolic groups whose hyperbolic boundary is the Sierpiński fractal. Therefore, one may use Ancona's theorem (27.1) to construct an irreducible random walk whose Martin boundary is $\mathbf{S}_2$. More recently, V. A. Kaimanovich has shown me a simple way to add edges to a tree in so that the resulting graph is hyperbolic and its hyperbolic boundary is $\mathbf{S}_d$, thus leading to another example where $\mathbf{S}_d$ arises as a Martin boundary.

Guivarc'h [158], in connection with the work on his book with Ji and Taylor [159], has developed a method for determining the Martin boundary on symmetric spaces that can also be adapted to random walks on symmetric spaces of p-adic Lie groups, that is, buildings like the ones studied in §12.C. I admit that I am not mature enough for a full understanding of this profound work.

Acknowledgments are due to many persons who have read through parts of the first versions of the four chapters, suggested better proofs, pointed out errors and/or misprints, references, and so on. Here is a list – I hope I did not forget anybody: Martine Babillot, Daniela Bertacchi, Donald Cartwright, Tullio Ceccherini-Silberstein, Thierry Coulhon, Manfred Einsiedler, Peter Grabner, Pierre de la Harpe, Vadim Kaimanovich, Russ Lyons, Röggi Möller, Yuval Peres, Nicolò Pintacuda, Laurent Saloff-Coste, Maura Salvatori, Klaus Schiefermayr, Oded Schramm, Mao Soardi, Ken Stephenson, Tatiana Nagnibeda, Fabio Zucca.

Book writing requires extended periods of undisturbed concentration, not like in everyday life at university, when students and colleagues knock at your door every other moment, exams have to be prepared, supervised and corrected, you have to attend numerous meetings, feed the bureaucrats with forms and reports, and teaching is the least disturbance. As a consequence, book writing is most efficient when you can flee from your own institute and go as a guest to a distant university, where all this does not occur (to you). I am therefore most grateful to the various institutions that I could visit for longer periods during recent years, and where about two thirds of this book were written: the Departments/Institutes of Mathematics at Universität Salzburg, Université de Rennes, Technische Universität Graz, Universität Linz, and in particular Universität Wien (with special thanks to Harald Rindler).

At the same time, I am also most grateful to the Department of Mathematics at my university in Milan, Italy, who gave me the freedom to go abroad several times, in particular for a whole semester in the final phase.

I thank Peter Gerl, who more than 20 years ago introduced me to the beautiful subject of random walks.

Finally, a special acknowledgment goes to Bélá Bollobas, who became "guilty" of giving rise to this book in 1994, when he came to Milan for a short visit before Easter and told me, "Wolfgang, I should have spoken with you ages ago. You should write a book on random walks."

Comments on the corrections (2020).

Besides the correction of various smaller misprints (others will still have been overlooked), there were three more substantial ones in the original (2000) edition.

1.) Local central limit theorem on $\mathbb{Z}^d$: for proving **Lemma 13.4** – which originally was just equation (13.4) – it had been stated that the function ψ was 2π-periodic, which is not true for the centred characteristic function unless the mean vector $\mathfrak{m}$ is in $\mathbb{Z}^d$. The mistake was pointed out to me by Laurent Saloff-Coste, who also provided input for the corrected proof. As a matter of fact, I found that I had copied this mistake from the classical paper [246] !

2.) In the proof of **Theorem 15.15** (random walks on lamplighter groups), I had used the random walk with law $\nu_0 * \mu_0$ (in the current notation) as if it were symmetric, but it is not, while $\nu_0 * \mu_0 * \nu_0$ *is* symmetric. This seems to be a popular error.

3.) In the proof of **Theorem 27.1**, I had incorporated a simplification (as compared with the original of [5]) taken from [188]. The late Martine Babillot pointed out that it comprised a mistake. The proof could be repaired by re-ordering the material.

BIBLIOGRAPHY

[1] Adyan, S. I.: Random walks on free periodic groups, *Math. USSR Izv.* **21** (1983), 425–434.

[2] Alexopoulos, G.: A lower estimate for central probabilities on polycyclic groups, *Canadian J. Math.* **44** (1992), 897–910.

[3] Amghibech, S.: Criteria of regularity at the end of a tree, *Séminaire de Probabilités XXXII*, Lecture Notes in Math. **1686**, Springer, Berlin, 1998, pp. 128–136.

[4] Ancona, A.: Negatively curved manifolds, elliptic operators, and the Martin boundary, *Ann. Math.* **125** (1987), 495–536.

[5] Ancona, A.: Positive harmonic functions and hyperbolicity, *Potential Theory, Surveys and Problems* (eds. J. Král et al.), Lecture Notes in Math. **1344**, Springer, Berlin, 1988, pp. 1–23.

[6] Ancona, A.: Théorie du potentiel sur les graphes et les variétés, *École d'Été de Probabilités de Saint-Flour XVIII - 1988* (ed. P. L. Hennequin), Lecture Notes in Math. **1427**, Springer, Berlin, 1990, pp. 4–112.

[7] Andrews, G.: *The Theory of Partitions*, Addison-Wesley, Reading, Massachusetts, 1976.

[8] Aomoto, K.: Spectral theory on a free group and algebraic curves, *J. Fac. Sci. Univ. Tokyo, Sect. I A* **31** (1984), 297–317.

[9] Avez, A.: Entropie des groupes de type fini, *C. R. Acad. Sci. Paris, Série A* **275** (1972), 1363–1366.

[10] Avez, A.: Limite de quotients pour des marches aléatoires sur des groupes, *C. R. Acad. Sci. Paris, Série A* **276** (1973), 317–320.

[11] Avez, A.: Thórème de Choquet–Deny pour les groupes à croissance non exponentielle, *C. R. Acad. Sci. Paris, Série A* **279** (1974), 25–28.

[12] Babai, L.: Automorphism groups, isomorphism, reconstruction, *Handbook of Combinatorics*, Chapter 27 (eds. R. L. Graham, M. Grötschel, L. Lovasz), North-Holland - Elsevier, Amsterdam, 1995, pp. 1447–1540.

[13] Babai, L.: The growth rate of vertex-transitive planar graphs, *Proc. 8th Ann. Symp. on Discrete Algorithms*, ACM–SIAM, New Orleans, 1997, pp. 564–573.

[14] Babillot, M.: *Le noyau potentiel des chaines semi-Markoviennes. Applications à l'étude du renouvellement des marches aléatoires*, Thèse 3ème cycle, Univ. Paris VII, 1985.

[15] Babillot, M.: Théorie du renouvellement pour des chaines semi-markoviennes transientes, *Ann. Inst. H. Poincaré Probab. Statist.* **24** (1988), 507–569.

[16] Babillot, M.: *Applications du renouvellement multi-dimensionel en géométrie et en systèmes dynamiques*, Thèse d'habilitation, Univ. Paris VI, 1997.

[17] Baldi, P.: Caractérisation des groupes de Lie connexes récurrents, *Ann. Inst. H. Poincaré Probab. Statist.* **17** (1981), 281–308.

[18] Baldi, P., Lohué, N., and Peyrière, J.: Sur la classification des groupes récurrents, *C. R. Acad. Sci. Paris, Série A* **285** (1977), 1103–1104.

[19] Ballmann, W.: On the Dirichlet problem at infinity for manifolds of nonpositive curvature, *Forum Math.* **1** (1989), 201–213.

[20] Ballmann, W.: *Lectures on Spaces of Nonpositive Curvature*, DMV Seminar **25**, Birkhäuser, Basel, 1995.

[21] Ballmann, W., and Ledrappier, F.: Discretization of positive harmonic functions on Riemannian manifolds and Martin boundary, *Actes de la Table Ronde de Géométrie Differentielle (Luminy, 1992), Semin. Congr.* **1**, Soc. Math. France, Paris, 1996, pp. 77–92.

[22] Barlow, M. T., and Bass, R. F.: Brownian motion and harmonic analysis on Sierpiński carpets, *Canadian J. Math* **51** (1999), 673–744.

[23] Barlow, M. T., and Bass, R. F.: Random walks on graphical Sierpiński carpets, *Symposia Math* **39** (1999), 26–55.

[24] Barlow, M. T., and Perkins, E. A.: Brownian motion on the Sierpiński gasket, *Probab. Th. Rel. Fields* **79** (1988), 543–623.

[25] Bartholdi, L., and Ceccherini-Silberstein, T.: Growth series and random walks on some hyperbolic graphs, *Monatsh. Math.* **136** (2002), 181–202.

[26] Bartholdi, L., Cantat, S., Ceccherini-Silberstein, T., de la Harpe, P.: Estimates for simple random walks on fundamental groups of surfaces, *Colloquium Math.* **72** (1997), 173–193.

[27] Bass, H.: The degree of polynomial growth of finitely generated groups, *Proc. London Math. Soc.* **25** (1972), 603–614.

[28] Beardon, A. F.: *The Geometry of Discrete Groups*, Springer, New York, 1983.

[29] Beardon, A. F.: *Iteration of Rational Functions*, Springer, Berlin, 1991.

[30] Beardon, A., and Stephenson, K.: The uniformization theorem for circle packings, *Indiana Univ. Math. J.* **39** (1990), 1383–1425.

[31] Benakli, N.: *Polyedres hyperboliques, passage du local au global*, Thèse 3ème cycle, Univ. Paris-Sud (Orsay), 1992.

[32] Bender, E. A., and Richmond, L. B.: Correlated random walks, *Ann. Probab.* **12** (1984), 274–278.

[33] Benjamini, I.: Instability of the Liouville property for quasi-isometric graphs and manifolds of polynomial volume growth, *J. Theoret. Probab.* **4** (1991), 631–637.

[34] Benjamini, I., Chavel, I., and Feldman, E. A.: Heat kernel lower bounds on Riemannian manifolds using the old ideas of Nash, *Proc. London Math. Soc.* **72** (1996), 215–240.

[35] Benjamini, I., Lyons, R., Peres, Y., and Schramm, O.: Uniform spanning forests, *Ann. Probab.* **29** (2001), 1–65.

[36] Benjamini, I., and Peres, Y.: Random walks on a tree and capacity in the interval, *Ann. Inst. H. Poincaré Probab. Statist.* **28** (1992), 557–592.

[37] Benjamini, I., and Schramm, O.: Harmonic functions on planar and almost planar graphs and manifolds, via circle packings, *Invent. Math.* **126** (1996), 565–587.

[38] Benjamini, I., and Schramm, O.: Random walks and harmonic functions on infinite planar graphs, using square tilings, *Ann. Probab.* **24** (1996), 1219–1238.

[39] Benjamini, I., and Schramm, O.: Every graph with a positive Cheeger constant contains a tree with a positive Cheeger constant, *GAFA* **7** (1997), 403–419.

[40] Bennett, C., and Sharpley, R.: *Interpolation of Operators*, Academic Press, Boston, Massachusetts, 1988.

[41] Berg, C., and Christensen, J.P.R.: On the relation between amenability of locally compact groups and the norms of convolution operators, *Math. Ann.* **208** (1974), 149–153.

[42] Berg, C., and Christensen, J.P.R.: Sur la norme des opérateurs de convolution, *Invent. Math.* **23** (1974), 173–178.

[43] Biggs, N. L., Mohar, B., and Shawe-Taylor, J.: The spectral radius of infinite graphs, *Bull. London Math. Soc.* **20** (1988), 116–120.

[44] Blackwell, D.: Extension of a renewal theorem, *Pacific J. Math.* **3** (1953), 315–320.

[45] Blackwell, D.: On transient Markov processes with a countable number of states and stationary transition probabilities, *Ann. Math. Statist.* **26** (1955), 654–658.

[46] Bowditch, B.: Notes on Gromov's hyperbolicity criterion for path-metric spaces, *Group Theory from a Geometrical Viewpoint* (eds. E. Ghys, A. Haefliger, A. Verjovsky), World Scientific, Singapore, 1991, pp. 64–167.

[47] Bowers, P. L.: Negatively curved graph and planar metrics with applications to type,

Michigan Math. J. **45** (1998), 31–53.

[48] Brown, K. S.: *Buildings*, Springer, Berlin, 1989.

[49] Calogero, A.: Strong isoperimetric inequality for the edge graph of a tiling of the plane, *Arch. Math.* **61** (1993), 584–595.

[50] Cannon, J. W.: The theory of negatively curved spaces and groups, *Ergodic Theory, Symbolic Dynamics and Hyperbolic Spaces* (eds. T. Bedford, M. Keane, C. Series), Oxford Univ. Press, Oxford, 1991, pp. 315–369.

[51] Carlen, E., Kusuoka, S., and Stroock, D.: Upper bounds for symmetric transition functions, *Ann. Inst. H. Poincaré Probab. Statist.* **23** (1987), 245–287.

[52] Carne, T. K.: A transmutation formula for Markov chains, *Bull. Sci. Math.* **109** (1985), 399–405.

[53] Cartier, P.: Fonctions harmoniques sur un arbre, *Symposia Math.* **9** (1972), 203–270.

[54] Cartier, P.: Géométrie et analyse sur les arbres, *Séminaire Bourbaki* **24**, Lecture Notes in Math. **317**, Springer, Berlin, 1973, pp. 123–140.

[55] Cartier, P.: Harmonic analysis on trees, *Proc. Symp. Pure Math. Amer. Math. Soc.* **26** (1973), 419–424.

[56] Cartwright, D. I.: Some examples of random walks on free products of discrete groups, *Ann. Mat. Pura Appl.* **151** (1988), 1–15.

[57] Cartwright, D. I.: On the asymptotic behaviour of convolution powers of probabilities on discrete groups, *Monatsh. Math.* **107** (1989), 287–290.

[58] Cartwright, D. I.: Singularities of the Green function of a random walk on a discrete group, *Monatsh. Math.* **113** (1992), 183–188.

[59] Cartwright, D. I., Kaimanovich, V. A., and Woess, W.: Random walks on the affine group of local fields and of homogeneous trees, *Ann. Inst. Fourier* (Grenoble) **44** (1994), 1243–1288.

[60] Cartwright, D. I., and Młotkowski, W.: Harmonic analysis for groups acting on triangle buildings, *J. Austral. Math. Soc.* **56** (1994), 345–383.

[61] Cartwright, D. I., and Sawyer, S.: The Martin boundary for general isotropic random walks in a tree, *J. Theoret. Probab.* **4** (1991), 111–136.

[62] Cartwright, D. I., and Soardi, P. M.: Harmonic analysis on the free product of two cyclic groups, *J. Funct. Anal.* **65** (1986), 163–180.

[63] Cartwright, D. I., and Soardi, P. M.: Random walks on free products, quotients, and amalgams, *Nagoya Math. J.* **102** (1986), 163–180.

[64] Cartwright, D. I., and Soardi, P. M.: A local limit theorem for random walks on the Cartesian product of discrete groups, *Boll. Un. Mat. Ital.* **1-A** (1987), 107–115.

[65] Cartwright, D. I., and Soardi, P. M.: Convergence to ends for random walks on the automorphism group of a tree, *Proc. Amer. Math. Soc.* **107** (1989), 817–823.

[66] Cartwright, D. I., Soardi, P. M., and Woess, W.: Martin and end compactifications of non locally finite graphs, *Trans. Amer. Math. Soc* **338** (1993), 679–693.

[67] Cassels, J. W. S.: *Local Fields*, Cambridge Univ. Press, Cambridge, 1986.

[68] Cassi, D.: Phase transitions and random walks on graphs: a generalization of the Mermin–Wagner theorem to disordered lattices, fractals, and other discrete structures, *Phys. Rev. Lett.* **68** (1992), 3621–3634.

[69] Cassi, D., and Regina, S.: Dynamical phase transitions on comb lattices, *Modern Phys. Lett. B* **6** (1992), 1887–1891.

[70] Cheeger, J.: A lower bound for the lowest eigenvalue of the Laplacian, *Problems in Analysis, a Symposium in Honour of Salomon Bochner* (ed. R. C. Gunning), Princeton University Press, Princeton, New Jersey, 1970, pp. 195–199.

[71] Chen, M. F.: Comparison theorems for Green functions of Markov chains, *Chinese Ann. Math.* **3** (1991), 237–242.

[72] Cherix, P. A., and Valette, A.: On spectra of simple random walks on one-relator groups, *Pacific J. Math.* **175** (1996), 417–438.

[73] Chihara, T. S.: *An Introduction to Orthogonal Polynomials*, Gordon and Breach, New York, 1978.

[74] Choquet, G., and Deny, J.: Sur l'équation de convolution $\mu = \mu * \sigma$, *C. R. Acad. Sci. Paris* **250** (1960), 799–801.

[75] Chung, K. L.: *Markov Chains with Stationary Transition Probabilities*, Springer, Berlin, 1960.

[76] Chung, K. L., and Fuchs, W. H. J.: On the distribution of values of sums of random variables, *Mem. Amer. Math. Soc.* **6** (1951).

[77] Chung, K. L., and Ornstein, D.: On the recurrence of sums of random variables, *Bull. Amer. Math. Soc.* **68** (1962), 30–32.

[78] Coornaert, M., Delzant, T., Papadopoulos, A.: *Géométrie et théorie des groupes: les groupes hyperboliques de Gromov*, Lecture Notes in Math. **1441**, Springer, Berlin, 1990.

[79] Coulhon, T.: Ultracontractivity and Nash type inequalities, *J. Funct. Anal.* **141** (1996), 510–539.

[80] Coulhon, T.: Analysis on infinite graphs with regular volume growth, *Symposia Math* **39** (1999), 165–187.

[81] Coulhon, T., and Grigor'yan, A.: On-diagonal lower bounds for heat kernels and Markov chains, *Duke Math. J.* **89** (1997), 133–199.

[82] Coulhon, T., and Grigor'yan, A.: Random walks on graphs with regular volume growth, *GAFA* **8** (1998), 656–701.

[83] Coulhon, T., and Ledoux, M.: Isopérimétrie, décroissance du noyau del la chaleur et transformations de Riesz: un contre-exemple, *Ark. Mat.* **32** (1994), 63–77.

[84] Coulhon, T., and Saloff-Coste, L.: Marches aléatoires non symmetriques sur les groupes unimodulaires, *C. R. Acad. Sci. Paris, Série I* **310** (1990), 627–630.

[85] Coulhon, T., and Saloff-Coste, L.: Puissances d'un opérateur régularisant, *Ann. Inst. H. Poincaré Probab. Statist.* **26** (1990), 419–436.

[86] Coulhon, T., and Saloff-Coste, L.: Isopérimétrie sur les groupes et les variétés, *Rev. Mat. Iberoamericana* **9** (1993), 293–314.

[87] Crotti, S.: *La frontiera di Martin del prodotto Cartesiano di un albero e di una retta*, Tesi di Laurea (Master's Thesis), Univ. Milano.

[88] Day, M. M.: Convolutions, means and spectra, *Illinois J. Math.* **8** (1964), 100–111.

[89] DeBaun, D. R, 1984: L^2-cohomology of noncompact surfaces, *Trans. Amer. Math. Soc.* **284** (1984), 543–565.

[90] Delmotte, T.: Estimations pour les chaines de Markov réversibles, *C. R. Acad. Sci. Paris, Série I* **324** (1997), 1053–1058.

[91] Delmotte, T.: Parabolic Harnack inequality and estimates of Markov chains on graphs, *Rev. Mat. Iberoamericana* **15** (1999), 181–132.

[92] Denker, M., and Sato, H.: Sierpiński gasket as a Martin boundary, *Potential Anal.* **14** (2001), 211–232.

[93] Derriennic, Y.: Marche aléatoire sur le groupe libre et frontière de Martin, *Z. Wahrscheinlichkeitsth. verw. Geb.* **32** (1975), 261–276.

[94] Derriennic, Y.: Quelques applications du théorème ergodique sous-additif, *Astérisque* **74** (1980), 183–201.

[95] Derriennic, Y., and Guivarc'h, Y.: Théorème de renouvellement pour les groupes non moyennables, *C. R. Acad. Sci. Paris, Série A* **277** (1973), 613–615.

[96] Dicks, W., and Dunwoody, M. J.: *Groups Acting on Graphs*, Cambridge Univ. Press, Cambridge, 1989.

[97] Dimca, A.: *Topics on Real and Complex Singularities*, Vieweg, Braunschweig, Germany, 1977.

[98] Dodziuk, J.: Difference equations, isoperimetric inequality, and transience of certain random walks, *Trans. Amer. Math. Soc.* **284** (1984), 787–794.

[99] Dodziuk, J., and Kendall, W. S.: Combinatorial Laplacians and isoperimetric inequality, *From Local Times to Global Geometry, Control and Physics* (ed. K. D. Elworthy), Pitman Res. Notes Math. Ser. **150**, 1986, pp. 68–74.

[100] Donsker, M., and Varadhan, S.: On the number of distinct sites visited by a random walk, *Comm. Pure Appl. Math.* **32** (1979), 721–747.

[101] Doob, J. L.: Discrete potential theory and boundaries, *J. Math. Mech.* **8** (1959), 433–458.

[102] Doob, J. L., Snell, J. L., and Williamson, R. E.: Application of boundary theory to sums of independent random variables, *Contributions to Probability and Statistics*, Stanford Univ. Press, Stanford, California, 1960, pp. 182–197.

[103] Doyle, P. G., and Snell, J. L.: *Random Walks and Electric Networks*, The Carus Math. Monographs **22**, Math. Association of America, 1984.

[104] van den Dries, L., and Wilkie, A. J.: On Gromov's theorem concerning groups with polynomial growth and elementary logic, *J. Algebra* **89** (1984), 349–374.

[105] Dubejko, T.: Random walks on circle packings, *Contemporary Math.* **211** (1997), 169–182.

[106] Duffin, R. J.: The extremal length of a network, *J. Math. Anal. Appl.* **5** (1962), 200–215.

[107] Dunford, N., and Schwartz, J. T.: *Linear Operators I–II*, Interscience, New York, 1963.

[108] Dunwoody, M. J.: Accessibility and groups of cohomological dimension one, *Proc. London Math. Soc.* **38** (1979), 193–215.

[109] Dunwoody, M. J.: Cutting up graphs, *Combinatorica* **2** (1982), 15–23.

[110] Dynkin, E. B.: Boundary theory of Markov processes (the discrete case), *Russian Math. Surveys* **24** (1969), 1–42.

[111] Dynkin, E. B., and Malyutov, M. B.: Random walks on groups with a finite number of generators, *Soviet Math. Doklady* **2** (1961), 399–402.

[112] Efremovic, V. A.: The proximity geometry of Riemannian manifolds, *Uspekhi Mat. Nauk.* **8** (1953), 189.

[113] Erickson, K. B.: The strong Liouville property for a class of random walks, *Monatsh. Math.* **109** (1990), 237–246.

[114] Faraut, J., and Picardello, M. A.: The Plancherel measure for symmetric graphs, *Annali Mat. Pura Appl.* **138** (1984), 151–155.

[115] Figà-Talamanca, A., and Picardello, M. A.: *Harmonic Analysis on Free Groups*, Lect. Notes Pure Appl. Math. **87**, Marcel Dekker, New York, 1983.

[116] Figà-Talamanca, A., and Steger, T.: Harmonic analysis for anisotropic random walks on homogeneous trees, *Memoirs Amer. Math. Soc.* **531** (1992).

[117] Flajolet, P., and Odlyzko, A. M.: Singularity analysis of generating functions, *SIAM J. Discrete Math.* **3** (1990), 216–240.

[118] Følner, E.: On groups with full Banach mean value, *Math. Scand.* **3** (1955), 243–254.

[119] Ford, L. R., and Fulkerson, D. R.: *Flows in Networks*, Princeton Univ. Press, Princetion, New Jersey, 1962.

[120] Freire, A.: On the Martin boundary of Riemannian products, *J. Diff. Geom.* **33** (1991), 215–232.

[121] Freudenthal, H.: Über die Enden diskreter Räume und Gruppen, *Comment. Math. Helvet.* **17** (1944), 1–38.

[122] Friedberg, R., and Martin, O.: Random walks on the Sierpiński gasket, *J. Physique* **47** (1986), 1661–1669.

[123] Fukushima, M.: *Dirichlet Forms and Markov Processes*, North-Holland, Amsterdam, 1980.

[124] Furstenberg, H.: Non commuting random products, *Trans. Amer. Math. Soc.* **108** (1963), 377–428.

[125] Furstenberg, H.: Intersections of Cantor sets and transversality of semigroups, *Problems in Analysis, Symposium in Honour of Salomon Bochner* (ed. R. C. Gunning), Princeton Univ. Press, Princeton, New Jersey, 1970, pp. 41–59.

[126] Furstenberg, H.: Random walks and discrete subgroups of Lie groups, *Advances in Probability and Related Topics* **1** (ed. P. Ney), M. Dekker, New York, 1971, pp. 1–63.

[127] Furstenberg, H.: Boundary theory and stochastic processes on homogeneous spaces, *Proc. Symp. Pure Math.* **26** (1973), 193–229.

[128] Furstenberg, H., and Kesten, H.: Products of random matrices, *Ann. Math. Statist.* **31** (1960), 457–469.

[129] Gehring, F. W., and Martin, G. J.: Discrete quasiconformal groups. I, *Proc. London Math. Soc.* **55** (1987), 331–358.

[130] Gerl, P.: Über die Anzahl der Darstellungen von Worten, *Monatsh. Math.* **75** (1971), 205–214.

[131] Gerl, P.: Irrfahrten auf F_2, *Monatsh. Math.* **84** (1977), 29–35.

[132] Gerl, P.: Wahrscheinlichkeitsmaße auf diskreten Gruppen, *Arch. Math.* **31** (1978), 611–619.

[133] Gerl, P.: A local central limit theorem on some groups, *The First Pannonian Symposium on Mathematical Statistics* (eds. P. Révész et al.), Lecture Notes in Statist. **8**, Springer, Berlin, 1981, pp. 73–82.

[134] Gerl, P.: Continued fraction methods for random walks on $\mathbb{N}$ and on trees, *Probability Measures on Groups* (ed. H. Heyer), Lecture Notes in Math. **1064**, Springer, Berlin, 1984, pp. 131–146.

[135] Gerl, P.: Rekurrente und transiente Bäume, *Séminaire Lotharingien de Combinatoire (IRMA Strasbourg)* **10** (1984), 80–87.

[136] Gerl, P.: Random walks on graphs, *Probability Measures on Groups VIII* (ed. H. Heyer), Lecture Notes in Math. **1210**, Springer, Berlin, 1986, pp. 285–303.

[137] Gerl, P.: Eine isoperimetrische Eigenschaft von Bäumen, *Sitzungsber. Österr. Akad. Wiss., Math.-naturw. Kl.* **195** (1986), 49–52.

[138] Gerl, P.: Natural spanning trees of $\mathbb{Z}^d$ are recurrent, *Discrete Math.* **61** (1986), 333–336.

[139] Gerl, P.: Random walks on graphs with a strong isoperimetric inequality, *J. Theoret. Probab.* **1** (1988), 171–187.

[140] Gerl, P., and Woess, W.: Local limits and harmonic functions for nonisotropic random walks on free groups, *Probab. Th. Rel. Fields* **71** (1986), 341–355.

[141] Geronimus, J.: On a set of polynomials, *Ann. Math.* **31** (1930), 681–686.

[142] Ghys, E., and de la Harpe, P. (eds.): *Sur les groupes hyperboliques d'après Mikhael Gromov*, Progress in Math. **83**, Birkhäuser, Basel, 1990.

[143] Goldman, J., and Rota, G.: The number of subspaces of a vector space, *Recent Progress in Combinatorics* (ed. W. Tutte), Academic Press, New York, 1969, pp. 75–83.

[144] Grabner, P. J., and Woess, W.: Functional iterations and periodic oscillations for simple random walk on the Sierpiński graph, *Stochastic Proc. Appl.* **69** (1997), 127–138.

[145] Greenleaf, F. P.: *Invariant Means on Topological Groups*, Van Nostrand, New York,

1969.

[146] Griffeath, D., and Liggett, T. M.: Critical phenomena for Spitzer's reversible nearest particle systems, *Ann. Probab.* **10** (1982), 881–895.

[147] Grigorchuk, R.I.: The growth degrees of finitely generated groups and the theory of invariant means, *Math. USSR Izv.* **25** (1985), 259–300.

[148] Grigorchuk, R. I., and de la Harpe, P: On problems related to growth, entropy and spectrum in groups theory, *J. Dynamical Control Syst.* **3** (1997), 55–89.

[149] Gromov, M.: Groups of polynomial growth and expanding maps, *Publ. Math. I. H. E. S.* **53** (1981), 53–73.

[150] Gromov, M.: Infinite groups as geometric objects, *Proceedings Int. Congr. Math. Warszawa 1983, Vol. 1*, 1984, pp. 385–392.

[151] Gromov, M.: Hyperbolic groups, *Essays in Group Theory* (ed. S. M. Gersten), Math. Sci. Res. Inst. Publ. **8**, Springer, New York, 1987, pp. 75–263.

[152] Grünbaum, B., and Shephard, G. C.: *Tilings and Patterns*, Freeman, New York, 1987.

[153] Guivarc'h, Y.: Groupes de Lie à croissance polynomiale, *C. R. Acad. Sci. Paris* **271** (1970), 237–239.

[154] Guivarc'h, Y.: Croissance polynomiale et périodes des fonctions harmoniques, *Bull. Soc. Math. France* **101** (1973), 333-379.

[155] Guivarc'h, Y.: Théorèmes quotients pour les marches aléatoires, *Astérisque* **74** (1980), 15–28.

[156] Guivarc'h, Y.: Sur la loi des grands nombres et le rayon spectral d'une marche aléatoire, *Astérisque* **74** (1980), 47–98.

[157] Guivarc'h, Y.: Application d'un théorème limite local à la transience et à la récurrence de marches de Markov, *Colloque de Théorie du Potentiel – Jacques Deny (Orsay 1983)*, Lecture Notes in Math. **1096**, Springer, Berlin, 1984, pp. 301–332.

[158] Guivarc'h, Y. (collaboration with J. C. Taylor and L. Ji): Compactifications of symmetric spaces and positive eigenfunctions of the Laplacian, *Institut de Recherches Mathématiques de Rennes Publ., Fascicule de Probabilités* (1994).

[159] Guivarc'h, Y., Ji, L., and Taylor, J. C.: *Compactifications of Symmetric Spaces*, Progress in Mathematics **156**, Birkhäuser, Boston, Massachusetts, 1998.

[160] Guivarc'h, Y., and Taylor, J. C.: The Martin compactification of the polydisc at the bottom of the spectrum, *Colloquium Math.* **50/51** (1990), 537–546.

[161] Gutkin, E.: Green's functions of free products of operators, with applications to graph spectra and to random walks, *Nagoya Math. J.* **149**, 93–116.

[162] Halin, R.: Über unendliche Wege in Graphen, *Math. Ann.* **157** (1964), 125–137.

[163] Halin, R.: Die Maximalzahl zweiseitig unendlicher Wege in Graphen, *Math. Nachr.* **44** (1970), 119–127.

[164] Halin, R.: Automorphisms and endomorphisms of infinite locally finite graphs, *Abh. Math. Sem. Univ. Hamburg* **39** (1973), 251–283.

[165] Hammersley, J. M.: Markovian walks on crystals, *Compositio Math.* **11** (1953), 171–186.

[166] de la Harpe, P., Robertson, A. G., and Valette, A.: On the spectrum of the sum of generators of a finitely generated group, *Israel J. Math.* **81** (1993), 65–96.

[167] de la Harpe, P., Robertson, A. G., and Valette, A.: On the spectrum of the sum of generators of a finitely generated group, II, *Colloquium Math.* **65** (1993), 87–102.

[168] He, Z.-X., and Schramm, O.: Fixed points, Koebe uniformization and circle packings, *Ann. Math.* **137** (1993), 369–406.

[169] He, Z.-X., and Schramm, O.: Hyperbolic and parabolic packings, *Discrete Comput. Geom.* **14** (1995), 123–149.

[170] Hebisch, W., and Saloff-Coste, L.: Gaussian estimates for Markov chains and random walks on groups, *Ann. Probab.* **21** (1993), 673–709.

[171] Hennequin, P. L.: Processus de Markoff en cascade, *Ann. Inst. H. Poincaré* **18** (1963), 109–196.

[172] Hewitt, E., and Ross, K. A.: *Abstract Harmonic Analysis I*, Springer, Berlin, 1963.

[173] Hille, E.: *Analytic Function Theory,* vols. I–II, Chelsea Publ. Comp., New York, 1962.

[174] Hopf, H.: Enden offener Räume und unendliche diskontinuierliche Gruppen, *Comment. Math. Helvet.* **16** (1943), 81–100.

[175] Hörmander, L.: *The Analysis of Linear Partial Differential Operators I.* 2nd ed., Springer, Berlin, 1990.

[176] Horn, R., and Johnson, Ch. A.: *Matrix Analysis*, Cambridge Univ. Press, Cambridge, 1985.

[177] Hughes, B. D.: *Random Walks and Random Environments, Volume 1: Random Walks*, Clarendon Press, Oxford, 1995.

[178] Hunt, G. A.: Markoff chains and Martin boundaries, *Illinois J. Math.* **4** (1960), 313–340.

[179] Imrich, W., and Seifter, N.: A survey on graphs with polynomial growth, *Discrete Math.* **95** (1991), 101–117.

[180] Ivanov, A. A.: Bounding the diameter of a distance-regular graph, *Soviet Math. Doklady* **28** (1983), 149–152.

[181] Jones, O. D.: Transition probabilities for the simple random walk on the Sierpiński graph, *Stochastic Proc. Appl.* **61** (1996), 4–69.

[182] Kaimanovich, V. A.: Spectral measure of the transition operator and harmonic functions connected with random walks on discrete groups, *J. Soviet Math.* **24** (1984), 550–555.

[183] Kaimanovich, V. A.: An entropy criterion for maximality of the boundary of random walks on discrete groups, *Soviet Math. Doklady* **31** (1985), 193–197.

[184] Kaimanovich, V. A.: Discretization of bounded harmonic functions on Riemannian manifolds and entropy, *Potential Theory* (ed. M. Kishi), de Gruyter, Berlin, 1991, pp. 213–223.

[185] Kaimanovich, V. A.: Poisson boundaries of random walks on discrete solvable groups, *Probability Measures on Groups X* (ed. H. Heyer), Plenum, New York, 1991, pp. 205–238.

[186] Kaimanovich, V. A.: Dirichlet norms, capacities and generalized isoperimetric inequalities for Markov operators, *Potential Anal.* **1** (1992), 61–82.

[187] Kaimanovich, V. A.: Measure-theoretic boundaries of Markov chains, 0–2 laws and entropy, *Harmonic Analysis and Discrete Potential Theory* (ed. M. A. Picardello), Plenum, New York, 1992, pp. 145–180.

[188] Kaimanovich, V. A.: Ergodicity of harmonic invariant measures for the geodesic flow on hyperbolic spaces, *J. reine angew. Math.* **455** (1994), 57–103.

[189] Kaimanovich, V. A.: The Poisson boundary for groups with hyperbolic properties, preprint, Univ. Rennes I (1998).

[190] Kaimanovich, V. A., and Vershik, A. M.: Random walks on discrete groups: boundary and entropy, *Ann. Probab.* **11** (1983), 457–490.

[191] Kaimanovich, V. A., and Woess, W.: The Dirichlet problem at infinity for random walks on graphs with a strong isoperimetric inequality, *Probab. Th. Rel. Fields* **91** (1992), 445–466.

[192] Kanai, M.: Rough isometries and combinatorial approximations of geometries of non-compact Riemannian manifolds, *J. Math. Soc. Japan* **37** (1985), 391–413.

[193] Kanai, M.: Rough isometries and the parabolicity of Riemannian manifolds, *J. Math. Soc. Japan* **38** (1986), 227–238.

[194] Kanai, M.: Analytic inequalities, and rough isometries between non-compact Riemannian manifolds, *Curvature and Topology of Riemannian Manifolds – Proceedings, Katata 1985*, Lecture Notes in Math. **1201**, Springer, Berlin, 1986, pp. 122–137.

[195] Karlin, S., and McGregor, J.: Random walks, *Illinois J. Math.* **3** (1959), 66–81.

[196] Katok, S.: *Fuchsian Groups*, Univ. of Chicago Press, Chicago, 1992.

[197] Kemeny, J. G., Snell, J. L., and Knapp, A. W.: *Denumerable Markov Chains*. 2nd ed., Springer, New York, 1976.

[198] Kesten, H.: Symmetric random walks on groups, *Trans. Amer. Math. Soc.* **92** (1959), 336–354.

[199] Kesten, H.: Full Banach mean values on countable groups, *Math. Scand* **7** (1959), 146–156.

[200] Kesten, H.: Ratio limits for random walks II, *J. Analyse Math.* **11** (1963), 323–379.

[201] Kesten, H.: The Martin boundary for recurrent random walks on countable groups, *Proc. 5th Berkeley Sympos. on Math. Statistics and Probability*, vol. 2, Univ. of California Press, Berkeley, California, 1967, pp. 51–74.

[202] Kesten, H.: Review of [225], *Math. Rev.* **36**-5269 (1968).

[203] Kesten, H., and Spitzer, F.: Random walk on countably infinite Abelian groups, *Acta Math.* **114** (1965), 237–265.

[204] Kindermann, R., and Snell, J. L.: *Markov Random Fields and their Applications*, Contemporary Math. **1**, American Math. Society, Providence, Rhode Island, 1980.

[205] Kingman, J. F. C.: The ergodic decay of Markov transition probabilities, *Proc. London Math. Soc.* **13** (1963), 337–358.

[206] Kingman, J. F. C.: The ergodic theory of subadditive processes, *J. Royal Stat. Soc., Ser. B* **30** (1968), 499-510.

[207] Koebe, P.: Kontaktprobleme der konformen Abbildung, *Ber. Ver. Sächs. Akad. Wiss. Leipzig Math.-Phys. Kl.* **88** (1936), 141–164.

[208] Krámli, A., and Szász, D.: Random walks with internal degrees of freedom, I. Local limit theorems, *Z. Wahrscheinlichkeitsth. verw. Geb.* **63** (1983), 85–95.

[209] Kuhn, G., and Soardi, P. M.: The Plancherel measure for polygonal graphs, *Ann. Mat. Pura Appl.* **134** (1983), 393–401.

[210] Kuich, W., and Salomaa, A.: *Semirings, Automata, Languages*, Springer, Berlin, 1985.

[211] Lalley, S. P.: Saddlepoint approximations and space-time Martin boundary for nearest neighbour random walk on a homogeneous tree, *J. Theoret. Probab.* **4** (1991), 701–723.

[212] Lalley, S. P.: Finite range random walks on free groups and homogeneous trees, *Ann. Probab.* **21** (1993), 2087–2130.

[213] Lebedev, N. N.: *Special Functions and their Applications*, Dover Publ., New York, 1972.

[214] Ledrappier, F.: Poisson boundaries of discrete groups of matrices, *Israel J. Math.* **50** (1985), 319–336.

[215] Levit, B. Ya., and Molchanov, S. A.: Invariant chains on a free group with a finite number of generators (in Russian), *Vest. Moscow Univ.* **26** (1971), 80–88.

[216] Losert, V.: On the structure of groups with polynomial growth, *Math. Z.* **195** (1987), 109–117.

[217] Lust-Piquard, F.: Lower bounds on $\|K^n\|_{1\to\infty}$ for some contractions K of $L^2(\mu)$, with applications to Markov operators, *Math. Ann.* **303** (1995), 699–712.

[218] Lyndon, R. C., and Schupp, P. E.: *Combinatorial Group Theory*, Springer, Berlin,

1977.

[219] Lyons, R.: Random walks and percolation on trees, *Ann. Probab.* **18** (1990), 931–958.

[220] Lyons, R.: Random walks and the growth of groups, *C. R. Acad. Sci. Paris, Série I* **320** (1995), 1361–1366.

[221] Lyons, R., Pemantle, R., and Peres, Y.: Random walks on the lamplighter group, *Ann. Probab.* **24** (1996), 1993–2006.

[222] Lyons, T.: A simple criterion for transience of a reversible Markov chain, *Ann. Probab.* **11** (1983), 393–402.

[223] Lyons, T.: Instability of the Liouville property for quasi-isometric Riemannian manifolds and reversible Markov chains, *J. Diff. Geom.* **26** (1987), 33–66.

[224] Lyons, T., and Sullivan, D.: Function theory, random paths and covering spaces, *J. Diff. Geom.* **19** (1984), 299–323.

[225] Margulis, G. A.: Positive harmonic functions on nilpotent groups, *Soviet Math. Doklady* **166** (1966), 241–244.

[226] Markvorsen, St., McGuinness, S., and Thomassen, C.: Transient random walks on graphs and metric spaces, with applications to hyperbolic surfaces, *Proc. London Math. Soc.* **64** (1992), 1–20.

[227] Martin, R. S.: Minimal positive harmonic functions, *Trans. Amer. Math. Soc.* **49** (1941), 137–172.

[228] McCaughan, G. J.: A recurrence/transience result for circle packings, *Proc. Amer. Math. Soc* **126** (1998), 3647–3656.

[229] McGuinness, S.: *Random walks on graphs and digraphs*, Ph. D. thesis, Univ. Waterloo, Ontario (1988).

[230] McGuinness, S.: Recurrent networks and a theorem of Nash-Williams, *J. Theoret. Probab.* **4** (1991), 87–100.

[231] McLaughlin, J. C.: *Random walks and convolution operators on free products*, Ph. D. thesis, New York Univ. (1986).

[232] Medolla, G., and Soardi, P. M.: Extensions of Foster's averaging formula to infinite networks with moderate growth, *Math. Z.* **219** (1995), 171–185.

[233] Merkl, F., and Wagner, H.: Recurrent random walks and the absence of continuous symmetry breaking on graphs, *J. Statist. Phys.* **75** (1995), 153–165.

[234] Mermin, N. D.: Absence of ordering in certain classical systems, *J. Math. Phys.* **8** (1967), 1061–1064.

[235] Mermin, N. D., and Wagner, H.: Absence of ferromagnetism or antiferromagnetism in one- or two-dimensional isotropic Heisenberg models, *Phys. Rev. Lett.* **17** (1966), 1133–1136.

[236] Milnor, J.: A note on curvature and fundamental group, *J. Diff. Geom.* **2** (1968), 1–7.

[237] Mohar, B.: Isoperimetric inequalities, growth and the spectrum of infinite graphs, *Linear Algebra Appl.* **103** (1988), 119–131.

[238] Mohar, B.: Some relations between analytic and geometric properties of inifinite graphs, *Discrete Math.* **95** (1991), 193–219.

[239] Mohar, B.: Isoperimetric numbers and spectral radius of some infinite planar graphs, *Math. Slovaca* **42** (1992), 411–425.

[240] Mohar, B., and Woess, W.: A survey on spectra of infinite graphs, *Bull. London Math. Soc.* **21** (1989), 209–234.

[241] Molchanov, S. A.: On the Martin boundaries for the direct products of Markov chains, *Th. Probab. Appl.* **12** (1967), 307–314.

[242] Möller, R. G.: Ends of graphs. II, *Math. Proc. Cambridge Phil. Soc.* **111** (1992),

455–460.

[243] Nagnibeda, T.: *On random walks and growth in groups with finitely many cone types*, Ph. D. Thesis, Univ. Genève (1997).

[244] Nagnibeda, T.: An estimate from above of spectral radii of random walks on surface groups, *Sbornik Seminarov POMI* **240** (1997), 154–166.

[245] Nash-Williams, C. St J. A.: Random walks and electric currents in networks, *Proc. Cambridge Phil. Soc.* **55** (1959), 181–194.

[246] Ney, P., and Spitzer, F.: The Martin boundary for random walk, *Trans. Amer. Math. Soc.* **121** (1966), 116–132.

[247] Northshield, S.: Circle boundaries of planar graphs, *Potential Anal.* **2** (1993), 299–314.

[248] Odlyzko, A. M.: Periodic oscillations of coefficients of power series that satisfy functional equations, *Adv. Math.* **44** (1982), 180–205.

[249] Odlyzko, A. M.: Asymptotic enumeration methods, *Handbook of Combinatorics*, Chapter 22 (eds. R. L. Graham, M. Grötschel, L. Lovasz) North-Holland – Elsevier, Amsterdam, 1995, pp. 1063–1229.

[250] Ol'shasnkii, A. Yu.: On the question of existence of an invariant mean on a group, *Russian Math. Surveys* **35** (1980), 180–181.

[251] Olver, F. W. J.: *Asymptotics and Special Functions*, Academic Press, San Diego, California, 1974.

[252] Papangelou, F.: Strong ratio limits, R-recurrence and mixing properties of discrete parameter Markov chains, *Z. Wahrscheinlichkeitsth. verw. Geb.* **8** (1967), 259–297.

[253] Parthasarathy, K.: *Probability Measures on Metric Spaces*, Academic Press, New York, 1967.

[254] Paschke, W. L.: Lower bound for the norm of a vertex-transitive graph, *Math. Z.* **213** (1993), 225–239.

[255] Paterson, A. T.: *Amenability*, Math. Surveys and Monographs **29**, American Math. Society, Providence, Rhode Island, 1988.

[256] Picardello, M. A.: Spherical functions and local limit theorems on free groups, *Ann. Mat. Pura Appl.* **33** (1983), 177–191.

[257] Picardello, M. A., and Woess, W.: Random walks on amalgams, *Monatsh. Math.* **100** (1985), 21–33.

[258] Picardello, M. A., and Woess, W.: Martin boundaries of random walks: ends of trees and groups, *Trans. Amer. Math. Soc.* **302** (1987), 185–205.

[259] Picardello, M. A., and Woess, W.: Harmonic functions and ends of graphs, *Proc. Edinburgh Math. Soc.* **31** (1988), 457–461.

[260] Picardello, M. A., and Woess, W.: A converse to the mean value property on homogeneous trees, *Trans. Amer. Math. Soc.* **311** (1989), 209–225.

[261] Picardello, M. A., and Woess, W.: Examples of stable Martin boundaries of Markov chains, *Potential Theory* (ed. M. Kishi), de Gruyter, Berlin, 1991, pp. 261–270.

[262] Picardello, M. A., and Woess, W.: Martin boundaries of Cartesian products of Markov chains, *Nagoya Math. J.* **128** (1992), 153–169.

[263] Picardello, M. A., and Woess, W.: The full Martin boundary of the bi-tree, *Ann. Probab.* **22** (1994), 2203–2222.

[264] Pier, J. P.: *Amenable Locally Compact Groups*, Wiley, New York, 1984.

[265] Pittet, Ch.: On the isoperimetry of graphs with many ends, *Colloq. Math.* **78** (1998), 307–318.

[266] Pittet, Ch., and Saloff-Coste, L.: A survey on the relationship between volume growth, isoperimetry, and the behaviour of simple random walk on Cayley graphs, with examples, available at `http://pi.math.cornell.edu/~lsc/papers/surv.pdf`.

[267] Pittet, Ch., and Saloff-Coste, L.: Amenable groups, isoperimetric profiles and random walks, *Geometric Group Theory Down Under, Canberra 1996* (eds. J. Cossey et al.), de Gruyter, Berlin, 1999, pp. 293–316.

[268] Pittet, Ch., and Saloff-Coste, L.: On the stability of the behaviour of random walks on groups, *J. Geometric Anal* **10** (2000), 713–737.

[269] Pólya, G.: Über eine Aufgabe der Wahrscheinlichkeitstheorie betreffend die Irrfahrt im Straßennetz, *Math. Ann.* **84** (1921), 149–160.

[270] Pólya, G.: *Orthogonal Polynomials*, American Math. Society, Providence, Rhode Island, 1939.

[271] Pruitt, W. E.: Eigenvalues of non-negative matrices, *Ann. Math. Statist.* **35** (1964), 1797–1800.

[272] Quenell, G.: Combinatorics of free product graphs, *Contemporary Math.* **206** (1994), 257–281.

[273] Raghunathan, M. S.: *Discrete Subgroups of Lie Groups*, Springer, Berlin, 1972.

[274] Rammal, R.: Spectrum of harmonic excitations on fractals, *J. Phys.* **45** (1984), 191–206.

[275] Rammal, R.: Random walk statistics on fractal structures, *J. Statist. Physics* **36** (1984), 547–560.

[276] Revuz, D.: *Markov Chains*, North-Holland, Amsterdam, 1975.

[277] Rigoli, M., Salvatori, M., and Vignati, M.: Subharmonic functions on graphs, *Israel J. Math.* **99** (1997), 1–27.

[278] Rodin, B., and Sullivan, D.: The convergence of circle packings to the Riemann mapping, *J. Diff. Geom.* **26** (1987), 349–360.

[279] Ronan, M.: *Lectures on Buildings*, Academic Press, New York, 1989.

[280] Rosenblatt, J.: Invariant measures and growth conditions, *Trans. Amer. Math. Soc.* **193** (1974), 33–53.

[281] Rudin, W.: *Fourier Analysis on Groups*, Interscience, New York, 1962.

[282] Rudin, W.: *Real and Complex Analysis*. 3rd ed., McGraw-Hill, New York, 1987.

[283] Rvačeva, E. L.: On domains of attraction of multi-dimensional distributions, *Selected Translations Math. Statist. Probab. Th.* **2** (1962), 183–205.

[284] Sabidussi, G.: Vertex transitive graphs, *Monatsh. Math.* **68** (1964), 427–438.

[285] Saloff-Coste, L.: Isoperimetric inequalities and decay of iterated kernels for almost-transitive Markov chains, *Combinatorics, Probab. Comput.* **4** (1995), 419–442.

[286] Saloff-Coste, L., and Woess, W.: Computing norms of group-invariant transition operators, *Combinatorics, Probab. Comput.* **5** (1996), 161–178.

[287] Saloff-Coste, L., and Woess, W.: Transition operators, groups, norms, and spectral radii, *Pacific J. Math.* **180** (1997), 333–367.

[288] Salvatori, M.: On the norms of group-invariant transition operators on graphs, *J. Theoret. Probab.* **5** (1991), 563–576.

[289] Salvatori, M.: Random walks on generalized lattices, *Monatsh. Math.* **1-2** (1996), 145–161.

[290] Sawyer, S.: Isotropic random walks in a tree, *Z. Wahrscheinlichkeitsth. verw. Geb.* **42** (1978), 279–292.

[291] Sawyer, S.: Martin boundaries and random walks, *Harmonic Functions on Trees and Buildings* (ed. A. Korányi), Contemporary Math. **206**, American Math. Society, Providence, Rhode Island, 1997, pp. 17–44.

[292] Sawyer, S., and Steger, T.: The rate of escape for anisotropic random walks in a tree, *Probab. Th. Rel. Fields* **76** (1987), 207–230.

[293] Schlichting, G.: Polynomidentitäten und Permutationsdarstellungen lokalkompakter Gruppen, *Invent. Math.* **55** (1979), 97–106.

[294] Schwarzc, A. S.: Volume invariants of coverings, *Doklady Akad. Nauk. SSSR* **105** (1955), 32–34.

[295] Segal, D.: *Polycyclic Groups*, Cambridge Univ. Press, Cambridge, 1983.

[296] Seifter, N., and Trofimov, V. I.: Automorphism groups of graphs with quadratic growth, *J. Combinatorial Th., Ser. B* **71** (1997), 205–210.

[297] Seneta, E.: *Non-negative Matrices and Markov Chains.* 2nd ed., Springer Series in Statist., New York, 1973.

[298] Series, C.: Martin boundaries of random walks on Fuchsian groups, *Israel J. Math.* **44** (1983), 221–242.

[299] Serre, J.-P.: *Local Fields*, Springer, New York, 1979.

[300] Short, H. (et al.): Notes on word hyperbolic groups, *Group Theory from a Geometrical Viewpoint (eds. E. Ghys, A. Haefliger, A. Verjovsky)*, World Scientific, Singapore, 1991, pp. 3–63.

[301] Soardi, P. M.: Simple random walks on $\mathbb{Z}^2 * \mathbb{Z}_2$, *Symposia Math.* **29** (1986), 303–309.

[302] Soardi, P. M.: The resolvent for simple random walks on the free product of discrete groups, *Math. Z.* **192** (1986), 109–116.

[303] Soardi, P. M.: Recurrence and transience of the edge graph of a tiling of the Euclidean plane, *Math. Ann.* **287** (1990), 613–626.

[304] Soardi, P. M.: *Potential Theory on Infinite Networks*, Lecture Notes in Math. **1590**, Springer, Berlin, 1994.

[305] Soardi, P. M., and Woess, W.: Amenability, unimodularity, and the spectral radius of random walks on infinite graphs, *Math. Z.* **205** (1990), 471–486.

[306] Soardi, P. M., and Yamasaki, M.: Classification of infinite networks and its applications, *Circuits, Syst. Signal Proc.* **12** (1993), 133–149.

[307] Spitzer, F.: *Principles of Random Walk.* 2nd edition, Springer, New York, 1976.

[308] Stallings, J.: *Group Theory and Three-Dimensional Manifolds*, Yale Univ. Press, New Haven, Connecticut, 1971.

[309] Steger, T.: *Harmonic analysis for an anisotropic random walk in a homogeneous tree*, Ph. D. Thesis, Washington Univ., St Louis, Missouri (1985).

[310] Stone, Ch.: On local and ratio limit theorems, *Proc. 5th Berkeley Sympos. on Math. Statistics and Probability*, vol. 2, Univ. of California Press, Berkeley, California, 1967, pp. 217–224.

[311] Taylor, J. C.: The Martin boundaries of equivalent sheaves, *Ann. Inst. Fourier* (Grenoble) **20** (1970), 433–456.

[312] Taylor, J. C.: The product of minimal functions is minimal, *Bull. London Math. Soc.* **22** (1990), 499–504; erratum: **24** (1991), 379–380.

[313] Telcs, A.: Spectra of graphs and fractal dimensions I, *Probab. Th. Rel. Fields* **85** (1990), 489–497.

[314] Telcs, A.: Spectra of graphs and fractal dimensions II, *J. Theoret. Probab.* **8** (1995), 77–96.

[315] Thomassen, C.: Isoperimetric inequalities and transient random walks on graphs, *Ann. Probab.* **20** (1992), 1592–1600.

[316] Thomassen, C.: Trees, ends, and transience, *Harmonic Analysis and Discrete Potential Theory* (ed. M. A. Picardello), Plenum, New York, 1992, pp. 259–266.

[317] Thomassen, C., and Woess, W.: Vertex-transitive graphs and accessibility, *J. Combinatorial Th., Ser. B* **58** (1993), 248–268.

[318] Trofimov, V. I.: Graphs with polynomial growth, *Math. USSR Sbornik* **51** (1985), 405–417.

[319] Trofimov, V. I.: Automorphism groups of graphs as topological groups, *Math. Notes (Transl. Mat. Zametki)* **38** (1986), 717–720.

[320] Uchiyama, K.: Green's functions for random walks on $\mathbb{Z}^N$, *Proc. London Math. Soc.* **77** (1988), 215–240.

[321] Varopoulos, N. Th.: Brownian motion and transient groups, *Ann. Inst. Fourier* (Grenoble) **33** (1983), 241–261.

[322] Varopoulos, N. Th.: Brownian motion and random walks on manifolds, *Ann. Inst. Fourier* (Grenoble) **34** (1984), 243–269.

[323] Varopoulos, N. Th.: Isoperimetric inequalities and Markov chains, *J. Funct. Anal.* **63** (1985), 215–239.

[324] Varopoulos, N. Th.: Semigroupes d'opérateurs sur les espaces L^p, *C. R. Acad. Sci. Paris, Série I* **301** (1985), 865–868.

[325] Varopoulos, N. Th.: Théorie du potentiel sur des groupes et des variétés, *C. R. Acad. Sci. Paris, Série I* **302** (1986), 203–205.

[326] Varopoulos, N. Th., Saloff-Coste, L. and Coulhon, T.: *Analysis and Geometry on Groups*, Cambridge Tracts in Math. **100**, Cambridge Univ. Press, Cambridge, 1992.

[327] Vere-Jones, D.: Geometric ergodicity in denumerable Markov chains, *Quart. J. Math. Oxford* **13** (1962), 7–28.

[328] Vere-Jones, D.: Ergodic properties of nonnegative matrices – I, *Pacific J. Math.* **26** (1967), 361–386.

[329] Vere-Jones, D.: Ergodic properties of nonnegative matrices – II, *Pacific J. Math.* **26** (1968), 601–620.

[330] Voiculescu, D.: Addition of certain non-commuting random variables, *J. Funct. Anal.* **66** (1986), 323–346.

[331] Voiculescu, D.: Noncommutative random variables and spectral problems in free product C^*-algebras, *Rocky Mountain J. Math.* **20** (1990), 263–283.

[332] Voiculescu, D., Dykema, K., and Nica, A.: *Free Random Variables*, CRM Monograph Series **1**, American Math. Society, Providence, Rhode Island, 1992.

[333] van der Waerden, B. L.: *Einführung in die algebraische Geometrie*, Springer, Berlin, 1939.

[334] Wagon, S.: *The Banach–Tarski paradox*, Encyclopedia of Math. **24**, Cambridge Univ. Press, Cambridge, 1985.

[335] Wall, H. S.: *Analytic Theory of Continued Fractions*, Van Nostrand, Toronto, 1948.

[336] Woess, W.: A local limit theorem for random walks on certain discrete groups, *Probability Measures on Groups* (ed. H. Heyer), Lecture Notes in Math. **928**, Springer, Berlin, 1982, pp. 468–477.

[337] Woess, W.: Puissances de convolution sur les groupes libres ayant un nombre quelconque de générateurs, *Inst. Élie Cartan* **7** (1983), 181–190.

[338] Woess, W.: A random walk on free products of finite groups, *Probability Measures on Groups* (ed. H. Heyer), Lecture Notes in Math. **1064**, Springer, Berlin, 1984, pp. 467–470.

[339] Woess, W.: Random walks and periodic continued fractions, *Adv. Appl. Probab.* **17** (1985), 67–84.

[340] Woess, W.: Transience and volumes of trees, *Arch. Math.* **46** (1986), 184–192.

[341] Woess, W.: Nearest neighbour random walks on free products of discrete groups, *Boll. Un. Mat. Ital.* **5-B** (1986), 961–982.

[342] Woess, W.: A description of the Martin boundary for nearest neighbour random walks on free products, *Probability Measures on Groups VIII* (ed. H. Heyer), Lecture Notes in Math. **1210**, Springer, Berlin, 1986, pp. 203–215.

[343] Woess, W.: Context-free languages and random walks on groups, *Discrete Math.* **67** (1987), 81–87.

[344] Woess, W.: Graphs and groups with tree-like properties, *J. Combinatorial Th., Ser.*

B **68** (1989), 271–301.

[345] Woess, W.: Boundaries of random walks on graphs and groups with infinitely many ends, *Israel J. Math.* **68** (1989), 271–301.

[346] Woess, W.: Topological groups and infinite graphs, *Discrete Math.* **95** (1991), 373–384.

[347] Woess, W.: Fixed sets and free subgroups of groups acting on metric spaces, *Math. Z.* **214** (1993), 425–440.

[348] Woess, W.: Random walks on infinite graphs and groups – a survey on selected topics, *Bull. London Math. Soc.* **26** (1994), 1–60.

[349] Woess, W.: Dirichlet problem at infinity for harmonic functions on graphs, *Potential Theory – ICPT94* (eds. J. Král et al.), de Gruyter, Berlin, 1996, pp. 189–217.

[350] Woess, W.: Topological groups and recurrence of quasi transitive graphs, *Rendiconti Sem. Mat. Fis. Milano* **64** (1994/96), 185–213.

[351] Woess, W.: *Catene di Markov e Teoria del Potenziale nel Discreto*, Quaderni dell'Unione Matematica Italiana **41**, Pitagora, Bologna, 1996.

[352] Woess, W.: A note on tilings and strong isoperimetric inequality, *Math. Proc. Cambridge Phil. Soc.* **124** (1998), 385–393.

[353] Wolf, J. A.: Growth of finitely generated solvable groups and curvature of Riemannian manifolds, *J. Diff. Geom.* **2** (1968), 421–446.

[354] Yamasaki, M.: Parabolic and hyperbolic infinite networks, *Hiroshima Math. J.* **7** (1977), 135–146.

[355] Yamasaki, M.: Discrete potentials on an infinite network, *Mem. Fac. Sci., Shimane Univ.* **13** (1979), 31–44.

[356] Zuk, A.: A remark on the norm of a random walk on surface groups, *Colloquium Math.* **72** (1997), 195–206.

[357] Zuk, A.: On the norms of the random walks on planar graphs, *Ann. Inst. Fourier* (Grenoble) **47** (1997), 1463–1490.

INDEX

Printed in the United States
By Bookmasters